Stochastic Analysis

Liber Amicorum for Moshe Zakai

Stochastic Analysis

Liber Amicorum for Moshe Zakai

Edited by

Eddy Mayer-Wolf
Department of Mathematics
Technion — IIT
Haifa, Israel

Ely Merzbach
Department of Mathematics
Bar Ilan University
Ramat Gan, Israel

Adam Shwartz
Department of Electrical Engineering
Technion — IIT
Haifa, Israel

ACADEMIC PRESS, INC.
Harcourt Brace Jovanovich, Publishers
Boston San Diego New York
London Sydney Tokyo Toronto

This book is printed on acid-free paper. ♾

ACADEMIC PRESS, INC.
1250 Sixth Avenue, San Diego, CA 92101

United Kingdom Edition published by
ACADEMIC PRESS LIMITED
24–28 Oval Road, London NW1 7DX

LCCN 91-070745

ISBN 0-12-481005-5

Printed in the United States of America

91 92 93 94 9 8 7 6 5 4 3 2 1

Contents

Invited Speakers to the Conference in Honor of Moshe Zakai are denoted by a star*.

Preface ... ix

Foreword ... xi

Publications by Moshe Zakai ... xv

Sergio Albeverio, Koichiro Iwata and Torbjörn Kolsrud, *Conformally invariant and reflection positive random fields in two dimensions* ... 1

John S. Baras*, *Real time architectures for the Zakai equation and applications* ... 15

Vaclav Beneš*, *Quadratic approximation by linear systems controlled from partial observations* ... 39

Alain Bensoussan, *A model of stochastic differential equation in Hilbert spaces applicable to Navier stokes equation in dimension 2* ... 51

Marc A. Berger, *Wavelets as attractors of random dynamical systems* ... 75

Etienne Carnal and John B. Walsh*, *Markov properties for certain random fields* ... 91

J. Martin C. Clark, *The anatomy of a low-noise jump filter: part I* ... 111

Mark H.A. Davis, Michael A.H. Dempster and Robert J. Elliott, *On the value of information in controlled diffusion processes* ... 125

Robert J. Elliott and Hans Föllmer, *Orthogonal Martingale representation* ... 139

Wendell H. Fleming and Qing Zhang,
Nonlinear filtering with small observation noise: piecewise monotone observations 153

G.J. Foschini and Larry A. Shepp,
Closed form characteristic functions for certain random variables related to Brownian motion 169

Nikos N. Frangos and Peter Imkeller,
Adaptedness and Existence of Occupation Densities for Stochastic Integral Processes in the Second Wiener Chaos . 189

Gopinath Kallianpur*, *A skeletal theory of filtering* 213

Ioannis Karatzas, Peter Lakner, John P. Lehoczky and Steven E. Shreve, *Equilibrium in a simplified dynamic, stochastic economy with heterogeneous agents* ... 245

Ioannis Karatzas and Daniel L. Ocone,
Feynman-Kac formula for a Degenerate Planar Diffusion and an application in stochastic control 273

Nicolai V. Krylov, *On the interior smoothness of harmonic functions for degenerate diffusion processes* 297

Hiroshi Kunita*, *The stability and approximation problems in nonlinear filtering theory* 311

Thomas G. Kurtz and Philip Protter,
Wong-Zakai corrections, random evolutions, and simulation schemes for SDE's 331

Harold J. Kushner*,
Nonlinear filtering for singularly perturbed systems 347

Paul Malliavin*, *Smooth σ-fields* 371

Annie Millet, David Nualart and Marta Sanz,
Composition of large deviation principles and applications 383

David Nualart*, *Nonlinear transformations of the Wiener measure and applications* 397

Etienne Pardoux and Marie Christine Roubaud,
Finite dimensional approximate filters
in the case of high signal-to-noise ratio 433

Boris Rozovskii, *A simple proof of*
uniqueness for Kushner and Zakai equations 449

Ichiro Shigekawa, *Itô-Wiener expansions*
of holomorphic functions on the complex Wiener space 459

Héctor J. Sussmann,
Limits of the Wong-Zakai type with a modified drift term .. 475

Shinzo Watanabe*,
Donsker's δ-functions in the Malliavin calculus 495

Eugene Wong*, *Implementing Boltzmann machines* 503

Ofer Zeitouni,
Infinite dimensionality results for MAP estimation 513

Preface

This book, dedicated to Moshe Zakai on the occasion of his 65th birthday, contains 29 papers in the fields of stochastic differential equations, nonlinear filtering, two-parameter martingales, Wiener space analysis and related topics. These areas reflect Moshe's interests and contributions. Some of the articles are surveys, others contain new results, methods or applications.

The publication of this volume has been timed to coincide with a three day conference on stochastic analysis at the Technion celebrating Moshe's birthday. As such, not only does it serve as the proceedings for the invited lectures at the meeting, but also provides an outlet for many of Moshe's other colleagues who, besides attending the conference, wished to dedicate an article to him. These latter contributions were refereed; unfortunately, not enough space was available to accommodate all the worthy papers.

We are grateful to the Technion and its Faculty of Electrical Engineering for having generously allocated the funds which made the conference and this book possible. Gitta Abraham and Lesley Price provided superb typing services and kept things under control, while a large number of colleagues have anonymously and graciously contributed to this volume as referees; special thanks are due to them. We are much indebted to Robert Adler for his foreword on Moshe Zakai the man and his contributions to-date. Robert's unique warmth and wit speak for us all. The authors themselves deserve a word of thanks: through their goodwill and efforts, a collection of papers has become a Liber Amicorum—a book of friends.

It is ultimately Moshe himself whom we wish to thank; once our teacher, now a real friend as well. Mazal Tov!

Eddy Mayer-Wolf
Ely Merzbach
Adam Shwartz

Haifa, January, 1991

Foreword

This collection of papers has been prepared to honour Moshe Zakai, on the occasion of his sixty-fifth birthday. That their authors represent only a few of Moshe's many friends, colleagues and students, will be obvious to anyone familiar with his many important contributions to Statistical Communication Theory and to the Theory of Probability and Stochastic Processes, or with his major impact on the development of Science and Engineering in Israel and, in particular, at the Technion.

We know that all of Moshe's friends will join us in dedicating this volume to him with deep affection and admiration, and in wishing him a future as fruitful, enjoyable, and productive as his past.

It is, of course, traditional and appropriate to now write a few words about Moshe's background and career. In keeping with Moshe's personality, and guided by the Talmudic dictum of not praising, to excess, a man to his face (We do rather hope that Moshe will read this book!), we shall attempt to do so without an overdose of superlatives. While this is not any easy task, we are aided by the list of Moshe's publications included in this volume, and we feel that we can rely on the reader to supply his own superlatives as he browses through some of the papers themselves. (However we warn the reader that this list is fast becoming obsolete!)

Moshe Zakai was born in Sokolka, Poland, on December 22, 1926. He came to Israel as a young child, where he found an environment more conducive to his personal growth and development. He received his B.Sc. from the Technion — Israel's primary engineering school and its first institution of higher learning — in 1951, and the Diplom Ingenieur degree in 1952, both in Electrical Engineering. His Ph.D. studies, which were funded by an Israel Government Fellowship, were undertaken at the University of Illinois, Urbana, from where he graduated in 1958.

Although it is often hard for Moshe's more mathematical friends to believe, his beginnings, like those of so many top mathematicians, were in extremely applied problems. Before leaving for Urbana Moshe spent five years with the Scientific Department of the Israel Defense Ministry, working on the research and development of radar systems. When he returned to Israel after his doctoral studies, he returned as the head of the Communications Research Group of the same department. During this period he also spent time with Sylvania Electronic Systems in Waltham, Massachusetts.

In 1965 Moshe joined the Faculty of Electrical Engineering at the Technion, where he now holds the William Fondiller Chair of Telecommunications. Since 1985 he has held the title of Technion Distinguished Professor, awarded not only to honour his scientific achievements but also in recognition of two decades of dedicated service to the Technion. For example, Moshe was dean of the Faculty of Electrical Engineering from 1970 to 1973, and Technion Vice President for Academic Affairs from 1976 to 1978.

The honours that have been extended to him over the years from international bodies are indicative of Moshe's wide range of academic interests. He is a fellow of both the Institute of Electrical and Electronic Engineers (1973) and the Institute of Mathematical Statistics (1988). He was recently (1989) elected as a Foreign Associate of the U.S. National Academy of Engineering, a comparatively rare honour for engineers outside the U.S.

There are a number of central threads that have run through Moshe's work over the years.

In the sixties, Moshe began working in the then new field of stochastic differential equations and Itô calculus. His primary motivation was the need to put the subjects of statistical communication and control in continuous time on a sound mathematical foundation. A basic problem here was that of relating the properties of Brownian motion, which was then beginning to take a central role as an idealised, mathematical model for stochastic noise, to those of more mundane, but physically realisable processes. In particular, together with Eugene Wong (thus starting a cooperation that has lasted a quarter of a century and produced 17 papers) Moshe studied the robustness properties of stochastic integrals, and clarified the relationships between the integrals of Stratonovich and Itô. They produced two papers [10,12][1] that remain classics to this day. Moshe was also interested in the long time stability of stochastic differential equations and properties of measure changes. It was this last theme that led him to develop (cf. [17]) what has since become known as the Zakai equation, one of the basic equations of non-linear filtering.

While the Kalman-Bucy filter of the early sixties had revolutionised the study of linear dynamical systems, the non-linear case remained an uncracked nut. Although other non-linear filtering equations were known before the Zakai equation, none had turned out to be amenable to analytic investigation. Moshe's way around this problem was based on the simple but insightful observation that it sufficed to obtain an unnormalized version of the filter's conditional measure and that, by comparing the given system to a simpler "reference measure", such a version arose as one factor in the underlying Radon-Nikodym derivative. Moreover it turned out that this unnormalized law evolved according to a comparatively simple (bilinear) stochastic partial differential equation. This was the equation that was to take on his name, and that was at the basis of most aspects of non-linear filtering theory during the seventies and eighties including the hardware implementation of filtering algorithms.

Today, the Zakai equation has obtained a new lease of life as one of the most frequently cited, and highly motivating, examples in the rapidly developing area of parabolic stochastic pde's.

In the early seventies Moshe turned to a new problem, that of multi-parameter martingales. This area, which was extremely active for over a decade, began with a fundamental paper of Moshe and Eugene Wong [31], in which they obtained a general representation of square integrable martingales by means of different stochastic integrals. This was a difficult theory, qualitatively different from that of its one-dimensional counterpart because of the lack of ordering inherent in the parameter space. Despite this, in a series of papers with Wong and others, Moshe developed a full and rich theory, covering topics as diverse as the characterisation of all multi-parameter Brownian martingales, stochastic integration, likelihood ratios and trans-

[1] References are to the list of publications given below.

formation of probability measures, Markov processes in the plane, martingale differential forms and random currents.

As the years passed by, although Moshe's motivation often still came from the world of Statistical Communication Theory, his technical expertise as a pure mathematician seemed to develop without bound, and in the early eighties he turned to one of the most mathematically demanding branches of stochastic process theory — the Malliavin calculus. He wrote a number of papers in the area including a superb exposition [52] of the Malliavin calculus, relating different approaches and introducing a number of novel insights.

As we enter the nineties, Moshe continues to expand his interests and increase even further his impact on the theory of stochastic processes. His current research centers around anticipative stochastic integrals and changes of measure under anticipative transformations, as seen with the perspective of the Malliavin calculus. Even without the benefit of hindsight, it is clear that his recent works with Nualart and Ustunel include a number of contenders to join his list of important, pathbreaking results.

In addition to the leadership that Moshe has provided to the international community through his published work, he has been a major force over the years in developing a new generation of Israeli probabilists, in particular at the Technion, where there are now close to twenty pure and applied probabilists, an active program of seminars and advanced courses in Probability, and a constant flow of distinguished visitors and postdocs. Much of this activity has its roots in Moshe's contagious enthusiasm, both as thesis supervisor and friend, and the fact that he was always more than happy to go out of his way to help young people along the academic path.

None of the above, though, really succeeds in capturing Moshe Zakai, the man. He is a delightful and unpretentious person, and the best of friends. He is a devoted family man, and he and his wife Shulamit (Mita) of 38 years are the proud parents of three children and (at last count) 7 grandchildren. All of us who have been privileged to know and to work with him have only the deepest admiration and affection for him.

Moshe – Happy birthday.

Publications by Moshe Zakai

1. M. Zakai, *A Note on the Impedance Transformation Properties of the Folded Dipole*, Proc. IRE, 41, 1061–1062, 1953.

2. M. Zakai, *On a Property of Wiener Filters*, IRE Trans. Inf. Th., IT-5, 15–17, 1959.

3. M. Zakai, *A Class of Definitions of Duration (or Uncertainty) and the Associated Uncertainty Relations*, Inf. Contr., 3, 101–115, 1960.

4. M. Zakai, *Second-Order Properties of the Pre-Envelope and Envelope Processes*, IRE Trans. Inf. Th., IT-6, 556–557, 1960.

5. M. Zakai, *The Representation of Narrow-Band Processes*, IRE Trans. Inf. Th., IT-8, 323–325, 1962.

6. M. Zakai, *On the First Order Probability Distribution of the Van der Pol Type Oscillator Output*, J. Elec. Contr., 14, 381–388, 1963.

7. Y. Ronen and M. Zakai, *The Maximum Likelihood Estimator for a Phase Comparison Angle Measuring System*, Proc. IEEE, 51, 1669–1670, 1963.

8. M. Zakai, *General Error Criteria*, IEEE Trans. Inf. Th., IT-10, 94–95, 1964.

9. M. Zakai, *Band Limited Functions and the Sampling Theorem*, Inf. Contr., 8, 143–158, 1965.

10. E. Wong and M. Zakai, *On the Relation Between Ordinary and Stochastic Differential Equations*, Int. J. Eng. Sci., 3, 213–229, 1965.

11. E. Wong and M. Zakai, *The Oscillation of Stochastic Integrals*, Z. Wahr. Verw. Geb. , 4, 103–112, 1965.

12. E. Wong and M. Zakai, *On the Convergence of Ordinary Integrals to Stochastic Integrals*, Ann. Math. Stat., 36, 1560–1564, 1965.

13. M. Zakai, *The Effect of Background Noise on the Operation of Oscillators*, Int. J. Elec., 19, 115–132, 1965.

14. E. Wong and M. Zakai, *On the Relation Between Ordinary and Stochastic Differential Equations—Applications to Stochastic Problems in Control Theory*, Proc. 3rd IFAC Conf., London, 3B-1–3B-8, 1967.

15. M. Zakai, *On the Ultimate Boundedness of Moments Associated with Solutions of Stochastic Differential Equations*, SIAM J. Contr., 5, 588–593, 1967.

16. M. Zakai, *Some Moment Inequalities for Stochastic Integrals and for Solutions of Stochastic Differential Equations*, Israel J. Math., 5, 170–176, 1967.

17. M. Zakai, *On the Optimal Filtering of Diffusion Processes*, Z. Wahr. Verw. Geb., 11, 230–243, 1969.

18. M. Zakai and J. Ziv, *On the Threshold Effect in Radar Range Estimation*, IEEE Trans. Inf. Th., IT-15, 167–170, 1969.

19. J. Ziv and M. Zakai, *Some Lower Bounds on Signal Parameter Estimation*, IEEE Trans. Inf. Th., IT–15, 386-391, 1969.

20. M. Zakai, *A Liapunov Criterion for the Existence of Stationary Probability Distributions for Systems Perturbed by Noise*, SIAM J. Contr., 7, 390–397, 1969.

21. E. Wong and M. Zakai, *Riemann-Stieltjes Approximations of Stochastic Integrals*, Z. Wahr. Verw. Geb., 12, 87–97, 1969.

22. M. Zakai and J. Snyders, *Stationary Probability Measures for Linear Differential Equations Driven by White Noise*, J. Diff. Eq., 8, 27–33, 1970.

23. J. Snyders and M. Zakai, *On Nonnegative Solutions of the Equation* $AD + DA' = -C^*$, SIAM J. Appl. Math., 18, 704–714, 1970.

24. J. Binia, A. Shenhar and M. Zakai, *Some Effects of Noise on the Operation of Oscillators*, Int. J. Nonl. Mech., 6, 593–606, 1971.

25. T. Kailath and M. Zakai, *Absolute Continuity and Radon-Nikodym Derivatives for Certain Measures Relative to Wiener Measure*, Ann. Math. Stat., 42, 130–140, 1971.

26. T. Kadota, M. Zakai and J. Ziv, *Mutual Information of the White Gaussian Channel With and Without Feedback*, IEEE Trans. Inf. Th., IT-17, 368–371, 1971.

27. T. Kadota, M. Zakai and J. Ziv, *The Capacity of a Continuous Memoryless Channel with Feedback*, IEEE Trans. Inf. Th., IT-17, 372–378, 1971.

28. M. Zakai and J. Ziv, *Lower and Upper Bounds on the Optimal Filtering Error of Certain Diffusion Processes*, IEEE Trans. Inf. Th., IT-18, 325–331, 1972.

29. M. Zakai and J. Ziv, *On Functionals Satisfying a Data-Processing Theorem*, IEEE Trans. Inf. Th., IT-19, 275–283, 1973.

30. J. Binia, M. Zakai and J. Ziv, *Bounds on the ϵ-Entropy of Wiener and* RC *Processes*, IEEE Trans. Inf. Th., IT-19, 359–362, 1973.

31. E. Wong and M. Zakai, *Martingales and Stochastic Integrals for Processes with a Multi-Dimensional Parameter*, Z. Wahr. Verw. Geb., 29, 109–122, 1974.

32. J. Binnia, M. Zakai and J. Ziv, *On the ϵ-Entropy and Rate Distortion Function of Certain Non-Gaussian Processes*, IEEE Trans. Inf. Th., IT-20, 517–524, 1974.

33. D. Chazan, M. Zakai and J. Ziv, *Improved Lower Bounds on Signal Parameter Estimation*, IEEE Trans. Inf. Th., IT-21, 90–93, 1975.

34. B.Z. Bobrovsky and M. Zakai, *A Lower Bound on the Estimation Error for Markov Processes*, IEEE Trans. Aut. Contr., AC-20, 785–788, 1975.

35. M. Zakai and J. Ziv, *A Generalization of the Rate-Distortion Theory and Applications*, in Information Theory—New Trends and Open Problems (G. Longo, ed.), Springer-Verlag, 87–123, 1975.

36. B.Z. Bobrovsky and M. Zakai, *A Lower Bound in the Estimation Error for Certain Diffusion Processes*, IEEE Trans. Inf. Th., IT-22, 45–52, 1976.

37. E. Wong and M. Zakai, *Weak Martingales and Stochastic Integrals in the Plane*, Ann. Prob., 4, 570–586, 1976.

38. E. Wong and M. Zakai, *Likelihood Ratios and Transformation of Probability Associated with Two-Parameters Wiener Processes*, Z. Wahr. Verw. Geb., 40, 283–308, 1977.

39. E. Wong and M. Zakai, *An Extension of Stochastic Integrals in the Plane*, Ann. Prob., 5, 770–778, 1977.

40. E. Wong and M. Zakai, *The Sample Function Continuity of Stochastic Integrals in the Plane*, Ann. Prob., 5, 1024–1027, 1977.

41. E. Wong and M. Zakai, *Differentiation Formulas for Stochastic Integrals in the Plane*, Stoch. Pr. Appl., 6, 339–349, 1978.

42. T. Kailath, A. Segall and M. Zakai, *Fubini-Type Theorems for Stochastic Integrals*, Sankhya, 40, Ser. A, Pt. 2, 138–143, 1978.

43. E. Merzbach and M. Zakai, *Predictable and Dual Predictable Projections of Two-Parameter Stochastic Processes*, Z. Wahr. Verw. Geb., 53, 263–269, 1980.

44. B.Z. Bobrovsky and M. Zakai, *On Lower Bounds for the Nonlinear Filtering Problem*, IEEE Trans. Inf. Th., IT-27, 131–132, 1981.

45. M. Zakai, *A Footnote to the Papers which Prove the Nonexistence of Finite Dimensional Filters*, in Stochastic Systems: The Mathematics of Filtering and Identification and Applications (M. Hazewinkel and J.C. Willems, eds.), D. Reidel Publishing Co., 649–650, 1981.

46. M. Zakai, *Some Remarks on Integration with Respect to Weak Martingales*, L.N. Math., 863, Springer-Verlag, 149–161, 1981.

47. M. Zakai, *Some Classes of Two-Parameter Martingales*, Ann. Prob., 9, 255–265, 1981.

48. B.Z. Bobrovsky and M. Zakai, *Asymptotic* a-Priori *Estimates for the Error in the Nonlinear Filtering Problem*, IEEE Trans. Inf. Th., IT-28, 371–376, 1982.

49. E. Wong and M. Zakai, *A Characterization of the Kernels Associated with the Multiple Integral Representation of Some Functionals of the Wiener Process*, Sys. Contr. Lett., 2, 94–98, 1982.

50. E. Wong and M. Zakai, *Some Results on Likelihood Ratios for Two Parameter Processes*, in Stochastic Differential Systems, L.N. Contr. Inf. Sc., 43, Springer-Verlag, 135–143, 1982.

51. E. Mayer-Wolf and M. Zakai, *On a Formula Relating the Shannon Information to the Fisher Information for the Filtering Problem*, L.N. Contr. Inf. Sc., 61, Springer Verlag, 164–171, 1984.

52. M. Zakai, *The Malliavin Calculus*, Acta Appl. Math., 3, 175–207, 1985.

53. M. Zakai, *Malliavin Derivatives and Derivatives of Functionals of the Wiener Process with Respect to a Scale Parameter*, Ann. Prob., 13, 609–615, 1985.

54. E. Wong and M. Zakai, *Markov Processes on the Plane*, Stochastics, 15, 311–333, 1985.

55. E. Merzbach and M. Zakai, *Bimeasures and Measures Induced by Stochastic Integrators in the Plane*, J. Mult. Anal., 19, 67–87, 1986.

56. D. Nualart and M. Zakai, *Generalized Stochastic Integrals and the Malliavin Calculus*, Prob. Th. Rel. Fields, 73, 255–280, 1986.

57. E. Wong and M. Zakai, *Multiparameter Martingale Differential Forms*, Prob. Th. Rel. Fields, 74, 429–453, 1987.

58. B.Z. Bobrovsky, E. Mayer-Wolf and M. Zakai, *Some Classes of Global Cramer-Rao Bounds*, Ann. Stat., 15, 1421–1438, 1987.

59. E. Merzbach and M. Zakai, *Stopping a Two Parameter Weak Martingale*, Prob. Th. Rel. Fields, 76, 499–507, 1987.

60. D. Nualart and M. Zakai, *Generalized Multiple Stochastic Integrals and the Representation of Wiener Functionals*, Stochastics, 23, 311–330, 1988.

61. E. Mayer-Wolf, M. Zakai and O. Zeitouni, *On the Memory Length of the Optimal Nonlinear Filter*, in Stochastic Differential Systems, Stochastic Control Theory and Applications (W. Fleming and P.L. Lions, eds.), IMA, Vol. 10, 311–322, Springer-Verlag, 1988.

62. B.Z. Bobrovsky, M. Zakai and O. Zeitouni, *Error Bounds for the Nonlinear Filtering of Signals with Small Diffusion Coefficients*, IEEE Trans. Inf. Th.,IT-34, 710–721, 1988.

63. A.S. Ustunel and M. Zakai, *Caractérisation géométrique de l'indépendance sur l'Espace de Wiener*, C.R. Acad. Sci. Paris, t.306, série I, 199–201, 1988.

64. A.S. Ustunel and M. Zakai, *Caractérisation géométrique de l'indépendance sur forte l'Espace de Wiener*, C.R. Acad. Sci. Paris, t.306, série I, 487–489, 1988.

65. D. Nualart, A.S. Ustunel and M. Zakai, *On the Moments of a Multiple Wiener-Itô Integral and the Space Induced by the Polynomials of the Integral*, Stochastics, 25, 233–240, 1988.

66. D. Nualart and M. Zakai, *Generalized Brownian Functionals and the Solution to a Stochastic Partial Differential Equation*, J. Func. Anal., 84, 279–296, 1989.

67. D. Nualart and M. Zakai, *The Partial Malliavin Calculus.* Séminaire de Probabilités XXIII, L.N. Math., 1372, Springer-Verlag, 362–381, 1989.

68. D. Nualart and M. Zakai, *A Summary of Some Identities of the Malliavin Calculus*, L.N. Math., 1390, Springer-Verlag, 192–196, 1989.

69. A.S. Ustunel and M. Zakai, *On Independence and Conditioning on Wiener Space*, Ann. Prob., 17, 1441–1453, 1989.

70. D. Nualart and M. Zakai, *On the Relation between the Stratonovich and Ogawa integrals*, Ann. Prob., 17, 1536–1540, 1989.

71. E. Wong and M. Zakai, *Spectral Representation of Isotropic Random Currents*, Séminaire de Probabilités XXIII, L.N. Math., 1372, Springer-Verlag, 503–526, 1989.

72. E. Wong and M. Zakai, *Isotropic Gauss-Markov Currents*, Prob. Th. Rel. Fields, 82, 137–154, 1989.

73. A.S. Ustunel and M. Zakai, *On the Structure of Independence on Wiener Space*, J. Func. Anal., 90, 113–137, 1990.

74. D. Nualart, M. Sanz and M. Zakai, *On the Relations Between the Increasing Processes Associated with Two Parameter Martingales*, Stoch. Pr. Appl., 34, 99–119, 1990.

75. D. Nualart, A.S. Ustunel and M. Zakai, *Some Relations among Classes of σ-Fields on Wiener Space*, Prob. Th. Rel. Fields, 85, 119–129, 1990.

76. D. Nualart and M. Zakai, *Multiple Wiener-Itô Integrals Possessing a Continuous Extension*, Prob. Th. Rel. Fields, 85, 131–145, 1990.

77. D. Nualart, A.S. Ustunel and M. Zakai, *Some Remarks on Independence and Conditioning on Wiener Space*, L.N. Math., 1444, Springer-Verlag, 122-127, 1990.

78. M. Zakai, *Stochastic Integration, Trace and the Skeleton of Wiener Functionals*, Stochastics and Stoch. Rep., 32, 93–108, 1990.

Conformally Invariant and Reflection Positive Random Fields in Two Dimensions

S. ALBEVERIO*#‡, K. IWATA*# , T. KOLSRUD**

* Ruhr-Universität Bochum, FRG
** Kungliga Tekniska Högskolan, Stockholm, Sweden
# SFB 237, Bochum-Essen-Düsseldorf, FRG, and
BiBo-S Research Centre, Universität Bielefeld, FRG
‡CERFIM, Locarno, Switzerland

Abstract. We present a construction of a class of random fields which are distributional sections of certain line bundles over a specific Riemann surface. These random fields A are invariant in law under the appropriate group of conformal mappings, and are obtained by solving the inhomogeneous conjugate Cauchy-Riemann equation $\partial A = F$, where F is a suitable Poisson- or Gauss-distributed noise.

1. Introduction.

In this article we are concerned with complex random fields A obtained as solutions of the inhomogeneous conjugate Cauchy-Riemann equation $\partial A = F$, where F is a given 'noise'. Our inspiration comes from quantum physics where one often considers physical systems obtained from classical equations of the form $LA = F$, where L is a differential (or pseudo-) operator of order one, in general non-scalar. We mention the Maxwell, Yang-Mills, and Dirac equations. The Cauchy-Riemann operators are related to the Maxwell and Dirac equations; see [4,6,7] for the four-dimensional case.

It is a fact, at least when L is linear, that free *Euclidean* quantised fields actually satisfy this equation with F a Gaussian noise. Letting F be a (generalised) Poisson noise, we obtain other random fields. These may be interpreted as models with self-interaction terms. Generally speaking, models of this kind are less singular than

Supported by the Swedish Natural Science Research Council, the Swedish National Board for Technical Development, the Royal Swedish Academy of Sciences, and the Sonderforschungsbereich 237 (Bochum-Essen-Düsseldorf).

ISBN 0-12-481005-5

those obtained using a Gaussian source F. This is particularly apparent in gauge field theory, where the non-Gaussian models need not be renormalised ([2, 7]).

In every physical system, invariance is a fundamental issue. Therefore, in trying to construct models in quantum field theory from equations of the form $LA = F$, invariance is a basic requirement. From a mathematical point of view this means that A and F must be invariant in law under appropriate transformation groups. Another important property, used when passing from Euclidean to Minkowski space-time, is 'reflection positivity'. In some cases, this can be obtained from reflection invariance if A is Markovian. In a series of papers, of which we mention [1-7, 9], we have studied these questions for different basic 'physical' equations. We shall not comment further on this work, but we would like to mention the related work carried out by Driver, Gross-King-Sengupta, Osipov, Tamura and others, to which appropriate references can be found in [1-7, 9]. Another type of random currents are treated in the article of Wong and Zakai [10].

In this article, invariance means conformal invariance, and our random fields can be viewed as distributional sections of line bundles over a particular Riemann surface M–the Riemann sphere minus two points. (A heuristic sketch of similar results is presented in [5].) To obtain invariance we first choose the measure defining the characteristic functional for F carefully. This means that we have started from a representation of the conformal group, G, say. To get A conformally invariant, we need a second representation of G, such that these two are intertwined by the $\bar{\partial}$-operator. These questions are taken care of in Section 2.

Random fields are most easily constructed through measures on co-nuclear spaces: they are indexed by certain nuclear test function spaces. The latter depend on the representations of G and have to be invariant. In fact, the choice of boundary conditions depend on certain parameters occurring in the representations in a rather delicate way. We deal with this question in Section 3, and the main existence result is formulated in Theorem 3.13. It should be mentioned here that in general, the case where L is strictly first-order (mass-zero case) is much more restrictive in this respect. Also, the particular choice of M, makes possible to invert $\bar{\partial}$. For general *compact* Riemann surfaces, a non-trivial kernel and cokernel of $\bar{\partial}$ must be taken into account. This is a consequence of the Riemann-Roch Theorem.

In the fourth and final section it is shown how to obtain reflection positivity by imposing a gauge condition on the test functions.

2. Representation theory, invariant measures.

We shall only consider random fields that are formally indexed by the Riemann sphere minus two points, and to this end we introduce the notation

$$M = \mathbf{CP}^1 \setminus \{0, \infty\} \cong \mathbf{C} \setminus \{0\}.$$

Let Γ_0 denote the orientation preserving, i.e. holomorphic, conformal transformations $M \to M$. Then Γ_0 consists of all Möbius transformation that leave $\{0, \infty\}$ invariant. Γ_0 is generated by the maps $z \to \gamma z$, $\gamma \neq 0$, (dilations and rotations) and $z \to 1/z$ (inversion). Identifying as usual the map $z \to (az+b)/(cz+d)$ with $\pm$ the matrix $\begin{pmatrix} a & b \\ c & d \end{pmatrix}$, where a, b, c, d are complex numbers satisfying $ad - bc = 1$, Γ_0 becomes a subgroup of $PSL(2, \mathbf{C}) = SL(2, \mathbf{C})/ \pm \mathrm{Id}$. Denote by G_0 the (simply connected) covering group of Γ_0. We introduce the function

$$J(g, z) = -cz + a, \quad z \in M, \; g \in G_0, \tag{2.1}$$

which is a square root of the derivative $dz/dg^{-1}z = (-cz + a)^2$ and which satisfies the cocycle condition

$$J(g_1 g_2, z) = J(g_1, z) J(g_2, g_1^{-1} z). \tag{2.2}$$

For every integer n we define two representations of G_0 by

$$(g \cdot \eta)(z) = J(g, z)^{-n} \eta(g^{-1} z) \equiv (\tau_n(g)\eta)(z), \tag{2.3}$$

and

$$(g \cdot \xi)(z) = \overline{J}(g, z)^{-2} J(g, z)^{-n} \xi(g^{-1} z) \equiv (\sigma_n(g)\xi)(z). \tag{2.4}$$

(Of course $\bar{J}$ is the complex conjugate of the function in Eq. (2.1).) This is done so that the Cauchy-Riemann operator $\bar{\partial} \equiv \frac{1}{2}(\partial/\partial x + i\partial/\partial y)$ intertwines the representations:

$$\bar{\partial} \tau_n = \sigma_n \bar{\partial}$$

i.e. $\bar{\partial}\eta = \xi$ implies $\bar{\partial}(g \cdot \eta) = g \cdot \xi$ for any $g \in G_0$.

(2.3) can be interpreted as the adjoint (cf. (2.10) below) of the following action of G_0 on $M \times \mathbf{C}$:

(2.5) $$(z, \alpha) \to (gz, \overline{J}(g, gz)^n \alpha).$$

We are interested in invariant measures for (2.5). Recall that viewing M as the punctured plane, the Riemannian volume $\mu(dxdy) = |z|^{-2}dxdy$ is G_0-invariant. Let now ν be a Lévy measure, i.e. a positive Radon measure on $\mathbf{C}^\times = \mathbf{C}\backslash\{0\}$ satisfying $\int \min(|\alpha|^2, 1)\,\nu(d\alpha) < \infty$. Assume also that ν is rotation invariant. Fix an integer n, and define a measure denoted $\mu \times_n \nu$ on $M \times \mathbf{C}^\times$ by

(2.6) $$\int_{M\times\mathbf{C}^\times} f\, d(\mu \times_n \nu) = \int_{M\times\mathbf{C}^\times} f(z, |z|^{n/2}\alpha)\,\mu(dxdy)\nu(d\alpha).$$

(2.7) PROPOSITION. *$\mu \times_n \nu$ is invariant under the action (2.5).*

PROOF: This result can be obtained directly by checking it for the generators of G_0. A shorter and more illuminating argument goes as follows: For any Riemann surface M the Riemannian volume μ can be expressed as $dxdy/\phi(z)^4$ (with $\phi \geq 0$) in local coordinates. μ is invariant under the group of Möbius transformations $M \to M$, i.e. under a certain subgroup of $PSL(2, \mathbf{C})$. Under such a transformation $z \to gz$, $\mu \to \phi(gz)^{-4}|cz+d|^{-4}dxdy$, so we must have $\phi(gz)|cz+d| = \phi(z)$, which, using $ad - bc = 1$, implies $\phi(z)|-cz+a|^{-1} = \phi(g^{-1}z)$, or

$$\phi(z)|J(g,z)|^{-1} = \phi(g^{-1}z).$$

Hence, with an action on $M \times \mathbf{C}^\times$ as in Eq. (2.5) and with ν rotation invariant,

$$\begin{aligned}
&\int_{M\times\mathbf{C}^\times} f(g^{-1}z, \phi(z)^n \bar{J}(g,z)^{-n}\alpha)\, d\mu(z)d\nu(\alpha) \\
&= \int_{M\times\mathbf{C}^\times} f(g^{-1}z, \phi(z)^n |J(g,z)|^{-n}\alpha)\, d\mu(z)d\nu(\alpha) \\
&= \int_{M\times\mathbf{C}^\times} f(g^{-1}z, \phi(g^{-1}z)^n\alpha)\, d\mu(z)d\nu(\alpha) \\
&= \int_{M\times\mathbf{C}^\times} f(z, \phi(z)^n\alpha)\, d\mu(z)d\nu(\alpha),
\end{aligned}$$

where we used that μ is invariant at the last step. For our choice of M, $\phi(z) = |z|^{1/2}$ and the assertion follows.

(2.8) REMARK: More general actions are possible if we restrict to subgroups of G_0. E.g. in the dilation/rotation case, when $g = \begin{pmatrix} a & 0 \\ 0 & 1/a \end{pmatrix}$ for non-zero complex a, we may let $g \cdot (z, a) = (a^n z, a^{m_1} \bar{a}^{m_2} \alpha)$, where n, m_1, m_2 are integers.

Define now

$$\psi(\beta) = \int_{\mathbf{C}^\times} \left(1 - e^{i\beta\cdot\alpha} + i\frac{\beta \cdot \alpha}{|\alpha|^2} \cdot 1_{\{|\alpha|\le 1\}}\right) \nu(d\alpha), \tag{2.9}$$

and put

$$\Psi(\eta) = \Psi_n(\eta) = \int_M \psi(|z|^{n/2}\eta(z))\mu(dxdy). \tag{2.10}$$

Then $\Psi(g \cdot \eta) = \Psi(\eta)$. Hence, defining $F = F_n$ through its characteristic functional

$$\mathbf{E}\{e^{i\langle F,\eta\rangle}\} = e^{-\Psi(\eta)}, \tag{2.11}$$

it is formally clear that F is invariant w.r.t. the representation (2.3) of G_0.

To make this more precise, we need an appropriate space of test functions indexing F in (2.11). Since we want to solve the random fields equation $\partial A = F$, further restrictions appear. We shall treat these questions in the next section.

(2.12) REMARK: The Gaussian case $\Psi(\eta) = \int_M \left||z|^{n/2}\eta(z)\right|^2 \mu(dxdy)$ can be obtained as a limit of (2.10). Although not explicitly stated, the results in this paper holds also in this case.

3. Invariant test functions and regularity. Existence.
Let $\mathcal{S}(M)$ denote the class of (test)functions

$$\begin{aligned} \mathcal{S}(M) \equiv \{\xi \in C^\infty(M \to \mathbf{C}) : \forall p > 0, \forall a \in \mathbf{Z}_+^2 \\ \lim_{|z|\to\infty} |z|^p D^a \xi(z) = 0 \text{ and } \lim_{|z|\to 0} |z|^{-p} D^a \xi(z) = 0\}. \end{aligned} \tag{3.1}$$

Then $\mathcal{S}(M)$ becomes a nuclear space using the natural seminorms. It is obvious that $\mathcal{S}(M)$ is invariant under composition with dilations and the map $z \to 1/z$. Furthermore, it is stable under multiplication with powers, including negative ones, of $|z|$. Hence $\mathcal{S}(M)$ is invariant for all the actions of G_0 through σ_n, $n \in \mathbf{Z}$.

Define another class $\mathcal{T}_1(M)$ of functions by

$$\mathcal{T}_1(M) \equiv \{\eta \in C^\infty(M \to \mathbf{C}): \lim_{|z|\to\infty} \eta(z) = 0, \lim_{z\to 0} z\eta(z) = 0 \text{ and } \bar{\partial}\eta \in \mathcal{S}(M)\}. \tag{3.2}$$

Then, as we shall see in Proposition (3.6) below, $\bar{\partial}\colon \mathcal{T}_1(M) \to \mathcal{S}(M)$ is injective. It is also surjective and the inverse map can be written

$$S\xi(\zeta) = S_1\xi(\zeta) = \frac{1}{\pi}\int_M \frac{\xi(z)}{\zeta - z}\, dxdy. \tag{3.3}$$

The only thing one needs to check surjectivity is the behaviour at the origin, but the formula shows that $S\xi(0)$ exists, and then $zS\xi(z) \to 0$ as $z \to 0$. Hence $\bar{\partial}\colon \mathcal{T}_1(M) \to \mathcal{S}(M)$ is in fact a bijection. We equip $\mathcal{T}_1(M)$ with the topology inherited from $\mathcal{S}(M)$ through the map $\bar{\partial}$.

To obtain similar spaces $\mathcal{T}_n(M)$ for general integers n, one is led to require that the boundary conditions are changed to $|z|^{(n+1)/2}\eta(z) \to 0$ at the origin and $|z|^{(n-1)/2}\eta(z) \to 0$ at infinity. For *odd* n the map

$$\mathcal{T}_n(M) \ni \eta \to z^{(n-1)/2}\eta \in \mathcal{T}_1(M) \tag{3.4}$$

is a bijection and the inverse to $\bar{\partial}$ is explicitly given by

$$S_n\xi(\zeta) = \frac{1}{\pi}\int_M \frac{z^{(n-1)/2}\xi(z)}{\zeta^{(n-1)/2}(\zeta - z)}\, dxdy, \quad \xi \in \mathcal{S}(M). \tag{3.5}$$

The reason for this is that the multipliers $\zeta^{(n\pm1)/2}$ appearing in the definition are holomorphic when n is odd (and only then). We give $\mathcal{T}_n(M)$ the induced topology, so that $\bar{\partial}\colon \mathcal{T}_n(M) \to \mathcal{S}(M)$ becomes a continuous mapping.

(3.6) PROPOSITION. *For odd integers n, the test functions spaces $\mathcal{S}(M)$ and $\mathcal{T}_n(M)$, defined by Eqs. (3.1,2,4), are invariant under the representations σ_n resp. τ_n of G_0 defined by (2.3-4). Furthermore*

$\bar{\partial}: \mathcal{T}_n(M) \to \mathcal{S}(M)$ is a continuous bijection whose inverse is given by (3.5).

PROOF: It remains to prove the injectivity of $\bar{\partial}$ and the invariance under the representation τ_n. We start with the latter issue for $n = 1$. Neither dilations nor rotations do affect the boundary conditions defining $\mathcal{T}_1(M)$. The inversion $z \to 1/z$ simply switches between the conditions at 0 and ∞, so invariance is clear for $n = 1$. Using the map (3.4) this extends to any odd integer.

To prove injectivity for $n = 1$, assume that η is mapped to 0, i.e. η is holomorphic off the origin. Then $z\eta$ is an entire function, vanishing at $z = 0$. It is an elementary exercise to deduce that then η is identically zero. Again, this extends to any odd integer by (3.4).

(3.7) REMARKS: Bijectivity is lost for even n. For $n = 0$ we see that constant functions are allowed, and in general rational functions will appear in the kernel of $\bar{\partial}$.
If one changes the condition at the origin in the definition of $\mathcal{T}_1(M)$ to $\eta(0) = 0$, then we get instead a map which is injective but not surjective. This shows the delicate situation when introducing boundary conditions.

The idea is now to define a random field $A = A_n$ on $\mathcal{S}(M)$ equipped with the representation σ_n of G_0 by

$$\langle A, \xi \rangle = -\langle F, S_n \xi \rangle, \quad \xi \in \mathcal{S}(M). \tag{3.8}$$

It follows that then

$$\langle A, \bar{\partial}\eta \rangle = -\langle F, \eta \rangle$$

so, by dualisation

$$\langle \partial A, \eta \rangle \equiv -\langle A, \bar{\partial}\eta \rangle = \langle F, \eta \rangle, \quad \eta \in \mathcal{T}_n(M). \tag{3.9}$$

Now for $\eta \in \mathcal{T}_n(M)$, $\langle F, \eta \rangle$ is defined by its characteristic functional as in Eq. (2.11). As we have already seen, A and F are invariant, but it remains to show that Ψ_n is well-defined on $\mathcal{T}_n(M)$. Of course this will not be true for general ψs. We therefore assume that ψ satisfies

$$\psi(\lambda) = O(|\lambda|^p), \quad \lambda \to 0, \tag{3.10}$$

for some $0 < p \leq 2$. It is always the case that $\psi(\lambda) = O(|\lambda|^2)$, $|\lambda| \to \infty$, see e.g. Berg-Forst [8]. We have

(3.11) PROPOSITION. *For odd integers n and for all $p > 0$*

$$\int_M \||z|^{n/2}\eta|^p \, d\mu < \infty, \quad \eta \in \mathcal{T}_n(M).$$

PROOF: As usual we need only prove this for $n = 1$. For $\eta \in \mathcal{T}_1(M)$, its value at the origin exists, as remarked above. Hence η is bounded and consequently

$$\int_{|z|\leq 1} \||z|^{1/2}\eta|^p \, d\mu = \int_{|z|\leq 1} |\eta|^p |z|^{p/2-2} \, dx dy < \infty, \tag{3.12}$$

since $p > 0$. Let $gz = 1/z$. Then g is represented by the off-diagonal matrix whose non-zero entries both are i. Hence the invariance of μ yields

$$\int_{|z|\geq 1} |\eta(z)|^p |z|^{p/2} \, d\mu = \int_{|z|\leq 1} |\eta(1/z)|^p |z|^{-p/2} \, d\mu$$

$$= \int_{|z|\leq 1} |(-iz)^{-1}\eta(1/z)|^p |z|^{p/2} \, d\mu = \int_{|z|\leq 1} |(g\cdot\eta)(z)|^p |z|^{p/2} \, d\mu,$$

and the latter is finite by invariance and (3.12).

This result shows that Ψ_n is a well-defined functional on $\mathcal{T}_n(M)$. We can now state our main result on the existence of conformally invariant random fields.

(3.13) THEOREM. *Assume that the Lévy measure ν in (2.9) is rotation invariant and satisfies condition (3.10). For each $n \in 2\mathbf{Z}+1$ there is a $\mathcal{T}_n'(M)$-valued random variable $F = (\langle F, \eta\rangle, \eta \in \mathcal{T}_n(M))$ defined through its characteristic functional by Eqs. (2.9)-(2.11). F is G_0-invariant in that $\langle F, g\cdot\eta\rangle$ and $\langle F, \eta\rangle$ are equal in law for each $g \in G_0$ and each $\eta \in \mathcal{T}_n(M)$. F defines an $\mathcal{S}'(M)$-valued random variable $A = (\langle A, \xi\rangle, \xi \in \mathcal{S}(M))$ by (3.8). A is also G_0-invariant through the representation σ_n, and satisfies $\partial A = F$ in the sense of (3.9).*

PROOF: As usual we only consider the case $n = 1$. Introduce the auxiliary Banach space

$$X \equiv \{f \in C(M \to \mathbf{C}) \colon \|f\| = \sup_z (1+|z|)|f(z)| < \infty\}.$$

Then X is invariant under the representation of G_0 on $\mathcal{T}_1(M)$, and the latter space is included in X. To see this, we first recall that $\sup_{|z|\leq 1} |f(z)| < \infty$ for any f in $\mathcal{T}_1(M)$. Hence $\sup_{|z|\geq 1} |zf(z)| = \sup_{|z|\leq 1} |z^{-1}f(z^{-1})| < \infty$ by invariance. These estimates combine to $\|f\| < \infty$, $f \in \mathcal{T}_1(M)$.

Now consider S as a map from (the whole of) $\mathcal{S}(M)$ into X. We note that $\mathcal{S}(M)$ and X are Fréchet spaces. Therefore, to show continuity it suffices to prove that S is closed, which follows immediately since both $\mathcal{S}(M)$ and X can be continuously embedded in the distribution space $\mathcal{D}'(\mathbf{R}^2)$.

The argument in Proposition 3.11 together with the assumption (3.10) shows in fact that $0 \leq \Psi(\eta) \leq C(\|\eta\|^p + \|\eta\|^2)$, so, as a functional on X, Ψ is continuous at $\eta = 0$, hence everywhere, $e^{-\Psi}$ being positive definite.

Combining these two continuous maps, we deduce that $\xi \to \Psi(S\xi)$ is continuous on $\mathcal{S}(M)$. Composing in turn this map with the by definition continuous map $\bar{\partial}$, we see finally that Ψ is continuous on $\mathcal{T}_1(M)$. At this stage we may invoke the Minlos Theorem.

4. Reflection positivity.

We shall now include the complex conjugation $\lambda\colon z \to \bar{z}$ into the transformation group. This means that we also allow antiholomorphic transformations, and in fact, the whole group of conformal mappings $M \to M$. Accordingly we extend G_0 to the semi-direct product

$$G = \mathbf{Z}_2 \times_{\mathbf{Z}_2} G_0,$$

where $\mathbf{Z}_2 = \{\lambda, 1\}$ (since $\lambda^2 = 1 = \text{Id.}$). Then

$$(\lambda, g) = (1, g)(\lambda, 1) = (\lambda, 1)(1, \bar{g}), \tag{4.1}$$

where of course $(1, g) = g$ as in the preceeding sections. (Note that if $g \in G_0$ then the matrix $\bar{g}$ obtained by conjugating all entries is also in G_0.) G is defined by this and the following relations

$$\begin{aligned}(1,g)(1,g') &= (1,gg') = (\lambda,g)(\lambda,\bar{g}'),\\ (\lambda,g)(1,g') &= (\lambda,g\bar{g}'),\\ (1,g)(\lambda,g') &= (\lambda,gg').\end{aligned} \tag{4.2}$$

We shall now extend our representations to G. First we define (identifying λ and $(\lambda, 1)$)

$$(4.3) \qquad \lambda \cdot \eta \colon z \to \bar{\eta}(\bar{z}).$$

The cocycle corresponding to λ in this case is not a complex number, rather it is the 2×2 matrix $\begin{pmatrix} 1 & 0 \\ 0 & -1 \end{pmatrix}$.

One is then lead to define the general action by

$$(4.4) \qquad ((\lambda, g) \cdot \eta)(z) \doteq J(g, z)^{-n} \bar{\eta}(\overline{g^{-1} z}) \equiv (\tau_n(\lambda, g)\eta)(z).$$

This defines the representation on $\mathcal{T}_n(M)$. To obtain the representation on $\mathcal{S}(M)$, i.e. to extend σ_n from G_0 to G, we first apply $\bar{\partial}$ to (4.3). Since λ and $\bar{\partial}$ commute, we can use the same action of λ on $\mathcal{S}(M)$. For the general case, we note that we must have $\bar{\partial}((\lambda, g) \cdot \eta) = \bar{\partial}((1, g)((\lambda, 1) \cdot \eta)) = (1, g) \cdot \bar{\partial}((\lambda, 1) \cdot \eta) = (1, g)(\lambda \cdot \bar{\partial}\eta)$, which we take as definition:

$$(4.5) \quad ((\lambda, g) \cdot \xi)(z) = \bar{J}(g, z)^{-2} J(g, z)^{-n} \bar{\xi}(\overline{g^{-1} z}) \equiv (\sigma_n(\lambda, g)\xi)(z).$$

We can now prove

(4.6) PROPOSITION. *For odd integers n, A and F are invariant under complex conjugation, and therefore under the whole conformal group G.*

PROOF: The boundary conditions used in the definitions of $\mathcal{S}(M)$ and $\mathcal{T}_n(M)$ are not affected by λ. It is obvious that the characteristic functional Ψ_n is invariant under reflection, hence F and λF have the same distribution. From $\lambda\bar{\partial} = \bar{\partial}\lambda$ follows $\lambda S_n = S_n \lambda$, so Eq. (3.8) shows that this invariance holds A too.

Thus we have a G-invariant measure, namely that of $\langle A, \cdot \rangle$, and this can be used to obtain a unitary representation of G. The existence of the involution $\theta : z \to 1/\bar{z}$ makes possible to construct a different representation. In the physics literature, and mostly in connection with simpler, (pseudo-)orthogonal, groups, this procedure is known as *reflection positivity.*

Consider first the σ-algebras

$$\mathcal{F}_{\pm} \equiv \sigma\{\langle A, \xi \rangle : \xi \in \mathcal{S}(M),\ \operatorname{supp} \xi \in M_{\pm}\}.$$

Here $M_- = \{z : |z| \leq 1\}$ and $M_+ = \{z : |z| \geq 1\}$. Clearly $\theta\colon M_\pm \to M_\mp$, and the unit circle is invariant. Furthermore, the underlying group actions also induce group actions on these σ-algebras and related classes of measurable functions.

Consider the bilinear form

$$(f, f') \equiv \mathbf{E}[f\, f' \circ \theta], \tag{4.7}$$

defined for (real-valued) bounded $\mathcal{F}_+$-measurable f and f'. Since θ is an involution and the underlying measure is invariant, this form is symmetric. Noting that for $g = \begin{pmatrix} a & 0 \\ 0 & a^{-1} \end{pmatrix}$, with $|a| \geq 1$ to preserve the support properties, $g\theta = \theta \bar{g}^{-1}$, we see that $f \circ g^{-1}$ is again $\mathcal{F}_+$-measurable and

$$(f \circ g^{-1}, f') = (f, f' \circ \bar{g}^{-1}),$$

i.e. (4.7) is a covariant form w.r.t. $g \to \bar{g}$.

Let us now try to find a subspace on which $(\cdot,\cdot)$ is non-negative definite. One way to do this is to introduce a gauge condition. This condition must be invariant under the action of the connected component of G, i.e. rotations and dilations. We shall only consider the case $n = 1$.

Let now π denote the map $M \ni z \to z^2 \in M$. Without further motivation we introduce the following (gauge) condition on ξ

$$\operatorname{Re}(\partial\pi^*(|z|\xi)) = 0. \tag{4.8}$$

We have the following auxiliary result.

(4.9) Lemma. *Let $g = \begin{pmatrix} a & 0 \\ 0 & a^{-1} \end{pmatrix}$. Then ξ satisfies (4.8) if and only if $g \cdot \xi$ satisfies (4.8). If instead $g = \begin{pmatrix} 0 & i \\ i & 0 \end{pmatrix}$, then ξ satisfies (4.8) if and only if $\operatorname{Im}(\partial\pi^*(|z|g \cdot \xi)) = 0$.*

Proof: First, let $g = \begin{pmatrix} a & 0 \\ 0 & a^{-1} \end{pmatrix}$. We have

$$|z|g \cdot \xi = \bar{a}^{-1}(|z|\xi) \circ g^{-1},$$

so

$$\pi^*(|z|g\cdot\xi)(z) = \bar{a}^{-1}(|z|\xi)\circ g^{-1}\circ\pi(z) = \bar{a}^{-1}(|z|\xi)\circ\pi(a^{-1}z),$$

whence

$$\partial\big(\pi^*(|z|g\cdot\xi)\big)(z) = \bar{a}^{-1}a^{-1}\big(\partial(|z|\xi)\circ\pi\big)(a^{-1}z).$$

If now $g = \begin{pmatrix} 0 & \mathrm{i} \\ \mathrm{i} & 0 \end{pmatrix}$, so that $gz = 1/z$, then using that g and π commute we obtain

$$\pi^*(|z|(g\cdot\xi)) = \overline{(-\mathrm{i}z^2)}^{-1}(|z|\xi)\circ\pi\circ g$$

from

$$|z|(g\cdot\xi) = \overline{(-\mathrm{i}z)}^{-1}(|z|\xi)\circ g,$$

so

$$\partial(\pi^*(|z|(g\cdot\xi))) = \overline{(-\mathrm{i}z^2)}^{-1}\partial((|z|\xi)\circ\pi\circ g)$$
$$=\overline{(-\mathrm{i}z^2)}^{-1}(-z^2)^{-1}\partial(\pi^*(|z|\xi))\circ g = \frac{\mathrm{i}}{|z|^4}\partial(\pi^*(|z|\xi))\circ g.$$

This proves the lemma.

We shall also need

(4.10) LEMMA. *Let $\eta \in \mathcal{T}_1(M)$ be defined by $\bar{\partial}\eta = \xi$, where ξ satisfies (4.8). Then $Re(z\pi^*\eta) \equiv 0$.*

PROOF: We have

$$\pi^*\xi = \pi^*\bar{\partial}\eta = \frac{1}{2\bar{z}}\bar{\partial}\pi^*\eta.$$

Using that $\overline{\partial}$ commutes with multiplication by z, we therefore get

$$0 = \mathrm{Re}\,\big(\partial(\pi^*(|z|\xi))\big) = \mathrm{Re}\,\big(\partial(z\bar{z}\cdot\frac{1}{2\bar{z}}\bar{\partial}\pi^*\eta)\big) = \frac{1}{8}\mathrm{Re}(\Delta(z\pi^*\eta)),$$

so $\mathrm{Re}\,(z\pi^*\eta)$ is harmonic in M. The conditions on η imply $|\pi^*\eta| \leq C/(1+|z|^2)$, so $z\pi^*\eta$ vanishes at $z=0$ and $z=\infty$. Hence its real part, being harmonic, is identically zero.

We can now state the main result of this section.

(4.11) PROPOSITION. *Define*

$$\tilde{\mathcal{F}}_+ = \sigma\{\langle A, \xi\rangle : \xi \in \mathcal{S}(M),\ supp\,\xi \subset M_+,\ \xi \text{ satisfies } (4.8)\}.$$

Then

$$\mathbf{E}[f\, f\circ\theta] \geq 0$$

for all bounded real-valued $\tilde{\mathcal{F}}_+$-measurable f.

PROOF: By the condition on the support of $\xi = \bar{\partial}\eta$, η is holomorphic in the interior of M_-, hence, so is $z\pi^*\eta$. Since the real part of this function vanishes, and its value at the origin is zero, $z\pi^*\eta \equiv 0$ in M_-, and so $\eta \equiv 0$ in M_-. Hence

$$\tilde{\mathcal{F}}_+ \subset \sigma\{\langle F, \eta\rangle : \operatorname{supp}\eta \subset M_+\}.$$

Since θ maps M_+ into M_-, and F is a noise, we deduce that $\tilde{\mathcal{F}}_+$ and $\theta\tilde{\mathcal{F}}_+$ are independent σ-algebras. By invariance we therefore may conclude

$$\mathbf{E}[f\, f\circ\theta] = \mathbf{E}[f]\mathbf{E}[f\circ\theta] = (\mathbf{E}[f])^2 \geq 0.$$

(4.12) REMARK: This result expresses 'reflection positivity'. As a consequence we have, for any real-valued $\tilde{\mathcal{F}}_+$-measurable f, $\mathbf{E}[f] = 0$ whenever $(f, f) = 0$, where $(\cdot,\cdot)$ is given by (4.7). Let $\mathcal{H}^0$ denote the linear space of such (real-valued $\tilde{\mathcal{F}}_+$-measurable) functions, equipped with the 'pre-inner' product (4.7), and denote by $\mathcal{H}$ the completion of the quotient space $\mathcal{H}^0/\{f \in \mathcal{H}^0 : (f,f) = 0\}$. The map $\mathcal{H}^0 \ni f \to \mathbf{E}[f] \in \mathbf{R}$ gives an identification of $\mathcal{H}$ with of the real line. Moreover the maps with $g = \begin{pmatrix} a & 0 \\ 0 & a^{-1} \end{pmatrix}$, $|a| \geq 1$, act trivially on $\mathbf{R}$. It would indeed be very interesting to find a Hilbert space which carries a non-trivial representation of the dilations. We leave this issue for further studies.

REFERENCES:

[1] Albeverio, S. and Høegh-Krohn, R.: *Construction of interacting local relativistic quantum fields in four space-time dimensions*, Phys. Lett. **B200**, 108-114 (1988). Erratum in *ibid.* **202**, p.621 (1988).

[**2**] Albeverio, S., Høegh-Krohn, R., Holden, H. and Kolsrud, T.: *Construction of quantised Higgs-like fields in two dimensions.* Phys. Lett. B. **222** (1989), 263-268.
[**3**] Albeverio, S., Høegh-Krohn, R. and Iwata, K.: *Covariant markovian random fields in four space-time dimensions with nonlinear electromagnetic interaction.* In 'Applications of self-adjoint extensions in quantum physics.' Proc. Dubna Conference 1987, Editors P. Exner, P. Seba, Springer Lect. Notes in Phys. **324**, 1989.
[**4**] Albeverio, S., Iwata., K. and Kolsrud., T.: *Random Fields as Solutions of the Inhomogeneous Quaternionic Cauchy-Riemann Equation. I. Invariance and Analytic Continuation.* Preprint Stockholm/BiBoS 1989. Commun. Math. Phys. (1990), .
[**5**] Albeverio, S., Iwata., K. and Kolsrud., T.: *Conformally Invariant Random Fields and Processes–Old and New.* Preprint TRITA/MAT-90/0009, Stockholm 1990. To appear in Proc. Lisboa Conf. 1989, Ed. A.B. Cruzeiro (Birkhäuser).
[**6**] Albeverio, S., Iwata., K. and Kolsrud., T.: *Homogeneous Markov Generalised Vector Fields and quantum Fields Over 4-dimensional space-time.* Preprint TRITA/MAT-series, Stockholm 1990. To appear in Proc. Trento Conf. Stochastic Partial Differential Equations and Applications III. Eds. G. DaPrato, L. Tubaro.
[**7**] Albeverio, S., Iwata., K. and Kolsrud., T.: *Reflection Positive Four-dimensional Euclidean Gauge Theory for Coupled Loop Variables.* Preprint TRITA/MAT-series, Stockholm 1990.
[**8**] Berg, C., and Forst, G.: Potential Theory on Locally Compact Abelian Groups. Springer, Berlin/Heidelberg/New York, 1975.
[**9**] Iwata., K.: On Linear Maps Preserving Markov Properties, and Applications to Multicomponent Generalised Random Fields. Ph. D. Thesis, Bochum University 1990.
[**10**] Wong, E. and Zakai, M.: *Isotropic Gauss-Markov currents.* Prob. Th. Rel. Fields **82** (1989), 137-154.

September 1990

Department of Mathematics, University of Bochum, D-4630, Bochum, FRG
Department of Mathematics, Royal Institute of Technology, S-100 44 Stockholm, Sweden

Real time architectures for the Zakai equation and applications

John S. Baras
Electrical Engineering Department
and Systems Research Center
University of Maryland at College Park

Abstract

We examine in detail real-time architectures for the sequential detection and/or estimation problems for diffusion type signals. We demonstrate the fundamental role played by the Zakai equation in defining candidate architectures. For scalar and two dimensional state models an architecture based on systolic arrays is derived. For higher dimensional problems a multilayer architecture based on an asychronous parallel implementation of the Multigrid algorithm is derived. Properties of the architectures and practical hardware implementation results are also reported.

1. Introduction

One of the basic activities of electrical engineering today is the processing of signals, be they in the nature of speech, radar, images, or of electromechanical or biological origin. By "processing" we generally mean conversion of the signals into some more acceptable format for analysis. Examples could be the reduction of noise content, parameter estimation, bandpass filtering, or the enhancement of contrast, as required for imaging systems. In feedback control systems analysis and synthesis, signal processing problems such as sequential detection and estimation are fundamental. The signal theorist develops algorithms for performing these functions by constructing mathematical models of signals and the operations conducted on them. The result of intensive research in this area over the last

ISBN 0-12-481005-5

twenty years has been a rather sophisticated theory, which utilizes advanced concepts from stochastic processes, differential equations and algebraic system theory. For a survey of such work the reader is referred to [2], where it can be seen that researchers have gone far beyond the classic work of Doob [3] and Wong [4].

One of the main reasons for the lack of impact of the more theoretical work on the engineering applications fields has been the failure to meet economic as well as real-time processing constraints imposed by the problems engineers face. The fundamental issue to be resolved for at least a wide class of signal processing problems involves meeting the time constraints implicit in the design. The problem is that such techniques will have much greater demands for their numerical analysis. As this translates to mean a greater number of arithmetic operations per second, meeting real-time processing conditions will be all the more difficult.

This is indeed the trend throughout much of signal processing: a greater volume of signals must be processed in a lesser amount of time, in addition to requiring more sophisticated analysis and relatively inexpensive electronics packaged on a small scale. To better understand these issues, we should examine in more detail the nature of some of these advanced techniques of signal processing. This type of analysis will become critical in the future as designers begin to assess model accuracy on device performance, cost of production and other factors.

A typical problem considered in this paper is described below. There are two hypotheses H_1, H_2 each representing that the observed data $y(t)$ originate from two different models. The decision maker receives the data $y(t)$ and has to decide which of the two hypotheses is valid. This problem is generic to a plethora of digital and analog signal processing problems, such as: pulse amplitude modulation, delta modulation, adaptive delta modulation, speech processing, direction finding receivers, digital phase lock loops, adaptive sonar and radar arrays, simultaneous detection and estimation.

As a matter of fact almost any sequential detection problem can be formulated in a similar manner. The underlying mathematical models can be diverse: diffusion processes, point processes, mixed processes, Markov chains etc. In this paper we shall concentrate on diffusion process models. That is to say, under each hypothesis the model for the observed data is

$$\begin{aligned} dx^i(t) &= f^i(x^i(t))dt + g^i(x^i(t))dw^i(t) \\ dy(t) &= h^i(x^i(t))dt + dv(t) \end{aligned} \tag{1.1}$$

where $i = 1, 2$ correspond to hypotheses H_1 or H_2. If we let

$$\hat{h}_i(t) = E_i \left\{ h^i(x^i(t)) | \mathcal{F}_t^y \right\} \tag{1.2}$$

the likelihood ratio for the problem is

$$\begin{aligned} \Lambda_t = \exp \Bigg(\int_0^t \left(\hat{h}_1(s) - \hat{h}_2(s) \right)^T dy(s) \\ - \frac{1}{2} \int_0^t \left(\|\hat{h}_1(s)\|^2 - \|\hat{h}_2(s)\|^2 \right) ds \end{aligned} \tag{1.3}$$

In [7] we showed that the optimal sequential detector utilizes threshold policies under both Neyman-Pearson and Bayes formulations and the likehood ratio Λ_t.

First it is clear that the detector has to select two things. A time τ, to stop collecting data, and a decision δ which declares one of the two hypotheses. Given the miss and false alarm probabilities α, β one computes thresholds A, B [7] and then the optimal detection strategy is given by

$$\tau^* = \inf \{ t \geq 0 | \Lambda_t \notin (A, B) \} \tag{1.4}$$

$$\delta^* = \begin{cases} 1, & \Lambda_{\tau^*} \geq B \\ 2, & \Lambda_{\tau^*} \leq A. \end{cases} \tag{1.5}$$

It is therefore clear that real time implementation of this rule is based on our ability to compute $\hat{h}_i(t)$ in real time. This is related to the Zakai equation of nonlinear filtering [1]. This is so because

$$\hat{h}^i(t) = \frac{\int u^i(x,t) h^i(x) dx}{\int u^i(x,t) dx} \tag{1.6}$$

where $u^i(x, t)$ is the unnormalized conditional probability density of $x(t)$ given $y(s), s \leq t$, under each model 1, or 2. This density satisfies the stochastic partial differential equation

$$
\begin{aligned}
du^i(x,t) &= L_i^* u^i(x,t)dt + u^i(x,t)h^{i^T}(x)dy(t) \\
u^i(x,0) &= p_0^i(x) \\
L_i^* u^i(x,t) &= \sum_{k,l} \frac{\partial^2}{\partial x_k \partial x_l}(\sigma_{k\,l}^i u^i(x,t)) - \\
&\quad - \sum_k \frac{\partial}{\partial x_k}(f_k^i(x)u^i(x,t)) \\
\sigma^i(x) &= \frac{1}{2} g^i(x) g^{i^T}(x).
\end{aligned}
\tag{1.7}
$$

It can be shown that the likelihood ratio (1.3) can be represented as

$$
\Lambda_t = \frac{\int u^1(x,t)dx}{\int u^2(x,t)dx}. \tag{1.8}
$$

As a consequence the real-time implementation issue, is reduced to the real time implementation of (1.7), (1.8), by a digital or analog circuit.

The Zakai equation (1.7), first indroduced by Moshe Zakai in [1], plays a fundamental role regarding the resolution of the real-time implementation problem for such algorithms, primarily due to its linearity! As is also well known the real-time implementation of the Zakai equation holds the keys to the solution of partially observed stochastic control problems [5], since it provides the natural state for the equivalent fully observed problem, which the feedback controller utilizes.

In section 4 we describe a special architecture, which can achieve real-time operation for many applications, utilizing systolic arrays, as first demonstrated in [7]. We emphasize that this architecture solves the problem for dimensions of x^i, less than or equal to 2. The higher dimensional problems were not addressed in [7]. In section 6 we provide a solution to the higher dimensional problem based on the so called multigrid method applied to (1.7).

The primary objective of the paper is to demonstrate that Moshe Zakai's fundamental contribution in [1], in addition to its well known theoretical value has paramount implications on the practical feasibility of real-time implementation of any sequential detector and/or estimator as well as feedback controller.

2. Review of Basic Sequential Detection Problems

The fundamental problem (1.1) can be easily reduced to two simpler problems of the following type. We are given a vector-valued signal $x_t \in R^n$ which satisfies the stochastic differential equation

$$\begin{aligned} dx_t &= f(x_t)\,dt + g(x_t)\,dw_t \\ x_0 &= \nu \end{aligned} \tag{2.1}$$

where w_t is a vector standard Brownian motion. Unfortunately, we cannot observe x_t directly, instead we only observe y_t, a vector-valued stochastic process $y_t \in R^p$. Under each hypothesis the observed data is the output of a stochastic differential equation, i.e.,

$$\begin{aligned} &\text{Under } H_1: \quad && dy_t = h(x_t)\,dt + dv_t \\ &\text{Under } H_0: \quad && dy_t = dv_t \end{aligned} \tag{2.2}$$

where v_t is another standard Brownian motion which is independent of w_t. Notice that if $f(\cdot)$ and $h(\cdot)$ are linear and $g(\cdot)$ is constant then this becomes a standard problem which can be solved by the Kalman filter.

Data are observed continuously starting at an initial time which is taken for convenience to be zero. We let $\mathcal{F}_t^y$ represent the information collected up to time t. At each time t, the decision-maker can either stop and declare one of the hypotheses to be true or can continue collecting data. We let τ represent the termination time and δ represent the decision. The decision-maker selects his decision based on the current information ,$\mathcal{F}_t^y$, so as to minimize an appropriate cost function. More precisely, an *admissible decision policy* is any pair $u = (\tau, \delta)$ of RV's where τ is an $\mathcal{F}_t^y$ stopping time and δ is a $\{0,1\}$-valued $\mathcal{F}_\tau^y$ measurable RV. An admissible policy $u = (\tau, \delta)$ is a *threshold policy* if there exist constants A and B, with $0 < A \leq 1 \leq B < \infty$ and $A \neq B$, such that

$$\tau = \inf(t \geq 0 \mid \Lambda_t \notin (A, B)) \tag{2.3}$$

$$\delta = \begin{cases} 1, & \Lambda_\tau \geq B \\ 0, & \Lambda_\tau \leq A \end{cases} \tag{2.4}$$

Here Λ_t is the likelihood ratio associated with this problem, namely

$$\Lambda_t = \exp\left(\int_0^t \hat{h}_s^T \, dy_s - \frac{1}{2} \int_0^t \|\hat{h}_s\|^2 \, ds \right), \tag{2.5}$$

and

$$\hat{h}_t = E_1(h(x) \mid \mathcal{F}_t^y). \tag{2.6}$$

Using Girsanov's theorem, it can be shown [6] that for threshold policies

$$P_0(\delta = 1) = \frac{1 - A}{B - A} \qquad P_1(\delta = 0) = \frac{A(B - 1)}{B - A}.$$

Some algebra gives the following result.

Let u be a threshold policy with A and B defined by

$$A = \frac{\beta}{1 - \alpha} \qquad B = \frac{1 - \beta}{\alpha} \tag{2.7}$$

where $\alpha + \beta < 1$, then

$$P_0(\delta = 1) = \alpha \qquad P_1(\delta = 0) = \beta. \tag{2.8}$$

Hence, given desired false alarm and miss probabilities, α, β, it is possible to find thresholds, (A, B), so that the resulting threshold policy has the required probabilities.

Given $0 < \alpha, \beta < 1$ with $\alpha + \beta < 1$, let $\mathcal{U}(\alpha, \beta)$ be the set of all admissible policies u such that

$$P_0(\delta = 1) \leq \alpha \qquad P_1(\delta = 0) \leq \beta. \tag{2.9}$$

The fixed probability of error formulation to the sequential hypothesis testing problem requires the solution of the following.

Problem ($\mathcal{P}_F$): *Find u^* in $\mathcal{U}(\alpha, \beta)$ such that for all u in $\mathcal{U}(\alpha, \beta)$,*

$$E_i\left(\int_0^T \|\hat{h}_s\|^2 \, ds \right) \geq E_i\left(\int_0^{T^*} \|\hat{h}_s\|^2 \, ds \right), \quad i = 0, 1. \tag{2.10}$$

The term in the expectation above represents the expected signal energy present. Usually, the observation time is minimized, subject to the error probability constraints (2.9). However, in this problem we cannot always be sure that we receive "good" data because $h(x_t)$ is itself a random process. It is clear that the longer we observe y the more energy signal we receive. Therefore, trying to decide "faster" is related to trying to decide while receiving the "minimum" signal energy. This intuitive idea is captured in (2.10). We can now state [7, 8]:

Theorem 1. *If u^* is the threshold policy with constants (A^*, B^*) defined by*

$$A^* = \frac{\beta}{1-\alpha}, \qquad B^* = \frac{1-\beta}{\alpha}, \tag{2.11}$$

then u^* solves problem $(\mathcal{P}_F)$.

For the Bayesian formulation, let H be a $\{0,1\}$-valued RV indicating the true hypothesis. By φ we denote the *a priori* probability that hypothesis H_1 is true.

We shall assume, two costs are incurred. The first cost is for observation and is accrued according to $k\int_0^t \|\hat{h}_s\|^2\,ds$, where $k > 0$ and $\{\hat{h}_t,\ t \geq 0\}$ is defined by (2.6). The second cost is associated with the final decision δ and is given by

$$C(H,\delta) = \begin{cases} c_1, & \text{when } H = 1 \text{ and } \delta = 0; \\ c_2, & \text{when } H = 0 \text{ and } \delta = 1; \\ 0, & \text{otherwise,} \end{cases} \tag{2.12}$$

where $c_1 > 0$ and $c_2 > 0$.

We are interested in minimizing the expected cost. If $u = (\tau, \delta)$ is any admissible policy, then the corresponding expected cost is

$$J(u) = E(\,k\int_0^\tau \|\hat{h}_s\|^2\,ds + C(H,\delta)). \tag{2.13}$$

The Bayesian approach to sequential detection seeks to find the solution to the following problem.

Problem $(\mathcal{P}_B)$: *Given $\varphi \in (0,1)$, find u^* such that,*

$$J(u^*) = \inf_{u\in\mathcal{U}} J(u). \tag{2.14}$$

It can be shown that to any admissible policy u there corresponds a threshold policy which has no greater cost, $J(u)$, therefore, the infimum in (2.14) need only be computed over threshold policies. In fact, it can be shown [7, 8] that the infimum is obtained and the following theorem results.

Theorem 2. *There exists an admissible threshold policy u^* that solves problem $(\mathcal{P}_B)$. The optimal thresholds $0 < A^* \leq 1 \leq B^* < \infty$ with $A^* \neq B^*$, are given by the relations*

$$A^* = \left(\frac{1-\varphi}{\varphi}\right)\left(\frac{a^*}{1-a^*}\right), \qquad B^* = \left(\frac{1-\varphi}{\varphi}\right)\left(\frac{b^*}{1-b^*}\right).$$

where a^* and b^* are the unique solutions of the transcendental equations

$$c_2 + c_1 = k(\Psi'(a^*) - \Psi'(b^*))$$

$$c_2(1-b^*) = c_1 a^* + (b^* - a^*)(c_1 - k\Psi'(a^*)) + k(\Psi(b^*) - \Psi(a^*)),$$

with

$$\Psi(x) = (1-2x)\log\frac{x}{1-x}.$$

satisfying $0 < a^* < b^* < 1$.

Here again the thresholds are unique functions of the cost parameters.

These results hold for the hypothesis testing problem described in (1.1) with the small modification of using Λ_t as defined by (1.3).

3. Numerical Solution for Scalar x

In this section we discuss the numerical method used to approximate $\hat{h}(t)$, Λ_t. As mentioned earlier, it can be shown that $\Lambda_t = \int_{R^n} u(x,t)\,dx$ where $u(x,t)$ is the solution to the *Zakai equation* (1.7). Our strategy will be to find an approximation to $u(x,t)$ which results in a good approximation to Λ_t.

In this section we consider in detail the case when x is scalar. Then in the *Zakai equation* (1.7)

$$\begin{aligned} L^* u(x,t) &= \frac{1}{2}\frac{d^2}{dx^2}[g^2(x)\,u(x,t)] - \frac{d}{dx}[f(x)\,u(x,t)] \\ &= a(x)\,u_{xx}(x,t) + b(x)\,u_x(x,t) + c(x)u(x,t) \\ &= A^* u(x,t) + c(x)u(x,t) \qquad (3.1) \end{aligned}$$

and

$$
\begin{aligned}
a(x) &= \frac{1}{2} g^2(x) \\
b(x) &= g(x)\, g'(x) - f(x) \\
c(x) &= g(x)\, g''(x) + (g'(x))^2 + g(x)\, g'(x) - f'(x)
\end{aligned}
\tag{3.2}
$$

with $p_0(x)$ the initial density of x.

In general, it is not possible to explicitly solve (1.7), (3.1). Therefore, we will approximate its solution using finite difference methods. Since (1.7) is a linear parabolic equation we use an implicit discretization for x-derivatives, in order to maintain stability. The general vector valued case is treated in [8, 9].

The solution is approximated on the interval $D = (a, b)$. Let $\Delta x > 0$ and define $x_k = a + k\,\Delta x$ and n such that $x_n \leq b$. Consider the collection of points $\{x_k\}_0^n$ in D. Let $\Delta t > 0$ and define $t_k = k\,\Delta t$.

The value of $u(x, t)$ at the grid point (x_i, t_k) is represented by v_i^k. We replace the x-derivatives in A^* with implicit finite difference approximations. To this end,

$$
\begin{aligned}
a(x)\, u_{xx}(x, t) &\sim a(x_i) \left(\frac{v_{i+1}^{k+1} - 2v_i^{k+1} + v_{i-1}^{k+1}}{(\Delta x)^2} \right) \\
b(x)\, u_x(x, t) &\sim b(x_i) \left(\frac{v_{i+1}^{k+1} - v_i^{k+1}}{\Delta x} \right) \quad \text{if } b(x_i) > 0 \\
& b(x_i) \left(\frac{v_i^{k+1} - v_{i-1}^{k+1}}{\Delta x} \right) \quad \text{if } b(x_i) < 0 \\
u_t(x, t) &\sim \frac{v_i^{k+1} - v_i^k}{\Delta t}
\end{aligned}
\tag{3.3}
$$

Let V^k represent the vector of mesh points at time $k\,\Delta t$. Then the above approximations result in a matrix A_n which approximates A^*. The approximation takes the form

$$
(I - \Delta t\, A_n) V^{k+1} = V^k + \text{ other terms.}
$$

Note that $a(x_i)$ is always positive and the special way of choosing the first derivative approximation guarantees that the matrix A_n is diagonally dominant and of the form

$$A_n = \begin{pmatrix} - & + & & \\ + & \ddots & \ddots & \\ & \ddots & \ddots & + \\ & & + & - \end{pmatrix}. \tag{3.4}$$

Therefore $(I - \Delta t\, A_n)$ is strictly diagonally dominant. In fact, it is also inverse positive, i.e., every element of the inverse is positive [10, Corollary 1.6b, p. 221].

The final step is to approximate the solution of (3.1) assuming $A^* = 0$. This leads to the matrix

$$D_k = \mathrm{diag}(e^{h(x_i)\,\Delta y_k + (c(x_i) - \frac{1}{2}h(x_i)^2)\,\Delta t}) \tag{3.5}$$

with $\Delta y_k = y_{(k+1)\Delta t} - y_{k\,\Delta t}$.

The overall approximation is

$$(I - \Delta t\, A_n)V^{k+1} = D_k\, V^k. \tag{3.6}$$

This approximation can be shown to be convergent [8, 9, 11].

A nice property of the above approximation is that it is positivity preserving. Regardless of the relationship of Δt and Δx, the solution V^k is always positive. This is important since we are approximating a probability density which we know can never be negative.

Schemes similar to (3.6) have been discussed in [9, 11]. Furthermore, numerical studies have been performed in [12] using these methods which have produced satisfactory results for approximations to $\hat{h}_t$.

Using V^k defined in (3.6) it is easy to construct a convergent approximation to the likelihood ratio via:

$$\Lambda_t^n = \int_a^b u_n(x,t)\,dx = \sum_{i=0}^{n} v_i^k\,\Delta x$$

Then Λ_t^n is a convergent approximation to Λ_t [8].

4. Real Time Architectures for Scalar and Two Dimensional x

The finite difference scheme used to approximate the solution of the Zakai equation involves solving the linear equation

$$(I - \Delta t A_n) V^{k+1} = D_k V^k, \tag{4.1}$$

for each time $t = k\Delta t$. Here D_k is a data dependent diagonal matrix. Our goal is to design a multiprocessor to efficiently solve (4.1). This means that:

(1) the time necessary to compute V^{k+1}, given V^k, A, and y_k, should be below a problem dependent threshold; and
(2) the control structure should be simple and regular. We have chosen the *systolic array architecture* of [13] to implement this scheme.

We are interested in systolic processors which perform linear algebra operations. The basic component of these arrays is the *inner product processor* (IPP). At each clock pulse the IPP takes the inputs x, y and a and computes $ax + y$ (the inner product step). This value is output on the y-output line, and the x and a values pass through to their respective output lines untouched. The number of processors in a systolic array solving the system in (4.1) depends only on the bandwidth of the matrix and not on its dimension. This makes a systolic design ideal for implementing *finite difference* approximations, where the number of mesh points is not known a priori but where the maximum bandwidth is determined by the specific scheme.

To solve (4.1), we use Gaussian elimination without pivoting. This method, which is stable since $(I - \Delta t\, A_n)$ is strictly diagonally dominant, results in matrices L and U such that

$$LU = (I - \Delta t\, A_n) \tag{4.2}$$

where U is an upper triangular matrix and L is a unit lower triangular matrix.

The matrices L and U will be bi-diagonal since Gaussian elimination without pivoting is used. As is standard with Gaussian elimination, once the factors L and U are found, (4.1) is solved by simple back substitution. It is well known that back substitution can be accomplished by systolic arrays [13]. The algorithm takes $2n + 2$ time units to operate. Here n is the dimension of the matrix. Hence we are able to solve (4.1) in $5n + 4$ time units. The extra n units result from a necessary buffering of the data between the two back substitution operations.

Notice that the matrix, $(I - \Delta t\, A_n)$, is independent of the received data, y_t, therefore, the LU factorization only has to be done at the time of filter design.

We have completed various implementations of the resulting processor using board level designs and employing current signal processing chips, with an IBM PC AT as the host. With current technology, such boards are capable of processing data at a rate of 20 kHz. We have also completed a special purpose VLSI chip design, which we named the *Zakai Chip* in honor of Moshe Zakai. We have recently completed a much improved special purpose VLSI chip design which we named the *Zakai II Chip.* Preliminary performance evaluation has been very good.

This programmable architecture solves a long standing problem for scalar state and observation models. It can be easily extended to vector observations of a scalar state. We have also shown that essentially the same architecture works for two dimensional state process x. However, the discovery of appropriate architectures for observations of vector states is much harder. For higher state dimensions, a different architecture is needed. In the rest of the paper we develop such multi-level architectures using numerical schemes based on multigrid methods.

5. Multigrid Algorithms for the Zakai Equation.

In this section we show how multigrid algorithms [14, 15] can be developed for the Zakai equation (1.7) in a systematic manner. As before we employ an implicit full discretization scheme that provides consistent time and space discretizations of the Zakai equation, in the sense that for each choice of discretization mesh, the problem can be interpreted as a nonlinear filtering problem for a discrete time, discrete state, hidden Markov chain.

To discretize (1.7), we choose a time step Δ and a space discretization mesh of size ϵ which determines a "grid" in R^d, the space where we wish to solve the Zakai equation. In other words $x \in R^d$. Using results on estimates of the tail behavior of $u(x,t)$ as $\|x\| \to \infty$ we can actually select a rectangular domain in R^d, D, where we are primarily interested in solving (1.7). Let us suppose that there are $n(\epsilon)$ points on each dimension of the grid of size ϵ. Let $G_d(\epsilon)$ denote the hypercube generated in R^d by the spatial mesh of size ϵ. We assume for simplicity here uniform grid spacing.

Given this set-up we can construct following the methods of Kushner [9] a matrix $A(\epsilon)$ which approximates the operator L^* in (1.7), in the sense that $A(\epsilon)$ defines an approximating Markov chain to the diffusion (1.1). Let

$$\begin{aligned} \Delta y(k) &= y((k+1)\Delta) - y(k\ \Delta) \qquad (5.1) \\ H_i(\epsilon) &= h(x_i(\epsilon)) \end{aligned}$$

where $x_i(\epsilon)$ is a generic point on the grid $G_d(\epsilon)$. Finally let $\mathbf{V}^{k+1}(\epsilon)$ be the vector of samples $u(x_i(\epsilon), (k+1)\Delta)$ of the unnormalized conditional density over the grid $G_d(\epsilon)$. In [8] the following was proved, using semigroup techniques.

Theorem 3: Let

$$\mathbf{D}(\epsilon, k) = \text{ diag } \{\exp(H_i^T(\epsilon)\Delta y(k) - \frac{1}{2}\|H_i(\epsilon)\|^2\Delta)\}$$

and consider the implicit iteration

$$\begin{aligned} \frac{(I - \Delta A(\epsilon))\mathbf{V}^{k+1}}{(\epsilon)} &= D(\epsilon, k)V^k(\epsilon) \qquad (5.2) \\ V^o(\epsilon) &= \{p_0(x_i(\epsilon))\}. \end{aligned}$$

Then as $k\Delta \to t$, with $\epsilon \to 0$ (along some sequence)

$$\lim_{\epsilon \to 0} \sup_{i \in G_d(\epsilon)} |V_i^k(\epsilon) - u(x_i(\epsilon), k\Delta)| = 0. \qquad (5.3)$$

In other words Theorem 3, provides a uniformly convergent scheme. Once we have this the likelihood ratio (1.8) can be easily approximated since

$$\int u(x,t)dx \sim \sum_{i \in G_d} V_i^k(\epsilon)\Delta x(\epsilon); \;\; \text{for } k\Delta \leq t < (k+1)\Delta \qquad (5.4)$$

where $\Delta x(\epsilon)$ is the approximation to the volume element in $G_d(\epsilon)$.

Therefore the real-time solution of (1.7) has been reduced to the analysis of the real-time computation of (5.2). We consider the matrix $I - \Delta A(\epsilon)$ on $G_d(\epsilon)$ and $G_{d+1}(\epsilon)$. There is a convenient way to label the states of the resulting Markov chain so as to have some recursion between these two representations. Indeed let the two matrices be denoted as Γ_d and Γ_{d+1} respectively. Then [16]

$$\Gamma_{d+1} = \begin{bmatrix} \Gamma_d & T & 0 & \cdots & 0 \\ T & \Gamma_d & T & \ddots & 0 \\ 0 & T & \ddots & \Gamma_d & T \\ 0 & \cdots & 0 & T & \Gamma_d. \end{bmatrix} \tag{5.5}$$

where T is a tridiagonal matrix with positive entries. Γ_{d+1} is an $n \times n$ block matrix. Furthermore it is straightforward to establish [16] that for any d, Γ_d is strongly diagonally dominant. Furthermore $I - A(\epsilon)\Delta$ has finite bandwidth [16]. The strong diagonal dominance of $I - \Delta A(\epsilon)$ implies that we will need no pivoting. As we shall see this property will help also in the selection of the relaxation scheme in the multigrid iteration.

The fundamental idea of multigrid (MG) algorithms is relatively easy to understand [14, 15]. The primary reasons for using MG methods are as follows. Direct solvers of discretized p.d.e's have computation time that grows linearly with n, the width of the finest grid, while MG methods can actually do much better than this as we shall see. As for relaxation schemes, slow convergence is a typical problem, although they are perfectly suited for parallel implementation, as they rely only on "local" information when a sweep is performed. Thus, relaxation schemes have a computation time that is independent of the size of the grid. Because of its naturally parallel properties, it turns out that the Multigrid method has a computation time that is essentially independent of the dimension of the problem. Because we wish to compute in real-time, such a numerical method is an ideal candidate for investigation.

We now described a one-cycle full Multigrid algorithm program. Let there be K point-grids which we will denote by $G_1, G_2, \ldots, G_K$ with the finest being G_K and the coarsest being G_1.

The finest grid contains the problem:

$$L^K U^K = f^K. \tag{5.6}$$

Smoothing Part I:

Given an initial approximation to the problem in (5.7), smooth j_1 times to obtain u^K.

Coarse-grid correction:

Compute the residual $d^K = f^K - L^K u^K$.

Inject the residual into the coarser grid G_{K-1},

$$d^{K-1} = I_K^{K-1} d^K.$$

Compute the approximate solution $\tilde{v}^{K-1}$ to the residual equation on G_{K-1}:

$$L^{K-1}\, v^{K-1} = d^{K-1}, \tag{5.7}$$

by performing $c \geq 1$ iterations of the Multigrid method, but this time we will be using the grids $G_{K-1}, G_{K-2}, \ldots, G_1$ applied to equation (5.7).

Interpolate the correction $\tilde{v}^K = I_{K-1}^K\, \tilde{v}^{K-1}$.

Compute the corrected approximation on G^K,

$$u^K + \tilde{v}^K. \tag{5.8}$$

Smoothing Part II:

Compute a new approximation to U^K by applying relaxation sweeps to $u^K + \tilde{v}^K$.

The recursive structure of the algorithm entering just after eq. (5.7) is apparent. Here the algorithm simply repeats itself, so in the case of $c = 1$ we have initial smoothing, computation of residual equation, injection to coarser grid, all until the coarsest grid is reached, where the equation is directly solved. Then we have interpolation upward through the grids, offering each finer grid an approximation for relaxation. This would be a "V-shape" structure as opposed to a "W-shape struture [16]. Only if $c > 1$, would we have a "W shape" structure. Of course it is clear that there can be many variants of the algorithm, depending on the number of interpolation and injection operations.

The convergence of the multigrid method is based on the following representation of the algorithm. For the general grid G_k we will have the equation

$$L^k U_k = f_k \quad , k = 1, \ldots, K. \tag{5.9}$$

The formula for obtaining a new approximation to the solution U_k from the old one u_k can be written as

$$\begin{aligned} \tilde{u}_k &= (I - ML^k)u_k + Mf_k \\ &= S_k u_k. \end{aligned} \tag{5.10}$$

Here S_k is the smoothing operation on the grid G_k, and we assume that M is invertible and the smoother is consistent. With this notation S_k^j denotes the smoother that uses j relaxation sweeps, or is applied j times.

As examples of smoothers, define D to be the matrix whose diagonal entries are equal to those of L^k, and which is zero everywhere else. Then $m = \omega D^{-1}$ is the modified Jacobi method. If T is the "upper triangular part" of L^k, and zero elsewhere, then we have the Gauss-Seidel method by setting $M = T^{-1}$. In fact, M is usually some approximation to the inverse of L^k, which forces $\rho(I - ML^k)$ to be close to zero.

By constructing the "Multigrid operator" we can show that, like any other iterative process, convergence is guaranteed under certain conditions.

Given u^k as the old approximation, the new approximation $\tilde{u}^k$ will be

$$\tilde{u}_k = M_k \; u_k + I_{k-1}^k (L^{k-1})^{-1} \; I_k^{k-1} f_k. \tag{5.11}$$

M_k will be the Multigrid operator on grid G_k we will concentrate on, for it is its spectral radius that determines whether the iteration converges or not. By M_k^c we will mean c multiples of the MG operator applied on the k grids. The following recursion defines this operator, which begins at grid level 2 and proceed up to $k = K - 1$ where we have K grid levels:

$$\begin{aligned} M_k^{k-1} &= S_k^{j_2}(I_k - I_{k-1}^k (L^{k-1})^{-1} \; I_k^{k-1} L^k) S_k^{j_1} \\ A_k^{k+1} &= S_{k+1}^{j_2} \; I_k^{k+1} : G_k \to G_{k+1} \\ A_{k+1}^k &= (L^k)^{-1} \; I_{k+1}^k (L^{k+1})^{-1} \; S_{k+1}^{j_1} : G_{k+1} \to G_k \end{aligned} \tag{5.12}$$

Thus we can write,

$$M_{k+1} = M_{k+1}^k + A_k^{k+1} \; M_k^c \; A_{k+1}^k. \tag{5.13}$$

Now if $\|M_{k+1}^k\|$, $\|A_k^{k+1}\|$ and $\|A_{k+1}^k\|$ for $k \leq K-1$ are known, then one can obtain an estimate of $\|M_K\|$, where $\|\cdot\|$ represents any reasonable operator norm; we refer to [17] for more detailed results on stability and convergence.

For the case of interest here, i.e. the discretization of the Zakai equation (5.2), following well known methodology for MG application we first identify a simpler, albeit characteristic problem. This is the problem with no input, i.e. when the right hand side of (5.2) becomes $V^k(\epsilon)$. In other words if one understands how MG is applied to the discretized Fokker-Planck equation

$$(I - \Delta A(\epsilon))\, V^{l+1}(\epsilon) = V^l(\epsilon), \tag{5.14}$$

then complete understanding of the application of MG to the Zakai equation is straightforward.

It follows [16] that the spectral radius of the associated MG operator (5.12) is determined entirely by the matrix $I - \Delta A(\epsilon)$, along with the choice of relaxation scheme. Also note that because $I - \Delta A(\epsilon)$ is not time dependent all *program parameters are precomputable.* In particular for the Zakai equation, they do not depend on the sample path $y(\cdot)$.

A highly recommended relaxation method in MG applications is the so called *successive overrelaxation method* (SOR). To define it suppose one wants to solve

$$Ax = b$$

with $a_{ii} \neq 0$. Then define B to be the $n \times n$ matrix

$$b_{ij} = \begin{cases} -a_{ij}/a_{ii}, & i \neq j \\ 0, & i = j \end{cases}$$

and define the vector c in R^n to have components, $c_i = b_i/a_{ii}$. Then let us consider the $L-U$ decomposition of B, $B = L + U$. Choose a real number ω, and define the iteration

$$x_{n+1} = \omega(Lx_{n+1} + Ux_n + c) + (1 - w)x_n. \tag{5.15}$$

This is the SOR method. If $\omega = 1$, the SOR method reduces to the Gauss-Seidel method, with $\omega > 1$ implying overcorrecting, and $\omega < 1$ implying undercorrecting.

Recall that the matrix $I - \Delta A(\epsilon)$, for the discretized Zakai equation, is strongly diagonally dominant. Furthermore this matrix is also an L-matrix, i.e. it has positive diagonal elements and non positive off diagonal elements. Finally this matrix is *consistently ordered* [16]. This is a consequence of the natural ordering on a rectangular grid. One can measure the properties of smoothing operators with a variety of measures [16]. So one can describe "optimal" smoothing operation. We thus have [16]:

Theorem 4: Because of the properties of $I - \Delta A(\epsilon)$, the MG operator converges and the optimal relaxation scheme for the Zakai equation is the SOR method. There is an optimal choice for ω in (5.15) with respect to convergence as well.

6. Architectures for Implementing MG in Real-Time.

In this section we analyze the complexity of the MG schemes described in section 5, in particular with respect to real time implementation. We shall see that the result is a multilayer processor network. Here the processors and the interconnections are more complicated than the ones used in the systolic architecture of [7]. So fabrication is a much harder problem.

The computing network will be a system of grids of identical processing elements. Therefore, we have two kinds of grids, one of points and one of processors, and these will be layered one on top of another. For each $1 \leq k \leq K$, processor grid P_k has $(n_k)^\gamma$ elements, where γ is a positive integer not greater than the problem dimension d. Also, we have $n_K = n$, and $n_i < n_j$, if $i < j$.

Similarly, in keeping with the above notation, there are, for each $1 \leq k \leq K$ a corresponding point grid G_k with $(n_k)^d$ points. (Note that the number of processors per grid is never greater than the number of points). Again we have $n_K = n$ while $n_i < n_j$ if $i < j$. A key assumption, which is quite realistic, is that for each step of the multigrid algorithm on point grid G_k, the processing grid P_k requires $O((n_k)^{d-\gamma})$ time to perform its computations.

To design a parallel machine capable of performing the MG algorithm, we assume our problem is in d dimensions over a rectangular domain using a regular point grid of n^d points. We further have

$$\begin{aligned} n_K &= n \\ n_{k+1} &= a(n_k + 1) - 1, \quad k = 1, 2, \ldots K - 1 \end{aligned} \tag{6.1}$$

for some integer $a \geq 2$. We map grid points in such a way that neighboring grid points reside in the same or neighboring processors.

Smoothing sweeps of at least some type can be accomplished in $O(n^{d-\gamma})$ time with this given connectivity. Let t be the time taken by a single processor to perform the operations at a single gridpoint that, done over the whole grid, constitute a smoothing sweep. Then, setting S as the time needed to perform the smoothing sweep over the whole of grid G_k on processor grid P_k, we have,

$$S = t\ n_k^{d-\gamma}. \tag{6.2}$$

Obviously, it is to our advantage to conduct as few smoothing sweeps as necessary and still assure sufficient accuracy.

Now processor grid P_k is connected to processor P_{k+1}. Processor $i \in P_k$ is connected to processor $a(i+1) - \mathbf{1} \in P_{k+1}$ where $\mathbf{1} = (1, 1, \ldots, 1)$. These connections allow any intergrid operations, such as interpolation, to be performed in $O(S)$ time. Now define the system of processor grids $\{P_1, P_2, \ldots, P_J\}$ as the machine M_J for $J = 1, 2, \ldots, K$. Then the execution of MG performed by M_k proceeds as follows:

1. First, j smoothing sweeps on grid G_k are done by P_k; all other processor grids idle.
2. The coarse grid equation is formed by P_k and transferred to P_{k-1}.
3. MG is iterated c times on grid G_{k-1} by M_{k-1}. P_k is idle.
4. The solution v^{k-1} is transferred to P_k by iterpolation: $I_{k-1}^k v^{k-1}$
5. The remaining m smoothing sweeps are done by P_k.

Now we let $W(n)$ be the time needed for steps $1, 2, 4, 5$ and find

$$W(n) = (j + m + s)\ t\ n^{d-\gamma}, \tag{6.3}$$

where s is the ratio of the time needed to perform steps 2 and 4 to the time needed for one smoothing sweep. Note that s is independent of n, d and γ.

We discuss now the time complexity of MG. We will denote by $T(n)$ the time complexity of the MG algorithm on a grid of n^d points. It turns out that $T(n)$ solves the recurrence:

$$T(an) = cT(n) + W(an), \tag{6.4}$$

where $W(an)$ denotes the work needed to pre-process and post-process the (an)-grid iterate before and after transfer to the coarser n-grid. In effect the term $W(an)$ includes the smoothing sweeps, the computation of the coarse grid correction equation (i.e., the right-hand side d^{k-1}) and the interpolation back to the fine grid $(I_{k-1}^k\ v^{k-1})$. Then we have

Theorem 5, [18]: Let $T_p(\cdot)$ be a particular solution of (6.4), i.e.,

$$T_p(an) = c\ T_p(n) + W(an).$$

Then the general solution of (6.4) is:

$$T(n) = \alpha\ n^{\log_a c} + T_p(n), \tag{6.5}$$

where α is an arbitrary constant. Using this result we have the general solution to (6.5),

$$T(n) = \begin{cases} \beta(a^p/(a^p - c))n^p & \text{if } c < a^p, \\ \beta n^p \log_a n + O(n^p) & \text{if } c = a^p, \\ O(n^{\log_a c}) & \text{if } c > a^p. \end{cases} \tag{6.6}$$

We see that it would take a single processor $O(n)$ steps to complete the above mentioned tasks on one dimension, while n processors could do the same for a two-dimensional problem in $O(n)$ time.

We say that the MG algorithm is of *optimal order* if $T(n) = O(n^{d-\gamma})$, a possibility that is sometimes precluded by some choices of c, a, γ and d, which in turn influence $T(n)$. Examination of (6.6) demonstrates the relations between the various parameters. As an example, in the one-processor case, with $\gamma = 0, d = 2$, we have $g(n) = n^2$. We then have an optimal scheme if $a = 2, c < 4$, for only then is $T(n) = O(n^2)$. But $c \geq 4$ is non-optimal, with $T(n) = O(n^2 \log n)$ for $c = 4$.

In general, we have an optimal scheme if and only if $c < a^d$.

There also exists a natural way to build a VLSI system to implement our algorithms. The $\gamma = 1$ machine can be embedded in two dimensions as a system of communicating rows of processors. The $\gamma = 2$ machine can be embedded in three dimensions as a system of communicating planes, and so on. Realizations in three-space will be possible in a natural way for any value of γ. Consider the case of $d = 2, \gamma = 2$. In this case, we have a set of homogeneous planar systolic arrays layered one on top of the other. If we let $a = 2, K = 3$, and $n_1 = 1, n_2 = 2(1+1) - 1 = 3, n_3 = 2(3+1) - 1 = 7$, we would have a 7×7 array on top of a 3×3 array which is then on top of a single processor corresponding to n_1. Unfortunately, this design differs from the classical systolic array concept of Kung [19] in that there exists no layout in which wire lengths are all equal. Also, each layer of the system is homogeneous while the entire machine is clearly not.

Now the four parameters c, a, γ, d are to be chosen with any implementation MG, and, of course, they are not unrelated to each other. Extending the earlier notation, we call any one choice of the four a design and denote its corresponding computing time by $T(c, a, \gamma, d)$. We will now begin with an examination of the trade-offs incurred by one choice over another. Following [18], an important issue is *efficiency*, E vs. *speedup* S in a particular design. We define,

$$S(c, a, \gamma, d) = T(c, a, 0, d)/T(c, a, \gamma, d) \tag{6.7}$$
$$E(c, a, \gamma, d) = T(c, a, 0, d)/(P(\gamma)T(c, a, \gamma, d))$$

Note that the speedup S corresponds to the gain in speed going from the one-processor system to that of the multiprocessor. Whereas the efficiency E reflects the trade-off between using more processors vs. time.

We say that a design $T(c, a, \gamma, d)$ is *asymptotically efficient* if E tends to a constant as $n \to +\infty$, and it will be *asymptotically inefficient* if $E \to 0$ asn $n \to +\infty$.

Theorem 6, [18]: Let $\gamma > 0$.

1) If $c < a^{d-\gamma}$ then $E(c,a,\gamma,d) = (a^\gamma - 1)(a^{d-\gamma} - c)/(a^d - c)$.
2) If $c = a^{d-\gamma}$ then $E(c,a,\gamma,d) = (a^\gamma - 1)a^{d-\gamma}/((a^d - c)\log_a n)$.
3) If $c > a^{d-\gamma}$ then

$$E(c,a,\gamma,d) = \begin{cases} O(1/n^{\log_a(c-d+\gamma)}) & \text{if } c < a^d \\ O((\log_a n)/n^\gamma) & \text{if } c = a^d \\ O(1/n^\gamma) & \text{if } c > a^d \end{cases}$$

We have at once that

1) *A design is asymptotically efficient if and only if* $c < a^{d-\gamma}$.

2) *The fully parallel design* $\gamma = d$, *is always asymptotically inefficient.*

3) "Halfway" between asymptotic efficiency and inefficiency is *logarithmic asymptotic efficiency*, with $E = O(\log n)$, as $n \to= \infty$. *A fully parallel design* ($\gamma = d$) *if logarithmically asymptotically efficient iff* $c = 1$.

4) If we start with a non-optimal design in the one processor case, then adding more processors will not make the design asymptotically efficient.

To get $T(n) = O(n)$ we have to select $c = 1$.

We also have considered the concurrent iteration schemes of Gannon and Van Rosendale [20].

Thus the fully parallel architecture has a computation time of at most $O(\log n)$ and so it is very competitive with the systolic direct solver. More importantly, this time is largely independent of dimension d, at least for small values of d. Of course, increases in d will result in large increases in circuit layout area, due to an increase in interconnections between grids, and thus a subsequent loss in computing speed.

We can implement the SOR method in a parallel fashion, using the "red-block" or "checker board" method [16]. In higher dimensions we need to utilize multicolor ordering. Employing the intrinsic locality of the SOR we can implement it asynchronously as well.

A detailed analysis of timing performed in [16] demonstrates that if the real-time constraint for the Zakai equation is 1 msec, then we can realistically achieve real-time implementation with the multi-layered networks of this section, only for dimension $d \leq 8$.

References

[1] M. Zakai, "On the Optimal Filtering of Diffusion Processes", *Z. Wahr. Verw. Geb.,* 11, pp. 230-243, 1969.

[2] M. Hazewinkel and J.C. Willems, edts, *Stochastic Systems: The Mathematics of Filtering and Identification*, Proc. of NATO Advanced Study Institute, Les Arcs, France, Dordrecht, The Netherlands: Reidel 1981.

[3] J. Doob, *Stochastic Processes*, Wiley, 1953.

[4] E. Wong and B. Hajek, *Stochastic Processes in Engineering Systems*, Springer-Verlag, 1985.

[5] A. Bensoussan, "Maximum Principle and Dynamic Programming Approaches to the Optimal Control of Partially Observed Diffusions", *Stochastics*, 9, pp. 169-222, 1983.

[6] R.S. Liptser and A.N. Shiryayev, *Statistics of Random Processes I: General Theory*, Springer-Verlag, New York, 1977.

[7] John S. Baras and Anthony LaVigna,"Real-Time Sequential Detection for Diffusion Signals", *Proc. 26th Conf. on Decision and Control*, pp. 1153-1157, 1987.

[8] A. LaVigna, "Real Time Sequential Hypothesis Testing for Diffusion Signals," M.S. Thesis, Univ. of Maryland, 1986.

[9] H.J. Kushner, *Probability Methods for Approximations in Stochastic Control and for Elliptic Equations*, Academic Press, New York, 1977.

[10] J. Schröder, "M-Matrices and Generalizations Using an Operator Theory Approach," *SIAM Review*, 20, pp. 213-244, 1978.

[11] E. Pardoux and D. Talay, "Discretization and Simulation of Stochastic Differential Equations," in *Publication de Mathematiques Appliquees Marseille-Toulon*, Université de Provence, Marseille, 1983.

[12] Y. Yavin, "Numerical Studies in Nonlinear Filtering," in *Lecture Notes in Control and Information Sciences 65*, Springer-Verlag, New York, 1985.

[13] H.T. Kung and C.E. Leiserson, "Algorithms for VLSI Processor Arrays," in *Introduction to VLSI Systems*, C. Mead and L. Conway, pp. 271-292, Addison-Wesley, Reading, Mass., 1980.

[14] S.F. McCormick, Edt, *Multigrid Methods*, Frontiers in Applied Mathematics, SIAM 1987.

[15] W.L. Briggs, *A Multigrid Tutorial*, SIAM, 1987.

[16] K. Holley, "Applications of the Multigrid Algorithm to Solving the Zakai Equation of Nonlinear Filtering With VLSI Implementation", Ph.D. Thesis, University of Maryland, December 1986.

[17] K. Stüben and V. Trottenberg, "Multigrid Methods: Fundamental Algorithms, Model Problem Analysis and Applications", in *Multigrid Methods*, W. Hackbusch and W. Trottenberg (edts), Springer Verlag, 1982.

[18] T. Chen and R. Schreiber, "Parallel Networks for Multi-grid Algorithms: Architecture and Complexity", *SIAM J. Sci. Stat. Comput.*, Vol. 6, No. 3, July 1985.

[19] H.T. Kung, "Systolic Algorithms", in *Large Scale Scientific Computation*, Academic Press, 1984.

[20] D. Gannon and J. Van Rosendale, "Highly Parallel Multigrid Solvers for Elliptic PDE's: An Experimental Analysis", ICASE Report No. 82-36, Nov. 1982.

Quadratic Approximation by Linear Systems Controlled From Partial Observations

V.E. Beneš
Dept. of Industrial Engineering
New Jersey Institute of Technology
Newark, NJ 07102
and
Dept. of Statistics
Columbia University
New York, NY 10027

Abstract

A class of quadratic feedforward control problems with linear dynamics forced by a semi-martingale is solved by recourse to the "predicted miss" and a Riccati equation.

I. Introduction

Many attempts at solving optimal control problems with partial observations have been based on using the conditional probability distribution of the unobserved state as a new state variable, moving according to the evolution equation of nonlinear filtering [3,4]. This approach typically leads to a Bellman–Hamilton–Jacobi equation with an

ISBN 0-12-481005-5

infinite–dimensional state. Haussmann's monograph [5] is an outstanding exception; it avoids the BHJ equation in favor of variational methods.

Recent work [6] by K. Helmes and R.W. Rishel presents examples of the former approach in which, thanks to a quadratic criterion and to the "feedforward" nature of the problems, the infinite–dimensional problem can be handled successfully: the concept of *predicted miss* provides a key for describing the optimal control without actually solving the hard part of the associated BHJ equation.

We show that the predicted miss approach also works for feedforward problems far more general than those based on Markov processes as in [6]. The novel feature is this: the Riccati equation arrived at in [6] (by guessing from BHJ?) can be derived from a variation and the properties of martingales, and so is a more pervasive property than was hitherto understood.

Our results concern controlled linear systems perturbed by outside influences which are semi–martingales with respect to the data available for control. They indicate that when the (final value) criterion is quadratic, the basic results remain the same as in [6]: linearity of the optimal control in the predicted miss, through a matrix satisfying a Riccati equation.

2. The Problem

Let (Ω, Y, P) be a probability space, on which is given a random variable η with values in R^n and $E|\eta|^2 < \infty$, a filtration $\{F_t, 0 \le t \le T\}$ satisfying the "usual" conditions, and a square–integrable semimartingale z taking values in R^n. The filtration represents the information available for control. The class H of admissible controls consists of R^m-valued processes u adapted to $\{F_t\}$ with

$$\| u \| = E\int_0^T |u_s|^2 ds < \infty$$

Let A,B be $n \times n$, $n \times m$ matrices, respectively. The task is this: to approximate η at the final time T by driving a linear system with dynamics

$$dx_t^u + Ax_t^u\, dt + Bu_t dt + dz_t\,, \qquad x_0 \text{ given}$$

with an admissible control $u \in H$, so that the functional

$$J(u) = E|\eta + x_T^u|^2 + \lambda E\int_0^T |u_s|^2 ds \qquad \lambda > 0$$

is a minimum.

This setup generalizes (and was inspired by) that of Helmes and Rishel [6], in which $F_t = \sigma\{x_s, 0 \le s \le t \le T\}$ is the filtration corresponding to a noisy controlled observation satisfying

$$dx_t = Ax_t dt + Bu_t dt + h(y_t)dt + dw_t \qquad x_0 = a$$

with y a diffusion independent of w. This case assumes the form above when we pass to the innovations representation

$$h(y_t)dt + dw_t = \hat{h}_t dt + d\nu_t$$

$$\hat{h}_t = E\{h(y_t)|F_t\}$$

and take $dz_t = \hat{h}_t dt + d\nu_t$.

The existence of an optimal u^* can be concluded from the facts that $J(u)$ is a convex functional on the Hilbert space H, lower semicontinuous in the norm topology, and coercive. We shall find this u^* by imposing the variational condition

$$\frac{d}{d\epsilon} J(u^* + \epsilon v)|_{\epsilon=0} = 0, \qquad \text{for every } v \in H.$$

One can calculate that

$$\frac{d}{d\epsilon} J(u + \epsilon v) = 2E\left\{ \epsilon\lambda \int_0^T |v_s|^2 ds + \lambda \int_0^T v_s' u_s ds \right.$$

$$+ \epsilon \left| \int_0^T e^{A(T-s)} B v_s ds \right|^2$$

$$+ \int_0^T v_s' (e^{A(T-s)} B)' E\{\eta + x_T^{u^*} | y_t\}$$

Hence the variational condition on u^* leads to

$$\lambda u_t^* + (e^{A(T-t)} B)' E\{\eta + x_T^{u^*} | y_t\} = 0$$

This extremal condition, together with the dynamical equation defining x^u from u, shall allow us to calculate u^*.

3. Predicted Miss

Thinking of η as a "target" random variable, and of x_t as a "pursuit" process partly under our control, we can entertain the amount $\eta + x_T$ by which we would miss η with x_T, if we do no control whatsoever in (t, T), i.e., if $u_s \equiv 0$ there. The *predicted miss* r_t is the conditional expectation of this miss, when the control is null from t until T, with respect to the data F_t at t:

$$r_t = E\{y + x_T^u | F_t\}, \quad \text{with } u_s \equiv 0 \text{ on } (t, T]$$

In this case of null control, at each t

$$x_T = e^{(T-t)A}x_t + \int_t^T e^{(T-s)A}dz_s$$

$$x_t = e^{tA}x_0 + \int_0^t e^{(t-s)A}Bu_s + \int_0^t e^{(t-s)A}dz_s$$

so

$$r_t = e^{tA}x_0 + \int_0^t e^{(T-s)A}Bu_s ds + E\{\eta + Z | F_t\}$$

where

$$Z = \int_0^T e^{(T-s)A}dz_s$$

It is known that the transformation to "predicted miss" usually has the virtue of suppressing or eliminating all dynamics except a (bounded variation) control term, and a martingale; we find in this case that

$$dr_t = e^{(T-t)A}Bu_t dt + dE\{\eta + Z | F_t\}$$

By construction, the predicted miss is defined for any control process $u \in H$, and is causal in u. The fundamental intuition behind introducing the predicted miss r_t is the guess that it alone is relevant to pushing x_t around so as to best approximate η at time T. This kind of guess has been proved correct in a number of examples [1,2], among them both fully and partially observed cases, some for more general criteria than the quadratic one assumed here. In the present case, by analogy with the classical linear regulator [5] and the Kalman filter, we can guess further that the optimal control u is to be a linear function of the predicted miss. It is possible, as in [6], to start with this guess and prove that it is right by showing that with its help one can "solve" the BHJ equation, using the predicted miss and the conditional density as state variables: these variables separate, and only the finite-dimensional one, the predicted miss, is pertinent to understanding the structure of the control problem. Computing the answer of course requires both. Our approach shows, under broad conditions not providing even a formal BHJ equation, that the guess can be proved from a variational condition by stochastic analysis. The point is that linearity of the optimal control in r_t, while intuitive and natural, comes directly from the variational argument and need not be adduced as an Ansatz.

4. Solution

The principal result can be stated thus:

Theorem: The unique extremal admissible control u* is given by the formula

$$u_t^* = -\frac{1}{\lambda} B'(e^{(T-t)A})' \Phi_t r_t^*$$

where

(i) Φ is the matrix solution of the Riccati equation

$$\Phi_T = I$$

$$\Phi_t = \Phi_t B(t)B(t)'\Phi_t$$

$$B(t) = e^{(T-t)A}B$$

(ii) r_t^* is the predicted miss corresponding to u^*, and satisfying

$$dr_t^* = B(t)u_t^* dt + dE\{\eta + Z|F_t\}$$

$$r_0^* = e^{TA}x_0 + E\{\eta + Z|F_0\}$$

(iii) the martingale $E\{\eta + x_T^{u^*}|F_t\}$ appearing in the variational condition has the representation

$$dM_t = dE\{\eta + x_T^{u^*}|F_t\} = \Phi_t dE\{\eta + Z|F_t\} = \Phi_t dm_t$$

Proof: We use the notations

$$M_t = E\{\eta + x_T^{u^*}|F_t\}$$

$$m_t = E\{\eta + Z|F_t\}$$

The arguments rest on an explicit representation of M_t in terms of m_t, stated in (iii) above. If the variational condition holds, then

$$\eta + x_T^{u^*} = \eta + e^{TA}x_0 + \int_0^T e^{(T-s)A}Bu_s^* ds + \int_0^T e^{(T-s)A}dz_s$$

$$= e^{TA}x_0 + \int_0^T B(s)u_s^* ds + \eta + Z$$

$$= e^{TA}x_0 - \frac{1}{\lambda}\int_0^T B(s)B(s)'E\{\eta + x_T^{u^*}|F_s\}ds + \eta + Z$$

Taking the conditional expectation of each side with respect to F_t, we find

$$M_t = e^{TA}x_0 - \frac{1}{\lambda}\int_0^t B(s)B(s)'M_s ds - \frac{1}{\lambda}E\{\int_t^T B(s)B(s)'M_s ds\,|\,F_t\} + m_t$$

$$= e^{TA}x_0 - \frac{1}{\lambda}\int_0^t B(s)B(s)'M_s ds - \frac{1}{\lambda}\int_t^T B(s)B(s)'ds M_t + m_t$$

Hence, applying the stochastic differential

$$dM_t = \frac{-1}{\lambda}\int_t^T B(s)B(s)'ds)^{-1}dm_t$$

$$= \left(I + \frac{1}{\lambda}\int_t^T B(s)B(s)'ds\right)^{-1} dm_t$$

Let Φ be the inverse matrix in the last expression. We can differentiate the identity

$$\Phi_t(I + \frac{1}{\lambda}\int_t^T B(s)B(s)'ds) = I$$

to obtain the Riccati DE

$$\lambda\dot{\Phi}_t = \Phi_t(B(t)B(t)'\Phi_t\ , \Phi_T = I$$

and the representations

$$M_t = \Phi_0(e^{TA}x_0 + m_0) + \int_0^t \Phi_s dm_s$$

$$u_t^* = -\frac{1}{\lambda}B(t)'M_t$$

The connection to the predicted miss r_t is now completed by showing that $M_t = \Phi_t r_t$. We calculate that $M_0 = \Phi_0 r_0$, and that

$$d(M_t - \Phi_t r_t) = dM_t - \dot{\Phi}_t r_t dt - \Phi_t B(t)u_t^* dt - \Phi_t dm_t$$

$$= \Phi_t B(t)B(t)'(M_t - \Phi_t r_t)dt$$

Therefore if $\Psi(t,s)$ is the matrix solution of

$$\dot{\Psi} = \Phi_t B(t) B(t)' \Phi \qquad \Phi(s,s) = I, \quad t > s$$

then

$$M_t - \Phi_t r_t = \Psi(t,0)(M_0 - \Phi_0 r_0) \equiv 0$$

5. Example: A Guidance Problem

The path of a target is represented by a stochastic process η_t taking values in R^n. Let $h(t,\eta)$ be causal R^m-valued functional of this path, square-integrable with probability one. We receive noisy data about the path η_t in the form of an observation equation

$$dy_t = h(t,\eta)dt + n_t dt + db_t \qquad y_0 = 0, \text{ say}$$

where n_t is an R^m-valued noise, assumed square-integrable, and db_t is an m-dimensional Brownian motion independent of the target process η_t. Our problem is to control an interceptor with coordinates $x_t \epsilon R^n$ to be near η_t at time T, while satisfying the dynamical equations

$$dx_t = (Ax_t + Bu_t)dt + dz_t, \qquad x_0 \text{ given}$$

with dz_t a martingale on the filtration generated by the data $Y_t = \sigma\{y_s, 0 \le s \le t\}, u_t$, a control process adapted to the data, and A and B are matrices of appropriate dimension. The criterion of performance is the quadratic functional

$$E|\eta_T + x_T|^2 + \lambda E \int_0^T |u_t|^2 dt,$$

We take $\eta = \eta_T$, $F_t = Y_t$, and discuss the martingale $m_t = E\{\eta + Z|Y_t\}$, that drives the predicted miss process. Under our assumptions this conditional expectation can be represented by a version of the Kallianpur–Striebel formula [7]:

$$E\{\eta + Z|Y_t\} = \frac{E^{\eta,n}\eta_T q(h + n, y)_t}{E^{\eta,n}q(h + n, y)_t} + E\{Z|Y_t\} \equiv F(t, y) + E\{Z|Y_t\}$$

where q is the Cameron–Martin–Girsanov functional:

$$q(f, g)_t = \exp\{\int_0^t f_s' dg_s - \frac{1}{2}\int_0^t |f_s|^2 ds\}$$

The guidance system can then be depicted by the following block diagram:

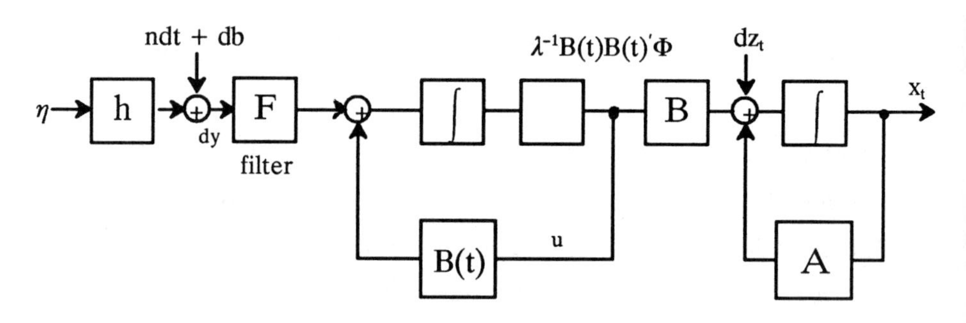

6. Example: A Disturbance Rejection Problem

Let ξ_t be a square-integrable "disturbance" process, and let w_t be an independent Brownian motion. We are to observe and control a process x_t solving

$$dx_t = Ax_t dt + Bu_t dt + \xi_t dt + dw_t$$

by a control process u_t adapted to $F_t = \sigma\{x_s, s \leq t\}$, so as minimize the criterion

$$E|x_T|^2 + \lambda E \int_0^T |u_s|^2 ds$$

More precisely, the admissible control processes are those for which there is a corresponding strong solution of (1) such that u_t is $\sigma\{x_s, 0 \leq s \leq t\}$–measurable. In this case [6] each F_t is simply

$$\sigma\{\int_0^s \xi_v dv + w_s, \quad 0 \leq s \leq t\}$$

To proceed, we cast equation (1) into the innovations form

$$dx_t = Ax_t dt + Bu_t dt + \hat{\xi}_t dt + d\nu_t$$

and "solve" the filtering problem of calculating $\hat{\xi}_t = E\{\xi_t | F_t\}$ by appeal to the formula (in [7])

$$\hat{\xi}_t = \frac{E^{\xi}\xi_t q(\xi, x)_t}{E^{\xi} q(\xi, x)_t}$$

The process $dz_t = \hat{\xi}_t dt + d\nu_t$ is now a semi–martingale on F_t and the theory of Section 4 is applicable. The case of Helmes and Rishel [6] is recovered upon taking $\xi_t = h(y_t)$, with y_t a diffusion: $dy_t = f(y_t)dt + dv_t$, v_t Brownian.

7. Acknowledgements

The value of conversation with I. Karatzas and R.W. Rishel is appreciated.

References

[1] V.E. Beneš, "Full Bang" to Reduce Predicted Miss is Optimal, SIAM J. Control and Optimiz., vol. 14, No. 1, January 1976, pp. 62–84.

[2] V.E. Beneš and R.W. Rishel, Optimality of "Full Bang to Reduce Predicted Miss" For Some Partially Observed Stochastic Control Problems, IMA Volume 10, Stochastic Differential Systems, Stochastic Control Theory and Applications, W. Fleming and P.L. Lions, Editors, Springer–Verlag, 1988.

[3] A. Bensoussan, Optimal Control of Partially Observed Diffusions, in "Advances in Filtering and Optimal Stochastic Control," W.H. Fleming and L.A. Gorostiza (Eds.), Proceeding of the IFIP–WG 7/1 Working Conference, Springer Lecture Notes in Control and Information Sciences, vol. 42, 1982.

[4] W.H. Fleming, Nonlinear Semigroup for Controlled Partially Observed Diffusions, SIAM J. Control and Optimiz. Vol. 20, No. 2, March 1982.

[5] G. Haussmann, A Stochastic Maximum Principle For Optimal Control of Diffusions, Pitman Research Notes in Mathematics 151, Longman 1986.

[6] K. Helmes and R.W. Rishel, An Optimal Control Depending on the Conditional Density of the Unobserved State, preprint, 1990.

[7] G. Kallianpur, Stochastic Filtering Theory, Applications of Mathematics 13, Springer–Verlag, 1980, p. 281.

A model of stochastic differential equation in Hilbert spaces applicable to Navier Stokes equation in dimension 2

A. Bensoussan
University Paris Dauphine and INRIA

1 Introduction

Zakai equation of non linear filtering has been an important source of motivation for studying linear and non linear stochastic partial differential equations (see in particular E. Pardoux [7]).

In a different context, there has been many articles presenting extensions of classical results of existence (or uniqueness) of solutions of deterministic P.D.E. to the stochastic case (see for instance A. Bensoussan, R. Temam [2], [3]). We consider here a model applicable to Navier Stokes equation, in dimension 2. The technique which is used is inspired from ideas of Nagase [6], and used previously by the author in connection with the splitting up approximation (see A. Bensoussan, R. Glowinski, A. Rascanu [1]).

2 Setting of the problem

2.1 Deterministic case

Let V, H be separable Hilbert spaces with

$$V \subset H, \quad \text{with dense and compact injection.} \tag{2.1}$$

ISBN 0-12-481005-5

We identify H and its dual, and call V' the dual of V, hence we have the sequence

$$V \subset H \subset V' \tag{2.2}$$

with continuous and dense injection. $| \ |$ and $|| \ ||$ denote the norms in H and V respectively.

let A, B be operators from V to V' such that

$$A \in L(V;V'), \quad \langle Av, v\rangle \geq \alpha ||v||^2, \quad \forall v \in V, \quad \alpha > 0 \tag{2.3}$$

$$\begin{aligned} &B \quad \text{is non linear and satisfies} \\ &||B(v_1) - B(v_2)||_{V'} \leq C(|v_1|^{1/2}||v_1||^{1/2} + \\ &\quad +|v_2|^{1/2}||v_2||^{1/2} + 1)|v_1 - v_2|^{1/2}||v_1 - v_2||^{1/2} \end{aligned} \tag{2.4}$$

$$\begin{aligned} &|\langle B(v), v\rangle| \leq \beta||v||^2 + C_\beta(|v|^2 + 1) \\ &\forall \beta > 0. \end{aligned} \tag{2.5}$$

Clearly B is continuous from V weakly to V'. This framework applies to Navier Stokes equation in dimension 2.

Let indeed $H =$ closure in $(L^2(\Omega))^2$, of the vector space of functions $\varphi \in C_0^\infty(\Omega)$, such that $\text{div}\varphi = 0$

$$V = \{v \in (H_0^1(\Omega))^2, \quad div \ v = 0\}$$

where Ω is a bounded domain of R^2.

Define for $u, v \in V$

$$\begin{aligned} \langle B(u), v\rangle &= \int_\Omega Duu.vdx \\ &= \sum_{i,j=1}^{2} \int_\Omega \frac{\partial u_i}{\partial x_j} u_j v_i dx = -\sum_{i,j=1}^{2} \int_\Omega u_i u_j \frac{\partial v_i}{\partial x_j} dx \end{aligned}$$

we have

$$\left|\int_\Omega Duu.vdx\right| \leq |Dv| \ |u|_{L^4}^2.$$

But for $n = 2$, one has the estimate

$$|u|_{L^4} \leq C||u||^{1/2}|u|^{1/2}$$

hence

$$|\langle B(u), v\rangle| \leq C\|u\| \, |u| \, \|v\|.$$

In particular $B(u) \in V'$. Next

$$\langle B(u) - B(u^*), v\rangle = -\sum_{i,j=1}^{2} \int_\Omega (u_i u_j - u_i^* u_j^*) \frac{\partial v_i}{\partial x_j} dx$$

hence

$$\begin{aligned} |\langle B(u) - B(u^*), v\rangle| &\leq C\|v\| \, |u - u^*|_{L^4} (|u|_{L^4} + |u^*|_{L^4}) \\ &\leq C\|v\| (|u|^{1/2}\|u\|^{1/2} + |u^*|^{1/2}\|u^*\|^{1/2}) \\ &\quad |u - u^*| \, \|u - u^*\|^{1/2} \end{aligned}$$

which yields (2.4). Finally

$$\langle B(u), u\rangle = \frac{1}{2} \sum_{i,j=1}^{2} \int_\Omega \frac{\partial}{\partial x_j} u_i^2 u_j dx = 0 \tag{2.7}$$

and (2.5) is trivially satisfied.

We first state the classical result (see J.L. Lions [5]).

Theorem 2.1 *We make the assumptions (2.3), (2.4), (2.5). Let* $y_0 \in H$, $f \in L^2(0,T;V')$, *then there exists one and only one solution* y *in* $W(0,T)$, *i.e.* $y \in L^2(0,T;V)$, $y' \in L^2(0,T;V)$ *such that*

$$\begin{aligned} & y' + Ay + B(y) = f \\ & y(0) = y_0 \end{aligned} \tag{2.8}$$

Proof

Let us consider an orthonormal basis of H, originated from the canonical isomorphism $\wedge$ of V into V'. Let w_j be such a basis, we have

$$((v, w_j)) = \lambda_j (v, w_j), \quad \forall v \in V$$

and λ_j is an increasing sequence, $\lambda_j \uparrow \infty$.

Moreover $\frac{w_j}{\sqrt{\lambda_j}}$ is an orthonormal basis of V, and $w_j\sqrt{\lambda_j}$ an orthonormal basis of V'. We denote by P_m the projector on $w_1 \cdots w_m$ in H. It is also a projector in V and in V' (on $\frac{w_j}{\sqrt{\lambda_1}} \cdots \frac{w_m}{\sqrt{\lambda_m}}$ and $w_1\sqrt{\lambda_j} \cdots w_m\sqrt{\lambda_m}$ respectively).

We look for $y_m(t)$ in the span of $w_1 \cdots w_m$ such that

$$\begin{aligned} &y_m' + P_m A y_m + P_m B(y_m) = P_m f \\ &y_m(0) = P_m y_0. \end{aligned} \tag{2.9}$$

This differential equation has a solution in $[0, T_m]$. A priori estimates will show that one can take $T_m = T$. We have

$$\frac{1}{2}\frac{d}{dt}|y_m|^2 + \langle Ay_m, y_m\rangle + \langle B(y_m), y_m\rangle = \langle f, y_m\rangle$$

hence

$$\frac{1}{2}\frac{d}{dt}|y_m|^2 + \alpha\|y_m\|^2 \leq \frac{\alpha}{4}\|y_m\|^2 + C_\alpha(|y_m|^2 + 1 + \|f\|^2) + \frac{\alpha}{4}\|y_m\|^2$$

and thus

$$y_m \text{ remains in a bounded set of } L^2(0,T;V) \cap L^\infty(0,T;H)$$

Now from the equation (2.9) we deduce, thanks to (2.4) which yields, in particular

$$\|B(v)\|_{V'} \leq C(1 + \|v\|\,|v|) \tag{2.10}$$

that y_m' remains in a bounded set of $L^2(0,T;V')$.

Let now y_m be a sequence which converges to y weakly in $L^2(0,T;V)$, strongly in $L^2(0,T;H)$, a.e. t in H, and y_m' converges weakly in $L^2(0,T;V')$. From (2.4) it follows that $B(y_m) \to B(y)$ in $L^2(0,T;V')$ weakly (not strongly). Indeed from (2.10) $B(y_m)$ remains in a bounded subset of $L^2(0,T;V')$. Let us show that $B(y_m) \to B(y)$ in $L^1(0,T;V')$ weakly. Since the dual of $L^1(0,T;V')$

is $L^\infty(0,T;V)$ (by the separability), we have to prove that if $z \in L^\infty(0,T;V)$ then

$$\left| \int_0^T \langle B(y_m) - B(y), z \rangle dt \right| \to 0.$$

We use (2.4) to assert that

$$\int_0^T |\langle B(y_m) - B(y), z \rangle| dt \leq C \int_0^T |y_m - y|^{1/2} (\|y_m\| + \|y\| + 1) dt$$

since y_m, y are in $L^\infty(0,T;H)$ and z in $L^\infty(0,T;V)$.

Since y_m remains in a bounded subset of $L^2(0,T;V)$ and $y_m \to y$ in $L^2(0,T;H)$, the left hand side of this inequality tends to 0 as $m \to \infty$.

Clearly y is a solution of (2.8).

Let us next prove uniqueness. Pick y, y^* two solutions, then setting $\tilde{y} = y - y^*$, we deduce

$$\frac{d\tilde{y}}{dt} + A\tilde{y} + B(y) - B(y^*) = 0$$
$$\tilde{y}(0) = 0,$$

and

$$\frac{1}{2}\frac{d}{dt}|\tilde{y}|^2 + \langle A\tilde{y}, \tilde{y} \rangle + \langle B(y) - B(y^*), \tilde{y} \rangle = 0.$$

Now from (2.4)

$$|\langle B(y) - B(y^*), \tilde{y} \rangle| \leq C \|\tilde{y}\|^{3/2} |\tilde{y}|^{1/2} (\|y\|^{1/2} + \|y^*\|^{1/2} + 1)$$

using the fact that y, $y^* \in L^\infty(0,T;H)$, hence

$$\leq \frac{\alpha}{2}\|\tilde{y}\|^2 + C_\alpha |\tilde{y}|^2 (\|y\|^2 + \|y^*\|^2 + 1).$$

Therefore

$$\frac{d}{dt}|\tilde{y}|^2 \leq C|\tilde{y}|^2 (\|y\|^2 + \|y^*\|^2 + 1)$$
$$\tilde{y}(0) = 0$$

which implies $\tilde{y}(t) = 0$.

2.2 Stochastic set up

Define next maps $g(v)$, $G(v)$ such that

$$\begin{aligned} &g \text{ is a nonlinear map from } H \text{ to } H, \text{ continuous} \\ &|g(v)| \leq \bar{g}(1+|v|) \end{aligned} \tag{2.11}$$

$$\begin{aligned} &G \text{ is a nonlinear map from } H \text{ to } H^m, \text{ continuous} \\ &|G(v)|_{H^m} \leq \bar{G}(1+|v|) \end{aligned} \tag{2.12}$$

We shall consider solutions of

$$\begin{aligned} &dy + (Ay + B(y) - g(y))dt = G(y)dw \\ &y(0) = y_0 \in H \end{aligned} \tag{2.13}$$

in the following sense.

A solution of (2.13) on $[0,T]$ is a system $(\Omega, \mathcal{A}, P, F^t, w, y)$ where

$$\begin{aligned} &\Omega, \mathcal{A}, P \text{ is a probability space, } F^t \text{ is a filtration on } (\Omega, \mathcal{A}, P), \\ &w(t) \text{ a } F^t \text{ standard Wiener process with values in } R^m, \\ &y(t) \text{ an element of } L^2_F(0,T;V) \cap L^2(\Omega, \mathcal{A}, P; C(0,T;H)) \end{aligned} \tag{2.14}$$

and (2.13) holds in the following sense

$$\begin{aligned} &\text{a.s. } y(t) + \int_0^t (Ay(s) + B(y(s)) - g(y(s))ds = \\ &= y_0 + \int_0^t G(y(s))dw(s) \\ &\forall t \in [0,T]. \end{aligned} \tag{2.15}$$

We recall that

$$L^2_F(0,T;V) = \{z(t;w) \in L^2((0,T) \times \Omega; dt \otimes dP; V| \\ \text{a.e. } t\ z(t) \in L^2(\Omega, F^t, V)\}$$

Our objective is to prove the

Theorem 2.2 *We make the assumptions (2.3), (2.4), (2.5) and (2.11), (2.12). Let $y_0 \in H$. Then for any $T > 0$, there exists a solution of (2.13) in the sense (2.14), (2.15).*

3 Approximation scheme

3.1 The algorithm

Let $k = \dfrac{T}{N+1}$. We define a map $\Psi_k : C(0,T;R^m) \rightarrow L^2(0,T;H)$ where $C(0,T;R^m)$ is equipped with the uniform convergence, as follows. We write

$$z_k(.) = \Psi_k(b(.)), \quad \text{where } b(.) \in C(0,T;R^m).$$

It is defined as follows. let $r = 0, \dots, N$, set

$$z_k(rk) = z_k^r.$$

Successively, we have

$$z_k' + Az_k + B(z_k) = 0, \qquad t \in [rk,(r+1)k[. \tag{3.1}$$

Given z_k^r, (3.1) defines $z_k((r+1)k-0)$. Then set

$$\begin{aligned} z_k^{r+1} &= z_k((r+1)k-0) + \int_{rk}^{(r+1)k} g(z_k(s))ds \\ &+G(z_k^r).(b((r+1)k) - b(rk)) \end{aligned} \tag{3.2}$$

and we start with

$$z_k^0 = y_0. \tag{3.3}$$

Successively (3.1), (3.2) defines z_k on $[0,T[$. We set for completeness

$$z_k(T) = z_k^{N+1} \tag{3.4}$$

The function $z_k(t)$ is continuous to the right, and has left limits, at discontinuity points $k \dots (N+1)k$. The map Ψ_k is continuous from $C(0,T;H)$ to $L^2(0,T;V)$.

Now let $\tilde{\Omega} = C(0,T;R^m)$, $\tilde{\mathcal{A}}$ the Borel σ-algebra, and $\tilde{P}$ the Wiener measure on $\tilde{\Omega}$, $\tilde{\mathcal{A}}$. We define a random variable in $L^2(0,T;V)$ by the setting

$$z_k(.;\tilde{\omega}) = \Psi_k(b(.;\tilde{\omega})), \text{ for } \tilde{\omega} \in \tilde{\Omega}. \tag{3.5}$$

3.2 A priori estimates

We shall prove the following a priori estimates

Lemma 3.1 *We have*

$$\tilde{E}\int_0^T ||z_k(t)||^2 dt \leq C \tag{3.6}$$

$$\tilde{E}|z_k(t)|^2 \leq C, \quad \tilde{E}|z_k(t)|^4 \leq C, \quad \forall t \in [0,T]. \tag{3.7}$$

Proof

We have from (3.1)

$$\begin{aligned} & |z_k(t)|^2 + 2\int_{rk}^t \langle Az_k, z_k\rangle ds + 2\int_0^t \langle B(z_k), z_k\rangle ds = |z_k^r|^2, \\ & t \in [rk,(r+1)k[. \end{aligned} \tag{3.8}$$

Using (1.5) we deduce

$$\begin{aligned} & |z_k(t)|^2 \leq e^{c_0 k}|z_k^r|^2 + c_1 k, \quad \forall t \in [rk,(r+1)k[\\ & \int_{rk}^{(r+1)k} ||z_k||^2 ds \leq c_2 k(e^{c_0 k}|z_k^r|^2 + c_1 k) \end{aligned} \tag{3.9}$$

Consider next the process

$$\begin{aligned} & \tilde{z}_k(t) = z_k((r+1)k - 0) + \int_{rk}^t g(z_k(s))ds \\ & + \mathbb{G}(z_k^r).(b(t) - b(rk)) \quad \text{for } t \in [rk,(r+1)k[. \end{aligned} \tag{3.10}$$

From Ito's formula, one gets

$$d|\tilde{z}_k(t)|^2 = (2(\tilde{z}_k(t), g(z_k(t))) + |G(z_k^r)|^2)dt + 2(\tilde{z}_k, G(z_k^r)).db$$

hence in particular, possibly modifying the constants

$$\tilde{E}|z_k^{r+1}|^2 \leq e^{c_0 k}\tilde{E}|z_k^r|^2 + c_1 k. \tag{3.11}$$

Iterating we get

$$\tilde{E}|z_k^r|^2 \leq e^{rc_0 k}|y_0|^2 + c_1 k(1 + e^{c_0 k} + \cdots + e^{(r-1)c_0 k}).$$

Therefore

$$\tilde{E}|z_k^r|^2 \le C,$$

independently of k, r.

From (3.9) we deduce the first estimate (3.7), and

$$\int_{rk}^{(r+1)k} ||z_k(t)||^2 dt \le Ck,$$

from which the first estimate (3.6) follows also. Let us prove the second estimate (3.7). We note that

$$\frac{d}{dt}|z_k(t)|^4 + 4|z_k(t)|^2(\langle Az_k, z_k\rangle + \langle B(z_k), z_k\rangle)dt = 0$$
$$\text{for } t \in [rk, (r+1)k[$$

hence again

$$|z_k(t)|^4 \le e^{c_0 k}|z_k^r|^4 + c_1 k.$$

Since also

$$\begin{aligned} d|\tilde{z}_k(t)|^4 &= 2|\tilde{z}_k(t)|^2(2(\tilde{z}_k(t), g(z_k(t))) + |Gz_k^r)|^2)dt \\ &+ |(\tilde{z}_k, G(z_k^r))|^2 dt + 4|\tilde{z}_k(t)|^2(\tilde{z}_k(t), G(z_k^r)).db \end{aligned}$$

we get again

$$\tilde{E}|z_k^{r+1}|^4 \le e^{c_0 k}\tilde{E}|z_k^r|^4 + c_1 k,$$

and the second estimate (3.7) is easily derived.

We now prove an estimate, which improves the first estimate (3.7).

Lemma 3.2 *We have*

$$\tilde{E} \sup_{t\in[0,T]} |z_k(t)|^2 \le C. \tag{3.12}$$

Proof

It is clear from (3.9) that

$$\sup_{t\in[0,T]} |z_k(t)|^2 \le C \max_{r=0\cdots N+1} |z_k^r|^2. \tag{3.13}$$

Now, noting that from (3.10)

$$\begin{aligned}|z_k^{r+1}|^2 &= |z_k((r+1)k-0)|^2 + \int_{rk}^{(r+1)k} (2(\tilde{z}_k, g(z_k)) + |G(z_k^r)|^2)dt \\ &+ 2\int_{rk}^{(r+1)k} (\tilde{z}_k, G(z_k^r)).db\end{aligned}$$

and combined with (3.8), applied at $t = (r+1)k - 0$, yields

$$\begin{aligned}&|z_k^{r+1}|^2 + 2\int_{rk}^{(r+1)k} \langle Az_k, z_k\rangle ds + 2\int_{rk}^{(r+1)k} \langle B(z_k), z_k\rangle ds \\ &= |z_k^r|^2 + \int_{rk}^{(r+1)k} (2(\tilde{z}_k, g(z_k)) + |G(z_k^r)|^2)dt \\ &+2\int_{rk}^{(r+1)k} (\tilde{z}_k, G(z_k^r)).db.\end{aligned} \tag{3.14}$$

Adding up these relations between, $r = 0$ and r we obtain

$$\begin{aligned}&|z_k^r|^2 + 2\int_0^{rk} \langle Az_k, z_k\rangle ds + 2\int_0^{rk} \langle B(z_k), z_k\rangle ds \\ &= |y_0|^2 + 2\int_0^{rk} (\tilde{z}_k, g(z_k))dt + k\sum_{j=0}^{r-1} |G(z_k^j)|^2 \\ &+2\sum_{j=0}^{r-1}\int_{jk}^{(j+1)k} (\tilde{z}_k, G(z_k^j)).db.\end{aligned} \tag{3.15}$$

It is convenient to define the process

$$\chi_k(t) = G(z_k^r), \qquad r_k \le t < (r+1)k.$$

Hence

$$\sum_{j=0}^{r-1}\int_{jk}^{(j+1)k} (\tilde{z}_k, G(z_k^j)).db = \int_0^{rk} (\tilde{z}_k, \chi_k(t)).db.$$

Therefore (3.15) yields

$$\begin{aligned}\tilde{E} \max_{r=0\cdots N+1} |z_k^r|^2 &\leq C + k \sum_{j=0}^{N} \tilde{E}|G(z_k^j)|^2 \\ &+2\tilde{E} \sup_{t\in[0,T]} \left| \int_0^t (\tilde{z}_k, \chi_k).db \right|.\end{aligned} \tag{3.16}$$

But

$$\begin{aligned}\tilde{E} \sup_{t\in[0,T]} \left| \int_0^t (\tilde{z}_k, \chi_k).db \right| &\leq \tilde{E} \left(\int_0^T |(\tilde{z}_k, \chi_k)|^2 dt \right)^{1/2} \\ &\leq \frac{1}{\sqrt{2}} \left(\tilde{E} \int_0^T |\tilde{z}_k(t)|^4 dt + \tilde{E} \int_0^T |\chi_k|^4 dt \right)^{1/2}.\end{aligned}$$

Noting that

$$\tilde{E} \int_0^T |\chi_k|^4 dt = k \sum_{j=0}^{N} \tilde{E}|G(z_k^j)|^4 \leq C$$

by the second estimate (3.7), and that for similar reasons

$$\tilde{E} \int_0^T |\tilde{z}_k(t)|^4 dt \leq C,$$

we deduce from (3.16) that

$$\tilde{E} \max_{r=0\cdots N+1} |z_k^r|^2 \leq C$$

and from (3.13), we obtain the desired estimate (3.12).

We state now our final estimate

Lemma 3.3 *We have*

$$\tilde{E} \sup_{|\theta|\leq\delta} \int_0^T ||z_k(t+\theta) - z_k(t)||_{V'}^{4/3} dt \leq C\delta^{1/3},\ \forall\delta \leq 1, \tag{3.17}$$

where z_k is extended by 0, outside $[0,T]$.

Proof

We assume $\theta > 0$. A similar calculation is done for $\theta < 0$. We write

$$\tilde{E} \sup_{0 \leq \theta \leq \delta} \int_0^T ||z_k(t+\theta) - z_k(t)||_{V'}^{4/3} dt \leq I_1 + I_2$$

where

$$\begin{aligned} I_1 &= \tilde{E} \sup_{0 \leq \theta \leq \delta} \int_0^{T-\delta} ||z_k(t+\theta) - z_k(t)||_{V'}^{4/3} dt \\ I_2 &= \tilde{E} \sup_{0 \leq \theta \leq \delta} \int_{T-\delta}^{T} ||z_k(t+\theta) - z_k(t)||_{V'}^{4/3} dt. \end{aligned}$$

Note that

$$I_2 \leq C\delta \tilde{E} \sup_{t \in [0,T]} |z_k(t)|^2 \leq C\delta$$

by lemma 2.2.

We consider now I_1. Note that we can write from the definition of z_k

$$\begin{aligned} & z_k(t+\theta) - z_k(t) + \int_t^{t+\theta} Az_k(s)ds + \int_t^{t+\theta} B(z_k)ds \\ &= \int_t^{t+\theta} g(z_k)ds + \int_{k[t/k]}^{k[t+\theta/k]} \chi_k(s).db \end{aligned} \tag{3.18}$$

where the process $\chi_k(t)$ has already been defined in the previous lemma.

We have next for $0 \leq \theta \leq \delta$

$$\left\| \int_t^{t+\theta} (Az_k + B(z_k) - g(z_k))ds \right\|_{V'} \leq C \int_t^{t+\delta} (||z_k|| + 1 + ||z_k|| |z_k|)ds.$$

Hence

$$\leq C\left[\delta+\delta^{1/2}\left(\int_0^T \|z_k\|^2 dt\right)^{1/2}+\int_t^{t+\delta}\|z_k\|^{3/2}ds+\int_t^{t+\delta}|z_k|^3 ds\right]$$

$$\leq C\left[\delta+\delta^{1/2}\left(\int_0^T \|z_k\|^2 dt\right)^{1/2}+\delta^{1/4}\left(\int_0^T \|z_k\|^2 dt\right)^{3/4}+\right.$$

$$\left.+\delta^{1/4}\left(\int_0^T |z_k|^4 dt\right)^{3/4}\right]$$

$$\leq C\delta^{1/4}\left(1+\int_0^T \|z_k\|^2 dt+\int_0^T |z_k|^4 dt\right)^{3/4}.$$

Hence

$$\|z_k(t+\theta)-z_k(t)\|_{V'}^{4/3}\leq C\left[\delta^{1/3}\left(1+\int_0^T \|z_k\|^2 dt+\int_0^T |z_k|^4 dt\right)\right.$$

$$\left.+\left|\int_{k[t/k]}^{k[t+\theta/k]}\chi_k(s).db\right|^{4/3}\right].$$

Therefore

$$I_1\leq C\delta^{1/3}+\tilde{E}\sup_{0\leq\theta\leq\delta}\int_0^T\left|\int_{k[t/k]}^{k[t+\theta/k]}\chi_k(s).db\right|^{4/3}dt.$$

But

$$\tilde{E}\int_0^T\sup_{0\leq\theta\leq\delta}\left|\int_{k[t/k]}^{k[t+\theta/k]}\chi_k(s).db\right|^{4/3}dt$$

$$\leq C\left(\tilde{E}\int_0^T\sup_{0\leq\theta\leq\delta}\left|\int_{k[t/k]}^{k[t+\theta/k]}\chi_k(s).db\right|^{2}dt\right)^{2/3}$$

$$\leq C\left(\int_0^T\left(\tilde{E}\int_{k[t/k]}^{k[t+\theta/k]}|\chi_k(s)|^2 ds\right)dt\right)^{2/3}.$$

Since $\tilde{E}|\chi_k(s)|^2\leq C$, we majorize by

$$\leq C\left(\int_0^T (k[t+\delta/k]-k[t/k])dt\right)^{2/3}\leq C\delta^{2/3}$$

and thus $I_1 \leq C\delta^{1/3}$.

The statement (3.17) has been proved.

4 Convergence

4.1 A compactness result

In the Hilbert space $L^2(0,T;H)$ we now consider a set Z depending on 3 constants K, L, M and two sequences μ_n, ν_n of numbers such that $\mu_n, \nu_n \geq 0$ and $\mu_n, \nu_n \to 0$ as $n \to \infty$.

The set is defined as follows

$$Z = \left\{ z \middle| \int_0^T ||z(t)||_V^2 dt \leq K, \quad |z(t)|_H^2 \leq L, \right. \\ \left. \sup_{|\theta| \leq \mu_n} \int_0^T ||z(t+\theta) - z(t)||_{V'}^{1/3} dt \leq \nu_n M, \quad \forall n \right\}. \tag{4.1}$$

Then the set Z is a compact subset of $L^2(0,T;H)$.

4.2 Tightness property

Let $S = C(0,T;R^m) \times L^2(0,T;H)$. We define on S the probability measure π_k image on S of the probability $\tilde{P}$ on $\tilde{\Omega}$, by the map

$$\tilde{\omega} \to (b(.,\tilde{\omega}), z_k(.,\tilde{\omega})) = (b(.,\tilde{\omega}), \psi_k(b(.,\tilde{\omega}))). \tag{4.2}$$

We have the

Lemma 4.1 *The family π_k is uniformly tight, i.e., $\forall \varepsilon$, $\exists W_\varepsilon \times Z_\varepsilon$ a compact subset of S, such that*

$$\pi_k(W_\varepsilon \times Z_\varepsilon) \geq 1 - \varepsilon, \quad \forall k. \tag{4.3}$$

Proof We define for given q_ε, r_ε

$$W_\varepsilon = \left\{ b(.) | \sup_{t \in [0,T]} |b(t)| \leq q_\varepsilon, \quad \sup_{\substack{t_1, t_2 \in [0,T] \\ |t_2 - t_1| < \frac{1}{n^6}}} |b(t_2) - b(t_1)| < \frac{r_\varepsilon}{n}, \quad \forall n \right\}$$

which is a compact subset of $C(0,T;R^m)$.

Let Z_ε be a set of the form (4.1) with constant K_ε, L_ε, M_ε and μ_n, ν_ε tending to 0, satisfying

$$\sum_n \frac{1}{\nu_n}(\mu_n)^{1/3} < \infty.$$

For instance pick $\nu_n = \frac{1}{n}$, and $\mu_n = \frac{1}{n^\rho}$, with $\rho > 6$. The set $W\varepsilon \times Z\varepsilon$ is compact, and the constants q_ε, r_ε, K_ε, L_ε, M_ε are easily adjusted.

4.3 Proof of Theorem 1.1

By Prokhorov's theorem, the family π_k is relatively compact in the set $\mathcal{P}(S)$ of probability measures on S, equipped with the weak convergence topology. Therefore, there exists a subsequence π_{k_j} such that

$$\pi_{k_j} \to \pi, \text{ i.e., } \forall \Phi : S \to R \tag{4.4}$$

continuous and bounded

$$\int \Phi(b(.), z(.))d\pi_{k_j} \to \int \Phi(b(.), z(.))d\pi$$

By Skorokhod's theorem, there exists a probability space $(\Omega, \mathcal{A}, P)$ and random variables $w_{k_j}(.,\omega)$, $y_{k_j}(.,\omega)$, $w(.,\omega)$, $y(.,\omega)$ on $(\Omega, \mathcal{A}, P)$ with values in S such that

$$\text{the probability law of } w_{k_j}(.),\ y_{k_j}(.) \text{ is } \pi_{k_j} \tag{4.5}$$

$$w_{k_j}(.),\ y_{k_j}(.) \to w(.),\ y(.) \text{ a.s. as } j \to \infty \tag{4.6}$$

$$\text{the probability law of } w(.),\ y(.) \text{ is } \pi. \tag{4.7}$$

We notice that, since ψ_k is continuous from $C(0,T;R^m)$ to $L^2(0,T;H)$ (in fact to $L^2(0,T;V)$), then

$$y_{k_j}(.,\omega) = \psi_{k_j}(w_{k_j}(.,\omega)). \tag{4.8}$$

Therefore we may write the relations

$$\begin{aligned}
&y'_{k_j} + Ay_{k_j} + B(y_{k_j}) = 0, \text{ for } t \in [rk_j, (r+1)k_j[\\
&y^{r+1}_{k_j} = y_{k_j}((r+1)k_j - 0) + \int_{rk_j}^{(r+1)k_j} g(y_{k_j}(s))ds + \\
&\quad + G(y^r_{k_j}).(w_{k_j}((r+1)k_j) - w_{k_j}(rk_j)) \qquad (4.9)\\
&y_{k_j}(rk_j) = y^r_{k_j} \\
&y^0_{k_j} = y_0 \\
&y_{k_j}(T) = y^{N+1}_{k_j}
\end{aligned}$$

with $r = 0 \cdots N_j$ where $N_j + 1 = T/k_j$.

Since w_k is a Wiener process, it is clear that Lemmas 2.1, 2.2, 2.3 will apply to give the estimates

$$E \int_0^T ||y_{k_j}(t)||^2_V dt \leq C \tag{4.10}$$

$$E \sup_{0 \leq t \leq T} |y_{k_j}(t)|^2 \leq C, \quad \sup_{0 \leq t \leq T} E|y_{k_j}(t)|^4 \leq C \tag{4.11}$$

$$E \sup_{|\theta| \leq \delta} \int_0^T ||y_{k_j}(t+\theta) - y_{k_j}(t)||^{4/3}_{V'} dt \leq C\delta^{1/3}, \ \forall \delta \leq 1. \tag{4.12}$$

Therefore we may assume, by extracting possibly a new subsequence, that

$$\begin{aligned}
y_{k_j} \to y \quad &\text{in} \quad L^2(\Omega, \mathcal{A}, P; L^2(0,T;V)) && \text{weakly} \\
&\text{in} \quad L^2(\Omega, \mathcal{A}, P; L^\infty(0,T;H)) && \text{weak star} \\
&\text{in} \quad L^\infty(0,T; L^4(\Omega, \mathcal{A}, P; H)) && \text{weak star}
\end{aligned} \tag{4.13}$$

But (4.6) implies

$$y_{k_j} \to y \text{ in } L^2(0,T;H) \text{ a.s.}$$

From the second estimate (4.11) we have then

$$y_{k_j} \to y \text{ in } L^2(\Omega, \mathcal{A}, P; L^2(0,T;H)) \tag{4.14}$$

and therefore, extracting a new subsequence,

$$y_{k_j} \to y \text{ in } H, \text{ a.e., a.s. (with respect to the measure } dP \otimes dt). \tag{4.15}$$

It follows easily from (4.14), (4.15) and Vitali's theorem, that

$$g(y_{k_j}) \to g(y) \text{ in } L^2(\Omega, \mathcal{A}, P; L^2(0,T;H)) \tag{4.16}$$

$$G(y_{k_j}) \to G(y) \text{ in } L^2(\Omega, \mathcal{A}, P; L^2(0,T;H^m)) \tag{4.17}$$

Moreover

$$B(y_{k_j}) \to B(y) \text{ in } L^3(\Omega, \mathcal{A}, P; L^{4/3}(0,T;V')) \text{ weakly.} \tag{4.18}$$

Let us check (4.18). From (1.10) we check that

$$\begin{aligned} E\int_0^T \|B(y_{k_j})\|_{V'}^{4/3} dt &\le CE\int_0^T (1 + \|y_{k_j}\||y_{k_j}|)^{4/3} dt \\ &\le \left(E\int_0^T (1 + \|y_{k_j}\|^2) dt\right)^{2/3} \left(E\int_0^T (1 + |y_{k_j}|^4) dt\right)^{1/3} \\ &\le C. \end{aligned}$$

Therefore to prove (4.18) it is enough to prove that

$$E\int_0^T \langle B(y_{k_j}) - B(y), z\rangle dt \to 0,$$

$\forall z \in L^\infty(\Omega \times (0,T); V)$. But then

$$E\int_0^T |\langle B(y_{k_j}) - B(y), z\rangle| dt \le CE\int_0^T |y_{k_j} - y|^{1/2} \|y_{k_j} - y\|^{1/2}$$

$$(1 + |y|^{1/2}\|y\|^{1/2} + |y_{k_j}|^{1/2}\|y_{k_j}\|^{1/2}) dt.$$

Take for instance the term

$$E\int_0^T |y_{k_j} - y|^{1/2} \|y_{k_j} - y\|^{1/2} |y_{k_j}|^{1/2} \|y_{k_j}\|^{1/2} dt.$$

From (4.10) and (4.14), it is clear that it goes to 0. Now if we set

$$F^t = \sigma(\mathbf{1}_t w(.;\omega); \mathbf{1}_t y(.,\omega)) \tag{4.19}$$

then one can check that

$$w(t) \text{ is a } F^t \text{ Wiener process.} \tag{4.20}$$

Let now define the process

$$\chi_{k_j}(t) = G(y^r_{k_j}), \quad t \in [rk_j, (r+1)k_j[.$$

We can write (4.9) in the following way

$$\begin{aligned} y_{k_j}(t) &+ \int_0^t (Ay_{k_j}(s) + B(y_{k_j}(s))ds = y_0 \\ &+ \int_0^{k_j[t/k_j]} g(y_{k_j}(s))ds + \int_0^{k_j[t/k_j]} \chi_{k_j}(s).dw_{k_j}(s). \end{aligned} \tag{4.21}$$

Let us prove that

$$\begin{aligned} &\int_0^{k_j[t/k_j]} \chi_{k_j}(s).dw_{k_j}(s) \to \int_0^T G(y(s)).dw(s) \\ &\text{in } L^2(\Omega, \mathcal{A}, P; H) \text{ weakly, for any } t. \end{aligned} \tag{4.22}$$

Write $\chi(s) = G(y(s))$.

We first notice that we can replace in the stochastic integral at the left hand side of (4.22), $k_j[t/k_j]$ by t, and thus we may as well take $t = T$. Thus we are going to prove that,

$$\begin{aligned} &\int_0^T \chi_{k_j}(s).dw_{k_j}(s) \to \int_0^T \chi(s).dw(s) \\ &\text{in } L^2(\Omega, \mathcal{A}, P; H) \text{ weakly.} \end{aligned} \tag{4.23}$$

From (4.17), it is easy to check that

$$\chi_{k_j} \to \chi \text{ in } L^2(\Omega, \mathcal{A}, P; L^2(0, T; H^m)). \tag{4.24}$$

Indeed let $\hat{y}_{k_j}(t) = y^r_{k_j}$, if $rk_j \leq t < (r+1)k_j$, $r = 0 \cdots N_j$, $(N_j + 1)k_j = T$, then let us check that

$$\hat{y}_{k_j} \to y \text{ in } L^2(\Omega, \mathcal{A}, P; L^2(0, T; H)). \tag{4.25}$$

If this is true, (4.24) follows, since $\chi_{k_j}(t) = G(\hat{y}_{k_j}(t))$. Now for $t \in [rk_j, (r+1)k_j[$, we have

$$\hat{y}_{k_j}(t) - y_{k_j}(t) = \int_{rk_j}^{t} (Ay_k(s)) + B(y_{k_j}(s))ds.$$

Hence $\hat{y}_{k_j} - y_{k_j} \rightarrow 0$ in $L^{4/3}(\Omega, \mathcal{A}, P; L^{4/3}(0,T;V'))$. Therefore $\hat{y}_{k_j} \rightarrow y$ in $L^2(\Omega, \mathcal{A}, P; L^2(0,T;H))$ weakly. But we have also, for $t \in [rk_j, (r+1)k_j[$,

$$\begin{aligned} |\hat{y}_{k_j}(t)|^2 &= |y_{k_j}(t)|^2 + 2\int_{rk_j}^{t} (\langle Ay_{k_j}, y_{k_j}\rangle + 2\langle B(y_{k_j}, y_{k_j}\rangle)ds \\ &\leq |y_{k_j}(t)|^2 + C\int_{rk_j}^{t} (|y_{k_j}(s)|^2 + 1)ds \\ &\leq |y_{k_j}(t)|^2 + Ck_j^{1/2}\left(\int_0^T |y_{k_j}(s)|^4\right)^{1/2} + Ck_j. \end{aligned}$$

Hence

$$E\int_0^T |\hat{y}_{k_j}(t)|^2 dt \leq E\int_0^T |y_{k_j}(t)|^2 dt + Ck_j^{1/2}.$$

Therefore

$$\limsup E\int_0^T |\hat{y}_{k_j}(t)|^2 dt \leq E\int_0^T |y(t)|^2 dt.$$

This estimate and the weak limit yields (4.25).

Define $\chi^\varepsilon(t) : \dfrac{1}{\varepsilon}\displaystyle\int_0^t e^{-\frac{t-s}{\varepsilon}}\chi(s)ds$, and a similar definition for $\chi^\varepsilon_{k_j}(t)$. We have

$$\begin{aligned} \chi^\varepsilon &\rightarrow \chi \quad &in \quad L^2(\Omega, \mathcal{A}, P; L^2(0,T;H^m)) \\ \chi^\varepsilon_{k_j} &\rightarrow \chi_{k_j} \quad &in \quad L^2(\Omega, \mathcal{A}, P; L^2(0,T;H^m)) \end{aligned} \tag{4.26}$$

as $\varepsilon \rightarrow 0$, for fixed k_j. Morever

$$\int_0^T |\chi^\varepsilon - \chi^\varepsilon_{k_j}|^2 dt \leq \int_0^T |\chi - \chi_{k_j}|^2 dt, \tag{4.27}$$

and

$$\int_0^T \chi_{k_j}^\varepsilon(t).dw_{k_j} = \frac{1}{\varepsilon} w_{k_j}(T).\int_0^T \chi_{k_j}(t)e^{-\frac{T-t}{\varepsilon}}dt$$
$$-\frac{1}{\varepsilon}\int_0^T w_{k_j}(t).(\chi_{k_j}(t) - \frac{1}{\varepsilon}\int_0^T \chi_{k_j}(s)e^{-\frac{t-s}{\varepsilon}}ds)dt \tag{4.28}$$

Now

$$E\left|\int_0^T \chi_{k_j}(t).dw_{k_j}\right|^2 = E\int_0^T |\chi_{k_j}(t)|^2 dt \leq C.$$

We consider any subsequence still denoted by k_j such that

$$\int_0^T \chi_{k_j}(t).dw_{k_j} \to \eta \text{ in } L^2(\Omega, \mathcal{A}, P; H) \text{ weakly.} \tag{4.29}$$

We may also assume that $\chi_{k_j} \to \chi$ for almost every ω, t. Now pick ξ any element in $L^2(\Omega, \mathcal{A}, P; H)$, then we can assert that

$$E\left(\xi, \int_0^T \chi_{k_j}^\varepsilon(t).dw_{k_j}(t)\right) \to E\left(\xi, \int_0^T \chi^\varepsilon(t).dw(t)\right) \tag{4.30}$$

as $k_j \to 0$, for fixed ε.

This follows from the facts that, thanks to the explicit formula (4.28),

$$\int_0^T \chi_{k_j}^\varepsilon(t).dw_{k_j}(t) \to \int_0^T \chi^\varepsilon(t).dw(t), \text{ a.s.}$$

as $k_j \to 0$, for fixed ε, and $E\left|\int_0^T \chi_{k_j}^\varepsilon(t).dw_{k_j}\right|^2 = E\int_0^T |\chi_{k_j}^\varepsilon(t)|^2 dt \leq C$ (independant of ε, k_j).

Now we write

$$E\left(\xi, \int_0^T \chi_{k_j}(t).dw_{k_j}(t)\right) - E\left(\xi, \int_0^T \chi(t).dw(t))\right)$$
$$= E\left(\xi, \int_0^T (\chi_{k_j} - \chi_{k_j}^\varepsilon).dw_{k_j}\right) + E\left(\xi, \int_0^T \chi_{k_j}^\varepsilon.dw_{k_j}\right)$$
$$-E\left(\xi, \int_0^T \chi^\varepsilon.dw\right) + E\left(\xi, \int_0^T (\chi^\varepsilon - \chi).dw\right) = I_{\varepsilon,k_j} + II_{\varepsilon,k_j} + III_\varepsilon.$$

Now by (4.27)

$$|I_{\varepsilon,k_j}| \leq C\left[E\int_0^T |\chi - \chi_{k_j}|^2 dt + E\int_0^T |\chi - \chi^\varepsilon|^2 dt\right].$$

Therefore letting k_j tend to 0, for fixed ε, we get

$$\limsup_{k_j\to 0}\left|E\left(\xi, \int_0^T \chi_{k_j}(t)dw_{k_j}(t)\right) - E\left(\xi, \int_0^T \chi(t) - dw(t)\right)\right|$$
$$\leq C\left(E\int_0^T |\chi - \chi^\varepsilon|^2 dt + III_\varepsilon\right).$$

Letting finally ε tend to 0, necessarily

$$\eta = \int_0^T \chi(t).dw(t).$$

Thus (4.22) holds for any t.

Clearly we also have the convergence in $L^2(\Omega, \mathcal{A}, P; L^2(0,T;H))$ weakly. Collecting results, we deduce from (4.21) passing to the limit in $L^{4/3}(\Omega, \mathcal{A}, P; L^{4/3}(0,T;V'))$ weakly, that

$$y(t)+\int_0^t (Ay(s)+B(y(s)))ds = y_0+\int_0^t g(y(s)))ds+\int_0^t G(y(s)).dw(s) \tag{4.31}$$

a.e. , a.s.

Now outside a set of probability 0, Ω_0, we have $\sup\limits_{t\in[0,T]} |y(t)| < \infty$. This implies $\int_0^t B(y(s))ds \in L^2(0,T;V')$. Define outside Ω_0 the process y_1 by solving pointwise

$$y_1(t) + \int_0^t Ay_1 ds + \int_0^t (B(y) - g(y))ds = y_0,$$

then $y_1 \in C(0,T;H)$, outside Ω_0.

Similarly there exists $y_2 \in L^2(\Omega, \mathcal{A}, P; L^2(0,T;V)) \cap L^2(\Omega, \mathcal{A}, P; C(0,T;H))$ such that outside Ω_1 of probability 0,

$$y_2(t) + \int_0^T Ay_2 ds = \int_0^t G(y(s)).dw(s), \quad \forall t \in [0,T].$$

Clearly

$$y = y_1 + y_2 \quad \text{a.e. } t, \text{ a.s. } \omega$$

and therefore we can replace in (4.31), y by $y_1 + y_2 = \tilde{y}$ which is a process with continuous trajectories in H, belonging to $L^2(\Omega, \mathcal{A}, P; L^2(0,T;V))$, and outside $\Omega_0 \cup \Omega_1$ satisfies

$$\tilde{y}(t) + \int_0^t (A\tilde{y} + B(\tilde{y})ds = y_0 + \int_0^t g(\tilde{y})ds + \int_0^t G(\tilde{y}(s)).dw(s)$$
$$\forall t \in [0,T].$$

Now since $E \sup_{t\in[0,T]} |y(t)|^2$ and $E \sup_{t\in[0,T]} |y_2(t)|^2$ are finite, necessarily $\tilde{y} \in L^2(\Omega, \mathcal{A}, P; C(0,T;H))$.

The proof of Theorem 1.1 has been completed.

Bibliography

[1] A. Bensoussan, R. Glowinski, A. Rascanu, *Approximation of Zakaï equation by the splitting up method,* to appear SIAM J.

[2] A. Bensoussan, R. Temam, *Equations aux dérivées partielles stochastiques non-linéaires,* Israël Journal of Mathematics, Vol. 11, n° 1, pp. 95-129, 1972.

[3] A. Bensoussan, R. Temam, *Equations stochastiques du type Navier Stokes* J. Funct. Anal., Vol. 2, pp. 195-222, Juin 1973.

[4] J.L. Lions, *Equations différentielles opérationnelles et problèmes aux limites,* Springer Verlag, Berlin, 1961.

[5] J.L. Lions, *Quelques méthodes de résolution de problèmes aux limites non linéaires,* Dunod - Gauthier-Villars, Paris 1969.

[6] Nagase, *On the Cauchy problem for nonlinear stochastic partial differential equation with continuous coefficients,* Existence Theorems, to be published.

[7] E. Pardoux, *Stochastic partial differential equations and filtering of diffusion processes,* Stochastics, 9, (1983).

Wavelets as Attractors of Random Dynamical Systems

Marc A. Berger
School of Mathematics
Georgia Institute of Technology
Atlanta, GA 30332

Abstract

Many algorithms for generating computer images today involve a recursive tree traversal. These include iterated function systems for generating fractals, sub-division refinement methods for generating B-splines and Beziér curves, line averaging methods for interpolants, and algorithms for wavelets and solutions to dilation equations. This article shows how ergodic theory can be used in a very general setting to produce random algorithms which generate the same images as the recursive ones. These images become attractors of random dynamical systems. In particular a random dynamical algorithm for generating Daubechies' wavelets is derived. The random dynamics even serves to shed new light on the structure of these wavelets.

Many algorithms in computer image generation today involve recursive tree traversal. This stems from the fact that they come from some underlying construction on "code space"

$$\Omega = (\mathbb{Z}_N)^\infty, \quad \text{where } \mathbb{Z}_N = \{0, \ldots, N-1\}.$$

The following three diverse examples illustrate this.

ISBN 0-12-481005-5

Example A: Fractals

Let $T(0),\ldots,T(N-1)$ be strictly contractive transformations $\mathbb{R}^d \to \mathbb{R}^d$. Then for every $\omega = (\omega_1, \omega_2, \ldots) \in \Omega$ the limit

$$\lim_{n\to\infty} T(\omega_1)\cdots T(\omega_n)x \tag{1}$$

exists, and this limit does not depend on x. (Hint: the sequence in (1) is a Cauchy sequence.) So we can define a map $X : \Omega \to \mathbb{R}^d$ by

$$X(\omega) = \lim_{n\to\infty} T(\omega_1)\cdots T(\omega_n)x;$$

and in fact X is continuous with respect to the "code space metric"

$$d(\omega, \sigma) = \sum_{k=1}^{\infty} \frac{|\omega_k - \sigma_k|}{(N+1)^k} \tag{2}$$

Hutchinson [6] discovered that the range $X(\Omega)$ is typically a fractal set, and Barnsley's text [1] is all about these sets. The image of $X(\Omega)$ can be computer-generated as the set of accumulation points of

$$\{T(\omega_1)\cdots T(\omega_n)x : n \in \mathbb{N}, (\omega_1, \ldots, \omega_n) \in (\mathbb{Z}_N)^n\} \tag{3}$$

(for some arbitrary x). It can also be generated as the closure of the set of fixed points

$$\{f(\omega_1, \ldots, \omega_n) : n \in \mathbb{N}, (\omega_1, \ldots, \omega_n) \in (\mathbb{Z}_N)^n\}, \tag{4}$$

where $f(\omega_1, \ldots, \omega_n)$ denotes the fixed point of $T(\omega_1)\cdots T(\omega_n)$. In practice one fixes a value of D and generates all points $T(\omega_1)\cdots T(\omega_D)x$, or all fixed points $f(\omega_1, \ldots, \omega_D)$ for $(\omega_1, \ldots, \omega_D) \in (\mathbb{Z}_N)^D$. These algorithms amount to an N-ary tree traversal. The nodes of this tree are points in $\mathbb{R}^d$ and the edges represent application of the transformations $T(\omega)$. The image drawn is the set of all leaves at the bottom (depth D) of this tree in the case (3), and the set of all nodes in the tree in case (4). The image in Diaconis and Shahshahani [5] was done this way.

Example B: Sub-Division Methods

Sub-division methods for curve generation and line averaging algorithms are analyzed in Micchelli and Prautzsch [7], [8]. In these schemes one is given $m \times m$ row stochastic matrices $P(0), \ldots,$ $P(N-1)$ (i.e., all row sums are one negative entries allowed!) and a set of "control points" $v_1, \ldots, v_m \in \mathbb{R}^d$. These points form the vertices of a "control polytope" $C \subseteq \mathbb{R}^d$, which we can associate with the $m \times d$ matrix

$$\begin{bmatrix} v_1^t \\ \cdots \\ v_m^t \end{bmatrix}$$

(This correspondence is well-defined if one thinks of C as an "ordered" polytope.) If $P = (p_{ij})$ is an $m \times m$ row stochastic matrix then one can identify an action of P on C; namely, PC. Equivalently PC is the polytope with vertices $v_t', \ldots, v_m'$ where

$$v_i' = \sum_j p_{ij} v_j'.$$

Under suitable (positivity and irreducibility-type) conditions on the $P(\omega)$'s it is shown in [8] that

$$P(\omega_n) \cdots P(\omega_1) C$$

converges to a singleton as $n \to \infty$ for any $\omega = (\omega_1, \omega_2, \ldots) \in \Omega$. By choosing the $P(\omega)$'s appropriately one generates a smooth curve $\mathcal{A}$ (the "attractor") as

$$\mathcal{A} = \bigcup \left(\lim_n P(\omega_n) \cdots P(\omega_1) C : \omega \in \Omega \right).$$

Here again we can construct a map $X : \Omega \to \mathbb{R}^d$ according to

$$X(\omega) = \lim_{n \to \infty} P(\omega_n) \cdots P(\omega_1) C.$$

(More precisely the term on the left should be $\{X(\omega)\}$, but we are identifying $X(\omega)$ here with the singleton containing it.) The image $X(\Omega)$ can be generated as the union of sets $P(\omega_D) \cdots P(\omega_1) C$ for some fixed depth D, and this is again a tree-traversal algorithm.

This time the nodes of the tree are polytopes, and the edges represent application of the matrices $P(\omega)$. The image generated is the union of all leaves at the bottom (depth D).

Example C: Wavelets

Wavelets are described in the survey article by Strang [10] and in the references therein. They are compactly supported functions $W(x)$, $x \in \mathbb{R}$, having the special feature that the family $\{W(2^j x - k) : j \leq 0, k \in \mathbb{Z}\}$ of translates and dilations are orthogonal (in the L_2-sense). These functions $W(2^j x - k)$ are local in both space and frequency, and thus afford a new type of basis for orthogonal expansions.

Wavelets are constructed through solutions $g(x)$, $x \in \mathbb{R}$, of certain special *dilation equations*

$$\begin{cases} g(x) = \sum_{k=-\infty}^{\infty} a_k g(2x - k) \\ \sum_{k=-\infty}^{\infty} g(x - k) = 1 \end{cases} \tag{5}$$

Typically in these dilation equations the coefficients a_k are finitely supported. The values of g at integers $x = \ell$ can be found through (5) as an eigenvector for the matrix (a_{2i-j}). Then using (5) one can find the values of g at the half-integers $x = \frac{\ell}{2}$, then at the quarter-integers $x = \frac{\ell}{4}$, etc.

The tree structure here is not as obvious as those above, yet in fact there is a simple binary tree structure. The 2^n nodes at depth n correspond to the sets of translate values

$$\left\{g\left(\frac{2\ell - 1}{2^{n+1}} - k\right) : k \in \mathbb{Z}\right\}; \quad \ell = 1, \ldots, 2^n.$$

They are arranged so that the translates

$$\left\{g\left(\frac{2\ell - 1}{2^{n+1}} - k\right)\right\}, \quad \left\{g\left(\frac{2\ell - 1}{2^{n+1}} + \frac{1}{2} - k\right)\right\}$$

are the sons of

$$\left\{g\left(\frac{2\ell - 1}{2^{n}} - k\right)\right\}.$$

The left and right edges correspond to multiplication by the matrices (a_{2i-j-1}) and (a_{2i-j}). To generate the curve for the solution of the dilation equations (5), draw the union of all the nodes down to depth D, for some fixed value of D.

The Daubechies' wavelets W_p presented in [4] can be constructed in terms of g as

$$W(x) = \sum_k (-1)^k a_{1-k} g(2x - k) \tag{6}$$

where (see [10]) $a_k = 0$ for $k < 0$ or $k > m = 2p - 1$, and the $2p$ coefficients $a_0, \ldots, a_m$ have to satisfy the $2p$ conditions

$$\sum a_{2k} = 1, \qquad \sum a_{2k+1} = 1$$

$$\sum (-1)^k k^\ell a_k = 0; \quad \ell = 1, \ldots, p-1$$

$$\sum a_k a_{k+2\ell} = 0; \quad \ell = 1, \ldots, p-1$$

Then $W = W_p$ given by (6) will have the desired orthogonality property. Other orthonormal wavelets with compact support can be constructed; these correspond to a choice $R \neq 0$ in [4, eq. (4.17)].

The more general dilation equation

$$g(x) = \sum_{k=-\infty}^{\infty} a_k g(Nx - k)$$

can be handled in the same fashion. The tree structure is now that of an N-ary tree. The nodes represent sets of translate values $\left\{g\left(\frac{\ell}{N^n} - k\right)\right\}$, N^{n-1} does not divide ℓ, and the edges represent multiplication by matrices

$$P(\omega) = (a_{N_{j-i-\omega}}), \quad 0 \leq \omega \leq N - 1.$$

This is described in [3].

Ergodic stochastic processes can provide a very convenient means of tree traversal . Let $\nu = \otimes\mu$ be product measure on Ω, $\mu(\{\omega\}) = \frac{1}{N}$, $0 \leq \omega \leq N - 1$. To run the image generation algorithms above one wants to simulate ν. That is, one wants to simulate a lot of points

$\omega \in \Omega$ (more precisely, a lot of points in $(\mathbb{Z}_N)^D$ for some depth D), according to ν. Below we construct a simple stochastic process (Z_n) in Ω for which a.s.

$$\frac{1}{n}\sum_{k=1}^{n} \delta_{Z_k} \Rightarrow \nu. \tag{7}$$

("$\Rightarrow$" denotes convergence in distribution.) The distributions on the left in (7) are the empirical distributions for $\{Z_1, \ldots, Z_n\}$. Thus by running a single trajectory of (Z_n) one is able to simulate ν.

Consider the right shift operators $R(\omega) : \Omega \to \Omega$, $0 \le \omega \le N-1$, given by

$$R(\omega) : \sigma \mapsto (\omega, \sigma) = (\omega, \sigma_1, \sigma_2, \ldots)$$

for $\sigma = (\sigma_1, \sigma_2, \ldots)$. This operator shifts σ to the right and inserts ω into the first slot. $R(\omega)$ is a contraction operator relative to the code space metric (1), in fact

$$d(R(\omega)\sigma, R(\omega)\sigma') = \frac{1}{N+1} d(\sigma, \sigma').$$

Any stationary ergodic measure ν on Ω extends to a measure $\tilde{\nu}$ on $\tilde{\Omega} = (\mathbb{Z}_N)^{\mathbb{Z}}$; that is, to two-sided infinite sequences $\tilde{\omega} = (\ldots, \omega_{-1}, \omega_0, \omega_1, \ldots)$ while keeping it stationary ergodic. Let $\tilde{\omega} \in \tilde{\Omega}$ and denote $\omega = (\omega_1, \omega_2, \ldots)$ for that $\tilde{\omega}$. For any $\sigma \in \Omega$

$$R(\omega_1) \cdots R(\omega_n)\sigma \to \omega$$

as $n \to \infty$. Thus for a given σ

$$Y_n(\tilde{\omega}) = R(\omega_1) \cdots R(\omega_n)\sigma \to \omega.$$

On the other hand

$$Z_n(\tilde{\omega}) = R(\omega_{-n}) \cdots R(\omega_{-1})\sigma \tag{8}$$

has, on account of stationarity, the same distribution as $Y_n(\tilde{\omega})$ under $\tilde{\nu}$, for each *fixed* n (*not* as processes). Thus $Z_n \Rightarrow \nu$. If we choose $\sigma = Z_0(\tilde{\omega}) = (\omega_0, \omega_1, \ldots)$ in (8) then (Z_n) will be stationary ergodic, and so by the Ergodic Theorem

$$\frac{1}{n}\sum_{k=1}^{n} \delta_{Z_k} \Rightarrow \nu \text{ a.s. } [\tilde{\nu}]. \tag{9}$$

Since for any $\sigma \in \Omega$

$$d(R(\omega_{-n})\cdots R(\omega_{-1})\sigma, R(\omega)_{-n})\cdots R(\omega_{-1})Z_0(\tilde{\omega})) \to 0$$

as $n \to \infty$, we see that (9) holds regardless of which σ is used in (8) to define Z_n.

In applications where $X : \Omega \to \mathbb{R}^d$ is some continuous function, running the random dynamical process (X_n) where

$$X_n = X(Z_n) \tag{10}$$

effectively accomplishes the tree traversal for generating $X(\Omega)$, since

$$\frac{1}{n}\sum_{k=1}^{n} \delta_{X_k} = \frac{1}{n}\sum_{k=1}^{n} \delta_{X(Z_k)} \Rightarrow \nu \circ X^{-1}$$

from (9). Heuristically the idea is to generate one long random path in the tree, rather than sample *all* paths of some fixed depth D.

Example A (cont'd):

Observe that

$$X \circ R(\omega) = T(\omega) \circ X \tag{11}$$

so that (X_n) given by (8), (10) evolves as

$$X_n = T(\omega_{-n})X_{n-1} \tag{12}$$

where $\tilde{\omega} = (\ldots, \omega_{-1}, \omega_0, \omega_1, \ldots)$ is distributed like $\tilde{\nu}$. Observe that when $\nu = \otimes\mu$ the *reversed-index process* $(\omega_{-1}, \omega_{-2}, \ldots)$ evolves as an i.i.d. sequence, since $\tilde{\nu}$ is reversible. Thus the random algorithm for generating $X(\Omega)$ via the orbit (X_n) is

Algorithm A (Fractals):

Initialize $x = f(0) \in \mathbb{R}^d$ (the fixed point of $T(0)$)

For $n = 1, 100000$
  plot x
  choose $\omega \in \{0, \ldots, N-1\}$ randomly
  $x \leftarrow T(\omega)x$

In recurrent iterated function systems [2] one is interested in generating only $X(\Omega_1)$, where Ω_1 is a special subset of Ω. Specifically

$$\Omega_1 = \{\omega \in \Omega : a_{\omega_1 \omega_{i+1}} = 1; i = 1, 2, \ldots\}$$

where $A = (a_{ij})$ is a zero/one matrix. Let $P = (p_{ij})$ be a transition probability matrix having A as sign pattern. If P is irreducible then we can take ν to be the measure on Ω corresponding to the stationary Markov chain evolving with transition probabilities p_{ij}. Then the reversed-index process $(\omega_{-1}, \omega_{-2}, \ldots)$ in (8) evolves as the reversed Markov chain, giving rise to an algorithm similar to Algorithm A.

Example B (cont'd):

Let $H_m \subseteq \mathbb{R}^m$ be the hyper-plane

$$H_m = \{x \in \mathbb{R}^m : \sum x_i = 1\}.$$

Given any $m \times d$ matrix C, there is a unique affine transformation $T : \mathbb{R}^{m-1} \to \mathbb{R}^d$ making the following diagram commute.

$$\begin{array}{ccc} H_m & \overset{Proj}{\longrightarrow} & \mathbb{R}^{m-1} \\ & \searrow C^t & \downarrow T \\ & & \mathbb{R}^d \end{array}$$

Here Proj is the projection $x \mapsto \tilde{x}$ from $\mathbb{R}^m \to \mathbb{R}^{m-1}$ which lops off the last component

$$\text{Proj} : \begin{bmatrix} x_1 \\ \cdots \\ x_m \end{bmatrix} \mapsto \begin{bmatrix} x_1 \\ \cdots \\ x_{m-1}. \end{bmatrix}$$

Denote $T = T(C)$. It can be written out in coordinate form

$$Tx = [v_1 - v_m | \cdots | v_{m-1} - v_m] x + v_m$$

where $v_1^t, \ldots, v_m^t$ are the row vectors of C. Similarly if P is an $m \times m$ row stochastic matrix then there is a unique affine transformation $T : \mathbb{R}^{m-1} \to \mathbb{R}^{m-1}$ making the following diagram commute.

$$\begin{array}{ccc} H_m & \overset{Proj}{\longrightarrow} & \mathbb{R}^{m-1} \\ \downarrow P^t & & \downarrow T \\ H_m & \overset{Proj}{\longrightarrow} & \mathbb{R}^{m-1} \end{array}$$

In fact $T = T(\widetilde{P})$ where $\widetilde{P}$ is the $m \times (m-1)$ matrix obtained from P by lopping off its last column. The $(m-1) \times (m-1)$ matrix A forming the linear part of T can be described via the diagram

$$\begin{array}{ccc} S_m & \xrightarrow{Proj} & \mathbb{R}^{m-1} \\ \downarrow P^t & & \downarrow A \\ S_m & \xrightarrow{Proj} & \mathbb{R}^{m-1} \end{array}$$

Here $S_m \subseteq \mathbb{R}^m$ is the subspace

$$S_m = \{x \in \mathbb{R}^m : \sum x_i = 0\}.$$

The matrix A can be written out in coordinate form as $A = (p_{ji} - p_{mi})_{i,j=1}^{m-1}$.

The following result is an easy consequence of the above diagrams and some elementary calculations.

Lemma 1 *Let $P(0), \ldots, P(N-1)$ be $m \times m$ row stochastic matrices, and define $T(0), \ldots, T(N-1)$ and $A(0), \ldots, A(N-1)$ in terms of them via the commutative diagrams above.*

i) The following are equivalent.

a) $lim_{n \to \infty} T(\omega_1) \cdots T(\omega_n)x$ exists, $\forall x \in \mathbb{R}^{m-1}$, for every $\omega \in \Omega$, and does not depend on x;

b) $\lim_{n \to \infty} P(\omega_n) \cdots P(\omega_1) = P_\infty(\omega)$ exists for every $\omega \in \Omega$, and $P_\infty(\omega)$ is a rank one row stochastic matrix (i.e., constant entries in each column).

ii) The following are equivalent, and imply a), b) above.

c) $\overline{\lim}_{n \to \infty} \frac{1}{n} \log \|A(\omega_1) \cdots A(\omega_n)\| < 0$ for every $\omega \in \Omega$;

d) $\overline{\lim}_{n \to \infty} \frac{1}{n} \log \operatorname{diam}(P(\omega_n) \cdots P(\omega_1)x) < 0$, $\forall x \in \mathbb{R}^m$, for every $\omega \in \Omega$, where $\operatorname{diam}(y) = \max_{1 \le i,j \le m}(y_i - y_j)$.

Using this Lemma and the process (X_n) from (8), (10) we arrive at the following random dynamical algorithm for generating $X(\Omega)$. The inputs are the row stochastic matrices $P(\omega)$ and the control polygon C. Set $T(\omega) = T(\widetilde{P}(\omega))$, according to the diagram above.

Algorithm B (Sub-Division Methods):
Initialize $x = f(0) \in \mathbb{R}^{m-1}$ (the fixed point of $T(0)$)

For $n = 1, 10000$
  plot $T(C)x \in \mathbb{R}^d$
  choose $\omega \in \{0, \ldots, N-1\}$ randomly
  $x \leftarrow T(\omega)x$

Observe how this algorithm parallelizes. Instead of running an orbit of length 10000 on one processor, one can run (say) 10 processors, each generating only 1000 points of an orbit.

Often one would like to identify Ω with [0,1] through

$$\omega \mapsto \sum_{k=1}^{\infty} \frac{\omega_k}{N^k} \tag{13}$$

where $\omega = (\omega_1, \omega_2, \ldots)$. In order for $X : \Omega \rightarrow \mathbb{R}^d$ to thereby induce a mapping $X' : [0,1] \rightarrow \mathbb{R}^d$ one needs to impose the "de Rham consistency conditions" [9]

$$X(R(\omega)\sigma(0)) = X(R(\omega - 1)(N-1))$$

where $\sigma(i)$ denotes the constant sequence $(i, i, \ldots)$. (This takes care of those numbers with two N-adic expansions.) In the sub-division methods this amounts to

$$P^t(\omega)\pi(0) = P^t(\omega - 1)\pi(N-1) \tag{14}$$

where $\pi(i) \in H_m$ is the normalized eigenvector of $P^t(i)$ corresponding to the eigenvalue $\lambda = 1$.

Example C (cont'd):

Assume that $a_k = 0$ for $k < 0$ or $k > m$, for some $m \in \mathbb{N}$, and that $\sum a_{2k} = \sum a_{2k+1} = 1$, as is typical for line averaging schemes. Thus, motivated by [7], this Example can be put into the framework of Example B by taking $N = 2$,

$$P(0) = (a_{2j-i-1})_{i,j=1}^{m}, \quad P(1) = (a_{2j-i})_{i,j=1}^{m};$$

and taking as control points

$$v_j = e_j \in \mathbb{R}^{m-1}, \quad 1 \leq j \leq m-1,$$

where e_j are the standard unit vectors, and

$$v_m = 0.$$

The condition (14) is automatically satisfied here, and so $X' : [0,1] \to \mathbb{R}^{m-1}$ is well-defined. It satisfies

$$X'\left(\frac{x+\omega}{N}\right) = T(\omega)X'(x), \quad x \in [0,1]$$

which is the *refinement equation* studied in [8]. Define

$$X'_m(x) = 1 - \sum_{k=1}^{m-1} X'_k(x)$$

and set $X'_k(x) = 0$ if $k \le 0$ or $k > m$. Then, as shown in [3], the solution $g(x)$ to the dilation equations (5) is given by

$$g(x) = X'_k(x+1-k), \quad k-1 \le x \le k$$

Thus the random algorithm for generating g is

Algorithm D (Solution of Dilation Equations):
Initialize $a = 0$, $x = \tilde{\pi}(0) \in \mathbb{R}^{m-1}$ (the projection of $\pi(0)$ — the normalized eigenvector for $P^t(0)$)

For $n = 1, 10000$
  $x_m = 1 - \sum_{k=1}^{m-1} x_k$
  plot $(a, x_1), (a+1, x_2), \ldots, (a+m-1, x_m)$
  choose $\omega \in \{0,1\}$ randomly
  $a \leftarrow 0.5(a+\omega)$
  $x \leftarrow T(\omega)x$

The corresponding random algorithm for generating wavelets is based on (6).

Algorithm C (Daubechies' Wavelets):
Initialize $a = 0$, $x = \tilde{\pi}(0) \in \mathbb{R}^{m-1}$ (the projection of $\pi(0)$ — the normalized eigenvector for $P^t(0)$)

For $n = 1, 10000$

$x_m = 1 - \sum_{k=1}^{m-1} x_k$

plot $\left(a + \frac{\ell}{2}, \sum_{k=\max(0,1-\ell)}^{\min(m,m-\ell)} (-1)^{k-1} a_k x_{k+\ell}\right), \quad -m+1 \leq \ell \leq m$

choose $\omega \in \{0, 1\}$ randomly

$a \leftarrow 0.5\,(a + \frac{\omega}{2})$

$x \leftarrow T(\omega)x$

This is illustrated in Figs. 1 and 2.

It is shown in [3] that when the coefficients of the dilation equations (6) are all non-negative, then $g(x)$ turns out to be the stationary density for a $1-D$ iterated function system. Identify $R'(\omega) : x \mapsto \frac{x+\omega}{N}$. (This is induced by the shift $R(\omega)$ under the identification (13).) Then g is the stationary density for the $1-D$ process

$$X_{n+1} = R'(\omega)X_n, \quad \text{with prob. } \frac{a_\omega}{2} \quad (0 \leq \omega \leq m).$$

(Observe that this is an overlapping IFS, in the sense of [1], if $m \geq 2$.) In this case of non-negative coefficients an equivalent algorithm to Algorithm D above is

Algorithm D′ (Solution of Dilation Equations with Non-Negative Coefficients):

Initialize $a = x = 0$, $y[0 : 1000] = 0$

For $n = 1, 250000$

choose $\omega \in \{0, \ldots, m\}$ according to the respective probabilities $\frac{1}{2}a_0, \ldots, \frac{1}{2}a_m$

$x \leftarrow 0.5(x + \omega)$

$i \leftarrow \text{int}\left(\frac{1000x}{m}\right)$

$y[i] \leftarrow y[i] + \frac{0.04}{m}$

endfor

For $i = 0, 1000$

plot $\left(\frac{im}{1000}, y[i]\right)$

endfor

The main theoretical concern in all of this is proving recurrence of the Markov chain (X_n) being used. For the sub-division methods this amounts to establishing condition c) of the Lemma above. If the $P(\omega)$'s have non-negative entries, then a mild irreducibility condition suffices to ensure this, as in [8]. For the general case of signed entries, sufficient conditions ensuring c) appearing in [3]. There the reader will find the full analysis of the results described herein.

Bibliography

[1] M. F. Barnsley, *Fractals Everywhere*, Academic Press, New York, 1988.

[2] M. F. Barnsley, J. H. Elton, and D. P. Hardin, *Recurrent iterated function systems*, Const. Approx. 5 (1989), 3–31.

[3] M. A. Berger, *Random affine iterated function systems: smooth curve generation*, to appear.

[4] I. Daubechies, *Orthogonal bases of compactly supported wavelets*, Comm. Pure Appl. Math. 41 (1988), 909-996.

[5] P. Diaconis, and M. Shahshahani, *Products of random matrices and computer image generation*, Contemp. Math. 50 (1986), 173-182.

[6] J. Hutchinson, *Fractals and self-similarity*, Indiana Univ. J. Math. 30 (1981), 713-747.

[7] C. A. Micchelli, and H. Prautzsch, *Refinement and subdivision for spaces of integer translates of a compactly supported function*, IBM Research Report #58951, October 1987.

[8] C. A. Micchelli, and H. Prautzsch, *Uniform refinement of curves*, IBM Research Report #60538, February 1988.

[9] G. de Rham, *Sur une courbe plane*, J. de Mathématiques Pures et Appliqués 35 (1956), 25–42.

[10] G. Strang, *Wavelets and dilation equations: a brief introduction*, SIAM Review 31 (1989), 614–627.

Initialize $r = x = 0$; $y = 1.366$

For $n = 1, 10000$

$b_1 = -0.183x; b_2 = -0.5x - 0.183y; b_3 = 0.866x - 0.317y - 0.183$

$b_4 = 0.5x + 1.183y - 0.5; b_5 = 0.683(1 - x - y)$

plot $(a - 1, b_1), (a - 0.5, b_2 - b_1), (a, b_3 - b_2), (a + 0.5, b_4 - b_3),$
$(a + 1, b_5 - b_4), (a + 1.5, -b_5)$

choose $k \in \{0, 1\}$ randomly

if $k = 0$

then $r \leftarrow 0.5r$

$y \leftarrow 0.366x + 0.5y + 0.683$

$x \leftarrow 0.683x$

else $r \leftarrow 0.5r + 0.25$

$z \leftarrow x$

$x \leftarrow 1.183x + 0.683y$

$y \leftarrow -1.366z - 0.866y + 1.183$

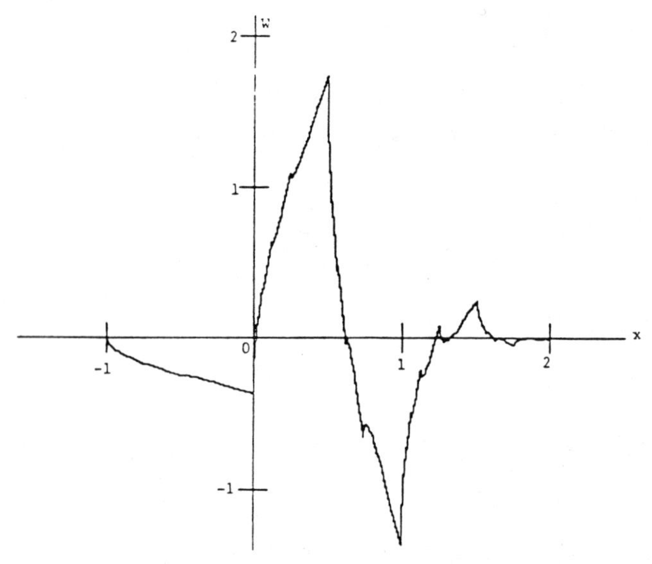

Figure 1: Random dynamical algorithm to generate Daubechies' wavelet W_4. The coefficients a_k are given by

$$4a_0 = 1 + \sqrt{3}, \quad 4a_1 = 3 + \sqrt{3}, \quad 4a_2 = 3 - \sqrt{3}, \quad 4a_3 = 1 - \sqrt{3}$$

Initialize $r = z_0 = z_6 = x_1 = 0$; $x_2 = 1.286$; $x_3 = -0.386$; $x_4 = 0.095$;

$$a_0 = 0.470;\ a_1 = 1.141;\ a_2 = 0.650;$$

$$a_3 = -0.191;\ a_4 = -0.121;\ a_5 = 0.050$$

For $n = 1, 10000$

   $x_5 = 1 - x_1 - x_2 - x_3 - x_4$; $a_0 \leftarrow -a_0$; $a_2 \leftarrow -a_2$; $a_4 \leftarrow -a_4$

   For $\ell = -4, 5$

      $m_1 = \max(0, 1 - \ell)$; $m_2 = \min(5, 5 - \ell)$; $y = 0$

      For $i = m_1, m_2$

         $y \leftarrow y + a_i x_{i+\ell}$

      endfor

      plot $\left(r + \frac{\ell}{2}, y\right)$

   endfor

   choose $k \in \{0, 1\}$ randomly

   $r \leftarrow 0.5r + 0.25k$; $a_0 \leftarrow -a_0; a_2 \leftarrow -a_2$; $a_4 \leftarrow -a_4$

   For $\ell = 1, 5$

      $z_\ell \leftarrow x_\ell; x_\ell \leftarrow 0$

   endfor

   For $\ell = 1, 4$

      $m_1 = \max(0, 2\ell - 6)$; $m_2 = \min(5, 2\ell - 1)$; $j = 2\ell + k - 1$

      For $i = m_1, m_2$

         $x_\ell \leftarrow x_\ell + a_i z_{j-1}$

      endfor

   endfor

endfor

Figure 2: Random dynamical algorithm to generate Daubechies' wavelet W_6. The coefficients a_k are given by

$$a_0 = 0.470,\ a_1 = 1.141,\ a_2 = 0.650,\ a_3 = -0.191,$$

$$a_4 = -0.121,\ a_5 = 0.050$$

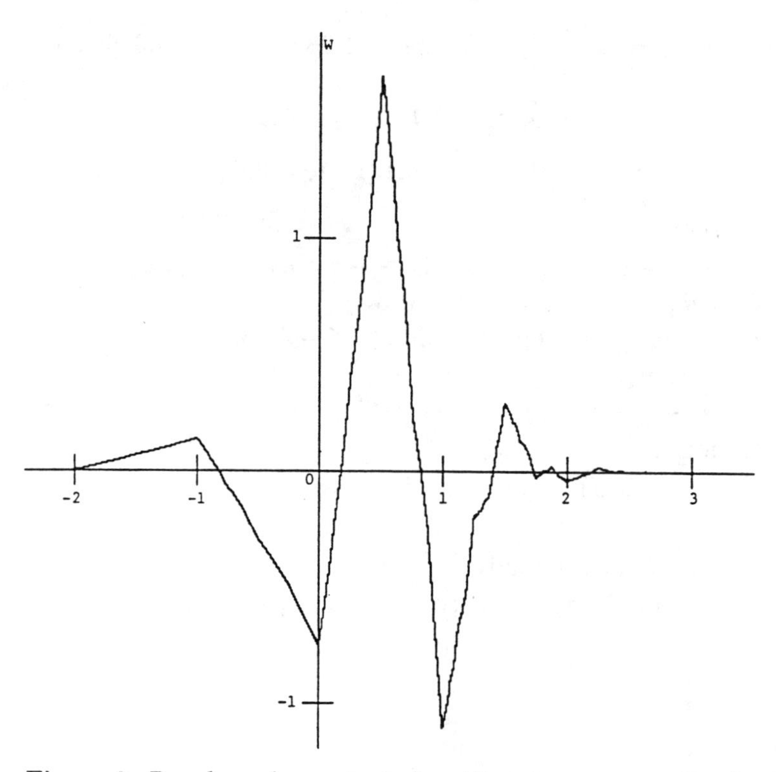

Figure 2: Random dynamical algorithm to generate Daubechies' wavelet W_6.

Markov Properties for Certain Random Fields

E. Carnal[1] and J. B. Walsh[2]

Abstract

Lévy's Markov and sharp Markov properties for random fields are studied, first in a general setting, and then in the context of two-parameter processes. It is shown that if Lévy's Markov property holds relative to finite unions of sets in some neighborhood base, then it holds for all bounded open sets. Two-parameter Gaussian processes which satisfy the usual Markov property along certain one-parameter curves are shown to satisfy Lévy's Markov property; they are in fact transforms of the Brownian sheet. Finally, a new proof is given that the Poisson sheet satisfies Lévy's sharp Markov property relative to all bounded relatively convex open sets.

1 Introduction

Let E be a topological space and let $\{X_t,\ t \in E\}$ be a stochastic process with parameter set E. There are a number of legitimate generalizations of the Markov property for X. One of the most appealing was suggested by P. Lévy. It is roughly this. If $A \subset E$, then the processes $\{X_t,\ t \in A\}$ and $\{X_t,\ t \in A^c\}$ are conditionally independent given $\{X_t,\ t \in \partial A\}$. If one thinks of A as the "past", of A^c as the "future", and of ∂A as the "present", this becomes the statement that the past and future are conditionally independent given

[1]This paper is based on a portion of the first author's doctoral dissertation. Tragically, E. Carnal died while this manuscript was still in a preliminary state. The present version has been assembled by the second author from the preliminary version and the thesis, with an additional section suggested by the first author's unfinished work.

[2]Department of Mathematics, University of British Columbia

ISBN 0-12-481005-5

the present, which is just the ordinary Markov property. We call this the *sharp Markov property* below; there are others which are equally relevant in certain situations, and even the sharp Markov property must be modified slightly if it is to apply to a large class of processes.

Our aim is to look at several of these Markov-like properties and the relations between them. The basic question addressed in the first section is this: if the Markov property holds for a certain class of sets, when can it be extended to a larger class? The second and third sections are largely influenced by the Brownian sheet, a process whose parameter set is $\mathbb{R}^2_+$ so the Markov properties we look at are connected with the order properties of the plane. We concentrate on Gaussian processes in section two, and the Poisson sheet in section three.

2 Markov Properties and Conditional Independence

Let (E,d) be a separable metric space, and let $(\Omega,\mathcal{F},P)$ be a probability space. Suppose that for each subset $A \subset E$ there exists a sub-σ-field of $\mathcal{F}$, denoted $\mathcal{F}(A)$, and that for any sequence $A_1, A_2, \ldots$ of sets, $\mathcal{F}(\bigcup_n A_n) = \vee_n \mathcal{F}(A_n)$. Let X denote the collection of these σ-fields: $X = \{\mathcal{F}(A) : A \subset E\}$. If the $\mathcal{F}(A)$ are generated by a stochastic process, say $\mathcal{F}(A) = \sigma\{Y_t,\ t \in A\}$, we say that X is the natural filtration of the process. However, there are many situations in which no process enters, so we will concentrate on the σ-fields rather than on the processes.

For $A \subset E$, define

$$\mathcal{G}(A) = \bigcap_{\varepsilon>0} \mathcal{F}(A_\varepsilon).$$

where $A_\varepsilon = \{x : d(x,A) < \varepsilon\}$. We call $\mathcal{G}(A)$ the **germ field** of A.

Let $\mathcal{A}$, $\mathcal{B}$, and $\mathcal{C}$ be σ-fields. We write $\mathcal{A} \perp \mathcal{B} \mid \mathcal{C}$ to mean that $\mathcal{A}$ and $\mathcal{B}$ are conditionally independent given $\mathcal{C}$, and we say that $\mathcal{C}$ is a **splitting field** for $\mathcal{A}$ and $\mathcal{B}$. Let us recall several useful properties of conditional independence. The first lemma is immediate and the second is due to Hunt [5].

Lemma 2.1 *(i)* $\mathcal{A} \perp \mathcal{B} \mid \mathcal{C}$ *if and only if for all* $A \in \mathcal{A}$,

$$P\{\mathcal{A} \mid \mathcal{B} \vee \mathcal{C}\} = P\{A \mid \mathcal{C}\};$$

(ii) if $\mathcal{A}' \subset \mathcal{A}$ *and* $\mathcal{B}' \subset \mathcal{B}$, *then*

$$\mathcal{A} \perp \mathcal{B} \mid \mathcal{C} \Longrightarrow \mathcal{A}' \perp \mathcal{B}' \mid \mathcal{C}.$$

Lemma 2.2 *Suppose* $\mathcal{D} \subset \mathcal{A} \cup \mathcal{B}$ *is a collection of subsets of* Ω. *(Caution:* $\mathcal{A} \cup \mathcal{B}$, *not* $\mathcal{A} \vee \mathcal{B}$.*) If* $\mathcal{A} \perp \mathcal{B} \mid \mathcal{C}$ *then*
(i) $\mathcal{A} \vee \mathcal{C} \perp \mathcal{B} \vee \mathcal{C} \mid \mathcal{C}$;
(ii) $\mathcal{A} \perp \mathcal{B} \mid \mathcal{C} \vee \sigma(\mathcal{D})$.

We will often use part (*ii*) in the following form: if $\mathcal{D}_1 \subset \mathcal{A}$ and $\mathcal{D}_2 \subset \mathcal{B}$ then $\mathcal{A} \perp \mathcal{B} \mid \mathcal{C} \Rightarrow \mathcal{A} \perp \mathcal{B} \mid \mathcal{C} \vee \sigma(\mathcal{D}_1) \vee \sigma(\mathcal{D}_2)$.

Lemma 2.3 *Suppose* $\mathcal{C}_1$, $\mathcal{C}_2$, $\mathcal{D}_1$, *and* $\mathcal{D}_2$ *are* σ*-fields, with* $\mathcal{C}_1 \cup \mathcal{C}_2 \subset \mathcal{B}$. *Then*
(i) if $\mathcal{A} \perp \mathcal{B} \mid \mathcal{C}_i$, $i = 1, 2$, *then* $\mathcal{A} \perp \mathcal{B} \mid \mathcal{C}_1 \cap \mathcal{C}_2$.
(ii) If $\mathcal{A} \perp \mathcal{B} \mid \mathcal{D}_1 \vee \mathcal{D}_2$ *and* $\mathcal{A} \perp \mathcal{D}_2 \mid \mathcal{D}_1$, *then* $\mathcal{A} \perp \mathcal{B} \vee \mathcal{D}_2 \mid \mathcal{D}_1$.

PROOF. *(i)* is due to McKean [7], and *(ii)* follows from Lemma 2.1:

$$P\{A \mid \mathcal{B} \vee \mathcal{D}_1 \vee \mathcal{D}_2\} = P\{A \mid \mathcal{D}_1 \vee \mathcal{D}_2\} = P\{A \mid \mathcal{D}_2\}$$

for $A \in \mathcal{A}$. ♣

If the σ-fields X are generated by a Gaussian process, we have an additional property whose proof is elementary.

Lemma 2.4 *Suppose* X *is the natural filtration of a Gaussian process. Let* A, B_1, B_2, *and* C *be subsets of* E. *If* $\mathcal{F}(A) \perp \mathcal{F}(B_i) \mid \mathcal{F}(C)$, $i = 1, 2$, *then* $\mathcal{F}(A) \perp \mathcal{F}(B_1 \cup B_2) \mid \mathcal{F}(C)$.

Definition 2.1 X *has the* **sharp Markov property** *(SMP) relative to a set* $A \subset E$ *if* $\mathcal{F}(A) \perp \mathcal{F}(A^c) \mid \mathcal{F}(\partial A)$. X *has the* **Markov property** *(MP) relative to a set* $A \subset E$ *if* $\mathcal{F}(A) \perp \mathcal{F}(A^c) \mid \mathcal{G}(\partial A)$.

Here are some elementary facts about the SMP and the MP which follow easily from the above lemmas.

Proposition 2.5 *The following three statements are equivalent:*
(i) X *has the SMP relative to* A*;*
(ii) $\mathcal{F}(\bar{A}) \perp \mathcal{F}(\bar{A^c}) \mid \mathcal{F}(\partial A)$*;*
(iii) $\mathcal{F}(A \setminus \partial A) \perp \mathcal{F}(A^c \setminus \partial A) \mid \mathcal{F}(\partial A)$*.*

PROOF. $\mathcal{F}(\bar{A}) = \mathcal{F}(A) \vee \mathcal{F}(\partial A)$ and $\mathcal{F}(\bar{A^c}) = \mathcal{F}(A^c) \vee \mathcal{F}(\partial A)$, so $(i) \Rightarrow (ii)$ by Lemma 2.2 *(i)*. Note that $(iii) \Rightarrow (ii)$ by the same reasoning, while both implications $(ii) \Rightarrow (i)$ and $(ii) \Rightarrow (iii)$ follow from Lemma 2.1 *(ii)*. ♣

Proposition 2.6 *The following are equivalent:*
(i) X *has the MP relative to* A*;*
(ii) if G *is a neighborhood of* ∂A*, then*

$$\mathcal{F}(A) \perp \mathcal{F}(A^c) \mid \mathcal{F}(G);$$

(iii) $\mathcal{G}(A) \perp \mathcal{G}(A^c) \mid \mathcal{G}(\partial A)$*.*

PROOF. Note that $\mathcal{F}(G) = \mathcal{F}(G \cap A) \vee \mathcal{F}(G \cap A^c)$, and let $\mathcal{D} = \mathcal{F}(G \cap A) \cup \mathcal{F}(G \cap A^c)$ in Lemma 2.2 *(ii)* to see that $(i) \Rightarrow (ii)$.

To see that $(ii) \Rightarrow (iii)$, let $\Lambda_1 \in \mathcal{G}(A)$, $\Lambda_2 \in \mathcal{G}(A^c)$, and let G_n be a $\frac{1}{n}$-neighborhood of ∂A. By *(ii)* and Lemma 2.2 *(i)*, $\mathcal{F}(A \cup G_n) \perp \mathcal{F}(A^c \cup G_n) \mid \mathcal{F}(G_n)$ so $P\{\Lambda_1 \cap \Lambda_2 \mid \mathcal{F}(G_n)\} = P\{\Lambda_1 \mid \mathcal{F}(G_n)\} P\{\Lambda_2 \mid \mathcal{F}(G_n)\}$. Note that $\bigcap_n \mathcal{F}(G_n) = \mathcal{G}(\partial A)$, so that the martingale convergence theorem can be applied to each term to see that $P\{\Lambda_1 \cap \Lambda_2 \mid \mathcal{G}(\partial A)\} = P\{\Lambda_1 \mid \mathcal{G}(\partial A)\} P\{\Lambda_2 \mid \mathcal{G}(\partial A)\}$.

Finally, $(iii) \Rightarrow (i)$ by Lemma 2.1 *(ii)*. ♣

Proposition 2.7 *If* X *has the SMP relative to* A*, it also has the MP relative to* A*.*

PROOF. By Lemma 2.2 *(ii)*, $\mathcal{F}(A) \perp \mathcal{F}(A^c) \mid \mathcal{F}(G)$ for any neighborhood G of ∂A, and the result follows from Proposition 2.6. ♣

Proposition 2.8 *Let* A_1 *and* A_2 *be subsets of* E *such that* $\partial(A_1 \cup A_2) = \partial A_1 \cup \partial A_2$*. If* X *has the SMP (respectively MP) relative to* A_i*,* $i = 1, 2$*, then* X *has the SMP (respectively MP) relative to* $A_1 \cup A_2$*.*

PROOF. Suppose X has the SMP relative to A_i, $i = 1,2$. Let $D_i = \partial A_i$. By Proposition 2.5, $\mathcal{F}(\bar{A}_i) \perp \mathcal{F}(\bar{A}_i^c) \mid \mathcal{F}(D_i)$, $i = 1,2$. Let $\mathcal{D} = \mathcal{F}(\bar{A}_2 \cap \bar{A}_1) \cup \mathcal{F}(\bar{A}_2 \cap \bar{A}_1^c)$. Then $\sigma(\mathcal{D}) = \mathcal{F}(\bar{A}_2)$, so by Lemma 2.2 *(ii)*,

$$\mathcal{F}(\bar{A}_1) \perp \mathcal{F}(\bar{A}_1^c) \mid \mathcal{F}(\mathcal{D}_1) \vee \mathcal{F}(\bar{A}_2).$$

Similarly, $\mathcal{F}(D_1) = \mathcal{F}(D_1 \cap A_2) \vee \mathcal{F}(D_1 \cap A_2^c)$, so a second application of Lemma 2.2 *(ii)* shows that $\mathcal{F}(\bar{A}_2) \perp \mathcal{F}(\bar{A}_2^c) \mid \mathcal{F}(D_1 \cup D_2)$.

By Lemma 2.1 *(ii)*,

$$\mathcal{F}(\bar{A}_1) \perp \mathcal{F}(A_1^c \cap A_2^c) \mid \mathcal{F}(D_1) \vee \mathcal{F}(\bar{A}_2)$$

and

$$\mathcal{F}(\bar{A}_2) \perp \mathcal{F}(A_1^c \cap A_2^c) \mid \mathcal{F}(D_1 \cup D_2).$$

By Lemma 2.3 *(ii)* with $\mathcal{D}_1 = \mathcal{F}(D_1 \cup D_2)$, and $D_2 = \mathcal{F}(\bar{A}_2)$,

$$\mathcal{F}(\bar{A}_1 \cup \bar{A}_2) \perp \mathcal{F}(A_1^c \cap A_2^c) \mid \mathcal{F}(D_1 \cup D_2).$$

But $D_1 \cup D_2 = \partial(A_1 \cup A_2)$, so this implies the SMP relative to $A_1 \cup A_2$.

If X has the MP relative to A, repeat the argument with the D_i replaced by neighborhoods G_i of ∂A_i, and apply Proposition 2.6. ♣

Proposition 2.9 *Let (G_n) be a sequence of disjoint open sets. If X satisfies the SMP (respectively MP) relative to each G_n, it satisfies the SMP (respectively MP) relative to $\bigcup_n G_n$.*

PROOF. Suppose X satisfies the SMP relative to each G_n. Since $\partial(\bigcup_1^N G_n) = \bigcup_1^N \partial G_n$ for each N, Proposition 2.8 and an induction argument imply that X has the SMP relative to $\bigcup_1^N G_n$. Fix N and let $\Lambda \in \mathcal{F}(\bigcup_1^N G_n)$. For $m > N$ the SMP implies that

$$P\{\Lambda \mid \mathcal{F}(\bigcap_1^m G_n^c)\} = P\{\Lambda \mid \mathcal{F}(\bigcup_1^m \partial G_n)\}.$$

Let $m \to \infty$. By the martingale convergence theorem,

$$P\{\Lambda \mid \bigcap_{m=1}^\infty \mathcal{F}(\bigcap_{n=1}^m G_n^c)\} = P\{\Lambda \mid \mathcal{F}(\bigcup_1^\infty \partial G_n)\}.$$

Let $G = \bigcup_1^\infty G_n$. Then $G^c = \bigcap_1^\infty G_n^c \supset \partial G \supset \bigcup_1^\infty \partial G_n$, so that

$$\bigcap_{m=1}^\infty \mathcal{F}(\textstyle\bigcap_1^m G_n^c) \supset \mathcal{F}(\bigcap_1^\infty G_n^c) = \mathcal{F}(G^c) \supset \mathcal{F}(\partial G) \supset \mathcal{F}(\bigcup_1^\infty \partial G_n).$$

Thus it follows that $P\{\Lambda \mid \mathcal{F}(G^c)\} = P\{\Lambda \mid \mathcal{F}(\partial G)\}$, since both sides equal $P\{\Lambda \mid \mathcal{F}(\bigcup_1^\infty \partial G_n)\}$. This is true for $\Lambda \in \mathcal{F}(\bigcup_1^N G_n)$ for any N, hence for $\Lambda \in \bigvee_1^\infty \mathcal{F}(G_n) = \mathcal{F}(G)$. By Lemma 2.1, we are done. The argument for the MP is similar. ♣

Corollary 2.10 *Suppose the space E is connected and locally connected and that the SMP (respectively the MP) holds for every open connected set whose complement is also connected. Then it holds for every open set.*

PROOF. If G is open, write $G = \bigcup_n G_n$, where the G_n are the connected components of G. Let $O_n = (\bar{G}_n)^c$ and let (O_{nj}) denote the connected components of O_n. If A is an open set containing $\bar{G}_n$, it must intersect each O_{nj}. Indeed, if it missed O_{nj}, say, then O_{nj} and $A \cup \left(\bigcup_{k \neq j} O_{nk}\right)$ would disconnect $\bar{G}_n \cup (\bigcup_k O_{nk}) = E$. It follows that $O_{nj}^c = \bar{G}_n \cup \left(\bigcup_{k \neq j} O_{nk}\right)$ is connected. Thus the hypothesis applies to the O_{nj}: the SMP (respectively MP) holds for each O_{nj}, hence for $O_n = \bigcup_n O_{nj}$ by Proposition 2.9. As $O_n^c = \bar{G}_n$, we have

$$\mathcal{F}(O_n) \perp \mathcal{F}(\bar{G}_n) \mid \mathcal{F}(\partial O_n),$$

(respectively, $\mathcal{F}(O_n) \perp \mathcal{F}(\bar{G}_n) \mid \mathcal{G}(\partial O_n)$), hence by Proposition 2.5 (respectively Proposition 2.6 and Lemma 2.1 *(ii)*) the SMP (respectively MP) holds for G_n. A final application of Proposition 2.9 shows that the SMP (respectively MP) holds for G. ♣

Let $A \triangle B = (A \setminus B) \cup (B \setminus A)$ denote the symmetric difference of A and B. For $x \in E$, let $d(x, B) = \inf\{d(s, y) : y \in B\}$. For any set B, let $N_\varepsilon(B) = \{x : d(x, B) < \varepsilon\}$ be the ε-neighborhood of B. The next results concern the MP rather than the SMP. We first show that if the MP holds for a class of sets which approximate A well enough, it also holds for A.

Theorem 2.11 *Let $A \in E$ and let $\{A_\alpha, \alpha \in T\}$ be a family of sets with the property that for each $\varepsilon > 0$ there is an $\alpha \in T$ such that $A \triangle A_\alpha \subset N_\varepsilon(\partial A)$. Then if X has the MP relative to A_α for each $\alpha \in T$, it also has the MP relative to A.*

PROOF. Let $\varepsilon > 0$ and let $G = N_\varepsilon(\partial A)$. There exists α such that $A_\alpha \triangle A \subset G$. Thus $A \setminus G = A_\alpha \setminus G$ and $A^c \setminus G = A_\alpha^c \setminus G$. It follows that $\partial A_\alpha \subset G$. By hypothesis and Proposition 2.6, $\mathcal{F}(A_\alpha) \perp \mathcal{F}(A_\alpha^c) \mid \mathcal{F}(G)$, hence by Lemma 2.2 *(i)*,

$$\mathcal{F}(A_\alpha \cup G) \perp \mathcal{F}(A_\alpha^c \cup G) \mid \mathcal{F}(G).$$

Since $A \subset A_\alpha \cup G$ and $A^c \subset A_\alpha^c \cup G$, Lemma 2.1 *(ii)* implies that $\mathcal{F}(A) \perp \mathcal{F}(A^c) \mid \mathcal{F}(G)$, and the result follows from Proposition 2.6. ♣

Notice that we gain nothing by assuming that X has the SMP—not just the MP—relative to each A_α, for the conclusion would still only be that X satisfied the MP relative to A. One can use this theorem when the A_α are sets which either increase or decrease to A. Here is a typical application.

Corollary 2.12 *Let $\mathcal{A}$ be a class of open subsets of E which contains a neighborhood basis of connected subsets of E. Let $\mathcal{A}_f$ be the class of all finite unions of sets in $\mathcal{A}$. Then if X has the MP relative to each set in $\mathcal{A}_f$, it has the MP relative to each relatively compact open subset of E.*

PROOF. Let G be open and relatively compact in E, and let $\varepsilon > 0$. Then $\bar{G}$ is compact, and can be covered by a finite number, say $A_1, \ldots, A_n$ of connected elements of $\mathcal{A}$ which have diameter less than $\varepsilon/2$. By throwing away some of the A_i if necessary, we may assume that each A_i intersects $\bar{G}$. Let $A = \bigcup_{i=1}^n A_i$. Now $G \subset \bar{G} \subset A$, so $\partial G \subset A \setminus G = A \triangle G$. If $z \in A \triangle G$, then $z \in A_j$ for some j; since $A_j \cap \bar{G} \neq \phi$, A_j must contain points of $\bar{G}$ and of G^c. Since A_i is connected by hypothesis, it must contain points of ∂G (otherwise, $A_i \cap G$ and $A_i \cap (\bar{G})^c$ would disconnect A_i). Thus $d(z, \partial G) < \varepsilon$, hence $A \triangle G \subset N_\varepsilon(\partial G)$, and the result follows from Theorem 2.11. ♣

3 Markov Properties in the Plane

Let I and J be intervals, not necessarily bounded or open, and let $E = I \times J$. By *rectangle*, we mean a bounded sub-rectangle of E of the form $I' \times J'$, where $I' \subset I$ and $J' \subset J$. We define the partial order $\prec$ and its complementary order $\overset{c}{\prec}$ by

$$(s,t) \prec (s',t') \text{ iff } s \leq s' \text{ and } t' \leq t\,;$$
$$(s,t) \overset{c}{\prec} (s',t') \text{ iff } s \leq s' \text{ and } t \geq t'\,.$$

Generalizations of the results of this section from $\mathbb{R}^2$ to $\mathbb{R}^n$ are straightforward for the most part, though they may be messy. For instance, one needs 2^{n-1} partial orders in $\mathbb{R}^n$ in place of the two above. We will confine our remarks to $\mathbb{R}^2$ for simplicity.

We will define several versions of the Markov property. Our main concern is with Gaussian processes, so we will consider the case where X is the natural filtration of a Gaussian process $Y = \{Y_z,\, z \in E\}$. If $z = (s,t)$, we will often write $Y(s,t)$ in place of Y_z.

Definition 3.1 *X satisfies the* **Markov property** *(resp.* **sharp Markov property***) if it has the MP (resp. SMP) relative to all open relatively compact subsets of E. X has the* **elementary Markov property** *if it has the MP relative to all finite unions of open rectangles. X has the* **order Markov property** *if, whenever $\gamma = \{\gamma(u),\, u \in [0,1]\}$ is a parameterized curve in E with the property that γ is increasing relative to either of the two partial orders $\prec$ or $\overset{c}{\prec}$, that $\{Y_{\gamma(u)},\, u \in [0,1]\}$ is a Markov process.*

There are many other Markov-type properties which have been suggested for two-parameter processes. Wong and Zakai [14] have studied a property which is related to both the MP and the order MP, and Nualart and Sanz [10] and Korezlioglu et al [6] have considered a closely related property.

Proposition 3.1 *The MP and the elementary MP are equivalent.*

PROOF. The MP clearly implies the elementary MP. The converse follows from Corollary 2.12 since the rectangles form a base for the topology of E. ♣

We will now specialize to centered Gaussian processes, for which Markov properties are more tractable. For a mean-zero Gaussian process $\{Y_z,\ z \in E\}$, the order MP reduces to the following:

if z_1, z_2, and z_3 are points in E for which either $z_1 \prec z_2 \prec z_3$ or $z_1 \overset{c}{\prec} z_2 \overset{c}{\prec} z_3$, then $Y_{z_1} \perp Y_{z_2} \mid Y_{z_3}$.

If $\Gamma(z, z')$ is the covariance function of Y, then it is easily seen that $Y_{z_1} \perp Y_{z_2} \mid Y_{z_3}$ if and only if

$$\Gamma(z_1, z_3) = \Gamma(z_1, z_2)\Gamma(z_2, z_3)(\Gamma(z_2, z_2))^{-1} \tag{1}$$

with the convention that $\frac{0}{0} = 0$.

The MP for a Gaussian process evidently reduces to a study of the covariance function. We say a covariance function $\Gamma(z, z')$ is of **Markov tensor product type**, or more simply, of **product type**, if there exist covariance functions $\Gamma_1(s, s')$, $s, s' \in I$ and $\Gamma_2(t, t')$, $t, t' \in J$, each of which is the covariance function of a one-parameter Gaussian Markov process, and a function f on E such that, if $z = (s, t)$ and $z' = (s', t')$, then

$$\Gamma(z, z') = f(z)f(z')\Gamma_1(s, s')\Gamma_2(t, t'). \tag{2}$$

For example, the Brownian sheet $\{W(s, t),\ s \geq 0, t \geq 0\}$ is a Gaussian process which has a covariance function of product type:

$$\Gamma((s, t'), (s', t')) = (s \wedge s')(t \wedge t'). \tag{3}$$

Theorem 3.2 *Suppose Y is a mean zero Gaussian process with a continuous covariance function which does not vanish on the diagonal. Then the following are equivalent.*

(i) Y has the order MP;

(ii) Γ is of product type;

(iii) there exist continuous increasing functions f on I and g on J, a continuous function h on E, and a Brownian sheet W, such that

$$Y(s, t) \equiv h(s, t)W(f(s), g(t)),$$

where "$\equiv$" means equivalence in distribution.

PROOF. $(i) \Rightarrow (ii)$. By taking $Z_z = \Gamma(z,z)^{-\frac{1}{2}} Y_z$ if necessary, we can assume that $\Gamma(z,z) = 1$. Let z_1, z_2, and z_3 be in E. The condition (2) that $Y_{z_1} \perp Y_{z_2} \mid Y_{z_3}$ becomes

$$\Gamma(z_1, z_3) = \Gamma(z_1, z_2)\Gamma(z_2, z_3) . \tag{4}$$

Let $z = (s,t) \prec (s',t') = z'$ and put $\bar{z} = (s,t')$ and $\underline{z} = (s',t)$. Both $z \prec \bar{z} \prec z'$ and $z \prec \underline{z} \prec z'$, so

$$\Gamma(z,z') = \Gamma(z,\bar{z})\Gamma(\bar{z},z') = \Gamma(z,\underline{z})\Gamma(\underline{z},z') . \tag{5}$$

Furthermore, both $\bar{z} \overset{c}{\prec} z' \overset{c}{\prec} \underline{z}'$ and $\bar{z} \overset{c}{\prec} z \overset{c}{\prec} \underline{z}$ so

$$\Gamma(\bar{z},\underline{z}) = \Gamma(\bar{z},z')\Gamma(z',\underline{z}) = \Gamma(\bar{z},z)\Gamma(z,\underline{z}) \tag{6}$$

Divide the second two terms of (5) by the corresponding terms of (6). Since Γ is symmetric, we see that $\Gamma(z,\bar{z})^2 = \Gamma(\underline{z},z')^2$, so that $\Gamma(z,\bar{z}) = \pm\Gamma(\underline{z},z')$ and hence that $\Gamma(z,\bar{z}) = \Gamma(\underline{z},z')$. Indeed, Γ is continuous and non-vanishing by (4) and the plus sign holds if $z = \bar{z} = \underline{z}$, so it holds everywhere.

In terms of s and t,

$$\Gamma((s,t),(s,t')) = \Gamma((s',t),(s',t'))$$

so that the left-hand side depends only on (t,t'), not on s. By symmetry,

$$\Gamma((s,t),(s',t)) = \Gamma((s,t'),(s',t')) ,$$

which depends only on (s,s'). Putting these together, we define $\Gamma_1(s,s') = \Gamma((s,t),(s',t))$ and $\Gamma_2(t,t') = \Gamma((s,t)(s,t'))$; then by (5)

$$\Gamma((s,t),(s',t')) = \Gamma_1(s,s')\Gamma_2(t,t') .$$

Now Γ_1 is the covariance function of the process $\{Y(s,t),\ s,t \in I\}$ for fixed t, which is a Gaussian Markov process since Y is Markovian along horizontal lines by hypothesis. By symmetry, Γ_2 is also a Markovian covariance function, so Γ is of product type.

(ii) ⇒ *(iii)*: By [7, §3.1] there exist continuous functions φ_i and ψ_i such that φ_i/ψ_i is increasing, and $\Gamma_i(u,v) = \varphi_i(u \wedge v)\psi_i(u \vee v)$. It follows that if $W(s,t)$ is a Brownian sheet, the process

$$Z(s,t) = \psi_1(s)\psi_2(t)\, W\left(\frac{\varphi_1(s)}{\psi_1(s)}, \frac{\varphi_2(t)}{\psi_2(t)}\right)$$

has covariance function $\Gamma_1\Gamma_2$. Thus we take $f(s) = \varphi_1(s)\psi_1(s)^{-1}$, $g(s) = \varphi_2(s)\psi_2(s)^{-1}$, and $h(s,t) = \psi_1(s)\psi_2(t)$.

(iii) ⇒ *(i)*: The map $(s,t) \mapsto (f(s), g(t))$ preserves both partial orders "$\prec$" and "$\overset{c}{\prec}$", so we need only show that the Brownian sheet satisfies the order MP. This follows by direct calculation using (3). ♣

Corollary 3.3 *Suppose that* Y *is a centered Gaussian process whose covariance is continuous and non-zero on the diagonal. If* Y *has the order MP, it also has the MP, and it has the SMP for finite unions of rectangles.*

PROOF. The mapping $(s,t) \mapsto (f(s), g(t))$ of E into $\mathbb{R}^2_+$ preserves order and set operations, and maps rectangles onto rectangles. By Theorem 3.2, it is enough to prove the Corollary for the Brownian sheet W. By [11,3,2][3] W satisfies the SMP for finite unions of rectangles. Thus it satisfies the MP by Proposition 3.1. ♣

Remark 3.1 The condition that the covariance function be continuous in Theorem 3.2 and Corollary 3.3 is not necessary, but it simplifies the proof enormously. The results are proved in [2] without the continuity condition.

[3]This result is difficult. A short and elegant proof has recently been found by Z-M Yang [15].

4 Markov Properties of the Poisson Sheet

Let Π be a homogeneous Poisson random measure on the Borel sets $\mathcal{B}$ of $\mathbb{R}^2_+$, that is

(i) $\Pi(A)$ is a Poisson r.v. with parameter $|A|$, $A \in \mathcal{B}$;

(ii) If $A_1, \ldots, A_n$ are disjoint, $\Pi(A_i)$, $i = 1, \ldots, n$ are independent and $\Pi(\bigcup_i A_i) = \sum_i \Pi(A_i)$,

where $|A|$ is the Lebesgue measure of A. The measure Π is a sum of point masses. The points form a Poisson point process on $\mathbb{R}^2_+$. We will use the symbol Π for both the random measure and the point process. "Points" below will refer to the points of this process.

If $z = (s,t) \in \mathbb{R}^2_+$, let R_z (or R_{st}) denote the rectangle $[0,s] \times [0,t]$. The **Poisson sheet** $\{X_z,\ z \in \mathbb{R}^2_+\}$ is defined by $X_z = \Pi(R_z)$.

It is not hard to see from (i) and (ii) that X satisfies the SMP for rectangles, but it is not obvious that X will satisfy the SMP relative to any much larger class of sets. The Brownian sheet, for instance, satisfies the SMP for finite unions of rectangles, but not for many other sets [11,12,13,3]. However, it turns out that the Poisson sheet satisfies the SMP for all bounded open sets.

The reason for the difference between the Poisson sheet and the Brownian sheet is in the global behavior: the Brownian sheet is continuous, while the Poisson sheet has discontinuities which propagate on lines. These discontinuities are the key to the SMP. In fact, all two parameter processes of independent increments with no Gaussian part satisfy the SMP with respect to bounded open sets, though this is a rather delicate fact in the case where the processes can have negative as well as positive jumps [4].

We will not prove this general theorem here, even for the Poisson sheet; we will limit ourselves to the special case of relatively convex open sets. (In fact, because of a delay in publishing this article—due entirely to the second author—it has been necessary to delete a conjecture in the original version, to the effect that the result should be true without the restriction of relative convexity. This conjecture has been proved in part by Merzbach and Nualart [8], who have shown that the SMP holds for a large class of point processes and for domains with piecewise-monotone boundaries; and by Dalang and the second author [4].) Both of the above proofs depend on earlier non-trivial results. By limiting ourselves to relatively convex sets, we can give a proof which is elementary and self-contained, and which avoids most of the cumbersome technicalities which come with increased generality.

The proper notion of convexity for the Poisson sheet is relative convexity. A set $A \subset I\!R_+^2$ is **relatively convex** if for each horizontal or vertical line L, $L \cap A$ is connected. Thus "relatively convex" means convex relative to horizontal and vertical lines. A convex set is relatively convex, but the converse is not true. The cross in the Swiss flag, for instance, is relatively convex but not convex. A relatively convex set need not be connected.

Theorem 4.1 *The Poisson sheet satisfies the sharp Markov property relative to all bounded open relatively convex sets, but not with respect to all unbounded open relatively convex sets. If A is a bounded open relatively convex set, then $\mathcal{F}(\partial A)$ is the minimal splitting field of $\mathcal{F}(A)$ and $\mathcal{F}(A^c)$.*

We will deal with the unbounded sets first. Let T be the triangular region $\{(s,t) : s > t\} \subset I\!R_+^2$. Then ∂T is the diagonal of the first quadrant, and $\mathcal{F}(\partial T) = \sigma\{X(t,t),\ t \geq 0\}$. Let

$$\begin{aligned}\Lambda_1 &= \{\exists \text{ exactly one point in } T \cap R_{11}\}\\ \Lambda_2 &= \{\exists \text{ exactly one point in } T^c \cap R_{11}\}\\ \Lambda_3 &= \{\exists \text{ exactly one point in } R_{11}\}.\end{aligned}$$

Now $\Lambda_3 = \{X(1,1) = 1\} \in \mathcal{F}(\partial T)$ and $\Lambda_1 \cap \Lambda_2 \cap \Lambda_3 = \phi$, so

$$P\left\{\textstyle\bigcap_1^3 \Lambda_i \mid \mathcal{F}(\partial T)\right\} \neq P\{\Lambda_1 \cap \Lambda_3 \mid \mathcal{F}(\partial T)\} P\{\Lambda_2 \cap \Lambda_3 \mid \mathcal{F}(\partial T)\}$$

since the first probability is zero and each of the last two is strictly positive. Thus $\mathcal{F}(T)$ is not conditionally independent of $\mathcal{F}(T^c)$ given $\mathcal{F}(\partial T)$.

Before dealing with the rest of the theorem we need some facts about relatively convex sets. The reader who is willing to limit himself to convex sets can skip points 1°–6°. In what follows, A

is an open, bounded, connected, relatively convex subset of $I\!R^2_+$. Let $I = \{s : \exists t \ni (s,t) \in A\}$. A is connected, so I is an interval, say $I = (\alpha, \beta)$. For $s \in I$, define $M(s) = \sup\{t : (s,t) \in A\}$ and $m(s) = \inf\{t : (s,t) \in A\}$.

1° If $M(u) < \lambda < M(v)$, there exists an s between u and v such that $(s, \lambda) \in A$.

PROOF. If not, the half planes $\{t < \lambda$ and $\{t > \lambda\}$ disconnect A.

2° There exist s_0 and s_1, not necessarily unique, such that M increases on (α, s_1) and decreases on (s_1, β); and m decreases on (α, s_0) and increases on (s_0, β).

PROOF. If the statement for M is false there exist λ and $u < v < w$ such that $M(v) < \lambda < M(u) \wedge M(w)$. By 1°, there exist $s' < v < s''$ such that the points (s', λ) and (s'', λ) are in A. By relative convexity, $(v, \lambda) \in A$ contradicting the definition of $M(v)$.

Extend M and m to α and β by continuity and define $\bar{M}(s) = \sup\{t : (s,t) \in \bar{A}\}$ and $\bar{m}(s) = \inf\{t : (s,t) \in \bar{A}\}$ for $s \in [\alpha, \beta]$. The following facts are now elementry, so we omit the proofs. We state them for M; analogous statements hold for m.

3° $\bar{M} \geq M$ and $\bar{M}(s) = M(s)$ at points of continuity of M. If M has a jump at s, the vertical line segment with endpoints $(s, M(s))$ and $(s, \bar{M}(s))$ forms part of ∂A. The set $N \equiv \{$ jumps of $\bar{M}$ $\}$ is countable.

4° ∂A is composed of points $\{(s, M(s)) :$ s a continuity point of $M\}$, $\{(s, m(s)) :$ s a continuity point of $m\}$, and vertical segments.

For a more graphical description of A, let $a = \bar{m}(s_0)$ and $b = \bar{M}(s_1)$. The points (s_0, a) and (s_1, b) are the "lowest" and "highest" points of ∂A. Similarly, there exist t_0 and t_1 such that (α, t_0) and (β, t_1) are in ∂A. We call these four points *relative extreme points* of A. In terms of these, we have the following picture of ∂A.

5° ∂A consists of four curves, (some of which may be degenerate):
the *upper left segment* UL with end points (α, t_0) and (s_1, b);
the *upper right segment* UR with end points (s_1, b) and (β, t_1);
the *lower right segment* LR with end points (s_0, a) and (β, t_1);

the *lower left segment* LL with end points (α, t_0) and (s_0, a).

For example, if A is a disc, the relative extreme points are at three, six, nine and twelve o'clock, while the curves UL, UR, LR, and LL are respectively from nine to twelve o'clock, from twelve to three o'clock, from three to six o'clock, and from six to nine o'clock.

6° Let $t = c$ be a horizontal line. Then there is a countable set N_1 such that if $c \notin N_1$ and $a < c < b$, then $t = c$ meets ∂A in exactly two points. Furthermore, if it meets, say, UL at the point (s, c), either

(i) (s, c) is in the interior of a vertical segment of UL, or

(ii) M is continuous at s and $M(s - \varepsilon) < M(s) < M(s + \varepsilon)$ for small $\varepsilon > 0$.

The Poisson sheet X has integer values. Each point of the point process gives rise to a jump discontinuity of X or, rather, since X has a two-dimensional parameter set, it gives rise to a pair of discontinuities which originate at the point. One runs horizontally, the other vertically (see figure 1).

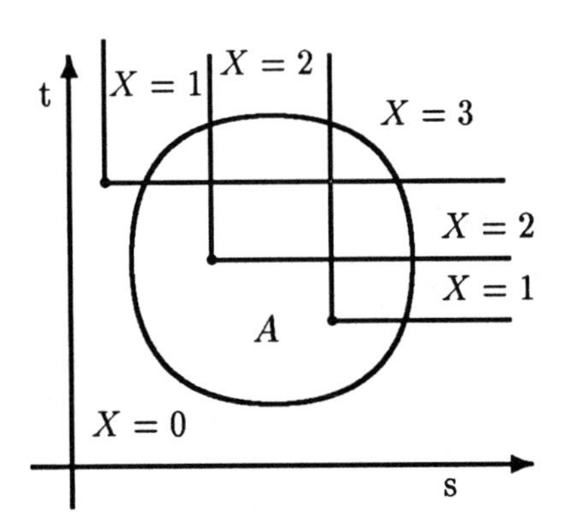

Figure 1: The discontinuities of X

When we speak of a discontinuity of X we mean one of these semi-infinite line segments, either horizontal or vertical, along which S has a jump discontinuity. If either discontinuity crosses ∂A, it causes a discontinuity of $X|_{\partial A}$. These jumps of $X|_{\partial A}$ give us information about the Poisson measure Π. To be more explicit, let us define

$$\Delta Y_U(s) = \lim_{r \downarrow\downarrow s, r \in Q} X(r, M(r)) - \lim_{r \uparrow\uparrow s, r \in Q} X(r, M(r))$$

$$\Delta Y_L(s) = \lim_{r \downarrow\downarrow s, r \in Q} X(r, m(r)) - \lim_{r \uparrow\uparrow s, r \in Q} X(r, m(r)) .$$

If we interchange the roles of s and t, we get analogous processes $\Delta \tilde{Y}_U(t)$ and $\Delta \tilde{Y}_L(t)$. All four are clearly $\mathcal{F}(\partial A)$-measurable.

We will use several facts about the Poisson point process without special mention, such as the fact that with probability one, no two points fall on any single horizontal or vertical line, and if K is any given set of measure zero, with probability one no point falls in K, and no two discontinuities of X can cross in K.

Lemma 4.2 *For a.e. ω, the following hold.*
(i) If $\Pi(\{(s,t)\}) = 1$ then
(a) $t < m(s)$ and $s \in (\alpha, \beta) \setminus N \Rightarrow \Delta Y_L(s) = 1$;
(b) $t < M(s)$ and $s \in (\alpha, \beta) \setminus N \Rightarrow \Delta Y_U(s) = 1$.
(ii) Conversely, suppose $s \in (s_1, \beta) \setminus N$ and $\Delta Y_U(s) = 1$, where s_1 is defined in 2°. Then $\exists\, t < M(s) \ni \Pi(\{(s,t)\}) = 1$.
(iii) If $s \in (\alpha, s_1) \setminus N$ and $\Delta Y_U(s) = 1$. Then either
(a) $\exists\, t < M(s) \ni \Pi(\{(s,t)\}) = 1$ or
(b) $\exists\, s' < s \ni \Pi(\{(s', M(s))\}) = 1$, and $\Delta \tilde{Y}_U(M(s)) \neq 0$.

PROOF. *(i)* See figure 1.
(ii) If $\Delta Y_U(s) \neq 0$ and $s \notin N$, there is a discontinuity of S passing through ∂A at $(s, M(s))$. Since $s > s_1$, the point is in UR, and $s \mapsto M(s)$ is strictly decreasing by 6°. If the discontinuity is horizontal, $\Delta Y_U(s) = -1$ (see figure 1), a contradiction. Thus it is vertical, and comes from a point located at (s,t) for some $t < M(s)$.
(iii) If *(a)* doesn't hold, the discontinuity through $(s, M(s))$ is horizontal. By 6° M is strictly increasing there, so the discontinuity must enter A at $(s, M(s))$, and it must then exit to the right at some point $(s'', M(s))$, causing a discontinuity of $X|_{\partial A}$, hence $\Delta \tilde{Y}_U(M(s)) \neq 0$.
♣

Let $\mathcal{D}_0$ be the collection of sets of the forms

$$V_U(c,d) \equiv \{(s,t) : c < s < d,\ 0 < t < M(s)\};$$

$$V_L(c,d) \equiv \{(s,t) : c < s < d,\ 0 < t < m(s)\};$$

along with their counterparts $H_U(c,d)$ and $H_L(c,d)$, gotten by interchanging s and t, and sets of the form R_z, for $z \in \partial A$. (V_U and V_L are vertical strips bounded above by portions of ∂A, and H_U and H_L are horizontal strips bounded to the right by portions of ∂A.) Let $\mathcal{D}'$ be the class of sets of the form $B \setminus A$, for sets $A \subset B$ such that $A, B \in \mathcal{D}_0$. Then let $\mathcal{D}$ be the class of finite disjoint unions of sets in $\mathcal{D}'$. The following lemma is the key to the SMP.

Lemma 4.3 *If $D \in \mathcal{D}$, then $\Pi(D)$ is $\mathcal{F}(\partial A)$-measurable.*

PROOF. It is enough to show measurability for the sets in $\mathcal{D}_0$. By Lemma 4.2 *(i)*,

$$\Pi(V_L(c,d)) = \#\{s \in (c,d) \setminus N : \Delta Y_L(s) \Delta Y_U(s) \neq 0\},$$

where N is the countable set of 3°. This is $\mathcal{F}(\partial A)$-measurable. If $c > s_1$,

$$\Pi(V_U(c,d)) = \#\{s \in (c,d) \setminus N : \Delta Y_U(s) = 1\}$$

by Lemma 4.2 *(i)* and *(ii)*, while if $d \leq s_1$, Lemma 4.2 *(i)* and *(iii)* imply that

$$\Pi(V_U(c,d)) = \#\{s \in (c,d) \setminus N : \Delta Y_U(s) = 1 \text{ and } \Delta \tilde{Y}_U(M(s)) = 0\}.$$

Thus these are $\mathcal{F}(\partial A)$-measurable.

The same is true of $H_U(c,d)$ and $H_L(c,d)$ by symmetry, and $\Pi(R_z) = X_z$ which is $\mathcal{F}(\partial A)$-measurable if $z \in A$. ♣

Let

$$\begin{aligned} X_z^1 &= \Pi(R_z \cap A) \\ X_z^2 &= \Pi(R_z \cap A^c). \end{aligned}$$

Then X^1 and X^2 are independent processes, and $X_z = X_z^1 + X_z^2$.

Lemma 4.4 *(i)* $z \in \bar{A} \Rightarrow X_z^2 \in \mathcal{F}(\partial A)$;
(ii) $z \in A^c \Rightarrow X_z^1 \in \mathcal{F}(\partial A)$.

PROOF. Let $z = (s,t) \in \bar{A}$. We claim $R_z \cap A^c \in \mathcal{D}$, which will prove *(i)* by Lemma 4.3. We can choose $s' \leq s$ and $t' \leq t$ such that $(s', t') \in$ LL. Then

$$R_{st} \cap A^c = R_{s't'} \cup V_L(s', s) \cup H_L(t', t) \in \mathcal{D}.$$

Similarly, $R_z \cap A \in \mathcal{D}$ if $z \in A^c$. We leave the verification to the reader. ♣

We can now prove the theorem.

PROOF of Theorem 4.1. Let A be a bounded open relatively convex set. Each connected component is necessarily relatively convex and bounded, so by Proposition 2.9 we may as well assume that A itself is open, connected, bounded and relatively convex. Notice that

$$\mathcal{F}(\bar{A}) = \sigma(X^1) \vee \mathcal{F}(\partial A), \qquad \mathcal{F}(A_c) = \sigma(X^2) \vee \mathcal{F}(\partial A). \quad (7)$$

Indeed, $X = X^1 + X^2$ and $\sigma(X^1) \in \mathcal{F}(\bar{A})$, so $\sigma(X^1) \vee \mathcal{F}(\partial A) \subset \mathcal{F}(\bar{A})$. On the other hand, $X^2|_{\partial A} \in \mathcal{F}(\partial A)$ by Lemma 4.4 *(i)* so if $z \in \bar{A}$, $X_z \in \sigma(X^1) \vee \mathcal{F}(\partial A)$. Thus $\mathcal{F}(\bar{A}) \subset \sigma(X_1) \vee \mathcal{F}(\partial A)$, proving the first half of (7). The second half is similar, using Lemma 4.4 *(ii)*.

Now both $X^1|_{\partial A}$ and $X^2|_{\partial A}$ are $\mathcal{F}(\partial A)$-measurable, so

$$\mathcal{F}(\partial A) = \sigma\left(X^1|_{\partial A}\right) \vee \sigma\left(X^2|_{\partial A}\right). \quad (8)$$

Let $\mathcal{C}$ be the trivial σ-field. X^1 and X^2 are independent, so certainly $\sigma(X^1) \perp \sigma(X^2) \mid \mathcal{C}$. By (8) and Lemma 2.2 *(i)*, we see that $\sigma(X^1) \perp \sigma(X^2) \mid \mathcal{F}(\partial A)$, hence $\sigma(X^1) \vee \mathcal{F}(\partial A) \perp \sigma(X^2) \vee \mathcal{F}(\partial A) \mid \mathcal{F}(\partial A)$, by Lemma 2.2 *(i)*, and we are done by (7).

Finally, to see that $\mathcal{F}(\partial A)$ is the minimal splitting field, let $\mathcal{S}$ be a splitting field and let $\Lambda \in \mathcal{F}(\partial A) \subset \mathcal{F}(A^c)$. Since $\mathcal{S}$ is a splitting field and Λ is in both $\mathcal{F}(\bar{A})$ and $\mathcal{F}(A^c)$

$$P\{\Lambda \mid \mathcal{S}\} = P\{\Lambda \mid \mathcal{F}(\bar{A})\} = I_\Lambda \Rightarrow \Lambda \in \mathcal{S}.$$

♣

Bibliography

[1] E. Carnal, *Propriétés markoviennes des produits tensoriels de covariances markoviennes,* C. R. Acad. Sci. Paris 288 sér. A, p. 149–152.

[2] E. Carnal, *Processus markoviens à plusieurs parametres,* Thesis, École Polytechnique Fédérale de Lausanne, 1979.

[3] R. Dalang and J. B. Walsh, *The sharp Markov property of the Brownian sheet and related processes,* to appear, Acta Math.

[4] R. Dalang and J. B. Walsh, *The sharp Markov property of Lévy sheets,* to appear, Ann. Prob.

[5] G. A. Hunt, *Martingales et Processus de Markov,* Dunod, Paris, 1966.

[6] H. Korezlioglu, P. Lefort and G. Mazziotto, *Une propriété markovienne et diffusions associées,* Processus Aléatoires à Deux Indices, Lect. N. Math. 863, Springer Verlag, (1981), p. 245–274.

[7] H. P. McKean, *Brownian motion with a several-dimensional time,* Th. Prob. and Appl. 8, p. 335–354.

[8] E. Merzbach and D. Nualart, *Markov Properties for point processes on the Plane,* Ann. Prob. 18, (1990), 342–358.

[9] J. Neveu, *Processus Aleatoires Gaussiens,* Presses de l'Université de Montréal, 1968.

[10] D. Nualart and M. Sanz, *A Markov property for two parameter Gaussian processes,* Stochastica 3 (1979), p. 1–16.

[11] F. Russo, *Étude de la propriété de Markov étroite en relation avec les processus planaires à accroissements indépendants,* Séminaire de Probabilités 18, Lect. N. Math 1059, Springer Verlag (1984), p. 353–378

[12] J. B. Walsh, *The propagation of singularities in the Brownian sheet,* Ann. Prob. 10, (1982), p. 279–288.

[13] J. B. Walsh, *Martingales with a multidimensional parameter and stochastic integrals in the plane,* Lect. N. Math. 1215, Springer Verlag, (1986), p. 229–491.

[14] E. Wong and M. Zakai, *Markov processes on the plane,* Stochastics 15, (1985) p. 311–333.

[15] Zhen-Ming Yang, *Equivalence between two Markov properties for processes with two parameters,* preprint.

The Anatomy of a Low-Noise Jump Filter: Part I

J.M.C. Clark
Imperial College, London

1. Introduction

This paper presents an asymptotic analysis of the filtering through low noise of a signal containing a distributed jump. The early paper of Zakai and Ziv [7] has led to an increasing interest in the use of perturbation methods for nonlinear filtering problems involving a small parameter, and in particular a number of authors, among them [1,2,3,5,7,8], have addressed the problem of filtering a diffusion from low observation noise. There seems, however, to have been little work on the corresponding problem for signal processes containing generally distributed jumps, though Kushner's treatment [4] of the filtering of perturbed signal processes of jump-diffusion type is certainly relevant to its 'fast' time-scale aspects. The non-local nature of the generators of discontinuous Markov processes ensures that the conditional laws, for low observation noise, behave quite differently from those for diffusion processes. In particular, in the discontinuous case the conditional laws are likely to have 'long' tails. Our motivating aim is the design of finite dimensional high-precision approximate filters for piecewise constant processes and, more generally, for the piecewise-deterministic processes espoused by M.H.A. Davis. There is no systematic way of doing this, except in the case of Markov chains with the jumps taking only a finite set of values, and it is natural to consider as a first step the asymptotic behaviour or 'shape' of the conditional laws of these processes. The behaviour not only determines the accuracy achievable by approximate

ISBN 0-12-481005-5

filters, but should also provide clues for their construction. This paper considers the conditional laws of a particular signal process containing a single, randomly occurring, generally distributed jump, at times prior to that jump. These 'pre-jump' laws are only one aspect of a rather special problem, but it is likely they are very similar to the 'inter-jump' conditional laws for more general piecewise-deterministic processes. The equally important 'fast time' behaviour of the laws at times just after a jump raises a separate set of questions and will, it is hoped, be treated elsewhere.

2. Preliminary definitions

Let $\{W_t : t \geq 0\}$, X and τ be a continuous Brownian motion and two random variables that are all strictly independent on a probability space $(\Omega, \mathcal{F}, \mathbb{P})$. Suppose that X is non-zero, and $\tau \geq 0$ possesses a strictly positive bounded continuous density g(t). F(t) and G(t) will denote the distribution functions of X and τ. The filtering model is as follows:

$$S_t = 0 \qquad \text{for } t < \tau$$
$$\quad = X \qquad \text{for } t \geq \tau$$

and for $\epsilon > 0$ $\{Y_t^\epsilon : t \geq 0\}$ is an observation process:

$$Y_t^\epsilon = \epsilon^{-1} \int_0^t S_s ds + W_t.$$

So, formally, ϵdY_t^ϵ represents a "signal plus low white noise". Let $\mathcal{Y}_t^\epsilon = \sigma(Y_t^\epsilon : 0 \leq s \leq t) \subset \mathcal{F}$. With a slight abuse of notation in the rewriting of S_t as $S_t(x,r)$, the Kallianpur-Striebel extension of Girsanov's formula allows us to express the conditional law of S_t as

$P(S_t \in dx|\mathcal{Y}_t^\epsilon) =$

$$N_t^{-1}\int_{\mathbb{R}}\int_0^\infty 1_{\{S_t(z,r)\in dx\}} \times \exp[\tfrac{1}{\epsilon}\int_0^t S_s(z,r)dY_s^\epsilon - \tfrac{1}{2\epsilon^2}\int_0^t S_s(z,r)^2 ds] \times g(r)dr\, dF(z)$$

which, on $\{\tau \geq t\}$, reduces to

$$P(S_t \in dx|\mathcal{Y}_t^\epsilon) = N_t^{-1}((1-G(t))\delta_0(dx)$$

$$\text{(1a)} \qquad + dF(x)\int_0^t \exp[\tfrac{x}{\epsilon}(W_t - W_r) - \tfrac{1}{2}\tfrac{x^2}{\epsilon^2}(t-r)]g(r)dr).$$

Here δ_0 is Dirac measure at 0, and N_t is the normalizing factor, which on $\{\tau \geq t\}$ is

$$\text{(1b)} \qquad N_t = 1 - G(t) + \int_{\mathbb{R}}\int_0^t \exp[\tfrac{x}{\epsilon}(W_t - W_r) - \tfrac{1}{2}\tfrac{x^2}{\epsilon^2}(t-r)]g(r)dr\, dF(x).$$

Equations (1a,b) are the basis of our analysis.

3. The conditional laws of S_t

If ϵ is small and the jump has not yet occurred then the conditional probability that $S_t = 0$ is obviously close to one. What is of more interest is the conditional probability of S_t being in a set excluding the origin, $P(S_t \in B\backslash\{0\}|\mathcal{Y}_t^\epsilon)$. Of particular interest is the case where $B=\mathbb{R}$; this is then just $P(\tau \leq t|\mathcal{Y}_t^\epsilon)$. It turns out precise answers can be given for two kinds of '$P(\cdot|\mathcal{Y}_t^\epsilon)$ – continuity' sets: Borel sets B containing an open interval containing $\{0\}$ (i.e. 'neighbourhoods' of $\{0\}$), and their complements.

Proposition 1 Suppose X has a continuous density on an open interval of the origin. If a Borel set B contains an open interval around the origin, then on the set $\{\tau \geq t\}$

$$\text{(2a)} \qquad \lim_{\epsilon \to 0} \epsilon^{-1} P(S_t \in B \backslash \{0\} | \mathcal{Y}_t^\epsilon)$$

$$= \frac{\sqrt{2\pi}\, f(0)}{1 - G(t)} \int_0^t \frac{1}{\sqrt{t-s}} \exp\left(\frac{(W_t - W_s)^2}{2(t-s)} \right) g(s)ds.$$

If B excludes an open interval around the origin, then on $\{\tau \geq t\}$

$$\text{(2b)} \qquad \epsilon^{-2} P(S_t \in B | \mathcal{Y}_t^\epsilon) \overset{\text{(law)}}{\rightarrow} \frac{g(t)}{1 - G(t)} \int_B \int_0^\infty \exp[x\beta_s - x^2 s/2] ds \, dF(x)$$

where $\{\beta_t : t \geq 0\}$ is a continuous Brownian motion. Furthermore the right members in (2a) and (2b) are finite (a.s).

Here $\overset{\text{(law)}}{\rightarrow}$ means weak convergence of the probability law of the left member of (2b), which, as a function of $\{W_t\}$ is of course random, to that of the right.

The proof will require the following lemma, which follows immediately from a joint application of the 'local' and 'global' forms of the law of the iterated logarithm for Brownian motion.

Lemma Suppose $\{\beta_t : t \geq 0\}$ is a continuous Brownian motion. Then

$$K := \sup\{\frac{\beta_t^2}{t \operatorname{loglog}(t + t^{-1} + e)} : 0 < t < \infty\} < \infty \text{ a.s.}$$

Proof of Proposition 1 Let

$$(3) \qquad I(B) = \int_0^t \left(\int_B \exp\left[\frac{x}{\epsilon} \bar{W}_s - \frac{x^2 s}{2\epsilon^2}\right] g(t-s)dF(x)\right)ds$$

where $\bar{W}_s$ is short for the reverse-time Brownian motion $W_t - W_{t-s}$. Decompose the range of integration of I(B) into

$$I(B) = \int_0^\alpha (\ldots)ds + \int_\alpha^t (\ldots)ds$$

$$=: I_0^\alpha + I_\alpha^t$$

Take α to be $\epsilon^{3/2}$. We will show that $I_0^\alpha = o(\epsilon)$. Now it follows from the identity

$$(4) \qquad \frac{x}{\epsilon} \bar{W}_s - \frac{x^2 s}{2\epsilon^2} = \frac{\bar{W}_s^2}{2s} - \frac{s}{2\epsilon^2}\left(x - \frac{\epsilon \bar{W}_s}{s}\right)^2$$

applied to (3) that

$$I_0^\alpha \leq \int_0^\alpha \exp\left(\frac{\bar{W}_s^2}{2s}\right) g(t\text{-}s)ds.$$

By the law of the iterated logarithm for Brownian motion

$$\frac{\bar{W}_s^2}{2s} \leq 2 \operatorname{loglog} \alpha^{-1}$$

on $0 \leq s \leq \alpha$ for all α sufficiently small. Set $\bar{g} = \max(g(s),\ 0 \leq s \leq t)$. Then it follows, again for sufficiently small α, that

$$I_0^\alpha \leq \bar{g}\,\alpha(\log \alpha^{-1})^2$$

As $\alpha=\epsilon^{3/2}$, this implies that $I_0^\alpha = o(\epsilon)$. To obtain the asymptotic estimate of I_α^t, first use the identity (4) to rewrite I_α^t as

$$(5)\quad I_\alpha^t = \epsilon\int_\alpha^t \sqrt{\frac{2\pi}{s}}\exp[\bar{W}_s^2/2s]\int_B \sqrt{\frac{s}{2\pi\epsilon^2}}\exp\left(-\frac{s}{2\epsilon^2}\left(x-\frac{\epsilon\bar{W}_s}{s}\right)^2\right)dF(x)g(t\text{-}s)ds$$

We see from the lemma that for $0 \leq s \leq t$

$$(6)\qquad \exp[\bar{W}_s^2/2s] \leq (\log(s^{-1}+s+e))^{K/2}$$

and that for $\epsilon^{3/2} \leq s \leq t$ and for ϵ sufficiently small

$$|\epsilon\bar{W}_s/s| \leq \epsilon^{1/4}\sqrt{K\,\log\log(\epsilon^{-3/2}+t+e)}$$

$$\leq \sqrt{K}\epsilon^{1/5}$$

Suppose $(-a,a)$ is an interval contained in B that supports the continuous density of f. It can easily be established by a standard inequality on the tails of the normal law that the inner integral in (5) restricted to $B\backslash(-a,a)$ satisfies

$$\int_{B\backslash(-a,a)} \ldots dF(x) \leq \sqrt{\frac{2t}{\pi\epsilon^2 a^2}}\exp\left(-\frac{1}{2\epsilon^{1/2}}\left(a-\sqrt{K}\epsilon^{1/5}\right)^2\right)$$

$$= o(\epsilon).$$

Regarding the corresponding integral over $(-a,a)$ as the integral of the continuous function f with respect to truncated Gaussian measures converging in weak* sense to the Dirac measure $\delta_0(dx)$ as $\epsilon \to 0$, we see that the integral converges to f(0). Consequently, it follows that

$$\lim_{\epsilon \to 0} \epsilon^{-1} I^t_\alpha = f(0) \int_0^t \sqrt{\frac{2\pi}{s}} \exp\left(\overline{W}_s^2/2s\right) g(t-s)ds.$$

The integral here is finite by (6). Since $I_0^\alpha = o(\epsilon)$ it follows that $\epsilon^{-1} I$ has the same limit and I is of order $0(\epsilon)$. (2a) then follows from the normalizing equation

$$(7) \qquad P(S_t \in B, \tau \le t | \mathcal{Y}_t^\epsilon) = \frac{I(B)}{1 - G(t) + I(\mathbb{R})}.$$

(2b) is included in this proposition for convenient comparison; but it is a corollary of the more detailed functional form which is stated in the next proposition: if we set

$$\begin{aligned} \phi(x) &= x^{-2} && \text{for } x \in B \\ &= 0 && \text{for } x \notin B, \end{aligned}$$

in Proposition 2, (2b) follows directly.

Notice that if $B = \mathbb{R}$ (2a) gives the asymptotic form for $P(\tau \le t | \mathcal{Y}_t^\epsilon)$ on $\{\tau \ge t\}$; so this is of order ϵ. A further distinctive feature of Proposition 1 is the fact that probabilities of sets outside an interval containing $\{0\}$ are all of the same order ϵ^2; this contrasts with the case of a Gaussian diffusion signal, where the corresponding tail probabilities are those of a normal law with vanishing variance ϵ^2, and also with the case of general diffusions, where estimates of 'large deviations' type apply [2].

In assessing the 'uncertainty' of S_t it is useful to have asymptotic forms for its conditional moments.

Proposition 2 Suppose the distribution F of the jump X has a continuous density on an open interval containing the origin. Suppose for some $\delta > 0$ a Borel function ϕ satisfies $E[((|X|^{\delta}+|X|^{-\delta})|\phi(X)|] < \infty$. Then for those values of $\tau \geq t$ $E[S_t^2\phi(S_t)|\mathcal{Y}_t^\epsilon]$ exists and, as $\epsilon \to 0$

$$(8)\quad \epsilon^{-2}E[S_t^2\phi(S_t)|\mathcal{Y}_t^\epsilon] \overset{(law)}{\rightarrow} \frac{g(t)}{1-G(t)} \int_{\mathbb{R}} x^2\phi(x) \int_0^\infty \exp[x\beta_s - x^2 s/2]ds\, dF(x)$$

where $\{\beta_t : t \geq 0\}$ is a continuous Brownian motion.

Proof First define for $s \geq 0$ $\beta_s^\epsilon = \epsilon^{-1}\overline{W}_{\epsilon^2 s \wedge t}$. Then $\{\beta_s^\epsilon;\ s \geq 0\}$ is a Brownian motion stopped at t/ϵ^2. Now the normalizing term N_t in (1b) is, by (3), $1-G(t) + I(\mathbb{R})$, and by the first part of Proposition 1 $I(\mathbb{R}) = 0(\epsilon)$. So, replacing $t - r$ in (1a) by $\epsilon^2 s$ and $\epsilon^{-1}(W_t - W_r)$ by β_s^ϵ we obtain

$$P(S_t^2\phi(S_t)|\mathcal{Y}_t^\epsilon) = -\epsilon^2(1-G(t))^{-1} \int_{\mathbb{R}} x^2\phi(x) \int_0^{t/\epsilon^2} \exp[x\beta_s^\epsilon - x^2 s/2]g(t-\epsilon^2 s)ds\,dF(x)$$

Given the identity of the laws of $\{\beta_s^\epsilon\}$ and $\{\beta_s\}$ on $[0,t/\epsilon^2]$, it is sufficient to establish the integrability with respect to ds dF(x), uniform in ϵ, of the integrand

$$J_\epsilon(s,x) := x^2\phi(x)\exp[x\beta_s - x^2 s/2]g(t-\epsilon^2 s)1_{[0,t/\epsilon^2]}(s)$$

for, as J_ϵ converges pointwise to J_0, its integral would then converge to the finite limit

$\int_{\mathbb{R}}\int_0^\infty J_0(s,x)ds\, dF(x)$ which is just the integral in the right member of (8).

Now by the lemma

$$\sup_s(x\beta_s - x^2s/4) \leq \sup_s\left(\sqrt{Ks\log\log(s+s^{-1}+e)}\,|x| - x^2s/4\right),$$

which is minimized, it turns out, when

$$|x| = 2\left(\sqrt{Ks^{-1}\log\log(s+s^{-1}+e)}\right)(1+o(1)),$$

where o(1) refers here to a bounded continuous function that vanishes at s=0 and ∞. The solution of this is

$$s = 4Kx^{-2}\log\log(x^2+x^{-2}+e)(1+o(1))$$

the o(1) now referring to a function of |x| vanishing at 0 and ∞. Consequently

$$\sup_s(x\beta_s - x^2s/4) \leq K_1\log\log(x^2+x^{-2}+e)$$

for some random K_1. So if $\overline{g} := \sup\{g(s) : s \geq 0\}$,

$$\int_0^{t/\epsilon^2} g(t-\epsilon^2 s)\exp[x\beta_s - x^2/2]ds$$

$$\leq \overline{g}\int_0^\infty \exp[K_1\log\log(x^2+x^{-2}+e) - x^2s/4]ds$$

$$= 4\overline{g}\, x^{-2}(\log(x^2+x^{-2}+e))^{K_1}.$$

Consequently

$$\int_0^{\epsilon} J_\epsilon(s,x)ds \leq 4\overline{g}\,\phi(x)\log(x^2+x^{-2}+e)^{K_1}$$

$$\leq K_2\phi(x)(x^{\delta}+x^{-\delta})$$

for some suitable random K_2. It follows from the integrability condition on ϕ that this last term is integrable with respect to $dF(x)$. So J_ϵ is integrable uniformly in ϵ and the proof of (8) is complete.

The 'gathering' around the true value of the underlying jump law that is produced by conditioning has the effect of increasing the order of integrable power moments by a factor of nearly two and reducing the order of inverse powers by a similar amount. More precisely

Corollary Suppose the density condition is satisfied. If $E[|X|^d] < \infty$ for some $d > 0$ then for all r where $1 < r < d+2$ $E[|S_t|^r|\mathcal{Y}_t^\epsilon]$ is of precise order ϵ^2 in the sense that

$$\epsilon^{-2}E[|S_t|^r|\mathcal{Y}_t^\epsilon] \overset{(law)}{\rightarrow} M_r$$

for some strictly positive finite random variable M_r.

Proof Since X has a continuous density near the origin, $E[|X|^{-q}]$ exists for $0<q<1$. Set $\delta=\frac{1}{2}\min\{1,r-1,d+2-r\}$ and $\phi(x)=x^{r-2}$. The integrability condition on ϕ is then satisfied and Proposition 2 applies. As the right member of (8) is a positive finite random variable, the order is precise.

A consequence of this is that the L^r norm of S_t

$$||S_t||_r := E[|S_t|^r|\mathcal{Y}_t^\epsilon]^{1/r}$$

is, for the range $r \in (1,2)$, of order $\epsilon^{2/r}$. This again contrasts with the case of Gaussian diffusion signals, for which the norms would all be of the same order of magnitude.

In circumstances where a family of random variables $\{Z_\epsilon : \epsilon > 0\}$ converges to zero in law with L^r norms of different orders of magnitude, it is natural to describe the rate of convergence in terms of orders of magnitude 'in law'. That is, Z_ϵ is $0(\alpha(\epsilon))$ in law if $\alpha(\epsilon)$ is a scaling function that ensures that the laws of the scaled variables $Z_\epsilon/\alpha(\epsilon)$ are ultimately tight. A necessary and sufficient condition for this is, of course, the existence of a strictly increasing function ϕ on $[0,\infty]$ with $\phi(0)=0$ and $\phi(\infty)=\infty$ such that $\limsup_{\epsilon\to 0} E[\phi(|Z_\epsilon|/\alpha(\epsilon))] < \infty$. If the limit is zero, then z_ϵ is $o(\alpha(\epsilon))$ in law.
Taking S_t with its conditional law $P(S_t \in dx|\mathcal{Y}_t^\epsilon)$ to be Z_ϵ, $\phi(x)$ to be x^r with $r=2/(2-\delta)$ for some $0<\delta < 1$, we see that under the conditions of the corollary S_t is $0(\epsilon^{2-\delta})$ in law. However it turns out this is a gross overestimate. Let ϕ be

$$\phi(x) = x \qquad \text{if } x \leq 1$$
$$= \log(x/e) \qquad \text{if } x > 1.$$

Then it can be estimated by arguments similar to those in the proof of Proposition 1 that under the conditions of the corollary $\limsup_{\epsilon\to 0} E[\phi(|S_t|e^{k/\epsilon})|\mathcal{Y}_t^\epsilon] < \infty$ for any $k>0$ and therefore that S_t is $0(e^{-k/\epsilon})$ in law!

However, not too much should be read into this remarkable rate of convergence; it merely reflects the features of the model that the 'pre-jump' value of S_t is known exactly and that $P(S_t \neq 0|\mathcal{Y}_t^\epsilon) = 0(\epsilon)$. If, for instance, S_0 were to have a prior positive density, this last conditional probability would be one, and it is likely that S_t would be of order ϵ in law.

4. The conditional relative density of S_t

The preceding results do not fully reveal the 'shape' of the tails of the conditional laws. (1a) shows that S_t possesses a conditional density relative to dF(x) on $\mathbb{R}\backslash\{0\}$:

$$(9)\qquad \begin{aligned}\hat{r}_t(x) &= \frac{d}{dF(x)}\,\mathbb{P}(S_t \leq x|\mathcal{Y}_t^\epsilon) \\ &= N_t^{-1}\int_0^t \exp[\tfrac{x}{\epsilon}(W_t - W_s) - \tfrac{x^2}{2\epsilon^2}(t-s)]g(s)ds\end{aligned}$$

where N_t is the normalizing factor given by (1b). $\hat{r}_t(x)$, depending as it does on $\{W_t\}$, is random. Thanks to a result of M. Yor [6] it is possible to describe precisely the asymptotic probability density of its scaled form.

Proposition 3 Suppose X has a continuous density (with respect to Lebesgue measure) on an interval around the origin. Suppose $x \neq 0$ is fixed. For those values of $\tau \geq t$

$$(10)\qquad \epsilon^{-2}\hat{r}_t(x) \overset{(law)}{\rightarrow} \frac{g(t)V}{(1-G(t))x^2}$$

as $\epsilon \rightarrow 0$, where V is a random variable with density $\frac{1}{v^2}e^{-1/v}$ (that is, with the density of the reciprocal of an exponential random variable).

Proof First, the proof of Proposition 1 established that $N_t = 1 - G(t) + 0(\epsilon)$, which gives the factor $(1-G(A))^{-1}$ in the right member of (10). Now, as before the process $\{\beta_s^{\epsilon/x} : 0 \le s \le x^2/\epsilon^2\}$, where $\beta_s^{\epsilon/x} = \frac{x}{\epsilon}(W_t - W_{t-\epsilon^2 s/x^2})$ is Brownian. The integral in the right member of (9) becomes with this substitution

$$\frac{\epsilon^2}{x^2} \int_0^{x^2 t/\epsilon^2} \exp[\beta_s^{\epsilon/x} - \tfrac{s}{2}] g(t - \epsilon^2 s/x^2) ds.$$

Since g is bounded and continuous the law of this integral converges weakly to that of $g(t)V$, where

$$V := \int_0^{\infty} \exp[\beta_s - \tfrac{s}{2}] ds$$

and $\{\beta_t : t \ge 0\}$ is any continuous Brownian motion. It is a corollary of a theorem of M. Yor [6] for exponential integrals of this type that V has the density of the reciprocal of an exponential random variable; that is, the density of V is $\frac{1}{v^2} e^{-1/v}$. The form of the right member of (10) then follows and the proof is complete.

There is a comparable result for the joint law of the densities $(\hat{r}_t(x_1), \hat{r}_t(x_2), \ldots, \hat{r}_t(x_n))$, the 'V' in (10) now being replaced by a vector $(V_{x_1}, V_{x_2}, \ldots, V_{x_n})$. The V_{x_i} all have the same given marginal law, but their joint law has at present no obvious characterization.

Suppose X has a density $f(x)$ and τ is exponential with density $\lambda\, e^{-\lambda t}$. Then $\{S_t\}$ is Markov and, for $t \le \tau$, the conditional density is asymptotically stationary and takes, asymptotically, the simple form

$\epsilon^2 \lambda V_x f(x)/x^2$. This demonstrates how the effect of conditioning on low-noise observations is to 'gather' the underlying law towards the true value of the signal by the introduction of a factor proportional to $\frac{\epsilon^2}{x^2}$.

5. References

[1] B.Z. Bobrovsky and M. Zakai, Asymptotic a priori estimates for the error in the nonlinear filtering problem, *IEEE Trans. Inform. Th.* **IT-28**(2)1982, 371-376.

[2] D. Ji, *Asymptotic Analysis of Nonlinear Filtering Problems,* Ph. Thesis, Brown University 1988.

[3] R. Katzur, B.Z. Bobrovsky and Z. Schuss, Asymptotic analysis of the optimal filtering problem for one-dimensional diffusions measured in a low noise channel, *SIAM J. Appl. Math.* **44** (1984), Part I: 591-604, Part II: 1176-1191.

[4] H.J. Kushner, *Weak Convergence Methods and Singularly Perturbed Stochastic Control and Filtering Problems*, Birkhauser Boston 1990.

[5] J. Picard, Nonlinear filtering of one-dimensional diffusions in the case of a high signal-to-noise ratio, *SIAM J. Appl. Math.* **46**(1986), 1098-1125.

[6] M. Yor, unpublished correspondence.

[7] M. Zakai, J. Ziv, Lower and upper bounds on the optimal filtering error of certain diffusion processes. *IEEE Trans Inf. Th.* **IT-18**(3) 1972, 325-331.

[8] O. Zeitouni, Approximate and limit results for nonlinear filters with small observation noise: the linear sensor and constant diffusions coefficient case, *IEEE Trans. Aut. Cont.* **33**(1988), 595-599.

On the Value of Information in Controlled Diffusion Processes

M.H.A. Davis
Imperial College of Science, Technology and Medicine

M.A.H. Dempster
Essex University

R.J. Elliott [1]
University of Alberta

Abstract

The requirement that the control process in a controlled diffusion be nonanticipative is formulated as an explicit equality constraint, following an idea of R.J.-B. Wets. The Lagrange multiplier associated with this constraint is identified, in a geometric setting.

1 Introduction

Stochastic optimization problems almost invariably take the following general form. Let $(\tilde{\Omega}, \tilde{\mathcal{F}}, \tilde{\mathbb{P}})$ be a finite (not necessarily probability) measure space and $\mathcal{U}$ be some class of measurable functions on $(\tilde{\Omega}, \tilde{\mathcal{F}}, \tilde{\mathbb{P}})$ taking values in another measurable space (U, Ξ). Let $c : \tilde{\Omega} \times U \to \mathbb{R}^+$ be a cost function, i.e., $c(\tilde{\omega}, u)$ is the cost incurred when state of nature $\tilde{\omega}$ is realized and decision u is taken. We then seek to compute the minimal "average cost" $\inf_{u \in \mathcal{U}} \tilde{\mathbb{E}}[c(\tilde{\omega}, u(\tilde{\omega})]$, (here $\tilde{\mathbb{E}}$ denotes integration with respect to $\tilde{\mathbb{P}}$) and to compute, if one exists, an optimal $u^0 \in \mathcal{U}$ which achieves the infimum. It is important to note at the outset that if $\mathcal{U}$ is equal to the set $\mathcal{U}^*$ of *all*

[1] Research Partially supported by NSERC grant 7964 and the U.S. Army Research Office under contract DAAL03-87-0102.

ISBN 0-12-481005-5

measurable function $u : \tilde{\Omega} \to U$ then there is nothing really "stochastic" about this problem at all. Indeed, it is evident in this case that (modulo some technical conditions)

$$\inf_{u \in \mathcal{U}^*} \tilde{\mathbb{E}}[c(\tilde{\omega}, u(\cdot))] = \tilde{\mathbb{E}}[\inf_{u \in U} c(\tilde{\omega}, u)].$$

The decision maker knows $\tilde{\omega}$ in advance, and the term inside the square brackets on the right is just a family of deterministic optimization problems indexed by $\tilde{\omega} \in \tilde{\Omega}$. A *selection theorem* is required to ensure that the solutions to these problems can be pieced together in a measurable way, giving an optimal $u^0 \in \mathcal{U}^*$. The role of the measure $\tilde{\mathbb{P}}$ is only to average the infima of the optimization problems to produce an average minimum cost. Thus a truly "stochastic" problem depends on $\mathcal{U}$ being some *strict subset* of $\mathcal{U}^*$.

Dynamic stochastic optimization problems are those in which $(\tilde{\Omega}, \tilde{\mathcal{F}}, \tilde{\mathbb{P}}) = (\Omega \times I,\ \mathcal{F} * \mathcal{J},\ \mathbb{P} * \mu)$, where $(\Omega, \mathcal{F}, \mathbb{P})$ is a probability space and $(I, \mathcal{J}, \mu)$ some "time" set. In this paper we will work in continuous time and I will be a finite interval, $I = [0, T]$, the measure μ being Lebesgue measure. The characteristic feature of these problems is that "information" is acquired as time proceeds. This is formulated by specifying a *filtration* $(\mathcal{F}_t)_{t \in [0,T]}$ contained in $\mathcal{F}$. A measurable function u on $\tilde{\Omega}$ is just a continuous time stochastic process on Ω, and the "information constraint" is that u be adapted to $(\mathcal{F}_t)$. In applications, $(\mathcal{F}_t)$ will be the natural filtration of some "observed" process (y_t) and the adaptedness requirement states, roughly, that the decision u_t should depend only on the past observations $(y_s,\ s \leq t)$. A nice way of formulating the adaptedness requirement is to introduce the predictable σ-field $\mathcal{P}$ in $\tilde{\Omega}$ and demand that u be $\mathcal{P}$-measurable, i.e. a predictable process. Of course this is slightly more restrictive than mere adaptedness, but it is the appropriate formulation in many applications. If $\mathcal{U}$ denotes the set of predictable U-valued processes, then evidently $\mathcal{U}$ is a linear subspace of $\mathcal{U}^*$. Thus *information constraint* that u be predictable is actually an *equality constraint* in $\mathcal{U}^*$ which we can write as $(I - \Pi)u = 0$, where Π denotes the predictable projection operator. Our general goal is to study the Lagrange multiplier associated with this constraint. The motivation for doing so is twofold: (a) by introducing the Lagrange multiplier we turn the optimization problem into a global minimization over $\mathcal{U}^*$; as pointed out earlier this is essentially a deterministic problem, and thus stochastic optimization becomes a special case of deterministic optimization! (b) the Lagrange multiplier has an interpretation as a price system for small violations of the constraint — in this case, small anticipative perturbations of the control. We are thus able to calculate an incremental "value of information" for the stochastic decision problem.

These questions have been studied in a series of papers Wets [17], Rockafellar and Wets [14], Dempster [6], Back and Pliska [1], in the stochastic programming literature (meaning that the function c is explicitly given and no 'dynamic system' intervenes). The first three of these are formulated in a discrete time setting, while [1] gives a general formulation in continuous time. The main results are that the Lagrange multiplier exists and has a certain "martingale" property. It will be clear from developments later in this paper why this martingale property holds.

Until recently no such results were available in stochastic control, where the function c is only implicitly defined since it generally depends on the solution of the controlled system's dynamic equations. However, the linear system/quadratic cost (LQG) case has now been completely resolved (Davis [3,5], Davis and Heunis [5]), and considerable progress has been made for controlled nonlinear stochastic differential equations (SDE), (Davis and Burstein [4]). In the latter case a foundational issue arises, in that the solutions of the SDEs must be defined when anticipative controls are used, so we must bring in — at least implicitly — some form of anticipative stochastic integral. This is why we hope this subject will appeal to Moshe Zakai. In fact the appropriate definition of the SDE solution is easily obtained using an extension of Kunita's decomposition of solutions of SDEs [10]; this is outlined in §2 below. The approach in [4] was via dynamic programming and this gives an explicit characterization of the Lagrange multiplier in terms of the solution of a certain backwards stochastic partial differential equation. It entails, however, the assumption that the optimal nonanticipative control $u^0(t,x)$, expressed as a feedback function of the state x, is a very smooth function of x, and this is difficult to check unless, as in the LQG case, u^0 is known explicitly.

In this paper we take a more abstract approach and obtain the Lagrange multiplier in terms of the Fréchet derivative of the pathwise cost with respect to the control. The "martingale property" is then immediate. Our result is however only a preliminary one, in that (a), no control constraints are allowed, and (b), we only establish the result when the diffusion coefficient in the system equations is state independent. We hope to remove these restrictions in future work.

The formulation of the problem and definition of solution are given in §2. In §3 we summarize the (almost trivial) theory of Lagrange multipliers associated with "subspace constraints", while §4 contains the main result of the paper, Theorem 4.1. By way of an example we derive the result for the LQG case [3,5] in §5.

2 Problem Formulation

Throughout this paper, $(\Omega, \mathcal{F}, (\mathcal{F}_t),\ t \in [0,T])$ will denote the canonical d-fold Wiener space, so that the components $(w_t^i,\ i = 1,\ldots,d)$ are independent Brownian motions and $\{\mathcal{F}_t\}$ is the (completed) natural filtration of (w_t), with $\mathcal{F} = \mathcal{F}_T$. Denote $\tilde{\Omega} = \Omega \times [0,T]$, $\tilde{\mathcal{F}} = \mathcal{F} * \mathcal{B}[0,T]$, $d\tilde{\mathbb{P}} = d\mathbb{P} \times dt$ and let $\mathcal{P}$ be the σ-field of $\mathcal{F}_t$-predictable subsets of $\tilde{\Omega}$. Now fix a number $p \geq 1$ and a positive integer m and define $\mathcal{U}_A := L_p^m(\tilde{\Omega}, \tilde{\mathcal{F}}, \tilde{\mathbb{P}})$, i.e., $\mathcal{U}_A$ is the set of measurable, (not necessarily adapted), functions $u : \tilde{\Omega} \to \mathbb{R}^m$ such that

$$\mathbb{E} \int_0^T |u(t,w)|^p dt < \infty.$$

Further, write $\mathcal{U}_{NA} := L_p^m(\tilde{\Omega}, \mathcal{P}, \tilde{\mathbb{P}})$, so that $\mathcal{U}_{NA}$ is the linear subspace of $\mathcal{U}_A$ consisting of those processes which are predictable. $\mathcal{U}_A$ and $\mathcal{U}_{NA}$ are the anticipative and nonanticipative control processes respectively.

Let $f : [0,T] \times \mathbb{R}^n \times \mathbb{R}^m \to \mathbb{R}^n$ and $\sigma : [0,T] \times \mathbb{R}^n \to \mathbb{R}^{n \times d}$ be functions satisfying the following conditions:

- A1 f is bounded and measurable, and for each t the map $(x,u) \to f(t,x,u)$ is C_b^1;
- A2 The elements σ^{ij} are bounded with bounded x-derivatives of all orders.

Under these conditions the following SDE has a unique solution (x_t) for any $u \in \mathcal{U}_{NA}$ and initial condition $x \in \mathbb{R}^n$

$$dx_t = f(t,x_t,u_t)dt + \sigma(t,x_t)dw_t, \quad x_0 = x \in \mathbb{R}^n. \tag{1}$$

This is our dynamic equation. We can also write it in Stratonovich form as

$$dx_t = \tilde{f}(t,x_t,u_t)dt + \sigma(t,x_t) \circ dw_t, \tag{2}$$

where "$\circ$" denotes the Stratonovich integral and

$$\tilde{f}^i(t,x,u) = f^i(t,x,u) - \frac{1}{2}\sum_{j=1}^{d}\sum_{k=1}^{n} \frac{\partial}{\partial x^k}\sigma_j^i(t,x)\sigma_j^k(t,x).$$

The *cost* corresponding to $u \in \mathcal{U}_{NA}$ is

$$J(u) = \mathbb{E}\theta(x_T)$$

where θ satisfies

- A3 $\theta : \mathbb{R}^n \to \mathbb{R}^+$ is bounded with bounded first derivatives.

These conditions are much more restrictive than strictly necessary, but save us from having to count exact orders of differentiability.

The control problem for nonanticipative controls is to find $u^0 \in \mathcal{U}_{NA}$ which minimized $J(u)$. We suppose that such a minimizing element exists:

- A4 There exists $u^0 \in \mathcal{U}_{NA}$ such that

$$J(u^0) \leq J(u) \quad \text{for all} \quad u \in \mathcal{U}_{NA}.$$

The normal approach to establishing (A4) is via dynamic programming. We denote by $\mathcal{U}_{FB}$ the set of measurable functions $u : [0,T] \times \mathbb{R}^n \to \mathbb{R}^m$, and by $\mathcal{U}_{FB}^L$ those functions u in $\mathcal{U}_{FB}$ which satisfy a linear growth condition $|u(t,x)| \leq K(1+|x|)$. Functions $u \in \mathcal{U}_{FB}$ are *feedback controls*, and by a result of Veretennikov [16] the SDE

$$dx_t = f(t, x_t, u(t, x_t))dt + \sigma(t, x_t)dw_t$$

has a unique, (strong), solution since the function $(t,x) \to f(t,x,u(t,x))$ is bounded and measurable. Further, $\sup_t \mathbb{E}|x_t|^q < \infty$ for any $q \geq 1$, by standard estimates, so that if $u^0 \in \mathcal{U}_{FB}^L$ then we can identify u^0 with $\hat{u} \in \mathcal{U}_{NA}$ defined by

$$\hat{u}(t, \omega) := u^0(t, x_t(\omega)). \tag{3}$$

Thus $\mathcal{U}_{FB}^L \subset \mathcal{U}_{NA}$. The "verification theorem" [9] of dynamic programming then states:

Theorem 2.1 *Suppose that the parabolic equation*

$$\begin{aligned} \frac{\partial V}{\partial t} + \inf_{u \in \mathbb{R}^n} \{\nabla_x V(t,x) \cdot f(t,x,u)\} &= 0 \\ V(T,x) &= \theta(x) \end{aligned} \tag{4}$$

has a $C^{1,2}$ solution V such that $|\nabla_x V| \leq K(1+|x|)$, and that the function $u^0(t,x) \in \mathcal{U}_{FB}^L$ satisfies

$$\nabla_x V(t,x) \cdot f(t,x,u^0(t,x)) = \min_u \nabla_x V(t,x) \cdot f(t,x,u).$$

Then u^0 is optimal in $\mathcal{U}_{FB}$ and the corresponding $\hat{u} \in \mathcal{U}_{NA}$ given by (3) is optimal in $\mathcal{U}_{NA}$. We have

$$V(0,x) = \inf_{u \in \mathcal{U}_{FB}} J(u) = \inf_{u \in \mathcal{U}_{NA}} J(u).$$

The proof is by a standard application of the Ito formula; see [9, Theorem VI.4.1]. Conditions can then be stated under which the Bellman equation (4) has a solution with the required smoothness.

The system equation (1) cannot, of course, be solved within the Ito theory when $u \in \mathcal{U}_A$, because $\sigma(t, x_t)$ will no longer be an adapted process. Instead we use the "decomposition" technique introduced by — for the adapted case — Kunita [10], and developed by Ocone and Pardoux [13]. It is closely related to the "pathwise solution" idea of Doss [7] and Sussmann [15].

Let $\xi_t(x)$ be the solution of the Stratonovich SDE

$$d\xi_t(x) = \sigma(t, \xi_t(x)) \circ dw_t, \qquad \xi_0(x) = x.$$

Under condition (A2) this has a unique solution $\xi_t(x)$ which is a flow of diffeomorphisms on $\mathbb{R}^n$. Further, for every $\delta > 0$ there exists $c(\delta) \in \bigcap_{q \geq 1} L_q(\Omega)$ such that

$$\sup_{t \leq T} \left| \left(\frac{\partial \xi_t(x)}{\partial x} \right)^{-1} \right| \leq c(\delta)(1 + |x|^2)^{\delta} \qquad \text{for all} \quad x \in \mathbb{R}^n, \quad \text{a.s.} \quad (5)$$

See [10]. Now define a function $F : [0, T] \times \Omega \times \mathbb{R}^n \times \mathbb{R}^m \to \mathbb{R}^n$ by

$$F(t, \omega, x, u) := \left(\frac{\partial \xi_t(x)}{\partial x} \right)^{-1} \cdot \tilde{f}(t, \xi_t(x), u) \qquad (6)$$

and let $\eta_t(x)$ be the solution of the following ordinary differential equation

$$\dot{\eta}_t(x) = F(t, \omega, \eta_t, u_t), \qquad \eta_0(x) = x \qquad (7)$$

for given $u \in \mathcal{U}$. This is well defined for almost all $\omega \in \Omega$, and by A1 and (5), using the argument of [10, Proposition II.3.1] it follows that (6) has a unique non-exploding solution. We now *define* the solution of (1) for $u \in \mathcal{U}_A$ by

$$X_t(x) = \xi_t \circ \eta_t(x). \qquad (8)$$

When $u \in \mathcal{U}_A$, a formal calculation shows that $X_t(x)$ defined by (2.7) satisfies (2); this is Kunita's decomposition of the solution of (1). The point here, of course, is that (7) is an ordinary differential equation and can be solved pathwise whether or not the control process (u_t) is adapted. The pathwise cost is expressed as

$$\mathcal{J}(u) = \theta(X_T) = \theta \circ \xi_T(\eta_T(x)). \qquad (9)$$

Note that minimizing $\mathcal{J}(u)$ subject to dynamics (7) is a deterministic optimal control problem, defined for almost every $\omega \in \Omega$.

In adopting (2.7) as our solution process we have implicitly defined a stochastic integral, (denoted here as "@"), by

$$\int_0^t \sigma(s, X_s)@dw_s := x_t - x - \int_0^t \tilde{f}(s, X_s, u_s)ds.$$

If (u_t) is sufficiently smooth, (in the sense of Malliavin calculus), then this will coincide with the anticipative Stratonovich integral of Nualart and Pardoux [12]. We will not, however, use this here.

3 Subspace Constraints and Lagrange Multipliers

This section summarizes the very simple geometric theory of Lagrange multipliers associated with subspace constraints. It is a special case of the theory in §9.3 of Luenberger [11].

Let X be a Banach space with dual space X^*, and let S be a closed linear subspace of X. We define

$$S^\perp = \{x^* \in X^* : \langle x^*, x\rangle = 0, \ \forall x \in S\},$$

where $\langle x^*, x\rangle$ denotes the pairing between $x \in X$ and $X^* \in X^*$.

Let $\phi : X \to \mathbb{R}$ be a Fréchet-differentiable functional and suppose that f achieves its minimum over S at $x_0 \in S$. The Fréchet derivative is a map $\phi' : X \to X^*$ such that for $h \in X$

$$\phi(x_0 + h) = \phi(x_0) + \langle \phi'(x_0), h\rangle + o(||h||).$$

Lemma 3.1 *If ϕ achieves its minimum over S at x_0, then $\phi'(x_0) \in S^\perp$.*

Proof. If $\phi'(x_0) \notin S^\perp$ then there exists $h \in S$ such that $\langle \phi'(x_0), h\rangle = \delta > 0$. But then $\phi(x_0 - \varepsilon h) = \phi(x_0) - \varepsilon(\delta + o(\varepsilon)/\varepsilon)$ so that $\phi(x_0 - \varepsilon h) < \phi(x_0)$ for small ε.

Theorem 3.1 *If ϕ is Fréchet differentiable and achieves its minimum over S at x_0 then there exists $\lambda \in S^\perp$ such that the Lagrange functional $L(x) = \phi(x) + \langle \lambda, x\rangle$ is stationary at x_0, i.e., $L'(x_0) = 0$.*

Proof. We have only to set $\lambda = -\phi'(x_0)$.

4 Application to the Control Problem

To apply the above results to our problem, we take $X = \mathcal{U}_A$, $S = \mathcal{U}_{NA}$ and $\phi(u) = J(u) = \mathbb{E}\theta(X_T)$, where (X_t) is given by (2.7). Then $X^* = L_q(\tilde{\Omega}, \tilde{\mathcal{F}}, \tilde{\mathbb{P}})$ where $q = p/(p-1)$, and

$$S^{\perp} = \left\{ \lambda \in X^* : \mathbb{E} \int_0^T \lambda_t \cdot u_t dt = 0 \qquad \text{for all} \quad u \in L_p(\tilde{\Omega}, \mathcal{P}, \tilde{\mathbb{P}}) \right\}.$$

Now

$$\mathbb{E} \int_0^T \lambda_t^T u_t dt = \mathbb{E} \int_0^T \hat{\lambda}_t^T u_t dt,$$

where $(\hat{\lambda}_t)$ is the *predictable projection* of (λ_t). (See [8, §6.40].) This is a modification of the process $\mathbb{E}[\lambda_t \mid \mathcal{F}_t]$, so roughly speaking elements of $S^{\perp}$ are (non-adapted) processes λ_t such that $\mathbb{E}[\lambda_t \mid \mathcal{F}_t] = 0$ for all t.

We now wish to show that $J(u)$ is Fréchet differentiable. Unfortunately we are presently able to do this only when the diffusion matrix σ does not depend on x. We will outline below what the obstruction is when σ does depend on x.

- A5 The diffusion matrix $\sigma(t)$ is a bounded measurable function of t (only).

Theorem 4.1 *Suppose that conditions A1, A3 and A5 hold. Then the function $J : \mathcal{U}_A \to \mathbb{R}$ is Fréchet differentiable. The Fréchet derivative is given by (11) below.*

Proof. When σ is independent of x the definition of solution simplifies as follows. We define $m_t = \int_0^t \sigma(s) dw_s$. Then $X_t(x) = \eta_t(x) + m_t$ where η_t satisfies the ODE

$$\dot{\eta}_t = f(t, \ \eta_t + m_t, \ u_t), \qquad \eta_0 = x.$$

Denote the solution process $X_t(x, u)$ to emphasize the dependence on $u \in \mathcal{U}_A$. We first compute the Gateaux derivative

$$\frac{d}{d\varepsilon} X_t(x, \ u + \varepsilon v) \Big|_{\varepsilon=0},$$

where $u, v \in \mathcal{U}_A$. Indeed, this is equal to the Gateaux derivative

$$z_t(v) := \frac{d}{d\varepsilon}\, \eta_t(u+\varepsilon v)\Big|_{\varepsilon=0}$$

and by standard results $z_t(v)$ is the solution of the linear ODE

$$\begin{aligned} \dot{z}_t(v) &= f_x(t,\ \eta_t + m_t,\ u_t)z_t(v) \\ &+ f_u(t,\ \eta_t + m_t, u_t)v_t, \qquad z_0(v) = 0, \end{aligned} \tag{10}$$

where $\eta_t = \eta_t(u)$. By (A1), f_λ and f_u are bounded. $z_T(v)$ is given explicitly by

$$z_T(v) = \int_0^T \Phi(T,t) f_u(t,\ \eta_t + m_t,\ u_t)v_t dt,$$

where $\Phi(t,s)$ is the transition matrix corresponding to the time-varying matrix $A(t) := f_x(t,\ \eta_t + m_t,\ u_t)$, and

$$\frac{d}{d\varepsilon} J(u+\varepsilon v)\Big|_{\varepsilon=0} = \mathbb{E}[\nabla\theta^T(\eta_T + m_T)z_T(v)].$$

The relationship between Gateaux and Fréchet derivatives is that if the Gateaux derivative takes the form $\mathbb{E}\int_0^T \lambda_t \cdot v_t dt$ for some $\lambda \in X^* = L_q(\tilde{\Omega}, \tilde{\mathcal{F}}, \tilde{\mathbb{P}})$ then J is Fréchet differentiable and $J'(u) = \lambda$. Thus it remains to show that

$$\lambda_t := \nabla\theta(\eta_T + m_T)\Phi(T,t) f_u(t,\ \eta_t + m_t,\ u_t) \tag{11}$$

belongs to L_q. In fact, λ_t is bounded under our assumptions. $\nabla\theta$ and f_u are bounded, and the i'th column of $\Phi(T,t)$ is equal to $y(T)$ where $y(t)$ satisfies

$$\dot{y}(s) = A(s)y(s), \qquad y(t) = e_i \tag{12}$$

with $e_i^T = (0,\ldots,0,1,0\ldots0)$ (1 in the i'th position). Now $\|A(s)\| \le M$ for some constant M, and an application of the Gronwall inequality gives $|y(T)| \le (1 + e^{MT})$. This completes the proof. Applying Theorem 3.2 now gives us our main result.

Theorem 4.2 *Suppose conditions A1, A3–A5 hold, so that in particular there is an optimal nonanticipative control $u^0 \in \mathcal{U}_{NA}$. Then there exists $\lambda \in L_q(\tilde{\Omega}, \tilde{\mathcal{F}}, \tilde{\mathbb{P}})$ for $q = p/(p-1)$ such that the Lagrange function L defined on $\mathcal{U}_A$ by*

$$L(u) = \mathbb{E}\Big\{\theta(x_T) + \int_0^T \lambda_t^T u_t dt\Big\}$$

is stationary at u_0. Further, $\hat{\lambda}_t \equiv 0$ where $(\hat{\lambda}_t)$ denotes the predictable projection of (λ_t), and λ_t is given explicitly by (11).

Remark 1. The property $\hat{\lambda}_t = 0$ is what has been called the "martingale property", actually an inappropriate term here. In a discrete time setting we can interpret λ_t as a *martingale difference* sequence.
Remark 2. If we assume A2 rather than A5, a difficulty arises in showing that the process $z_t(v)$ of (10) is in L_q. f_x in (10) must be replaced by F_x, where F is defined by (6). Even though it is known that $(\partial\xi_t(x)/\partial x)^{-1}$ belongs to all L_q it is not clear that the solution of the linear equation (10) (or (12)) is in L_q. For example in the scalar case if $A = Z^2$ where Z is a standard normal random variable then clearly A is in all L_q, but the solution of the linear ODE $\dot{y} = Ay$, $y(0) = 1$ is $y(t) = e^{At}$ which is not in L_1 for $t \geq \frac{1}{2}$.

5 Example: the LQG Problem

This was treated from a different angle in Davis [3,5]. The dynamics are given by $dx_t = Ax_t dt + Bu_t + Cdw_t, \quad x_0 = x$ and the objective is to minimize the quadratic cost

$$J(u) = \mathbb{E}\Big\{\int_0^T (x_t^T Q x_t + u_t^T R u_t)dt + x_t^T F x_T\Big\}. \tag{13}$$

Here Q, R, F are non-negative definite symmetric matrices with R positive definite. The nonanticipative optimal control is

$$u_t^0 = -R^{-1}B^T S(t)x_t, \tag{14}$$

where $S(t)$ is the solution of the Riccati equation

$$-\dot{S}(t) = S(t)A + Q - S(t)R^{-1}B^T S(t), \qquad S(T) = F.$$

Thus the dynamics of the optimally controlled system are

$$dx_t = (A - BR^{-1}B^T S(t))x_t dt + C dw_t, \qquad x_0 = x. \tag{15}$$

Introduce an extra state variable x^{n+1} defined by $dx_t^{n+1} = (x_t^T Q x_t + u_\alpha^T R u_t)dt$ and let $\bar{x}^T = (x^T, x^{n+1})$. The cost (13) is then given by $J(u) = \mathbb{E}\theta(\bar{x}_T)$, where

$$\theta(\bar{x}) = x^T F x + x^{n+1}. \tag{16}$$

In the notation of §4 (but using $\bar{x}$ in place of x) we have

$$f(\bar{x}, u) = \begin{bmatrix} Ax + Bu \\ x^T Q x + u^T R u \end{bmatrix}$$

so that

$$f(\bar{x}, u) = \begin{bmatrix} A & 0 \\ 2x^T Q & 0 \end{bmatrix}, \qquad f_u(\bar{x}, u) = \begin{bmatrix} B \\ 2u^T R \end{bmatrix} \tag{17}$$

and

$$\Phi(T, t) = \begin{bmatrix} e^{A(T-t)} & 0 \\ 2\int_t^T x_s^T Q \Omega^{A(s-t)} ds & 1 \end{bmatrix}.$$

Using (14), (16), (17) we then find that λ_t of (11) is given by $\lambda_t = 2B^T \hat{\beta}_t$ where

$$\hat{\beta}_t = e^{A^T(\tau - t)} F x_t + \int_t^T e^{A^T(s-t)} Q x_s ds - S(t) x_t.$$

We note that $\hat{\beta}_t = 0$, and, using the Riccati equation and the dynamic equation (15) it can be checked that $\hat{\beta}(t)$ satisfies the linear stochastic equation

$$d\hat{\beta}_t = -A^T \hat{\beta}_t dt - S(t) C dw_t, \qquad \hat{\beta}_T = 0.$$

The Lagrange multiplier for the LQG problem is thus $\lambda_t = -2B^T \hat{\beta}_t$, in agreement with the result of [3].

Bibliography

[1] K. Back and S.R. Pliska, *The shadow price of information in continuous-time decision problems*, Stochastics 22 (1987), 151–186.

[2] M.H.A. Davis, *Anticipative LQG control*, IMA J. Math Control and Information 6 (1989), 259–265.

[3] M.H.A. Davis, *Anticipative LQG control II*, in Applied Stochastic Analysis, eds. M.H.A. Davis and R.J. Elliott, Gordon and Breach, London 1990.

[4] M.H.A. Davis and G. Burstein, *On the relation between deterministic and stochastic optimal control*, Proc. 3rd CIRM Symposium on Stochastic Partial Differential Equations, Trento, January 1990, Lecture Notes in Mathematics, Springer–Verlag, Berlin (to appear).

[5] M.H.A. Davis and A.J. Heunis, *Stochastic processes and Linear System Theory*, Chapman and Hall, London (forthcoming).

[6] M.A.H. Dempster, *The expected value of perfect information in the optimale evoluation of stochastic systems*, in Stochastic Differential Systems, ed. M. Kohlmann, Lecture Notes in Control and Information Sciences 36, Springer–Verlag, Berlin 1981.

[7] H. Doss, *Liens entre équations différentielles stochastiques et ordinaires*, Ann. Inst. H. Poincaré 13 (1977), 99–125.

[8] R.J. Elliott, *Stochastic Calculus and Applications*, Springer–Verlag, Berlin 1982.

[9] W.H. Fleming and R.W. Rishel, *Deterministic and Stochastic Optimal Control*, Springer–Verlag, New York 1975.

[10] H. Kunita, *Stochastic differential equations and stochastic flows of diffeomorphisms*, in Ecole d'Eté de Saint–Flour XII, ed. P.L. Hennequin, Lecture Notes in Mathematics 1097, Springer–Verlag, Berlin 1984, pp. 1434–303.

[11] D.G. Luenberger, *Optimization by Vector Space Methods*, Wiley, New York 1969.

[12] D. Nualart, *Noncausal stochastic integrals and calculus* in Stochastic Analysis and Related Topics, eds. H. Korezlioglu and S. Ustunel, Lecture Notes in Mathematics 1316, Springer–Verlag, Berlin 1988.

[13] D. Ocone and E. Pardoux, *A generalized Ito–Ventzell formula. Application to a class of anticipating stochastic differential equations*, Ann. Inst. H. Poincaré 25 (1989), 39–71.

[14] R.T. Rockafellar and R.J.-B. Wets, *Nonanticipativity and L_1 martingales in stochastic optimization problems*, in Stochastic Systems: Modelling, Identification and Optimization II, Mathematical Programming Study 6, North–Holland, Amsterdam 1976, pp. 179–187.

[15] H.J. Sussmann, *On the gap between deterministic and stochastic ordinary differential equations*, Ann. Prob. 6 (1978), 19–41.

[16] A.Y. Veretennikov, *On the criteria for existence of a strong solution to a stochastic equation*, Theory Probab. Appl. 27 (1982), 441–449.

[17] R.J.-B. Wets, *On the relation between stochastic and deterministic optimization*, in Control Theory, Numerical Methods and Computer Systems Modelling, eds. A. Bensoussan and J.L. Lions, Lecture Notes in Economics and Mathematical Systems 107, pp. 350–361, Springer–Verlag, Berlin 1975.

Orthogonal Martingale Representation

Robert J. Elliott
University of Alberta
Hans Föllmer
Universität Bonn

Abstract

Stochastic integrals with respect to a martingale X often involve a predictable process integrated against the continuous martingale component X^c together with terms which are integrals of the compensated random measures associated with the jumps. The latter are related to 'optional' stochastic integrals. The main result of this paper relates such a stochastic integral with the sum of a predictable stochastic integral of X and an orthogonal martingale. The result has applications in the hedging of contingent claims in finance.

Acknowledgments

Research partially supported by NSERC Grant A-7964, the Air Force Office of Scientific Research, United States Air Force, under contract AFOSR-86-0332, and the U.S. Army Research Office under contract DAAL03-87-0102.

1 Introduction

For a real local martingale

$$X_t = X_0 + M_t^c + M_t^d$$

write $\mu = \mu^X$ for the random measure associated with the jumps of X (see Jacod [7]), and $\nu = \mu^P$ for its predictable compensator.

ISBN 0-12-481005-5

Consider a local martingale of the form

$$N_t = N_0 + \int_0^t \phi_s dM_s^c + \int_0^t \int_R \psi_s(y)(\mu(dy, ds) - \nu(dy, ds)) \qquad (1)$$

where ϕ and ψ are suitable integrands. Projecting N on X we can write

$$N_t = N_0 + \int_0^t \gamma_s dX_s + \Gamma_t \qquad (2)$$

where Γ is a local martingale orthogonal to X, in the sense that the product ΓX is a local martingale. In this paper our purpose is to determine explicit formulae for γ in terms of ϕ and ψ, both in the general case and for Markov diffusions with jumps.

These results have applications in the hedging of contingent claims in finance. See [5] and [6]. For example, suppose that the martingale X_t represents the price of some asset at time t and that for $T > 0$ a contingent claim is given by $H(X_T)$, where H is a function such that $H(X_T)$ is a real, square integrable random variable. Suppose at time t we invest amount ξ_t in the asset and an amount η_t in a riskless bond with zero interest rate and price $Y = 1$. Then the value of our portfolio at time t is

$$V_t = \xi_t X_t + \eta_t Y = \xi_t X_t + \eta_t.$$

We assume ξ is predictable with respect to the filtration $\{F_t\}$ generated by X, and η is adapted. The accumulated gain from the asset price fluctuations up to time t is the stochastic integral $\int_0^t \xi_s dX_s$. Then the cost accumulated to time t by using the investment strategy (ξ_t, η_t) is

$$C_t = V_t - \int_0^t \xi_s dX_s, \qquad 0 \leq t \leq T.$$

We want our investment strategy to duplicate the contingent claim, so for a strategy (ξ, η) to be admissible we also require

$$V_T = \xi_T X_T + \eta_T = H(X_T).$$

Now suppose $H(X_T)$ can be represented as a (predictable) stochastic integral

$$H(X_T) = E[H(X_T)] + \int_0^T \xi_s^H dX_s \quad \text{a.s.} \tag{3}$$

for some (predictable) integrand ξ^H. Then let us take an investment strategy (ξ, η) and value process V defined by:

$$\xi_t = \xi_t^H, \qquad \eta_t = V_t - \xi_t X_t$$

and

$$V_t = E[H(X_T)] + \int_0^t \xi_s^H dX_s.$$

We have $V_T = H(X_T)$, so the strategy is admissible and for all $t \in [0, T]$

$$C_t = C_T = C_0 = E[H(X_T)].$$

That is the strategy is self-financing because, apart from the initial cost $C_0 = E[H(X_T)]$, no additional costs arise and no risks are involved.

Conversely, if there is a self-financing strategy (ξ, η),

$$V_t = C_t + \int_0^t \xi_s dX_s = C_0 + \int_0^t \xi_s dX_s,$$

so V_t is a martingale. Therefore,

$$V_t = E[V_T \mid F_t] = E[H(X_T) \mid F_t]$$

and

$$V_0 = C_0 = E[H(X_T)],$$

so the martingale V has the representation

$$V_t = E[H(X_T)] + \int_0^t \xi_s dX_s. \tag{4}$$

The existence of a self-financing strategy is, therefore, equivalent to the representation of the martingale $E[H(X_T) \mid F_t]$ in the form (4) for some predictable integrand ξ. In general, a representation in this form is not available. In this case we can proceed as follows (cf. [5] and [6]).

Definition 1.1 *An admissible investment strategy* (ξ, η) *is said to be mean self-financing if the corresponding cost process* C *is a martingale. That is, for* $t \leq T$,

$$E[C_T - C_t \mid F_t] = 0 \quad \text{a.s.}$$

In this case, by definition

$$V_t = C_t + \int_0^t \xi_s dX_s = E[C_T \mid F_t] + \int_0^t \xi_s dX_s.$$

Therefore, V is a martingale for an admissible mean self-financing strategy. Consequently,

$$V_0 = C_0 = E[H(X_T)]$$

and

$$V_t = E[H(X_T) \mid F_t] = E[H(X_T)] + K_t + \int_0^t \xi_s dX_s$$

where K_t is the martingale $E[C_T \mid F_t] - C_0$. Note that, if (ξ, η) is an admissible, mean self-financing strategy, $V_t = E[H(X_T) \mid F_t]$ is independent of ξ. However,

$$C_t = C_t^\xi = V_t - \int_0^t \xi_s dX_s$$

does depend on ξ, as does K above. Therefore, $K_t^\xi = E[C_T^\xi \mid F_t] - C_0^\xi$ and each admissible, mean self-financing strategy (ξ, η) gives rise to a decomposition:

$$V_t = E[H(X_T) \mid F_t] = E[H(X_T)] + K_t^\xi + \int_0^t \xi_s dX_s. \tag{5}$$

Definition 1.2 *For each admissible mean self-financing strategy the remaining risk is defined to be*

$$R_t^\xi = E[(C_T^\xi - C_t^\xi)^2 \mid F_t].$$

Consider the unique Kunita–Watanabe decomposition

$$V_t = E[H(X_T)] + \Gamma_t + \int_0^t \xi_s^* dX_s$$

where Γ is a martingale orthogonal to X and ξ^* is a predictable integrand. Now define an investment strategy (ξ^*, η^*) by putting

$$V_t = E[H(X_T)] + \Gamma_t + \int_0^t \xi_s^* dX_s \quad \text{and} \quad \eta_t^* = V_t - \xi_t^* X_t. \tag{6}$$

Then (ξ^*, η^*) is an admissible, mean self-financing strategy which minimizes the remaining risk R_t. To see this note that for any other admissible, mean self-financing strategy (ξ, η):

$$C_T^{\xi} - C_t^{\xi} = K_T^{\xi} - K_t^{\xi}.$$

However, from (5) and (6)

$$K_T^{\xi} - K_t^{\xi} + \int_t^T \xi_s dX_s = \Gamma_T - \Gamma_t + \int_t^T \xi_s^* dX_s.$$

Therefore, because Γ is orthogonal to X,

$$E[(C_T^{\xi} - C_t^{\xi})^2 \mid F_t] = E[(\Gamma_T - \Gamma_t)^2 \mid F_t] + E\Big[\int_0^T (\xi_s^* - \xi_s)^2 d\langle X, X\rangle_s \mid F_t\Big]$$

and this is minimized when $\xi = \xi^*$. Consequently, the unique admissible, risk minimizing investment strategy is (ξ^*, η^*), where ξ^* is the predictable integrand arising in the representation (6).

This discussion indicates why decompositions such as (6), together with an explicit formula for the integrand, are of interest in finance. Representations such as (1) arise when the asset price $X_t = X_0 + M_t^c + M_t^d$ also involves random disturbances of jump type. In that case, a contingent claim typically admits a representation as in (1)

$$H(X_T) = E[H(X_T)] + \int_0^T \phi_s dM_s^c + \int_0^T \int_R \psi_s(y)(\mu(dy, ds) - \nu(dy, ds)). \tag{7}$$

Then ϕ_s and $\psi_s(y)$ for each $y \in R$, (or, at least, for each y in the support of $\mu = \mu^X$), must represent amounts invested in different assets in order to duplicate (i.e., represent) the claim $H(X_T)$. However, if the only assets available are X and $Y = 1$, we must consider the alternative representation (2)

$$H(X_T) = E[H(X_T)] + \int_0^T \gamma_s dX_s + \Gamma_t.$$

Then γ will generate the risk minimizing mean self-financing investment strategy described above.

In particular, even though (after a Girsanov change of measure) the Markov diffusion process considered by Aase in [1] is complete in a mathematical sense, that is, contingent claims have a representation of the form (7), it is not complete in the financial sense; that is, they do not necessarily admit a decomposition (3). To replicate the claims in Aase's model an uncountable number of additional artificial assets would be required. Clearly this is not realistic.

Orthogonal martingale representation after a Girsanov change of measure will be discussed in another paper.

2 Orthogonal Projection

Consider a real local martingale

$$X_t = X_0 + M_t^c + M_t^d.$$

Suppose

$$\mu = \mu^X(dy, dt) = \sum_{s>0} I_{\{\Delta X_s \neq 0\}} \delta_{(s, \Delta X_s)}(dy, dt)$$

and

$$\nu = \nu(dy, dt) = \mu^P.$$

Write $\{F_t\}$ for the right continuous, complete filtration generated by X. Consider a process N which is a stochastic integral of the form

$$N_t = N_0 + \int_0^t \phi_s dM_s^c + \int_0^t \int_R \psi_s(y)(\mu(dy, ds) - \nu(dy, ds)) \tag{8}$$

for suitable integrands ϕ and ψ. What we wish to do is write

$$N_t = N_0 + \int_0^t \gamma_s dX_s + \Gamma_t \tag{9}$$

where γ is a predictable integrand and Γ is a local martingale orthogonal to X. From Jacod [7] we know that the stochastic integral $\int\int \psi(d\mu - d\nu)$ in (8) is related to an optional stochastic integral with respect to $M^d = \int_0^t \int_R y(\mu(dy, ds) - \nu(dy, ds))$; consequently we are relating the optional integrals in (8) to the predictable integral in (9).

Proposition 2.1 *Assume there is a reference measure $v = v(w, ds)$ such that $\langle M^c, M^c\rangle$ and λ are absolutely continuous with respect to v, where*

$$\nu(dy, ds) = m(s, dy)\lambda(ds).$$

Write $\rho_s = d\langle M^c, M^c\rangle / dv$ and $\lambda_s = d\lambda/dv$. Then if N_t is the martingale given by (8), the process γ in (9) is

$$\gamma_s = \frac{\rho_s\phi_s + \lambda_s \int_R y\psi_s(y)m(s, dy)}{\rho_s + \lambda_s \int_R y^2 m(s, dy)}. \tag{10}$$

Proof. Note if γ is the predictable integrand of (9)

$$\begin{aligned} \Lambda_t &= \int_0^t \gamma_s dX_s = \int_0^t \gamma_s dM_s^c + \int_0^t \gamma_s dM_s^d \qquad (11) \\ &= \int_0^t \gamma_s dM_s^c + \int_0^t \int_R \gamma_s y(\mu(dy, ds) - \nu(dy, ds)). \end{aligned}$$

From (8) and (9)

$$\Gamma_t = \int_0^t (\phi_s - \gamma_s)dM_s^c + \int_0^t \int_R (\psi_s(y) - \gamma_s y)(\mu(dy, ds) - \nu(dy, ds)).$$

The martingales Γ and X are orthogonal if $[\Gamma, X]$ is a martingale (see Dellacherie and Meyer VIII.41, [3]). However, writing X as

$$X_t = X_0 + M_t^c + \int_0^t \int_R y(\mu(dy, ds) - \nu(dy, ds))$$

we have

$$[\Gamma, X]_t = \int_0^t (\phi_s - \gamma_s) d\langle M^c, M^c \rangle_s + \int_0^t \int_R y(\psi_s(y) - \gamma_s y)\mu(dy, ds)$$

$$= \int_0^t (\phi_s - \gamma_s)\rho_s ds + \int_0^t \int_R y(\psi_s(y) - \gamma_s y)(\mu(dy, ds) - \nu(dy, ds))$$

$$+ \int_0^t \int_R y(\psi_s(y) - \gamma_s y) m(s, dy)\lambda_s ds.$$

This is a martingale if and only if

$$\int_0^t (\phi_s - \gamma_s)\rho_s ds + \int_0^t \int_R y(\psi_s(y) - \gamma_s y) m(s, dy)\lambda_s ds$$

is the null process, which is the case if and only if the integrand is zero. Therefore,

$$(\phi_s - \gamma_s)\rho_s + \lambda_s \int_R y(\psi_s(y) - \gamma_s y) m(s, dy) = 0$$

and (10) follows.

Remarks 2.2. 1) If $M^d = 0$, so that $\mu = \nu = 0$, then $\gamma = \phi$.

2) Note

$$\begin{aligned} \int_R y\psi_s(y) m(s, dy) &= E[\Delta X_s \Delta N_s \mid F_{s-}] \\ &= {}^p(\Delta X \Delta N)_s \end{aligned}$$

and

$$\int_R y^2 m(s, dy) = E[\Delta X_s^2 \mid F_{s-}] = {}^p(\Delta X^2)_s,$$

where ${}^p(\quad)$ denotes the predictable projection.

3) If $M^c = 0$,

$$\gamma_s = \frac{\int_R y\psi_s(y) m(s, dy)}{\int_R y^2 m(s, dy)} = \frac{E[\Delta X_s \Delta N_s \mid F_{s-}]}{E[\Delta X_s^2 \mid F_{s-}]} = \frac{{}^p(\Delta X \Delta N)_s}{{}^p(\Delta X^2)_s}$$

so γ can be interpreted as the regression of the jumps of N on the jumps of X.

4) With appropriate interpretation of products as tensor products in R^m, the same expression for γ is valid when X is an m-dimensional martingale.

3 Representation Results

Consider a real, local martingale

$$X_t = X_0 + M_t^c + M_t^d$$

and let $\mu = \mu^X$, $\nu = \mu^p$. Suppose $F \in C^{1,2}$, the space of functions continuously differentiable in t and twice continuously differentiable in x. Then the differentiation rule gives (see Jacod [7])

$$\begin{aligned} F(t, X_t) = F(0, X_0) &+ \int_0^t \frac{\partial F}{\partial s}(s, X_{s-})ds + \int_0^t \frac{\partial F}{\partial x}(s, X_{s-})dM_s^c \\ &+ \int_0^t \int_R (F(s, X_{s-} + y) - F(s, X_{s-}))(\mu(dy, ds) - \nu(dy, ds)) \\ &+ \frac{1}{2}\int_0^t \frac{\partial^2 F}{\partial x^2}(s, X_{s-})d\langle M^c, M^c\rangle_s \qquad (12) \\ &+ \int_0^t \int_R \Big(F(s, X_{s-} + y) - F(s, X_{s-}) - \frac{\partial F}{\partial x}(s, X_{s-})\Big)\nu(dy, ds). \end{aligned}$$

Suppose X is Markov. For a time $T > 0$ and an integrable C^2 function $H(\cdot)$, consider the random variable $H(X_T)$ and the martingale

$$N_t = E[H(X_t) \mid F_t].$$

Because X is Markov

$$N_t = E[H(X_T) \mid X_t] = V(t, X_t), \quad \text{say.}$$

The following representation result appears to be part of the folklore.

Proposition 3.1 *Suppose V is $C^{1,2}$, that is continuously differentiable in t and twice continuously differentiable in x. Then the martingale N is given by the stochastic integral representation*

$$\begin{aligned} N_t = E[H(X_T)] &+ \int_0^t \frac{\partial V}{\partial x}(s, X_{s-})dM_s^c \\ &+ \int_0^t \int_R (V(s, X_{s-} + y) - V(s, X_{s-})) \\ &\times(\mu(dy, ds) - \nu(dy, ds)). \qquad (13) \end{aligned}$$

Furthermore, suppose

$$\langle M^c, M^c \rangle_t = \int_0^t \rho_s ds$$

and

$$\nu(dy, ds) = m(s, dy)\lambda_s ds.$$

Then V is the solution of the backward Kolmogorov equation

$$\begin{aligned}
&\frac{\partial V}{\partial s} + \frac{1}{2}\frac{\partial^2 V}{\partial x^2}\rho_s \\
&\quad + \int_R \Big(V(s, X_{s-} + y) - V(s, X_{s-}) - \frac{\partial V}{\partial x}(s, X_{s-})y\Big) m(s, dy)\lambda_s \\
&\quad = 0 \qquad (14)
\end{aligned}$$

with terminal condition

$$V(T, X_T) = H(X_T).$$

Proof. The result follows by expanding $V(t, X_t)$ by the Ito rule (12) and observing that, because $V(t, X_t) = N_t$ is a martingale, the sum of the bounded variation terms must be the null process.

Remarks 3.2. Often, (see Example 3.4), the differentiability of V follows from flow properties. In the pure jump case only increments of V enter in (13).

Corollary 3.3 *Write $\Delta V_s(y) = V(s, X_{s-} + y) - V(s, X_{s-})$.*

Then from Proposition 2.1 N_t can be written

$$N_t = E[H(X_T)] + \int_0^t \gamma_s dX_s + \Gamma_t$$

where

$$\gamma_s = \frac{\rho_s \frac{\partial V}{\partial x} + \lambda_s \int_R y \Delta V_s(y) m(s, dy)}{\rho_s + \lambda_s \int_R y^2 m(s, dy)} \qquad (15)$$

and Γ is a martingale orthogonal to X.

Example 3.4. Suppose X is a Markov diffusion:

$$X_t = X_0 + \int_0^t g(s, X_{s-})dB_s + \int_0^t \int_R h(s, X_{s-}, y)(\tilde{\mu}(dy, ds) - \tilde{\nu}(dy, ds)). \quad (16)$$

Here $\tilde{\mu}$ is a random measure and $\tilde{\nu} = \tilde{\mu}^p$; note $\tilde{\mu}$ is not μ^X. Suppose $\tilde{\nu}(dy, ds) = m(s, X_{s-}, dy)\lambda_s(X_{s-})ds$. Write $\xi_{r,t}(x)$ for the solution of (16) starting at time r in position x, so that

$$\xi_{r,t}(x) = x + \int_r^t g(s, \xi_{r,s-}(x))dB_s + \int_r^t \int_R h(s, \xi_{r,s-}(x), y)(\tilde{\mu}(dy, ds) - \tilde{\nu}(dy, ds)). \quad (17)$$

Suppose g, h, m and λ and their first two derivatives in x are measurable with linear growth in the x variable. Then from the theory of stochastic flows, see [2], it is known there is a set $A \subset \Omega$ of measure zero such that the map $(r, t, x) \to \xi_{r,t}(x)$ is twice differentiable in x with derivative $\frac{\partial \xi_{r,t}(x)}{\partial x} = D_{r,t}$. Again write $\{F_t\}$ for the right continuous σ-field generated by X and suppose H is a C^2, integrable function. For $T > 0$ consider the right continuous martingale

$$\begin{aligned} N_t &= E[H(\xi_{0,T}(x_0)) \mid F_t] \\ &= E[H(\xi_{t,T}(X_t)) \mid X_t] = V(t, X_t). \end{aligned}$$

From the differentiability of the flow

$$\frac{\partial V}{\partial x}(t, X_t) = E\Big[\frac{\partial H}{\partial x}(X_T)D_{t,T} \mid F_t\Big].$$

Substituting in (13) and (15) we have the representations

$$\begin{aligned} N_t = \; & E[H(X_T)] + \int_0^t E\Big[\frac{\partial H}{\partial x}(X_T)D_{0,T} \mid F_{s-}\Big] D_{0,s-}^{-1} g(s, X_{s-})dB_s \\ & + \int_0^t \int_R (E[H(\xi_{s-,T}(X_{s-} + y)) - H(\xi_{s-,T}(X_{s-})) \mid X_{s-}] \\ & \times h(s, \xi_{0,s-}(x_0), y)(\tilde{\mu}(dy, ds) - \tilde{\nu}(dy, ds)) \end{aligned}$$

and from Corollary 3.3 this is

$$= E[H(X_T)] + \int_0^t \gamma_s dX_s + \Gamma_t$$

where

$$\gamma_s = \Big\{ g(s, X_{s-})^2 E\Big[\frac{\partial H}{\partial x}(X_T) D_{0,T} \mid F_{s-}\Big] D_{0,s}^{-1}$$
$$+\lambda_s(X_{s-}) \int_R y E[H(\xi_{s-,T}(X_{s-} + y)) - H(\xi_{s-,T}(X_{s-})) \mid X_{s-}\Big]$$
$$\times h(s, \xi_{0,s-}(x_0), y) m(s, \xi_{0,s-}(x_0), dy)\Big\}$$
$$\times \Big[g(s, X_{s-})^2 + \lambda_s(X_{s-})$$
$$\times \int_R y^2 h(s, \xi_{0,s-}(x_0), y)^2 m(s, \xi_{0,s-}(x_0), dy)\Big]^{-1}.$$

Example 3.5. The random measure in (16) could be a Poisson random measure. However, for simplicity suppose it is a finite sum of independent Poisson processes N_t^i, $i = 1, \dots, n$ with time varying jump sizes a_t^i and intensities λ_s^i. Suppose $g(s, X_{s-}) = \sigma X_{s-}$ and $h(s, X_{s-}, y) = X_{s-}$, so that X_t, (representing an asset price under a 'risk neutral' measure), is given by the following "log Poisson plus log normal" equation

$$X_t = X_0 + \sigma \int_0^t X_{s-} dB_s + \sum_{i=1}^n \int_0^t X_{s-} a_s^i (dN_s^i - \lambda_s^i ds). \tag{18}$$

Suppose for an integrable C^2 function H, $H(X_T)$ represents a contingent claim depending on the asset price at time $T > 0$. Then

$$\begin{aligned} N_t &= E[H(X_T) \mid F_t] \\ &= E[H(X_T)] + \int_0^t \gamma_s dX_s + \Gamma_t \end{aligned}$$

where

$$\gamma_s = \frac{\sigma^2 X_{s-}^2 E[\frac{\partial H}{\partial x}(x_T) D_{0,T} \mid F_{s-}] D_{0,s-}^{-1} + \sum_{i=1}^n \lambda_s^i a_s^i \Delta_s^i V}{\sigma^2 X_{s-}^2 + \sum_{i=1}^n \lambda_s^i (a_s^i)^2}$$

and

$$\Delta_s^i V = E[H(\xi_{s-,T}(X_{s-} + a_s^i)) - H(\xi_{s-,T}(X_{s-})) \mid X_{s-}].$$

From (18)

$$\begin{aligned} X_t = \xi_{s-,t}(X_{s-}) = \\ & X_{s-} \exp\Big(\sigma(B_t - B_s) - \frac{\sigma^2}{2}(t-s) - \sum_{i=1}^{n} \int_s^t a_s^i \lambda_s^i ds\Big) \\ & \times \prod_{s \le r \le t} (1 + a_r^i \Delta N_r^i). \end{aligned} \tag{19}$$

so

$$\begin{aligned} D_{s-,T} = D_{0,T} \cdot D_{0,s-}^{-1} = \\ & \exp\Big(\sigma(B_t - B_s) - \frac{\sigma^2}{2}(t-s) - \sum_{i=1}^{n} \int_s^t a_s^i \lambda_s^i ds\Big) \\ & \times \prod_{s \le r \le t} (1 + a_r^i \Delta N_r^i). \end{aligned}$$

Suppose $H(X_T)$ is a call option of the form $H(X_T) = (X_T - K)^+$. Then H is not C^2 but is the limit of the smooth functions $H_\varepsilon(X_T) = \frac{1}{2}(X_T - K + \sqrt{(X_T - K)^2 + \varepsilon})$; using approximation arguments it is shown, for example in [4], that the above theory applies to H. Now $\frac{\partial H}{\partial x}(X_T) = I_{X_T \ge K}$, so

$$E\Big[\frac{\partial H}{\partial x}(X_T) D_{0,T} \mid F_s\Big] D_{0,s-}^{-1} = E[I_{X_T \ge K} D_{s-,T} \mid X_{s-}]$$

and because the B and N^i are independent this can be evaluated as in [1], giving a Black–Scholes type formula. Similarly, with $\xi_{s-,T}(X_{s-})$ given by (19) $\Delta_s^i V$ can be calculated.

Bibliography

[1] K.K. Aase, *Contingent claims valuation when the security price is a combination of an Itô process and a random point process*, Stochastic Processes and App. 28, 1988, p. 185–220.

[2] J.M. Bismut, *Martingales, the Malliavin calculus and hypoellipticity under general Hörmander's conditions*, Zeits. für Warsch. 56, 1981, p. 469–505.

[3] C. Dellacherie and P.A. Meyer, *Probabilités et Potentiel, B: Théorie des Martingales*, Hermann, Paris, 1980.

[4] D. Duffie, *Security Markets. Stochastic Models,* Academic Press, Boston, New York, London, 1988.

[5] H. Föllmer and M. Schweizer, *Hedging of contingent claims under incomplete information*, Workshop on Stochastic Processes. Imperial College, London, April 1989. Gordon and Breach, to appear.

[6] H. Föllmer and D. Sondermann, *Hedging of non-redundant contingent claims*, in W. Hildenbrandt and A. Mas–Colell (Eds.) Contributions to Mathematical Economics, 1986, p. 205–223.

[7] J. Jacod, *Calcul stochastique et problèmes de martingales*, Lecture Notes in Math. Vol. 714, Springer–Verlag, Berlin, Heidelberg, New York, 1979.

Nonlinear Filtering with Small Observation Noise: Piecewise Monotone Observations

Wendell H. Fleming
Division of Applied Mathematics
Brown University
Providence, RI 02912

Qing Zhang
Faculty of Management
University of Toronto
Toronto, Ontario, Canada M5S 1V4

Abstract

A discrete time model for filtering with small observation noise is considered in this paper. The observation function is assumed to be a piecewise monotone function with a finite number of intervals of monotonicity. Under a certain detectability hypothesis, a sequential quadratic variation test is proposed to detect the intervals of monotonicity of the observation function. An upper bound for the mean time of reaching a decision is given in terms of the observation noise level. Then based on the quadratic variation test, accurate approximate finite dimensional filters can be used to construct an asymptotically optimal filter as the observation noise tends to zero.

1 Introduction

There is a substantial literature on the problem of optimal nonlinear filtering. In continuous time an unobserved state X_t and observation

ISBN 0-12-481005-5

Y_t are modelled according to

$$\begin{cases} dX_t = f(X_t)dt + g(X_t)dW_t, \ 0 \le t \le T \\ dY_t = h(X_t)dt + \varepsilon dV_t, \ Y_0 = 0, \end{cases} \tag{1}$$

where W_t, V_t are independent brownian motions and T is a finite number.

To find the mean square optimal estimate $\hat{X}_t$ for X_t given Y_s for $0 \le s \le t$ requires knowing the conditional distribution of X_t. Since the dynamics of the conditional distribution are governed by the nonlinear functional-partial differential equation of nonlinear filtering, the problem is inherently infinite dimensional [7].

If X_t and Y_t are of the same dimension and h is one-to-one, then X_t would be known exactly if $\varepsilon = 0$. For small $\varepsilon > 0$ good finite dimensional approximate filters have been described in [5] and [9]. References [2] [4] are concerned with the case when h is piecewise one-to-one. Under a certain 'detectability' condition, one can perform a hypothesis test based on observations on Y_t to decide that X_t belongs to a region on which h is one-to-one. Once this is done, an approximate filter of the type in [9] is used to estimate $\hat{X}_t$.

In this paper we consider the following discrete time analogue of (1):

$$\begin{cases} x_{k+1} = x_k + \varepsilon f(x_k) + \sqrt{\varepsilon} g(x_k) w_k \\ y_k = h(x_k) + \sqrt{\varepsilon} v_k, \ y_0 = 0, \end{cases} \tag{2}$$

where w_k, v_k, $k = 0, 1, 2, ..., [T/\varepsilon]$ are i.i.d. standard normal random variables, x_0 is a random variable independent of w_k, v_k with $E \exp \theta_0 |x_0|^2 < \infty$ for some $\theta_0 > 0$, and $[T/\varepsilon]$ is the largest integer not greater than T/ε.

Actually, (2) approximates (1) in the following way. One discretizes (1) with time step size ε and replaces $X_{k\varepsilon}$ by x_k, $W_{(k+1)\varepsilon} - W_{k\varepsilon}$ by $\sqrt{\varepsilon} w_k$, $V_{(k+1)\varepsilon} - V_{k\varepsilon}$ by $\sqrt{\varepsilon} v_k$, and $\varepsilon^{-1}(Y_{(k+1)\varepsilon} - Y_{k\varepsilon})$ by y_k.

The purpose of this paper is to provide a sequential statistical test to discriminate among the intervals of monotonicity of the observation function h during time intervals in which the state x_k does not cross critical points of h. Then we use accurate approximate finite dimensional filters during such time intervals as an approximation to the state x_k.

To simplify the exposition, in most of the paper, we take the single critical point of h to be a global minimum at $x = 0$ with $h(0) = 0$. Extensions to multiple critical points of h and to dimension larger than one can be carried out similarly as in [2, Sec. 8]. We begin in Sections 2 and 3 with estimates which correspond to estimates in [2] for the continuous time filter model. After some introductory material in Section 2, two possible approximations m_k^+ and m_k^- are introduced. Lemma 3.1 provides a hypothesis test to guarantee that x_k does not cross zero during a time interval $K_0 \leq k \leq K_1$ with small probability of error. Next, we propose a sequential test based on quadratic variations to decide whether $x_k > 0$ on $K_0 \leq k \leq K_1$ or $x_k < 0$ on $K_0 \leq k \leq K_1$. Then we turn to estimates of error probabilities and mean decision times. If the positive alternative is chosen then m_k^+ is used as an approximation to x_k; and m_k^- is used if the negative alternative is chosen. Finally estimates for the error $\hat{x}_k - m_k^{\pm}$ are given in Section 5.

2 Problem formulation

We make the following assumptions about the functions f, g, h in (2).

(A1) There exists f_0 such that $|f(x) - f_0 x|$ is bounded. $g(x)$ is Lipschitz and bounded, and $g(x) \geq c_1'$ for some $c_1' > 0$.

(A2) $h(x)$ is a C_b^2 function with $h(0) = 0$ and $h(x) > 0$ for $x \neq 0$. Moreover, $h(x) \to \infty$ as $|x| \to \infty$.

(A3) There exist $\bar{y} > 0$ and $\bar{c} > 0$ such that $h(x^+) = h(x^-) = y \geq \bar{y}$ and $x^- < 0 < x^+$ imply

$$|(gh')^2(x^+) - (gh')^2(x^-)| \geq \bar{c}. \tag{3}$$

Let $\delta_0 > 0$ be such that $h(x) \geq \bar{y}$ implies $|x| \geq 2\delta_0$. Assume that there exist C^2 functions h_+ and h_- on $(-\infty, \infty)$ such that

$$h(x) = \begin{cases} h_+(x) & \text{if } x \geq \delta_0 \\ h_-(x) & \text{if } x \leq -\delta_0 \end{cases}$$

and

$$c_1 \leq h_+'(x) \leq c_2, \; c_1 \leq -h_-'(x) \leq c_2, \; \forall x \in R^1 \tag{4}$$

for some constants $c_1 > 0$ and $c_2 > 0$. Then, (3) implies one of the following:

$$(gh'_+)^2(x^+) - (gh'_-)^2(x^-) \geq \bar{c}, \tag{5}$$

$$(gh'_+)^2(x^+) - (gh'_-)^2(x^-) \leq -\bar{c}. \tag{6}$$

In what follows, we assume (6) only, the discussion for the case of (5) being similar.

First of all, we start with an elementary lemma concerning with the exponential boundness of the state x_k.

Lemma 2.1. *For all $m = 0, 1, 2, \ldots$, there exists a constant C such that*

$$E \sup_{k \leq T/\varepsilon} |x_k|^{2m} \leq C^{2m}(2m-1)!!, \tag{7}$$

where $(2m-1)!! = (2m-1)\cdot(2m-3)\cdots 3\cdot 1$. Moreover, for some $\theta > 0$, we have

$$E \exp \theta \sup_{k \leq T/\varepsilon} |x_k|^2 < \infty. \tag{8}$$

Proof. Suppose (7) holds, then for $0 < \theta < (2C^2)^{-1}$,

$$E \exp \theta \sup_{k \leq T/\varepsilon} |x_k|^2 \leq (1 - 2\theta C^2)^{-1} < \infty.$$

We now show (7). Let $a_\varepsilon = 1 + f_0 \varepsilon$. Then we have

$$\begin{aligned} x_k = \; & a_\varepsilon^{k-1}(x_0 + \varepsilon f(x_0)) \\ & + \textstyle\sum_{i=1}^{k-1} a_\varepsilon^{k-i-1}(x_i + \varepsilon f(x_i) - a_\varepsilon x_i) + \sqrt{\varepsilon} a_\varepsilon^{k-1} M_{k-1}, \end{aligned} \tag{9}$$

where $M_{k-1} = \sum_{i=0}^{k-1} a_\varepsilon^i g(x_i) w_i$ is a $\sigma_{k-1} = \sigma\{w_0, \ldots, w_{k-1}\}$ martingale. Under assumption (A1), we have for some constant C_1

$$|x_k| \leq C_1[|x_0| + \sqrt{\varepsilon}|M_{k-1}|].$$

Thus,

$$E \sup_{k \leq T\varepsilon} |x_k|^{2m} \leq C_1^{2m} 2^{2m-1}[E|x_0|^{2m} + \varepsilon^m E \sup_{k \leq T/\varepsilon} |M_k|^{2m}]. \tag{10}$$

Moreover, by the elementary martingale property,

$$E \sup_{k \leq T/\varepsilon} |M_k|^{2m} \leq C_2 \sup_{k \leq T/\varepsilon} E|M_k|^{2m}, \tag{11}$$

where $C_2 = \sup_m (2m/(2m-1))^{2m} < \infty$.

Similarly as in [6, pp. 128], we can show that for some constant C_3,

$$\varepsilon^m E \sup_{k \le T/\varepsilon} |M_k|^{2m} \le C_3(2m-1)!!. \tag{12}$$

Then, we combine (10), (11), and (12) to complete the proof. □

We consider the two filters $m^{\pm}$.

$$m_{k+1}^{+} = m_k^{+} + q(y_{k+1} - h_{+}(m_k^{+})) \tag{13}$$

where $(1-\rho_1)/c_1 \le q \le 1/c_2$ for some $0 \le \rho_1 < 1$; and

$$m_{k+1}^{-} = m_k^{-} - q(y_{k+1} - h_{-}(m_k^{-})) \tag{14}$$

with $m_0^{\pm} = Ex_0$. With filters $m_k^{\pm}$ so defined we show that $h(x_k)$ can be well approximated by $h_{\pm}(m_k^{\pm})$.

Lemma 2.2. *For each $a > 0$ and $m = 0, 1, 2, \ldots$ there exists $\varepsilon_m > 0$ such that for $0 < \varepsilon \le \varepsilon_m$*

$$\sup_{a \le \varepsilon k \le T} E|h(x_k) - h_{+}(m_k^{+})|^{2m} \le C_m \varepsilon^m, \tag{15}$$

$$\sup_{a \le \varepsilon k \le T} E|h(x_k) - h_{-}(m_k^{-})|^{2m} \le C_m \varepsilon^m, \tag{16}$$

where $C_m = C^{2m}(2m-1)!!$ for some constant C.

Proof. We only show (15), (16) being similar. Let $z_k = h(x_k) - h_{+}(m_k^{+})$, and let $\bar{a}_k = (h_{+}(m_k^{+} + qz_k) - h_{+}(m_k^{+}))/qz_k \in [c_1, c_2]$. Define $\xi_k = (h(x_{k+1}) - h(x_k)) - (h_{+}(m_{k+1}^{+}) - h_{+}(m_k^{+} + qz_k))$. Then we have

$$z_{k+1} = (1 - \bar{a}_k q) z_k + \xi_k.$$

Recall that $(1-\rho_1)/c_1 \le q \le 1/c_2$ where c_1, c_2 are given in (4), $0 \le 1 - a_k q \le \rho_1 < 1$. This implies that $|z_{k+1}| \le \rho_1 |z_k| + |\xi_k|$. By induction over k, we obtain,

$$|z_k| \le \rho_1^{k-1}|\xi_0| + \rho_1^{k-2}|\xi_1| + \cdots + |\xi_{k-1}| + \rho_1^k |z_0|.$$

Let $v_k = 1 + \rho_1 + \cdots + \rho_1^k \le 1/(1-\rho_1) := C_1$. Then, for all $m = 0, 1, 2 \ldots$

$$\begin{aligned} E|z_k|^{2m} \le\ & v_k^{2m-1} E(\rho_1^{k-1}|\xi_0|^{2m} + \rho_1^{k-2}|\xi_1|^{2m} \\ & + \cdots + |\xi_{k-1}|^{2m} + \rho_1^k |z_0|^{2m}) \\ \le\ & C_1^{2m}(\sup_k E|\xi_k|^{2m} + \rho_1^k E|z_0|^{2m}). \end{aligned}$$

By (A2) and the assumption on x_0, we have there exists C_0 such that $E|z_0|^{2m} \le C_0^{2m} m!$. We choose $0 < \varepsilon_m < 1$ such that $\varepsilon_m \log \varepsilon_m = (a \log \rho_1)/m$. Then for $0 < \varepsilon \le \varepsilon_m$, $\rho_1 \le \rho_1^{a/\varepsilon} \le \varepsilon^m$.

Moreover, under (A1) and (A2), there exists a constant C_2 such that $|\xi_k| \le C_2(\varepsilon + \varepsilon|x_k| + \sqrt{\varepsilon}|w_k| + \sqrt{\varepsilon}|v_{k+1}|)$. Thus

$$E|\xi_k|^{2m} \le 4^{2m-1} C_2^{2m}[\varepsilon^{2m}(1 + E|x_k|^{2m}) + 2\varepsilon^m(2m-1)!!].$$

Applying Lemma 2.1, we obtain $\sup_k E|\xi_k|^{2m} \le C_3^{2m}(2m-1)!!\varepsilon^m$ for a constant C_3. Let $C = C_1(C_3 + C_0)$, then

$$|z_k|^{2m} \le C^{2m}(2m-1)!!\varepsilon^m. \square$$

3 Sequential quadratic variation test

Let δ_0 be as in Section 2. Given $0 < K_0 < K_1$, let

$$\begin{aligned}
B^+ &= \{x_k \ge \delta_0,\ K_0 \le k \le K_1\}\\
B^- &= \{x_k \le -\delta_0,\ K_0 \le k \le K_1\}\\
C_\varepsilon^+ &= \{h_+(m_k^+) \ge \bar{y},\ K_0 \le k \le K_1\}\\
C_\varepsilon^- &= \{h_-(m_k^-) \ge \bar{y},\ K_0 \le k \le K_1\}\\
B_\varepsilon &= B^+ \cup B^-,\ C_\varepsilon = C_\varepsilon^+ \cup C_\varepsilon^-.
\end{aligned}$$

We assume that $\varepsilon K_0 = a_1$, $\varepsilon K_1 = a_2$ with $0 < a_1 < a_2 \le T$, and $a_2 - a_1$ bounded away from 0. If C_ε occurs, then it is very probable that B_ε should also occur (see Lemma 3.1). We wish to decide between B^+ and B^-, based on observations y_k for $K_0 \le k \le K_0 + n^*$, where the "decision time" n^* may be random. Typically, n^* is considerably less than $K_1 - K_0$, and in fact is of order $\log \varepsilon$ (Theorem 4.1).

Let H_+ and H_- denote the hypotheses B^+ and B^- respectively. We define a test statistic:

$$S_n = \sum_{k=K_0}^{K_0+n-1} [(y_{k+1} - y_k)^2/\varepsilon - c(k)] \tag{17}$$

where $c(k) = [(gh'_+)^2(m_k^+) + (gh'_-)^2(m_k^-)]/2 + 2$ and $n \le K_1 - K_0$.

Description of the test: For pre-chosen $A > 0$ and $B > 0$, we keep accumulating of the process S_n until it is bigger than A or less than $-B$, then we stop and make a decision. We accept H_+ if $S_n \leq -B$ and accept H_- if $S_n \geq A$. More precisely, let $n^* = \inf\{n : S_n \notin [-B, A]\}$. Then

$$\begin{array}{lll} \text{if} & S_{n^*} \leq -B & \text{we accept } H_+ \\ \text{if} & S_{n^*} \geq A & \text{we accept } H_-. \end{array}$$

Lemma 3.1. *There exist $\varepsilon_m > 0$ and C_m, $m = 1, 2, \cdots$, such that for $0 < \varepsilon \leq \varepsilon_m$, $P(B_\varepsilon^c \cap C_\varepsilon) \leq C_m \varepsilon^m$, $\inf_{0<\varepsilon\leq\varepsilon_m} P(C_\varepsilon) > 0$.*

Therefore, $P(B_\varepsilon^c | C_\varepsilon) \leq (\inf_{0<\varepsilon\leq\varepsilon_m} P(C_\varepsilon))^{-1} C_m \varepsilon^m := C'_m \varepsilon^m$.

Proof. Observe that there exists $\alpha > 0$ such that for all x,

$$h(x) \geq \bar{y} - \alpha \text{ implies } |x| \geq \delta_0.$$

For this fixed α, we have

$$\begin{aligned} P(B_\varepsilon^c \cap C_\varepsilon) \leq & \; P(\sup_{K_0 \leq k \leq K_1} |h(x_k) - h_+(m_k^+)| \geq \alpha) \\ \leq & \; \alpha^{-2m} C_m \varepsilon^m. \end{aligned}$$

The last inequality is due to Lemma 2.2.

To show $P(C_\varepsilon)$ is bounded below by a positive constant, it suffices to show $P(C_\varepsilon^+)$ is so. We first show that for any $a_1 \leq t \leq a_2$, $k_0 = [t/\varepsilon]$ and for any $\bar{x} > 0$ there exists $c_0'' > 0$ such that

$$P(|x_{k_0}| \geq \bar{x}) \geq c_0''. \tag{18}$$

In fact, by (9), there exists c such that $|x_{k_0}| \geq \sqrt{\varepsilon}|M_{k_0-1}| - c$. This implies

$$\begin{aligned} P(|x_{k_0}| \geq \bar{x}) \geq & \; P(\sqrt{\varepsilon}|M_{k_0-1}| \geq \bar{x} + c) \\ = & \; P(|M_{k_0-1}|/s_{k_0} \geq (\bar{x} + c)/(\sqrt{\varepsilon} s_{k_0})), \end{aligned}$$

where $s_{k_0}^2 = E[g^2(x_0) + \cdots + g^2(x_{k_0-1})]$. By (A1), we have $s_{k_0} \geq \sqrt{k_0} c_1'$. Let $\beta = (\bar{x} + c)/(\sqrt{t} c_1')$. Then $(\bar{x} + c)/(\sqrt{\varepsilon} s_{k_0}) \leq \beta$. Thus,

$$P(|M_{k_0-1}|/s_{k_0} \geq (\bar{x} + c)/(\sqrt{\varepsilon} s_{k_0})) \geq P(|M_{k_0-1}|/s_{k_0} \geq \beta).$$

Then apply [1, Theorem 9.3.2],

$$\begin{aligned} & P(|M_{k_0-1}|/s_{k_0} \geq \beta) \\ \rightarrow \; & (\sqrt{2\pi})^{-1} \int_{|x| \geq \beta} e^{-x^2/2} dx := 2c_0'' > 0, \text{ as } \varepsilon \rightarrow 0. \end{aligned}$$

Therefore there exists $\varepsilon_m > 0$ such that for $\varepsilon \leq \varepsilon_m$, (18) holds. Then

$$\begin{aligned} P(C_\varepsilon^+) \geq\ & P(\sup_{K_0 \leq k \leq K_1} h(x_k) \geq \bar{y} + \alpha) \\ & - P(\sup_{K_0 \leq k \leq K_1} |h(x_k) - h_+(m_k^+)| \geq \alpha) \\ \geq\ & P(\sup_{K_0 \leq k \leq K_1} h(x_k) \geq \bar{y} + \alpha) - \alpha^{-2} C_2 \varepsilon^2 \\ \geq\ & P(|x_{k_0}| \geq \bar{x}) - \alpha^{-2} C_2 \varepsilon^2 \\ \geq\ & c_0'' - \alpha^{-2} C_2 \varepsilon^2 \end{aligned}$$

where $\bar{x}$ is so chosen that $|x| \geq \bar{x}$ implies $h(x) \geq \bar{y}$.

Therefore, there exists $\varepsilon_m > 0$ such that for $0 < \varepsilon \leq \varepsilon_m$, $c_1' := c_0'' - \alpha^{-2} C_2 \varepsilon^2 > 0$ and $P(C_\varepsilon^+) \geq c_1' > 0$. □

Lemma 3.2. *For any $\alpha > 0$, there exist $C > 0$ and $0 < \rho < 1$ such that*

$$\begin{aligned} P(|\textstyle\sum_{k=K_0}^{K_0+n-1} & [((gh')(x_k)w_k + v_{k+1} - v_k)^2 \\ & -((gh')^2(x_k) + 2)]| \geq n\alpha) \leq C\rho^n, \end{aligned} \tag{19}$$

for all $n = 1, 2, ..., [T/\varepsilon]$.

Proof. It suffices to show that there exist C and $0 < \rho < 1$ such that the following four inequalities hold for all $\alpha > 0$ and n:

i) $P(|\sum_{k=K_0}^{K_0+n-1}((gh')^2(x_k)(w_k^2 - 1)| \geq n\alpha) \leq C\rho^n$
ii) $P(|\sum_{k=K_0}^{K_0+n-1}[((gh')(x_k)w_k v_{k+1}| \geq n\alpha) \leq C\rho^n$
iii) $P(|\sum_{k=K_0}^{K_0+n-1}[((gh')(x_k)w_k v_k| \geq n\alpha) \leq C\rho^n$
iv) $P(|\sum_{k=K_0}^{K_0+n-1}[(v_{k+1} - v_k)^2 - 2]| \geq n\alpha) \leq C\rho^n$.

By the standard results of large deviation theory, it can be shown easily that iv) holds. We now prove the first three inequalities.

For any fixed $|\theta| < (2C_1)^{-1}$ (here C_1 is the upper bound for $(gh')^2$), we consider a function

$$\phi_n(\theta) = E \exp[\theta \sum_{k=K_0}^{K_0+n-1} (gh')^2(x_k)(w_k^2 - 1)]. \tag{20}$$

Let $\sigma_k = \sigma\{w_0, ..., w_k\}$, then $\{x_i, i \leq k\}$ is σ_{k-1} measurable and is independent of $\{w_k, w_{k+1}, ...\}$. Thus,

$$\begin{aligned}\phi_{n+1}(\theta) = & \ E\exp[\theta \textstyle\sum_{k=K_0}^{K_0+n-1}(gh')^2(x_k)(w_k^2-1)] \\ & \exp[\theta(gh')^2(x_n)(w_n^2-1)] \\ = & \ E\textstyle\sum_{m=0}^{\infty} E[\exp[\theta\sum_{k=K_0}^{K_0+n-1}(gh')^2(x_k)(w_k^2-1)] \\ & [\theta(gh')^2(x_n)(w_n^2-1)]^m/m!|\sigma_{K_0+n-1}] \\ = & \ \textstyle\sum_{m=0}^{\infty} E\exp[\theta\sum_{k=K_0}^{K_0+n-1}(gh')^2(x_k)(w_k^2-1)] \\ & [\theta(gh')^2(x_n)]^m E(w_n^2-1)^m/m!.\end{aligned}$$

By direct calculation, it can be shown that $E(w_n^2-1)^m \geq 0$ for all m. Thus,

$$\begin{aligned}\phi_{n+1} \leq & \ \phi_n(\theta)\textstyle\sum_{m=0}^{\infty}((|\theta|C_1)^m/m!)E(w_1^2-1)^m \\ = & \ \phi_n(\theta)E\exp|\theta|C_1(w_1^2-1).\end{aligned}$$

Therefore, we have by induction $\phi_n(\theta) \leq [E\exp|\theta|C_1(w_1^2-1)]^n$, which is finite provided $|\theta| < (2C_1)^{-1}$. For $0 < \theta < (2C_1)^{-1}$,

$$\begin{aligned} & P(|\textstyle\sum_{k=K_0}^{K_0+n-1}((gh')^2(x_k)(w_k^2-1)| \geq n\alpha) \\ \leq & \ P(\textstyle\sum_{k=K_0}^{K_0+n-1}((gh')^2(x_k)(w_k^2-1) \geq n\alpha) \\ & +P(\textstyle\sum_{k=K_0}^{K_0+n-1}((gh')^2(x_k)(w_k^2-1) \leq -n\alpha) \\ \leq & \ \exp(-n\alpha\theta)[\phi_n(\theta)+\phi_n(-\theta)] \\ \leq & \ 2[\exp(-\alpha\theta)E\exp(\theta C_1(w_1^2-1))]^n \\ := & \ 2[\Phi(\theta)]^n.\end{aligned} \tag{21}$$

Since $\Phi(0) = 1$, and $(d/d\theta)\Phi(0) = -\alpha < 0$, there exists $0 < \theta < (2C_1)^{-1}$ such that $\rho := \Phi(\theta) < 1$. Thus,

$$P(|\sum_{k=K_0}^{K_0+n-1}((gh')^2(x_k)(w_k^2-1)| \geq n\alpha) \leq 2\rho^n.$$

i) is proved. We now show ii). Let $\sigma_n' = \sigma\{w_0, ..., w_{n-1}, v_0, ..., v_n\}$. Then $\{x_k, v_{k+1} : k \leq n\}$ is σ_n' measurable and σ_n' is independent of w_n and v_{n+1}. We define a function

$$\psi_n(\theta) = E\exp[\theta \sum_{k=K_0}^{K_0+n-1}(gh')(x_k)w_kv_{k+1}]. \tag{22}$$

Then we have the following:

$$\begin{aligned}\psi_{n+1}(\theta) = \; & E\{\exp[\theta \textstyle\sum_{k=K_0}^{K_0+n-1}(gh')(x_k)w_k v_{k+1}] \\ & \exp[\theta(gh')(x_n)w_n v_{n+1}]\} \\ = \; & E[\textstyle\sum_{m=0}^{\infty} E\{\exp[\theta \sum_{k=K_0}^{K_0+n-1}(gh')(x_k)w_k v_{k+1}] \\ & \cdot[\theta(gh')(x_n)w_n v_{n+1}]^m/m! | \sigma'_{K_0+n}\} \\ = \; & E\textstyle\sum_{m=0}^{\infty} E\{\exp[\theta \sum_{k=K_0}^{K_0+n}(gh')(x_k)w_k v_{k+1}] \\ & (\theta(gh')(x_n))^{2m} | \sigma'_{K_0+n}\} E w_n^{2m} E v_{n+1}^{2m}/(2m)!.\end{aligned}$$

Thus,

$$\begin{aligned}\psi_{n+1}(\theta) \leq \; & \psi_n(\theta)\textstyle\sum_{m=0}^{\infty}(\theta^2 C_1)^m((2m-1)!!)^2/(2m)! \\ \leq \; & \psi_n(\theta)/(1-\theta^2 C_1).\end{aligned}$$

Thus, $\psi_n(\theta) \leq (1-\theta^2 C_1)^{-n}$. Observe that there exists $0 < \theta$ such that $0 < e^{-\theta\alpha}/(1-\theta^2 C_1) < 1$. Then by applying the same procedure as in (21), we prove ii). iii) can be shown similarly. □

Lemma 3.3. *For $n \leq [T/\varepsilon]$, let*

$$\begin{aligned}E_1 = \; & \{n^{-1}\textstyle\sum_{k=K_0}^{K_0+n-1}((gh')^2(x_k) \\ & -[(gh'_+)^2(m_k^+) + (gh'_-)^2(m_k^-)]/2) \geq -\bar{c}/8\}, \\ E'_1 = \; & \{n^{-1}\textstyle\sum_{k=K_0}^{K_0+n-1}((gh')^2(x_k) \\ & -[(gh'_+)^2(m_k^+) + (gh'_-)^2(m_k^-)]/2) \leq -\bar{c}/8\},\end{aligned}$$

where $\bar{c}$ is the constant given in (A3). Then for all $m = 0, 1, 2, \ldots$, there exist C_m and $\varepsilon_m > 0$ such that for $0 < \varepsilon \leq \varepsilon_m$,

$$P(E_1 \cap B^+ \cap C_\varepsilon) \leq C_m \varepsilon^m. \tag{23}$$

$$P(E'_1 \cap B^- \cap C_\varepsilon) \leq C_m \varepsilon^m. \tag{24}$$

Proof. We only show (23). By (A2) and (A3) there exists an $\alpha_0 > 0$ such that for $\xi_+ \geq 0$ and $\xi_- < 0$ the conditions

$$|h(x) - h(\xi^\pm)| < \alpha_0, \;\; \text{and } h(x) \geq \bar{y} - \alpha_0$$

imply that $\xi^+ \geq \delta_0, \xi^- \leq -\delta_0$ and $(gh')^2(\xi^+) - (gh')^2(\xi^-) < -\bar{c}/2$. Let

$$\begin{aligned}E_2 = \; & \{n^{-1}\textstyle\sum_{k=K_0}^{K_0+n-1}((gh'_+)^2(m_k^+) - (gh'_-)^2(m_k^-)) < -\bar{c}/2\} \\ E_3 = \; & \{n^{-1}\textstyle\sum_{k=K_0}^{K_0+n-1}|(gh')^2(x_k) - (gh'_+)^2(m_k^+)| \geq \bar{c}/8\} \\ E_4 = \; & \{\sup_{K_0 \leq k \leq K_0+n-1}|h(x_k) - h_+(m_k^+)| < \alpha\} \\ & \cap\{\sup_{K_0 \leq k \leq K_0+n-1}|h(x_k) - h_-(m_k^-)| < \alpha\}.\end{aligned}$$

Note that

$$P(E_1 \cap B^+ \cap C_\varepsilon) \le P(E_4^c) + P(E_1 \cap E_4 \cap B^+ \cap C_\varepsilon). \tag{25}$$

By Lemma 2.2, $P(E_4^c)$ is bounded by $C_m \varepsilon^m$ for small ε. Now we are to show that for α small enough $E_1 \cap E_4 \cap B^+ \cap C_\varepsilon = \emptyset$. In fact,

$$\begin{aligned} E_1 \cap E_4 \cap B^+ \cap \quad C_\varepsilon \subset & \, E_3 \cap E_4 \cap B^+ \cap C_\varepsilon \\ & \cup E_2^c \cap E_4 \cap B^+ \cap C_\varepsilon. \end{aligned}$$

Observe the following:
i) By the definition of δ_0, we can show that there exists $\alpha_0' > 0$ such that

$$\begin{aligned} h_+(m_k^+) \ge \bar{y} - \alpha_0' \text{ implies } m_k^+ \ge \delta_0 \text{ and} \\ h_-(m_k^-) \ge \bar{y} - \alpha_0' \text{ implies } m_k^- \le -\delta_0. \end{aligned}$$

Thus, on $E_4 \cap C_\varepsilon$, for $\alpha \le \alpha_0'$, $m_k^+ \ge \delta_0$ and $m_k^- \le -\delta_0$.
ii) On the set $E_3 \cap E_4 \cap B^+ \cap C_\varepsilon$, there exist a constant $C' > 0$ such that

$$|(gh')^2(x_k) - (gh'_+)^2(m_k^+)| \le C'|h(x_k) - h_+(m_k^+)| < C'\alpha.$$

Thus, for $\alpha < \bar{c}/(8C')$, $E_3 = \emptyset$.
iii) By our earlier claim, there exists $\alpha' > 0$ such that for all $\alpha < \alpha'$, $E_2^c = \emptyset$.

We take $\alpha = \min\{\alpha_0, \alpha_0', \alpha', \bar{c}/(8C')\}$ in (25), then there exists C_m and ε_m such that $E_1 \cap E_4 \cap B^+ \cap C_\varepsilon = \emptyset$. Thus,

$$P(E_1 \cap B^+ \cap C_\varepsilon) \le P(E_4^c) \le C_m \varepsilon^m.$$

We complete the proof of this lemma. □

Lemma 3.4. *For each* $m, N = 0, 1, 2...$ *there exist constants* $C_1' > 0$ *and* $C_2' > 0$ *such that for* $A \ge C_1' N + C_2'(m+1)|\log \varepsilon|$

$$P((|S_{n^*}| \ge A) \cap (n^* \le N)) \le C\varepsilon^m.$$

Proof. First of all,

$$P((|S_{n^*}| \ge A) \cap (n^* \le N)) \le \sum_{n=0}^{N} P(|S_n| \ge A).$$

Now we are to find an upper bound for S_n. Recall that gh' is a bounded function. We have for some constant C

$$\begin{aligned}|S_n| \le\ & Cn + \textstyle\sum_{k=K_0}^{K_0+n-1}(y_{k+1}-y_k)^2/\varepsilon \\ \le\ & Cn + 2\textstyle\sum_{k=K_0}^{K_0+n-1}(h(x_{k+1})-h(x_k))^2/\varepsilon + (v_{k+1}-v_k)^2) \\ \le\ & Cn + C\textstyle\sum_{k=K_0}^{K_0+n-1}(\varepsilon f^2(x_k) + (g(x_k)w_k)^2 + (v_{k+1}-v_k)^2) \\ \le\ & Cn + C\textstyle\sum_{k=K_0}^{K_0+n-1}(\varepsilon x_k^2 + w_k^2 + (v_{k+1}-v_k)^2) \\ \le\ & C_1[n + \sup_{k\le K_0+n-1} x_k^2 + \textstyle\sum_{k=K_0}^{K_0+n-1}(w_k^2 + (v_{k+1}-v_k)^2)].\end{aligned}$$

Choose $A > 4C_1N \ge 4C_1 n$. Then we have

$$\begin{aligned}P(|S_n| \ge A) \le\ & P(C_1 \sup_{k\le K_0+n-1} x_k^2 \ge A/4) \\ & + P(C_1 \textstyle\sum_{k=K_0}^{K_0+n-1} w_k^2 \ge A/4) \\ & + P(C_1 \textstyle\sum_{k=K_0}^{K_0+n-1}(v_{k+1}-v_k)^2 \ge A/4).\end{aligned}$$

By Lemma 2.1, there exists C'' such that

$$P(C_1 \sup_{k\le K_0+n-1} x_k^2 \ge A/4) \le C' \exp(-c_0 A)$$

where $c_0 = 1/(4C_1) > 0$. Moreover, there exists a θ such that

$$\begin{aligned}& P(C_1 \textstyle\sum_{k=K_0}^{K_0+n-1} w_k^2 \ge A/4) \\ \le\ & \exp(-A/(4C_1\theta)) E \exp(\theta \textstyle\sum_{k=K_0}^{K_0+n-1} w_k^2) \\ \le\ & \exp(-c_0' A + c' n),\end{aligned}$$

where $c_0' = 1/(4C_1\theta)$ and $c' = \log(1-2\theta)^{-1/2}$. Similarly,

$$P(C_1 \sum_{k=K_0}^{K_0+n-1} (v_{k+1}-v_k)^2 \ge A/4) \le \exp(-c_0'' A + c'' n),$$

for some constants $c_0'' > 0, c'' > 0$.

Let $C_1' = \max(c'/c_0', c''/c_0'')$ and $C_2' = \max(1/c_0, 1/c_0', 1/c_0'')$. Then for $A \ge C_1' N + C_2'(m+1)|\log\varepsilon|$,

$$P(|S_n| \ge A) \le (3 + C'')\varepsilon^{m+1}.$$

Therefore,

$$\begin{aligned}P(|S_{n^*}| \ge A) \le\ & (3 + c'')(T/\varepsilon)\varepsilon^{m+1} \\ =\ & (3 + C'')T\varepsilon^m := C\varepsilon^m. \square\end{aligned}$$

Theorem 3.1. *There exist constants $a_0 > 0$ $\varepsilon_m > 0$ and C_m, $m = 1, 2, \cdots$, such that for*

$$\min\{A, B\} \geq a_0(m+1)|\log \varepsilon|,\ 0 < \varepsilon \leq \varepsilon_m$$

$$P((S_{n^*} \geq A) \cap B^+ \cap C_\varepsilon) \leq C_m \varepsilon^m; \tag{26}$$

$$P((S_{n^*} \leq -B) \cap B^- \cap C_\varepsilon) \leq C_m \varepsilon^m. \tag{27}$$

Proof. We only show (26). Note that for any $N = 0, 1, 2, \ldots$,

$$\begin{aligned} & P((S_{n^*} \geq A) \cap B^+ \cap C_\varepsilon) \\ \leq\ & P((S_{n^*} \geq A) \cap (n^* \leq N)) \\ & + P((S_{n^*} \geq A) \cap (n^* \geq N) \cap B^+ \cap C_\varepsilon) \\ \leq\ & P((S_{n^*} \geq A) \cap (n^* \leq N)) \\ & + \textstyle\sum_{n \geq N} P((S_n \geq A) \cap B^+ \cap C_\varepsilon). \end{aligned}$$

By Lemma 3.4, the first term is bounded by $C_m \varepsilon^m$ provided $A \geq C_1' N + C_2'(m+1)|\log \varepsilon|$. So, it suffices to show

$$P((S_n \geq A) \cap B^+ \cap C_\varepsilon) \leq C_m \varepsilon^{m+1} \tag{28}$$

for all $n \geq N \geq c_0(m+1)|\log \varepsilon|$ and for some $c_0 > 0$.

Let $\zeta_k = h(x_{k+1}) - h(x_k) - h'(x_k)(x_{k+1} - x_k)$. Then $|\zeta_k| \leq C|x_{k+1} - x_k|^2$ for some $C > 0$.

We define a sequence ζ_k'

$$\begin{aligned} \zeta_k' =\ & \zeta_k^2/\varepsilon + \varepsilon (fh')^2(x_k) + 2((gh')(x_k) w_k + v_{k+1} - v_k)\zeta_k/\sqrt{\varepsilon} \\ & + 2\sqrt{\varepsilon}((gh')(x_k) w_k + v_{k+1} - v_k)(fh')(x_k) + 2(fh')(x_k)\zeta_k. \end{aligned}$$

Then we may rewrite S_n as follows: $S_n = I_n + II_n + III_n$ where

$$\begin{aligned} I_n &= \textstyle\sum_{k=K_0}^{K_0+n-1} \{[(gh')(x_k) w_k + (v_{k+1} - v_k)]^2 - [(gh')^2(x_k) + 2]\} \\ II_n &= \textstyle\sum_{k=K_0}^{K_0+n-1} \{(gh')^2(x_k) - [(gh'_+)^2(m_k^+) + (gh'_-)^2(m_k^-)]/2\} \\ III_n &= \textstyle\sum_{k=K_0}^{K_0+n-1} \zeta_k'. \end{aligned}$$

Applying Lemma 2.1, we can show that there exists C_m', C_m'', C_m''' such that

$$\begin{aligned} E|\zeta_k|^{2m} &\leq C_m' |x_{k+1} - x_k|^{4m} \\ &\leq C_m'' \varepsilon^{2m} E[\varepsilon^{2m}(1 + |x_k|^{4m}) + w_k^{4m}] \\ &\leq C_m''' \varepsilon^{2m} (4m-1)!!. \end{aligned}$$

Thus we obtain $E|\zeta_k'|^{2m} \le C_m \varepsilon^m$ for some C_m. Let $\alpha \le \bar{c}/32$, then

$$\begin{aligned} & (S_n \ge A) \subset (S_n \ge -n(\bar{c}/16)) \\ \subset \ & (|I_n| \ge n\alpha) \cup (II_n \ge -n(\bar{c}/8)) \cup (|III_n| \ge n\alpha). \end{aligned} \tag{29}$$

Thus, $P(|III_n| \ge n\alpha) \le \alpha^{-2m+2} E|\zeta_k'|^{2m+2} \le C_m \varepsilon^{m+1}$ for suitable C_m. Moreover, by Lemmas 3.2 and 3.3, the probabilities of the remaining two terms are bounded by $C_m \varepsilon^{m+1}$ for $n \ge N = c_0(m+1)|\log \varepsilon|$ and $0 < \varepsilon \le \varepsilon_m$ for suitable $\varepsilon_m > 0$. Let $a_0 = c_0 C_1' + C_2'$. Then when $\min\{A, B\} \ge a_0(m+1)|\log \varepsilon|$, (26) holds. □

Remark. Theorem 3.1 gives bounds for Type I and II error probabilities for the sequential hypothesis test.

4 Mean decision time

Let us fix m, and define $\bar{n} = \bar{n}(m, \varepsilon)$ by

$$\bar{n} = \max\{16A/\bar{c}, 16B/\bar{c}, c_0(m+1)|\log \varepsilon|\} + 1$$

where c_0 is given in (28). Then $\bar{n} = \max\{16a_0/\bar{c}, c_0\}(m+1)|\log \varepsilon| + 1$ provided $A = B = a_0(m+1)|\log \varepsilon|$.

Theorem 4.1. *For each m, there exist $\varepsilon_m > 0$ and C_m such that for $0 < \varepsilon \le \varepsilon_m$, i) $P(n^* \ge \bar{n}, C_\varepsilon) \le C_m \varepsilon^m$ and ii) $E[n^*; C_\varepsilon] \le \bar{n} + 1$.*

Proof. First of all,

$$\begin{aligned} & P(n^* \ge \bar{n}, C_\varepsilon) \\ \le \ & P(n^* \ge \bar{n}, B^+ \cap C_\varepsilon) + P(n^* \ge \bar{n}, B^- \cap C_\varepsilon) + P(B_\varepsilon^c \cap C_\varepsilon) \\ \le \ & P(n^* \ge \bar{n}, B^+ \cap C_\varepsilon) + P(n^* \ge \bar{n}, B^- \cap C_\varepsilon) + C_m' \varepsilon^m, \end{aligned}$$

for some ε_m', C_m' and $0 < \varepsilon \le \varepsilon_m'$. The last inequality is because Lemma 3.1. Note that

$$\begin{aligned} & P(n^* \ge \bar{n}, B^+ \cap C_\varepsilon) \\ \le \ & P(-B < S_{\bar{n}-1} < A, B^+ \cap C_\varepsilon) \\ \le \ & P(S_{\bar{n}-1} \ge -(\bar{n}-1)(\bar{c}/16), B^+ \cap C_\varepsilon) \\ & + P(-B < S_{\bar{n}-1} < -(\bar{n}-1)(\bar{c}/16), B^+ \cap C_\varepsilon). \end{aligned}$$

In (29), we take $\alpha = \bar{c}/32$. Then we obtain that there exist ε_m'', C_m'' such that

$$P(S_{\bar{n}-1} \ge -(\bar{n}-1)(\bar{c}/16), B^+ \cap C_\varepsilon) \le C_m'' \varepsilon^m, \ 0 < \varepsilon \le \varepsilon_m''.$$

Moreover, by the definition of $\bar{n}$, $\{-B < S_{\bar{n}-1} < -(\bar{n}-1)(\bar{c}/16)\} = \emptyset$. Thus, $P(n^* \geq \bar{n}, B^+ \cap C_\varepsilon) \leq C''_m \varepsilon^m$. Similarly, there exist ε'''_m and C'''_m such that $P(n^* \geq \bar{n}, B^- \cap C_\varepsilon) \leq C'''_m \varepsilon^m$. for $0 < \varepsilon \leq \varepsilon'''_m$.

Let $C_m = C'_m + C''_m + C'''_m$ and let $\varepsilon_m = \min\{\varepsilon'_m, \varepsilon''_m, \varepsilon'''_m\}$. Then i) holds for $0 < \varepsilon \leq \varepsilon_m$.

Now we show ii). In fact,

$$\begin{aligned} E[n^*; C_\varepsilon] = & \textstyle\sum_{n=1}^{[T/\varepsilon]} P(n^* \geq n, C_\varepsilon) \\ \leq & \bar{n} + \textstyle\sum_{n=\bar{n}}^{[T/\varepsilon]} P(n^* \geq n, C_\varepsilon) \\ \leq & \bar{n} + (T/\varepsilon) 3 C_m \varepsilon^m \\ \leq & \bar{n} + 3 T C_m \varepsilon^{m-1}. \end{aligned}$$

We take $m = 2$ and ε_m small enough so that $3TC_2\varepsilon \leq 1$. Then we obtain $E[n^*; C_\varepsilon] \leq \bar{n} + 1$. □

5 Asymptotic optimal filters

Let $\hat{B}^+ = \{S_{n^*} \leq -B\}$ and $\hat{B}^- = \{S_{n^*} \geq A\}$. Then Theorem 3.1 implies $P(\hat{B}^+ \cap B^- \cap C_\varepsilon) \leq C_m \varepsilon^m$, $P(\hat{B}^- \cap B^+ \cap C_\varepsilon) \leq C_m \varepsilon^m$.

When $\hat{B}^+ \cap C_\varepsilon$ holds, we use m_k^+ as an approximation to the optimal discrete time filter $\hat{x}_k$; and when $\hat{B}^- \cap C_\varepsilon$ holds we use m_k^- as an approximation. We state without proof the following result.

Theorem 5.1. *Let $0 < p < 1$. Then, for $m = 1, 2, \cdots$, there exist $\varepsilon_m > 0$, C_m such that, for $K_0 \leq k \leq K_1$, $0 < \varepsilon \leq \varepsilon_m$,*

$$P(|\hat{x}_k - m_k^+| > \varepsilon^p | \hat{B}^+ \cap C_\varepsilon) \leq C_m \varepsilon^m, \tag{30}$$

$$P(|\hat{x}_k - m_k^-| > \varepsilon^p | \hat{B}^- \cap C_\varepsilon) \leq C_m \varepsilon^m, \tag{31}$$

where $\hat{x}_k$ is the conditional mean of x_k given the observations up to k.

A proof can be given along the lines of the corresponding continuous time result [2, Theorem 7.1]. For this purpose a discrete time analogue (Milheiro [8]) of the Picard estimates [9] for the case of one-one observation function h is needed, together with a suitable change of probability measure under which h_+ (or h_-) becomes the observation function.

Bibliography

[1] Y.S. Chow and H. Teicher, Probability Theory, Springer-Verlag, New York, (1978).

[2] W. Fleming and E. Pardoux, Piecewise monotone filtering with small observation noise, SIAM J. on Control and Optimiz., Vol. 27, No. 5, pp. 1156-1181, (1989).

[3] W. Fleming, D. Ji and E. Pardoux, Piecewise linear filtering with small observation noise, Proc. 8th Int. Conf. on Analysis and Optimization of Systems (Antibes Conf. 1988) Lect. N. in Control and Info. Sci, No. 111, pp. 725-739, Springer (1988).

[4] W.H. Fleming, D. Ji, P. Salame, and Q. Zhang, Piecewise monotone filtering in discrete time with small observation noise, to appear in IEEE Trans. Auto. Control, (1991).

[5] R. Katzur, B.Z. Bobrovsky, and Z. Schuss, Asymptotic analysis of the optimal filtering problem for one-dimensional diffusions measured in a low noise channel, Part I and II, SIAM J. Appl. Math., 44, pp. 591-604, and pp. 1176-1191, (1984).

[6] G. Kallianpur, Stochastic Filtering Theory, Springer-Verlag, New York, (1980).

[7] R.S. Liptser and A.N. Shiryayev, Statistics of Random Processes, Springer-Verlag, New York, (1977).

[8] P. Milheiro, Etudes asymptotiques en filtrage nonlineare avec petit bruit d'observation, These Universite de Provence, (1990).

[9] J. Picard, Nonlinear filtering of one-dimensional diffusions in the case of a high signal-to-noise ratio, SIAM J. Applied Math, 46, pp. 1098-1125, (1986).

Closed Form Characteristic Functions for Certain Random Variables Related to Brownian Motion

G. J. Foschini and L. A. Shepp
AT&T Bell Laboratories

Abstract

In what follows let $W(t) = (W_1(t), W_2(t), W_3(t))$ be a three-dimensional Wiener process (Brownian motion) starting at the origin. P. Lévy has shown that the characteristic function of the random variable $X = \int_0^1 [W_1(t) - tW_1(1)] d[W_2(t) - tW_2(1)]$, which measures the signed area of a random planar loop, is $Ee^{izX} = (z/2)\operatorname{csch}(z/2)$. (The corresponding density, of importance in polymer physics, is $(\pi/2)\operatorname{sech}^2 \pi x$). Lévy's derivation uses N. Wiener's simple sine-cosine Fourier expansion of white Gaussian noise. This expansion is used here as the starting point in showing that the six-dimensional vector $(W(1), \int_0^1 W(t) \times dW(t))$, needed to characterize polarization dispersion effects for high speed communications over optical fibers, also has an elementary closed form characteristic function.

Employing the same Fourier expansion we prove the following result, generalizing the two examples: Let V be any finite dimensional vector whose components are arbitrary linear combinations of random variables of the form $W_k(1)$, $W_k(1)W_l(1)$, $W_k(1)\int_0^1 W_l(t)dt$, $\int_0^1 W_k(t)dW_l(t)$ $(1 \le k \le 3, 1 \le l \le 3)$. The joint characteristic function of V can be expressed in closed form using elementary functions.

ISBN 0-12-481005-5

1 Introduction

Let $W(t) = (W_1(t), W_2(t), W_3(t))$ be a standardized Wiener process (Brownian motion) on [0,1]. By standardized, we mean that for each t, $W(t)$ is distributed as $N(0,tI)$. Every time we write an integral sign in this paper the integral is understood to be over the unit interval. In the next section we derive characteristic functions as noted in the following two examples, the first of which uses only two dimensions of $W(t)$ and is a known result.

Example I: Signed Area of a Random Planar Loop, X. Let $\overline{W}_m(t)$ denote the pinned Wiener process [1-3] (also called the Brownian bridge)

$$\overline{W}_m(t) \triangleq W_m(t) - tW_m(1) . \tag{1}$$

Since $\overline{W}_m(t)$ starts and ends at 0, the curve $\{(\overline{W}_1(t), \overline{W}_2(t))\}_{0 \le t \le 1}$ is a random planar loop. Using Green's Theorem (see for example, [4]) the signed loop area is given by

$$X = \int \overline{W}_1(t) \cdot d\overline{W}_2(t) . \tag{2}$$

In this case the characteristic function is

$$Ee^{izX} = (z/2)/\sinh(z/2) . \tag{3}$$

(The corresponding density is $p(x) = \frac{\pi}{2}\frac{1}{\cosh^2 \pi x}$.) This was first obtained by P. Lévy [5] using a series representation for $W_i(t)$ that was introduced by Wiener in [6] (see equation (7) below). Sparked by interest for studying topologically constrained polymers, (3) was recently rediscovered using different methods, in [7]. Results related to (3) appear in [8-13].

Example II: Fiber Optical Polarization Dispersion Vector Y. Let $\times$ denote vector cross-product

$$Y \triangleq (W(1), \int W(t) \times dW(t)) . \tag{4}$$

Let $z = (z_1, z_2, z_3)$, $\zeta = (\zeta_1, \zeta_2, \zeta_3)$ be the transform variates with denoting transpose and vector multiplication such as $z'\zeta$ denoting the scalar product. We have

$$Ee^{iz'W(1)+i\zeta'\int W(t)\times dW(t)} =$$

$$\operatorname{sech}|\zeta|\exp-1/2\left\{|z|^2\frac{\tanh|\zeta|}{|\zeta|}+\frac{(z'\zeta)^2}{|\zeta|^2}\left(1-\frac{\tanh|\zeta|}{|\zeta|}\right)\right\}. \quad (5)$$

See [14] for a discussion of the importance of (5) for the theory of polarization dispersion in single mode fibers.

As we shall see, the following simple Fourier expansions [1,3] of the white Gaussian noise processes $\{dW_m(t)/dt\}_{m=1}^3$ are useful in deriving the characteristic functions for the examples

$$\dot{W}_m(t) = \xi_0^m + 2^{1/2}\sum_{j=1}^{\infty}(\xi_j^m\cos 2\pi jt+\eta_j^m\sin 2\pi jt). \quad (6)$$

Each element of $\{\xi_k^m, 0\le k<\infty, 1\le m\le 3\}\cup\{\eta_k^m, 1\le k\le\infty, 1\le m\le 3\}$ is distributed as $N(0,1)$ and statistically independent of all the other elements. As remarked, this was first used by Wiener [6]. A similar expansion holds for any complete orthonormal sequence [1]. The sines and cosines have a crucial advantage here in that the expansion is orthogonal in more than one way as we shall see in (9) below. In particular, the Karhunen-Loeve expansion of the Wiener process is not the right expansion to use.

Employing the formal expansion (6) we prove the following:

Theorem. *For $W(t)$ a three-dimensional Wiener Process, the joint characteristic function of any random variables of the form*

$$W_k(1),\ W_k(1)W_l(1),\ W_k(1)\int W_l(t)dt,\ \int W_k(t)dW_l(t)$$

can be expressed in closed form using elementary functions.

The theorem is valid for either Ito or Stratonovich [15,16] integrals.

The above result, has the immediate implication that any finite dimensional vector, V, whose components, $(V_1, V_2, \ldots, V_K)$, are linear combinations of the random variables mentioned in the theorem, also has a joint characteristic function expressible in closed form using elementary functions. So this generalization covers the second example. To see that the theorem itself covers the first example, use integration by parts on the $-\int tW_1(1)dW_2(t)$ term that appears when expanding $\int \overline{W}_1(t)d\overline{W}_2(t)$ out into the four integrals. The curve $\{\overline{W}_1(t),\overline{W}_2(t),\overline{W}_3(t)\}$ $0 \le t \le 1$ is a three dimensional loop. One could use the constructive proof of the theorem to find the joint characteristic function of the signed areas of linear mappings of this loop onto a finite set of planes.

The very basic example $\int W_1(t)dW_2(t)$ is clearly covered by the theorem. The characteristic function turns out to be $\text{sech}^{1/2} z$ (the corresponding density is $\left((2^{3/2}\pi)^{-1}\left|\Gamma\left(\frac{1}{4}+i\frac{x}{2}\right)\right|^2\right)$. We will not derive the characteristic function explicitly in this case since the derivation is very much along the lines of the derivation of Example II, but simpler. The last component of Y in Example II is $\int W_1(t)dW_2(t) - \int W_2(t)dW_1(t)$. From (5) the characteristic function is $\text{sech} z$, so $\int W_1(t)dW_2(t)$ and $-\int W_2(t)dW_1(t)$ add like independent random variables. (The probability density function corresponding to $\text{sech} z$ is $1/2\,\text{sech}(\pi x/2)$).

All densities mentioned thus far can be obtained from their characteristic function using the tables [17]. As of this writing (5) has not been inverted in closed form. We note that the six univariate marginals densities of Y have precisely the same functional form as their characteristic functions.

Lévy's formula ((1.3.4) of [5]) is more general than (3). It is a formula for the characteristic function of the area included by the loop defined by a planar Brownian motion and its (origin-to-endpoint) chord, conditioned on the chord length. This celebrated formula is a straight-forward consequence of our equation (5).

2 Derivation of Characteristic Functions for the Examples

The expansion (6) is formal, the integrated form

$$W_m(t)=\xi_0^m t + 2^{1/2}\sum_{j=1}^{\infty}(2\pi j)^{-1}(\xi_j^m \sin 2\pi jt + \eta_j^m(1-\cos 2\pi jt)) \quad (7)$$

converges uniformly with probability one [1,3,18]. Note $W_m(1) = \xi_0^m$.

Although the theorem generalizes the examples, the derivation of Example I is short and serves to very simply illustrate the use of (6) and (7). We will also derive Example II as it provides a concrete illustration of a somewhat elaborate case and provides much of what is needed for the theorem.

The following expansions will be needed. They follow using (6) and (7)

$$W_k(1)\int_0^1 W_l(t)dt = \xi_0^k\Big((\xi_0^l/2)+2^{1/2}\sum_{j=1}^{\infty}(2\pi j)^{-1}\eta_j^l\Big) \tag{8a}$$

$$\int_0^1 W_k(t)dW_l(t) = (\xi_0^k\xi_0^l/2)+\sum_{j=1}^{\infty}(2\pi j)^{-1}[\eta_j^l(\xi_j^k-2^{1/2}\xi_0)^k-\eta_j^k(\xi_j^l-2^{1/2}\xi_0^l)]\,. \tag{8b}$$

Example I. Using the definitions of $\overline{W}_1(t)$ and $\overline{W}_2(t)$ in $X=\int \overline{W}_1(t)d\overline{W}_2(t)$, and then substituting using (8a) and (8b), we obtain on account of the orthogonality of the terms in the expansion (7) *and their derivatives*,

$$Ee^{izX} = \prod_{j=1}^{\infty} E\exp\Big\{iz(\xi_j^1\eta_j^2-\xi_j^2\eta_j^1)/(2\pi j)\Big\} \tag{9a}$$

$$= \prod_{j=1}^{\infty} E_{\xi_j}(e^{-1/2(z\xi_j/(2\pi j))})^2 = \prod_{j=1}^{\infty}\Big(1+(z/(2\pi j))^2\Big) \tag{9b}$$

$$= (z/2)\sinh(z/2)\,. \tag{9c}$$

See [19] for the last equality.

Example II. In deriving the characteristic function of the six dimensional random vector $(W(1),\int W(\tau)\times dW(\tau))$ we will use the following representation of the last three components constituting the integrated vector cross product

$$\int W(\tau) \times dW(\tau) = (2\int W_2(\tau)dW_3(\tau) - W_2(1)W_3(1),$$
$$2\int W_3(\tau)dW_1(\tau) - W_3(1)W_1(1),$$
$$2\int W_1(\tau)dW_2(\tau) - W_1(1)W_2(1)) . \quad (10)$$

We have used integration by parts, so one integral, not two, appears in each component.

Let (p,q,r) index the cyclic permutations $\{(1,2,3), (3,1,2), (2,3,1)\}$. Form the six dimensional characteristic function as a product over the three cycles

$$\phi(z_1,z_2,z_3,\zeta_{23},\zeta_{13},\zeta_{12}) =$$
$$E \prod_{\{(p,q,r)\}} e^{i\{W_p(1)z_p + [2\int W_q(\tau)dW_r(\tau) - W_q(1)W_r(1)]\zeta_{qr}\}} . \quad (11)$$

The dual subscripting of the last three transform variables is a temporary notational convenience. Drawing on (6) and (7) and doing the elementary integrals we rewrite (11) as

$$\phi(z_1,z_2,z_3,\zeta_{23},\zeta_{13},\zeta_{12}) =$$
$$E \prod_{\{p,q,r\}} e^{iz_p\xi_o^p + i\zeta_{qr}\sum_{j=1}^{\infty}\left[\frac{\eta_j^r(\xi_j^q - \sqrt{2}\,\xi_o^q) - \eta_j^q(\xi_j^r - \sqrt{2}\,\xi_o^r)}{\pi j}\right]} \quad (12)$$

where the $\zeta_{qr}W_q(1)W_r(1)$ terms cancelled out of each sum. We will next proceed to take the expectation in three stages: first with respect to the $\eta_j^{(\cdot)}$ variates, second with respect to the $\xi_j^{(\cdot)}$ $(j \ge 1)$ variates, and last with respect to the three $\xi_o^{(\cdot)}$ variates.

Expectation over the η Variates

In preparation for taking the expectation over the η_j^p, η_j^q, η_j^r variates we collect $\eta_j^{(\cdot)}$ terms with the same superscript. Collecting and taking the expectation

$$\phi(z_1, z_2, z_3, \zeta_{23}, \zeta_{13}, \zeta_{12})$$

$$= Ee^{z'\xi_0} \prod_{j=1}^{\infty} \prod_{\{r,q,p\}} e^{\frac{i}{\pi j}\left\{\eta_j^r [z_{qr}(\xi_j^q - \sqrt{2}\,\xi_0^q) - z_{rp}(\xi_j^p - \sqrt{2}\,\xi_0^p)]\right\}} \tag{13a}$$

$$= Ee^{z'\xi_0} \prod_{j=1}^{\infty} \prod_{\{r,q,p\}} e^{-(\frac{1}{\pi j})^2 [z_{qr}(\xi_j^q - \sqrt{2}\,\xi_0^q) - z_{rp}(\xi_j^p - \sqrt{2}\,\xi_0^p)]^2} \quad . \tag{13b}$$

For this example dual subscripting on the three ζ variables is no longer useful. We will employ the following notation

$$z \triangleq (z_1, z_2, z_3) \tag{14a}$$

$$\zeta \triangleq (\zeta_1, \zeta_2, \zeta_3) \text{ in place of } (\zeta_{23}, \zeta_{31}, \zeta_{12}) \tag{14b}$$

$$|\zeta| \triangleq \sqrt{\zeta_1^2 + \zeta_2^2 + \zeta_3^2} \tag{14c}$$

$$\xi_j \triangleq (\xi_j^1, \xi_j^2, \xi_j^3) \qquad j \geq 0 \, . \tag{14d}$$

The next step is, for each $j \geq 1$, to collect the quadratic forms in each vector $\xi_j - 2^{1/2}\xi_0$. We express $\phi(z, \zeta)$ using an infinite product of exponentials of quadratic forms

$$\phi(z, \zeta) = Ee^{iz'\xi_0} \prod_{j=1}^{\infty} e^{-1/2(\xi_j - 2^{1/2}\xi_0)' A_j (\xi_j - 2^{1/2}\xi_0)} \tag{15}$$

where A_j is the 3×3 matrix

$$A_j = (\pi j)^{-2} [|\zeta|^2 I - \zeta\zeta'] \tag{16a}$$

$$= (|\zeta| / \pi j)^2 \, (I - \zeta\zeta' / |\zeta|^2) \, . \tag{16b}$$

For short we are writing A_j instead of $A_j(\zeta)$. Notice

$$A_j^k = (|\zeta| / \pi j)^{2k} \, (I - \zeta\zeta' / |\zeta|^2) \, . \tag{17}$$

Expectation over the ξ_j Variates

To take the expectation over each ξ_j $(j \geq 1)$, a triple integral is required where the integrand of the exponent is

$$-1/2[(\xi_j - 2^{1/2}\xi_0)' A_j(\xi_j - 2^{1/2}\xi_0) + \xi_j' I \xi_j] . \tag{18}$$

To evaluate the triple integral we rewrite (17) as a single quadratic form in ξ_j. Then the triple integral, with ξ_0 fixed, can be viewed as the integral over all of three dimensional space of an *unnormalized* trivariate Gaussian density. The triple integral can be easily evaluated, by comparing the integrand with the corresponding trivariate Gaussian probability density (same mean and variance-covariance matrix) that is properly normalized (integrates to unity).

We begin this evaluation process by rewriting (17) as

$$-1/2[(\xi_j - \mu_j)'(A_j + I)(\xi_j - \mu_j) + 2\xi_0' A_j \xi_0 - \mu_j'(A_j + I)\mu_j] \tag{19}$$

where we solve for μ_j so (17) and (18) are equal. The solution is

$$\mu_j = 2^{1/2}(A_j + I)^{-1} A_j \xi_0 \tag{20}$$

which, when used in the last term of (19), gives

$$-1/2\left\{(\xi_j - \mu_j)'(A_j + I)(\xi_j - \mu_j) + 2\xi_0'[A_j - A_j(I + A_j)^{-1} A_j]\xi_0\right\}. \tag{21}$$

Conditional on the value of ξ_0 we use (21) to get

$$Ee^{-1/2(\xi_j - 2^{1/2}\xi_0)' A_j(\xi_j - 2^{1/2}\xi_0)} = |I + A_j|^{-1/2} e^{-1/2\xi_0' 2[A_j - A_j(I + A_j)^{-1} A_j]\xi_0} . \tag{22}$$

Here the $|I + A_j|$ means the determinant of $I + A_j$.

The formula for $\phi(z, \zeta)$ can now be written

$$\phi(z, \zeta) = Ee^{iz\xi_0 - 1/2\xi_0' B \xi_0} \prod_{j=1}^{\infty} |I + A_j|^{-1/2} \tag{23}$$

where

$$B(\zeta) = 2\sum_{j=1}^{\infty} [A_j - A_j(I + A_j)^{-1} A_j] . \tag{24}$$

For short we will write B instead of $B(\zeta)$.

Closed Form Representation for Infinite Product and Infinite Sum

To express (23) is closed form we need to evaluate an infinite product and an infinite sum. We do the infinite product first. The matrix $I + A_j$ has eigenvalues of 1 with multiplicity one and $1 + (|\zeta|/\pi j)^2$ of multiplicity 2. Therefore

$$\prod_{j=1}^{\infty} |I+A_j|^{-1/2} = \prod_{j=1}^{\infty} (1+(|\zeta|/\pi j)^2)^{-1} \tag{25a}$$

$$= |\zeta|/\sinh|\zeta| . \tag{25b}$$

Now we turn our attention to finding B in closed form. We evaluate B by rewriting (24) as

$$B = 2\sum_{j=1}^{\infty} \{A_j - A_j^2 + A_j^3 - A_j^4 + \cdots\} \tag{26a}$$

$$= 2\sum_{j=1}^{\infty} \sum_{k=1}^{\infty} (-1)^{k+1} (|\zeta|/\pi j)^{2k} [I-(\zeta\zeta'/|\zeta|^2)] \tag{26b}$$

$$= 2\sum_{j=1}^{\infty} (|\zeta|/\pi j)^2 \frac{1}{1+(|\zeta|/\pi j)^2} [I-(\zeta\zeta'/|\zeta|^2)] \tag{26c}$$

$$= 2\sum_{j=1}^{\infty} \frac{|\zeta/\pi|^2}{j^2+|\zeta/\pi|^2} [I-(\zeta\zeta'/|\zeta|^2)] \tag{26d}$$

$$= \left(\frac{|\zeta|}{\tanh|\zeta|} - 1\right) [I-(\zeta\zeta'/|\zeta|^2)] . \tag{26e}$$

The passage from (26d) to (26e) uses the following well known series representation which appears in reference [20]

$$2|\zeta/\pi|^2 \sum_{j=1}^{\infty} \frac{1}{j^2+|\zeta/\pi|^2} = \frac{|\zeta|}{\tanh|\zeta|} - 1 . \tag{27}$$

Expectation over ξ_0

We use the same sort of trick to evaluate the last expectation as we used to evaluate the previous expectation. Here the evaluation is much easier since we have no need to complete the square. We get

$$Ee^{iz'\xi_0 - \frac{1}{2}\xi_0' B\xi_0} = |B+I|^{-\frac{1}{2}} e^{-\frac{1}{2}z'(I+B)^{-1}z} \tag{28}$$

From (26e)

$$I + B = \frac{|\zeta|}{\tanh|\zeta|} I + \left(1 - \frac{|\zeta|}{\tanh|\zeta|}\right)\frac{\zeta\zeta'}{|\zeta|^2}. \tag{29}$$

It is straightforward verification to show

$$(I+B)^{-1} = \frac{\tanh|\zeta|}{|\zeta|} I + \left(1 - \frac{\tanh|\zeta|}{|\zeta|}\right)\left(\frac{\zeta\zeta'}{|\zeta|^2}\right). \tag{30}$$

This inverse has eigenvalues 1, with multiplicity one, and $|\zeta|^{-1}\tanh|\zeta|$, with multiplicity two, so

$$|B+I|^{-\frac{1}{2}} = |\zeta|^{-1}\tanh|\zeta|. \tag{31}$$

Therefore

$$\phi(z,\zeta) =$$

$$\frac{|\zeta|}{\sinh|\zeta|} \cdot \frac{\tanh|\zeta|}{|\zeta|} e^{-\frac{1}{2}\left\{|z|^2 \frac{\tanh|\zeta|}{|\zeta|} + \frac{(z'\zeta)^2}{|\zeta|^2}\left(1 - \frac{\tanh|\zeta|}{|\zeta|}\right)\right\}}. \tag{32}$$

Finally

$$\phi(z,\zeta) = \operatorname{sech}|\zeta| e^{-\frac{1}{2}\left\{|z|^2 \frac{\tanh|\zeta|}{|\zeta|} + \frac{(z'\zeta)^2}{|\zeta|^2}\left(1 - \frac{\tanh|\zeta|}{|\zeta|}\right)\right\}}. \tag{33}$$

3 Proof of the Theorem

3.1 Paring the Listed Variates

It is easy to see that if the theorem is true, it remains true for new variates defined in terms of the original variates V_j listed in the theorem,

$$\tilde{V}_j = \alpha_j V_j + \beta_j \quad (\alpha_j \text{ and } \beta_j \text{ constants}) . \tag{34}$$

So if the theorem is true for either of the Ito or Stratonovich interpretation, it is true for the other, since the difference in interpretation amounts to a constant offset. The only terms sensitive to the choice of interpretation, are the $\int W_k(t)dW_k(t)$ terms, and since linear combinations of 1 and $W_k^2(1)$, span both interpretations it suffices to prove the theorem with $\int_0^1 W_k(t)dW_k(t)$ omitted from the variates listed in the theorem. For $l \neq k$, it follows from the integration by parts formula

$$\int W_l(t)dW_k(t) = W_l(1)W_k(1) - \int W_k(t)dW_l(t) \tag{35}$$

that terms of the form $\int W_l(t)dW_k(t)$ can also be dropped from the list. So it suffices to prove the theorem with all terms of the form $\int W_l(t)dW_k(t)$ with $k \geq l$ omitted from the list. It is therefore enough to prove the theorem with only the 21 variates $\{W_k(1)\ 1 \leq k \leq 3;$ $W_l(1)W_k(1) \quad 1 \leq l \leq k \leq 3; \quad W_l(1)\int W_k(t)dt \quad 1 \leq l, \quad k \leq 3;$ $\int W_l(t)dW_k(t)\ 1 \leq l < k \leq 3\}$ on the list.

Let $\hat{V}$ be a 21 dimensional random vector with each of the 21 listed variates the sole occupant of one component. Let U be the 12 dimensional vector with the $W_k(t)\int W_l(t)dt$ variates occupying the first nine positions in the order $(k,l)=(1,1), (1,2) \cdots (3,3)$ and the $\int_0^1 W_k(t)dW_l(t)$ variates occupying the last three coordinates in the order $(k,l)=(2,3), (1,3), (1,2)$. We shall see that we can essentially limit our consideration to U.

Before we show how to use U to complete the proof of the theorem we need the following notation:

The twelve dimensional vector transform variate:

$$Z = (z_{11}, z_{12}, \ldots, z_{33}, \zeta_{23}, \zeta_{13}, \zeta_{23}) \tag{36a}$$

and random vectors, the first three dimensional, and the last two infinite dimensional with all components i.i.d. standard normal:

$$\xi_0 = (\xi_0^1, \xi_0^2, \xi_0^3) \tag{37b}$$

$$\xi = (\xi_1^1, \xi_1^2, \xi_1^3, \xi_2^1, \xi_2^2, \xi_2^3, \cdots) \tag{37c}$$

$$\eta = (\eta_1^1, \eta_1^2, \eta_1^3, \eta_2^1, \eta_2^2, \eta_2^3, \cdots). \tag{37d}$$

Suppose, upon taking the two inner expectations on the right-hand side of

$$E^{iZU} = E_{\xi_0} E_\xi E_\eta e^{iZU} \tag{38}$$

we have an equation of the form

$$Ee^{iZU} = E_{\xi_0} e^{\xi_0' C \xi_0 + c' \xi_0 + \gamma} \tag{39}$$

with the entries of the matrix C, the vector c, and the scalar γ are deterministic closed form expressions involving the transform variables. Then the expectation in (39) can clearly be carried out to express Ee^{iZU} in closed form. Consequently, the characteristic function of $\hat{V}$ is also expressible in closed form. So we need only establish the form of (39) and that C,c and γ are expressible in closed form. We refer to the exponent on the right-hand-side of (39) as a quadratic form in ξ_0 (meaning second order *and* lower).

Expectation With Respect to η

We proceed now to take the expectation with respect to η and then with respect to ξ. To begin this process for η, we look at iZU and use (8) and the (p,q,r) cycle notation of Example II to express the η_j^r term

$$i\left\{\frac{\eta_j^r}{2\pi j}\,[\zeta_p(\xi_j^q - 2^{1/2}\xi_0^q) - \zeta_q(\xi_j^p - 2^{1/2}\xi_0^p) + 2^{1/2}(z_{1r}\xi_0^1 + z_{2r}\xi_0^2 + z_{3r}\xi_0^3)]\right\}. \tag{40}$$

For $s = 1,2,3$ let

$$\hat{\xi}_j^s \triangleq (\xi_j^s - 2^{½}\xi_0^s) \tag{41a}$$

$$\hat{\xi}_j \triangleq (\hat{\xi}_j^1, \hat{\xi}_j^2, \hat{\xi}_j^3) \tag{41b}$$

$$z_s \triangleq (z_{1s}, z_{2s}, z_{3s})' \tag{41c}$$

$$\Theta_j^r \triangleq \text{term in } \{\} \text{ in equation (40)} \tag{41d}$$

$$\mu_s \triangleq 2^{½} z_s' \xi_0 \tag{41e}$$

$$\mu^2 \triangleq \mu_1^2 + \mu_2^2 + \mu_3^2 \,. \tag{41f}$$

Using (41d), independence of the η_j, and the characteristic function formula for a univariate Gaussian we get

$$E_\eta e^{iZU} = E_\eta e^{i\sum_{j=1}^{\infty}(\Theta_j^1 + \Theta_j^2 + \Theta_j^3)} \tag{42a}$$

$$= \prod_{j=1}^{\infty} E_{\eta_j} e^{i(\Theta_j^1 + \Theta_j^2 + \Theta_j^3)} \tag{42b}$$

$$= \prod_{j=1}^{\infty} \exp - ½ (2\pi j)^{-2} \{(\Theta_j^1)^2 + (\Theta_j^2)^2 + (\Theta_j^3)^2\} \,. \tag{42c}$$

Working out this exponent in detail, we have

$$-1/2(2\pi j)^{-2}\{(\Theta_j^1)^2 + (\Theta_j^2)^2 + (\Theta_j^3)^2\} = \tag{43a}$$

$$-1/2(2\pi j)^{-2} \sum_{\{(p,q,r)\}} \left\{\zeta_p \hat{\xi}_j^q - \zeta_q \hat{\xi}_j^p + 2^{½}(z_r'\xi_0)\right\}^2 = \tag{43b}$$

$$-1/2(2\pi j)^{-2} \sum_{\{(p,q,r)\}} \Big[\zeta_p^2(\hat{\xi}_j^q)^2 - 2\zeta_p\zeta_q\hat{\xi}_j^q\hat{\xi}_j^p$$

$$+ \zeta_q^2(\hat{\xi}_j^p)^2 + 2\zeta_p\hat{\xi}_j^q\mu_r - 2\zeta_q\hat{\xi}_j^p\mu_r + \mu_r^2\Big] \,. \tag{43c}$$

Expectation With Respect to ξ

To proceed with the proof that the form of $E_\xi E_\eta e^{iZU}$ is as indicated in (39), collect terms in (43c) that involve ξ_j with j an arbitrary fixed index strictly greater than zero. In carrying out the integration required for taking the expectation with respect to ξ_j the exponent of the integrand is obtained by subtracting $-1/2(\xi_j' I \xi_j)$ from (43c). Employing A_j from Example II, we rewrite the integrand exponent as

$$-1/2\Big\{(\xi_j - 2^{1/2}\xi_0)' A_j (\xi_j - 2^{1/2}\xi_0) + \xi_j'\xi_j +$$

$$(2\pi j)^{-2}[2(\zeta_3\mu_2 - \zeta_2\mu_3,\ \zeta_1\mu_3 - \zeta_3\mu_1,\ \zeta_2\mu_1 - \zeta_1\mu_2)'(\xi_j - 2^{1/2}\xi_0)]$$

$$+|\mu/2\pi j|^{-2}\Big\}. \tag{44}$$

Let

$$\alpha = 2(\zeta_3\mu_2 - \zeta_2\mu_3, \zeta_1\mu_3 - \zeta_3\mu_1, \zeta_2\mu_1 - \zeta_1\mu_2). \tag{45}$$

We complete the square for (44) rewriting it as

$$-1/2\Big\{(\xi_j - \omega_j)'(A_j + I)(\xi_j - \omega_j) - \omega_j'(A_j + I)\omega_j$$

$$+ 2\xi_0' A_j \xi_0 - (2\pi j)^{-2} 2^{1/2} \alpha' \xi_0 + (2\pi j)^{-2} |\mu|^2\Big\} \tag{46}$$

where ω_j is defined to be the vector required for (44) and (46) to be equal.

If we sum the last three terms in (46)

$$-1/2 \sum_{j=1}^{\infty} [2\xi_0' A_j \xi_0 - (2\pi j)^{-2} 2^{1/2} \alpha' \xi_0 + (2\pi j)^{-2} |\mu(\xi_0)|^2] \tag{47}$$

it is evident that we obtain a closed form quadratic form in ξ_0. Summing the first term in (46) over $j = 1, 2, \ldots$ gives rise to a closed form multiplier that does not depend on ξ_0 in the expression for the characteristic function (just as in the examples).

Checking for Closed Form

From the above paragraph, the only term we need to consider further to see if it is a closed form quadratic form in ξ_0 is

$$-1/2\sum_{j=1}^{\infty}\omega_j'(A_j+I)\omega_j \tag{48}$$

where the vector ω_j', which gives equality between (44) and (46), is

$$\omega_j' = (2^{1/2}\xi_0'A_j - 1/2(2\pi j)^{-2}\alpha'(I+A_j)^{-1} . \tag{49}$$

Writing out (46), with the ω_j expression from (49) substituted, gives

$$-1/2\left\{(2^{1/2}A_j\xi_0 - 1/2(2\pi j)^{-2}\alpha)'(I+A_j)^{-1}(2^{1/2}A_j\xi_0 - 1/2(2\pi j)^{-2}\alpha)\right\}. \tag{50}$$

Multiplying out, we get four terms in the brackets, two of which are the same. The distinct terms are

$$2\,\xi_0^1\,A_j(I+A_j)^{-1}A_j\,\xi_0 \tag{51a}$$

$$-2^{1/2}(2\pi j)^{-2}\alpha'(I+A_j)^{-1}A_j\,\xi_0 =$$

$$-2^{1/2}(2\pi j)^{-2}\alpha'[I-(I+A_j)^{-1}]\xi_0 \tag{51b}$$

$$1/4(2\pi j)^{-4}\alpha^1(I+A_j)^{-1} . \tag{51c}$$

The middle equality comes from replacing A_j by $I+A_j-I$. From Example II, upon summing over $j \geq 1$, expression (51a) leads to a closed form expression.

Looking at (51b) and (51c) it is enough to check that the following two matrices can be expressed in closed form

$$\sum_{j=1}^{\infty}(2\pi j)^{-2}(I+A_j)^{-1} \tag{52a}$$

and

$$\sum_{j=1}^{\infty}(2\pi j)^{-4}(I+A_j)^{-1} . \tag{52b}$$

A simple check shows

$$(I+A_j)^{-1} = \frac{I}{1+|\zeta/\pi j|^2} + \frac{\zeta\zeta'}{(\pi j)^2(1+|\zeta/\pi j|^2)}. \tag{53}$$

From (53), closed form for (52a) and (52b) turns on closed form for the left-hand sides of

$$\sum_{j=1}^{\infty}(|\zeta/\pi|^2+j^2)^{-1} = 1/2|\pi/\zeta|^2\left(\frac{|\zeta|}{\tanh|\zeta|}-1\right) \tag{54a}$$

$$\sum_{j=1}^{\infty}j^{-2}(|\zeta/\pi|^2+j^2)^{-1} = |\pi/\zeta|^2[(\pi^2/6)+1/2|\pi/\zeta|^2] - (\pi/2)|\pi/\zeta|^3\coth|\zeta| \tag{54b}$$

$$\sum_{j=1}^{\infty}j^{-4}(|\zeta/\pi|^2+j^2)^{-1} = (\pi/|\zeta|)^2\sum_{j=1}^{\infty}(j^{-4}-j^{-2}((\zeta/\pi)^2+j^2)^{-1}). \tag{54c}$$

The right-hand side of (54a) comes from (27), which comes from reference [20] as does (54b). Equation (54c) is just elementary algebra. The right-hand side of (54c) is closed form from (54b) and the result that $\sum_{j=1}^{\infty} j^{-4} = (\pi^4/90)$ as given in reference [19].

4 Closing Remarks

By generalizing beyond the examples, we demonstrated that closed form joint characteristic functions for certain Wiener functionals are not just a fluke. No attempt was made to find a maximal list in the statement of the theorem and we would be extremely surprised if the list could not be greatly expanded. From [21], we see that the joint characteristic function of $W_m(1)$, $\int W_m(t)dt$, $\int W_m^2 dt$ can be expressed in closed form. While there are many ways to seek to generalize the theorem, perhaps the next logical step is to check and see if the theorem remains true if all random variables of the form

$$\int W_k(t)\,dt,\ \int W_k^2(t)\,dt,\ \int W_k(t)\,dt \int W_l(t)\,dt$$

are added to the list.

Acknowledgements

J. M. Steele posed the question of finding the distribution of the area integral solved in (3) to one of us. C. L. Mallows has given a different derivation of (3) as a corollary of other results. Neal Madras and Jim Pitman pointed out some of the references. The problem of finding the characteristic function in Example II arose in (joint) work with Craig Poole [14].

Bibliography

[1] L. A. Shepp, "Radon-Nikodym derivatives of Gaussian measures," *Ann. Math. Statist.*, vol. 37, pp. 321-354, 1966.

[2] S. Karlin and H. M. Taylor, "A Second Course in Stochastic Process," New York: Academic, ch. 15, 1981.

[3] F. B. Knight, "Essentials of Brownian Motion and diffusion," *American Mathematical Society*, pp. 12-20, 1981.

[4] W. Kaplan, "Advanced Calculus," *Addison-Wesley*, Massachusetts, Chapter IV, 1959.

[5] P. Lévy, Wiener's random function and other Laplacian random functions, Proceedings Second Berkeley Symposium, Mathematical Statistics, Probability Volume II, University of California, 171-186, 1950.

[6] N. Wiener, "Une probleme de probabilities denombrables," Bullitin Societe Mathematique France, Vol 52, pp. 569-578, 1924.

[7] D. C. Khandekar and F. W. Wiegel, Distribution of the area enclosed by a plane random walk, J. Phys. A: Math. Gen. 21 (1988), L 563-566.

[8] Jim Pitman and Marc Yor, A decomposition of Bessel bridges, Z. Wahr. verw. Gebiete 59, 425-457, 1982.

[9] J. F. Le Gall and Marc Yor, Enlacements du mouvement Brownian autour des combes de l'espace, Trans. AMS, 317, 687-722, 1990.

[10] P. Biane and M. Yor, Variations sur une formule de Paul Lévy, Annals Institute Henri Poincare 23, 359-377, 1987.

[11] M. Yor, On stochastic areas and averages of planar Brownian motion, J. Phys. A: Math. Gen. 22, 3049-3057, 1989.

[12] B. Duplantier, Areas of planar Brownian curves, J. Phys. A: Math. Gen. 22, 3033-3048, 1989.

[13] P. Biane and M. Yor, A relation between Lévy's stochastic area formula, Legendre polynomials and some continued fractions of Gauss, Technical Report no. 74, Department of Statistics, Berkeley, 1986.

[14] G. J. Foschini and C. D. Poole, "Statistical theory of polarization dispersion in single mode fibers," to be published.

[15] A. H. Jazwinski, "Stochastic processes and filtering theory," *Academic Press*, New York, Chapter IV, 1970.

[16] L. Arnold, "Stochastic differential equations theory and application," John Wiley and Sons, New York, Chapters 4 and 10, 1974.

[17] A. Erdelyi, W. Magnus, F. Oberhettinger, F. G. Tricomi, "Tables of integral transforms," *Bateman Manuscript Project*, vol. I, McGraw-Hill, 1954.

[18] J. Delporte, "Fonctions aléatories presque sûrement continues sur un intervalle fermé," *Ann. Inst. Henri Poincare´ Sec. B1* 111-215,1964.

[19] I. S. Gradshteyn and I. M. Ryzhik, "A table of integrals series and products," *Academic Press*, 1980.

[20] A. D. Wheelon, "Tables of summable series and integrals involving Bessel functions," *Holden-Day Inc.*, Chapters 2 and 5, 1968.

[21] G. J. Foschini and G. Vannucci, "Characterizing filtered light waves corrupted by phase noise," *IEEE Trans. Inform. Theory*, vol. 34, pp. 1437-1448, Nov. 1988.

Adaptedness and Existence of Occupation Densities for Stochastic Integral Processes in the Second Wiener Chaos

Nikos N. Frangos
Dept. of Mathematics
Hofstra University

Peter Imkeller
Math. Institut der LMU Muenchen

Abstract

Let U be a Skorohod integral process in the second Wiener chaos associated with a square integrable function f on the unit square. We derive a necessary and sufficient integral criterion for the existence of a square integrable occupation density of U in terms of the Hilbert-Schmidt operator associated with f. This criterion trivializes in case U is adapted and thus exhibits the special role played by adaptedness in this area.

1 Introduction

In [2], Berman's idea to employ Fourier analysis for the study of local times was used to give a necessary and sufficient criterion for the existence of occupation densities of Skorohod integral processes U in the second Wiener chaos. U is just a (renormalized) double Ito integral of a square integrable function f on the unit square. The main instrument of the approach consists in a method for translating from stochastics to analysis. The property of the samples of U to oscillate fast enough to "distribute" the places visited smoothly

ISBN 0-12-481005-5

enough on the real axis is translated into a purely analytic language using determinants and resolvents related to the Hilbert-Schmidt operator belonging to f. In this process of transcription stochastically intuitive facts such as the independence of increments of the Wiener process W are replaced at a very early stage by the more analytically oriented orthogonality of Gaussian components. Therefore the integral criterion obtained appears less readily accessible to stochastic intuition. This adds to the fact that it looks, in its interpretation with Fredholm's theory (see [3]), rather delicate: it might just reflect the delicacy of the problem itself. In [3] we were able to apply the criterion successfully in the case of finitely many interacting Gaussian components, and obtained results which seem to be beyond the reach of the stochastic techniques of enlargement of filtration. Yet, surprisingly, it did not present the expected easy answer in the simple and well known case of adapted integrands.

This paper was written to resolve this mystery and shed some light on the special role played by adaptedness or backwards adaptedness. How are these properties reflected analytically? It soon became clear that to answer this question the translation into analysis in [2] was probably done in a too early stage. To counterbalance this, we chose to proceed roughly in the following way.

If u is the integrand, U its integral process, the criterion to be established reads

$$\int_{[-1,1]^c} \int_0^1 \int_0^1 E(\exp(iy(U_t - U_s))u_s^2 u_t^2) ds dt dy < \infty.$$

To be able to take advantage of the independence of increments of W in the most efficient way, in the first step $\int_0^1 \exp(iyU_s)u_s^2 ds$ is described in a formula which in some way is related to the Ito type formula presented in Nualart, Pardoux [6], by some kind of stochastic integral and a random variable. But after having exhausted the stochastic means, we also have to be able to translate into operator theory. For this purpose it is most convenient to start working with matrices instead of their infinite dimensional counterparts. We therefore approximate the stochastic integral mentioned by Riemann sums in which only finite dimensional processes appear. The resulting representation quickly leads to an isolation of the "adapted" contribution and consequently to an integral condition, which is put into analytical terms in the second step.

More formally, we realize the program just sketched in the following way. In section 1, which is purely preparatory, we introduce the finite dimensional integrands u^n and their integral processes U^n. We show that, given a 0-sequence of partitions $(\mathbf{J}_n)_{n\in\mathbf{N}}$ by intervals $J = [s^J, t^J[$, the integral $\int_0^1 \exp(iyU_s)u_s^2 ds$ can be approximated by

$$\frac{2i}{y}X_n + \frac{2}{y^2}(1 - \exp(iyU_1^N)),$$

where

$$X_n = \sum_{J\in\mathbf{J_n}} \exp(iyU_{s^J}^n)u_{s^J}^n(W_{t^J} - W_{s^J}).$$

In section 2, using this description, we single out the essential parts of Berman's condition. This way it is seen to be equivalent to the integrability of $y \mapsto y^{-2}\lim_{n\to\infty} E(|X_n|^2)$ on $[-1,1]^c$. We then isolate the adapted contribution of $E(|X_n|^2)$ and thus obtain in stochastic terms a necessary and sufficient criterion for the existence of occupation densities, which is trivial in the adapted case (theorem 2.1).

After the translation based on [2] is performed (section 3), it is given in purely operator theoretic terms in theorem 3.1.

We expect no essential problem in the extension of the first two sections to the whole Wiener space. It is, however, not yet clear what can be done about section 3. It seems likely that the direct arguments concerning the whole information on the distribution of Skorohod integral processes can be replaced by techniques related to partial integration on Wiener space.

2 Notations and Conventions

We shall deal with the Wiener process W indexed by $[0,1]$, defined on some fixed probability space $(\Omega, \mathbf{F}, P)$, and its stochastic integrals in the second chaos. More precisely, for $f \in L^2([0,1]^2)$ we consider integrands $u_t = \int_0^1 f(s,t)dW_s$ and their "Skorohod integral processes" $U_t = \delta(1_{[0,t]}u), \quad t \in [0,1]$. Here $\delta(1_{[0,t]}u)$ is simply the double Ito integral of the function $1_{[0,t]}f$. As for δ, we will refer to Nualart, Pardoux [6] for some more relevant notions and their basic properties we adopt from the calculus of Skorohod's integral. For example, we will need some elementary facts about the Malliavin derivative D, confined to the second chaos, and the space

$$\mathbf{L}_{2,1} = \{u \in L^2(\Omega \times [0,1]), \int_0^1 \| u_t \|_2^2 \, dt + \int_0^1 \int_0^1 \| D_s u_t \|_2^2 \, ds dt < \infty\}$$

where $\| \, . \, \|_p$ denotes the p-norm in $L^p(\Omega, \mathbf{F}, P), p \geq 1$.

A sequence of partitions $(\mathbf{J}_n)_{n \in \mathbf{N}}$ of $[0,1]$ by intervals $J = [s^J, t^J]$ is called "0-sequence" if it increases with respect to fineness and its mesh converges to 0. For the increment of a process X over an interval J we write $\Delta_J X$. For functions f on $[0,1]^2$, matrices A, operators T, transposes are denoted by f^*, A^*, T^*, respectively. The symbol I is used for both the unit matrix and the identity operator. If f,g are L^2-kernels on $[0,1]^2$, we write

$$fg(u,v) = f(u,.)g(.,v) = \int_0^1 f(u,w)g(w,v)dw, \quad (u,v) \in [0,1]^2,$$

for their product kernel. At places, Lebesgue measure on the Borel sets of $[0,1]$ is written λ. For the more sophisticated notions of operator theory, such as the trace (denoted by tr), the regularized Fredholm determinant (denoted by det_2) or the second Fredholm "minor" (denoted by D_2) of an operator, we refer to Simon [9]. If z is a complex number, Re(z) (Im(z)) are its real (imaginary) parts.

3 A Finite Dimensional Approximation

Let $f \in L^2([0,1]^2)$, u and U according to the preceding section. In case f is zero off the lower triangle in $[0,1]^2$, i.e. $f(u,v) = 1_{[0,v]}(u)f(u,v), \quad u,v \in [0,1]$, u is adapted and consequently U an Ito integral process in the classical sense. In [2], a necessary and sufficient criterion for the existence of a square integrable occupation density of U was derived. It states that such a density exists iff

$$\int_{\mathbf{R}} \int_0^1 \int_0^1 E(\exp(iyU_t - U_s)u_s^2 u_t^2) ds dt dy < \infty. \tag{1}$$

To tackle (1), we shall approximate u and U by "finite dimensional" processes u^n, U^n so that (1) can essentially be checked for the approximates instead of u,U. This will provide a good basis for the use of stochastic arguments as well as for the transcription into operator theory later.

For simplicity, from now on we will make the assumption that f is continuous. Only at the very end we shall make rather trivial extensions to non-continuous functions as well. To give the finite dimensional approximations of u,U, we let $(\mathbf{J}_n)_{n\in\mathbf{N}}$ be a 0-sequence of partitions of $[0,1]$. Define

$$f^n(u,v) = \sum_{J,K\in\mathbf{J}_n, J\neq K} f(s^J,s^K)1_J(u)1_K(v), \quad u,v\in[0,1], n\in\mathbf{N}.$$

and $u_t^n = \sum_{J,K\in\mathbf{J}_n} f^n(s^J,s^K)\Delta_J W 1_K(t)$,
$U_t^n = \sum_{J,K\in\mathbf{J}_n)} f^n(s^J,s^K)\Delta_J W \Delta_K W 1_{[0,t]}(t^K), \quad t\in[0,1], n\in\mathbf{N}$.
Some consequences of the calculus of Skorohod's integral follow.

Proposition 3.1 *For any $p\geq 2$ there exists a constant c_p such that for any continuous g on $[0,1]^2$, $v_t = \int_0^1 g(s,t)dW_s$,
$V_t = \delta[1_{[0,t]}v), \quad t\in[0,1]$, we have*

$$\| v_t \|_p^p \leq c_p \| g \|_p^p, \qquad \| V_t \|_p^p \leq c_p \| g \|_2^p, \quad t\in[0,1].$$

Proof: The first inequality is a consequence of Burkholder's inequalities, whereas the second follows from Nualart, Pardoux [6], p. 544, and the simple equation $D_s v_t = g(s,t), \quad s,t\in[0,1]$.

We next compute Malliavin derivatives of the functions we will mainly be concerned with.

Proposition 3.2 *For $y\in\mathbf{R}, s,t\in[0,1], g,v,V$ as in prop. 3.1.*

$$D_s \exp(iyV_t) = iy\exp(iyV_t)[1_{[0,t]}(s)v_s + \int_0^t g(s,v)dW_v].$$

Proof: According to proposition 5.5 of Nualart, Pardoux [6], we have

$$D_s V_t = 1_{[0,t]}(s)v_s + \int_0^t g(s,v)dW_v.$$

Now apply the chain rule for Malliavin derivatives (cf. for example Nualart, Pardoux [6], p. 539).

For the Malliavin derivatives needed we have the following inequalities.

Proposition 3.3 *For $p \geq 2$ there exists a constant c_p such that for all g, v, V as in proposition 3.1, all $y \in \mathbf{R}$, all measurable $b : [0,1] \to [0,1]$ we have*

$$\int_0^1 \int_0^1 \| D_s \exp(iyV_{b(t)}) \|_p^p \, dsdt \leq c_p \mid y \mid^p \| g \|_p^p,$$

$$\int_0^1 D_t \exp(iyV_{b(t)}) dt \leq c_p \mid y \mid^p \| g \|_p^p .$$

Proof: By proposition 3.2 and boundedness of exp on the unit circle we have

$$\int_0^1 \int_0^1 \| D_s \exp(iyV_{b(t)}) \|_p^p \, dsdt$$

$$\leq 2 \mid y \mid^p [\int_0^1 \| v_{b(t)} \|_p^p \, dt + \int_0^1 \int_0^1 \| \int_0^{b(t)} g(s,v) dW_v \|_p^p \, dsdt].$$

Now apply proposition 3.1 twice, to g and to $1_{[0,b(t)]} g$. The second inequality is proved in the same way.

Remark: In proposition 3.3, we are only interested in the choice $b(t) = s^J$ for $t \in J, J \in \mathbf{J}_n, n \in \mathbf{N}$.

Proposition 3.4 *For $p \geq 2$ there exists a constant c_p such that for all g, v, V as in proposition 3.1 and g',v', V' correspondingly, all $y \in \mathbf{R}$, all measurable $b : [0,1] \to [0,1]$ we have*

$$\int_0^1 \int_0^1 \| D_s[\exp(iyV_{b(t)}) - \exp(iyV'_{b(t)})] \|_p^p \, dsdt$$

$$\leq c_p[\mid y \mid^p \| g - g' \|_p^p + \mid y \mid^{2p} \| g - g' \|_2^p (\| g \|_{2p}^p + \| g' \|_{2p}^p)],$$

$$\int_0^1 \| D_t[\exp(iyV_{b(t)}) - \exp(iyV'_{b(t)})] \|_p^p \, dt$$

$$\leq c_p[\mid y \mid^p \| g - g' \|_p^p + \mid y \mid^{2p} \| g - g' \|_2^p (\| g \|_{2p}^p + \| g' \|_{2p}^p)].$$

Proof: Proposition 3.2 gives for $s, t \in [0,1]$

$$\begin{aligned}
D_s[\exp(iyV_t) - \exp(iyV_t')] \\
&= iy\exp(iyV_t)[1_{[0,t]}(s)v_s + \int_0^t g(s,v)dW_s] \\
&-iy\exp(iyV_t')[1_{[0,t]}(s)v_s' + \int_0^t g'(s,v)dW_v] \\
&= iy\exp(iyV_t)[1_{[0,t]}(s)(v_s - v_s') + \int_0^t (g-g')(s,v)dW_v] \\
&+iy[\exp(iyV_t) - \exp(iyV_t')][1_{[0,t]}(s)v_s' + \int_0^t g'(s,v)dW_v]. \quad (2)
\end{aligned}$$

Let us estimate the two expressions on the right hand side of (2) for b(t) instead of t separately. The estimate for the first one is given by proposition 3.3. More precisely, for a universal constant c_p'

$$\int_0^1 \int_0^1 \| iy\exp(iyV_{b(t)})[1_{[0,t]}(s)(v_s - v_s') + \int_0^t (g-g')(s,v)dW_v] \|_p^p \, dsdt \le c_p' \mid y \mid^p \| g - g' \|_p^p .$$

To estimate the second expression, we make use of the mean value theorem applied to the function $z \mapsto \exp(iyz)$, and the inequality of Cauchy-Schwarz. This combination yields

$$\begin{aligned}
\int_0^1 \int_0^1 \| iy[\exp(iyV_{b(t)}) - \exp(iyV_{b(t)}')](1_{[0,b(t)]}(s)v_s' \\
+ \int_0^{b(t)} g'(s,v)dW_v) \|_p^p \, dsdt \\
\le 2 \mid y \mid^{2p} [\int_0^1 \int_0^1 \||| V_{b(t)} - V_{b(t)}' \| v_s' \|||_p^p \\
+ \||| V_{b(t)} - V_{b(t)}' \| \int_0^{b/t)} g'(s,v)dW_v \|||_p^p \, dsdt \\
\le 2 \mid y \mid^{2p} [(\int_0^1 \| V_{b(t)} - V_{b(t)}' \|_{2p}^{2p} dt \int_0^1 \| v_s' \|_{2p}^{2p} ds)^{1/2} \\
+(\int_0^1 \| V_{b(t)} - V_{b(t)}' \|_{2p}^{2p} dt \int_0^1 \int_0^1 \| \int_0^{b(t)} g'(s,v)dW_v \|_{2p}^{2p} dsdt)^{1/2}] \\
\le c_p'' \mid y \mid^{2p} \| g - g' \|_2^p (\| g \|_{2p}^p + \| g' \|_{2p}^p) \qquad (\textit{proposition } 3.1)
\end{aligned}$$

with a universal constant c_p''.

These two estimates combine to yield the first one of the asserted inequalities. The proof of the second one is analogous and easier.

In the situations of interest we get the following corollaries.

Corollary 3.1 *For $p \geq 2$ there exists a constant c_p such that for all $n \in \mathbf{N}, y \in \mathbf{R}$ we have*

$$\int_0^1 \int_0^1 \sum_{J \in \mathbf{J}_n} 1_J(t) \parallel D_s \exp(iyU^n_{s^J}) \parallel_p^p dsdt \leq c_p \mid y \mid^p \parallel f^n \parallel_p^p,$$

$$\int_0^1 \sum_{J \in \mathbf{J}_n} 1_J(t) \parallel D_t \exp(iyU^n_{s^J}) \parallel_p^p dt \leq c_p \mid y \mid^p \parallel f^n \parallel_p^p .$$

Proof: Apply proposition 3.3 to $g = f^n, b(t) = s^J$ for $t \in J, J \in \mathbf{J}_n$.

Corollary 3.2 *For $p \geq 2$ there exists a constant c_p such that for all $n \in \mathbf{N}, y \in \mathbf{R}$ we have*

$$\int_0^1 \int_0^1 \sum_{J \in \mathbf{J}_n} 1_J(t) \parallel D_s[\exp(iyU^n_{s^J}) - \exp(iyU_{s^J}) \parallel_p^p dsdt$$
$$\leq c_p[\mid y \mid^p \parallel f - f^n \parallel_p^p + \mid y \mid^{2p} \parallel f - f^n \parallel_2^p (\parallel f^n \parallel_{2p}^p + \parallel f \parallel_{2p}^p)],$$
$$\int_0^1 \sum_{J \in \mathbf{J}_n} 1_J(t) \parallel D_t \exp(iyU^n_{s^J}) - \exp(iyU_{s^J}) \parallel_p^p dt$$
$$\leq c_p[\mid y \mid^p \parallel f - f^n \parallel_p^p + \mid y \mid^{2p} \parallel f - f^n \parallel_2^p (\parallel f^n \parallel_{2p}^p + \parallel f \parallel_{2p}^p)].$$

Proof: Apply proposition 3.4 to $g = f^n, g' = f, b(t) = s^J$ for $t \in J, J \in \mathbf{J}_n$.

From corollary 3.2 we get the following asymptotic equivalence.

Proposition 3.5 *Let $y \in \mathbf{R}$. Then (in the L^2- sense)*

$$\lim_{n \to \infty} \exp(iyU_1^n) = \exp(iyU_1),$$
$$\lim_{n \to \infty} \sum_{J \in \mathbf{J}_n} [\exp(iyU^n_{s^J}) u^n_{s^J} \Delta_J W - \exp(iyU_{s^J}) \Delta_J U] = 0.$$

Proof: By the mean value theorem and proposition 3.1 we get

$$\| \exp(iyU_1^n) - \exp(iyU_1) \|_2^2 \leq | y |^2 \| U_1^n - U_1 \|_2^2 \leq c_2 | y |^2 \| f^n - f \|_2^2,$$

which converges to 0 by continuity of f. For the more difficult second assertion, note that for $n \in \mathbf{N}, J \in \mathbf{J}_n$ due to the trace-free definition of f^n, $u^n_{s^J} \Delta_J W = \delta(1_J u^n)$ and by Nualart, Pardoux [6] p. 542,

$$\begin{aligned}
&\sum_{J \in \mathbf{J}_n} [\exp(iyU^n_{s^J}) u^n_{s^J} \Delta_J W - \exp(iyU_{s^J}) \Delta_J U] \\
&\quad = \sum_{J \in \mathbf{J}_n} [\exp(iyU^n_{s^J}) - \exp(iyU_{s^J})] \delta(1_J u) \\
&\quad + \sum_{J \in \mathbf{J}_n} \exp(iyU^n_{s^J}) \delta(1_J(u^n - u)) \\
&\quad = \delta(v^n) + \sum_{J \in \mathbf{J}_n} \int_J u_t D_t[\exp(iyU^n_{s^J}) - \exp(iyU_{s^J})] dt \\
&\quad + \delta(w^n) + \sum_{J \in \mathbf{J}_n} \int_J (u^n - u)_t D_t[\exp(iyU^n_{s^J})] dt, \qquad (3)
\end{aligned}$$

where

$$\begin{aligned}
v^n &= \sum_{J \in \mathbf{J}_n} 1_J u [\exp(iyU^n_{s^J}) - \exp(iyU_{s^J})], \\
w^n &= \sum_{J \in \mathbf{J}_n} 1_J (u^n - u) \exp(iyU^n_{s^J}).
\end{aligned}$$

Let us take care of the second and fourth terms in (3) first. To abbreviate, we denote $Y^n = \exp(iyU^n), Y = \exp(iyU), b(t) = s^J$ for $t \in J \in \mathbf{J}_n$. The inequality of Cauchy-Schwarz gives

$$\begin{aligned}
\| \int_0^1 u_t D_t[Y^n_{b(t)} - Y_{b(t)}] dt \|_2 &\leq \| \int_0^1 u_t^2 \, | D_t[Y^n_{b(t)} - Y_{b(t)}] |^2 \, dt \|_1 \\
&\leq (\int_0^1 \| u_t \|_4^4 \, dt \quad \int_0^1 \| D_t[Y^n_{b(t)} - Y_{b(t)}] \|_4^4 \, dt)^{1/2}, \quad (4)
\end{aligned}$$

$$\| \int_0^1 (u^n - u) D_t Y^n_{b(t)} dt \|_2$$

$$\leq \| \int_0^1 (u^n - u)_t^2 \mid D_t Y_{b(t)}^n \mid^2 dt \|_1$$

$$\leq \left(\int_0^1 \| (u^n - u) \|_4^4 \, dt \int_0^1 \| D_t Y_{b(t)}^n \|_4^4 \, dt\right)^{1/2}. \tag{5}$$

The last line of (4) converges to 0 as $n \to \infty$ by corollary 3.2, proposition 3.1 and the continuity of f, whereas for the convergence of the last line of (5) we may appeal to corollary 3.1, proposition 3.1 and continuity. This takes care of the second and fourth terms in (3). To treat the first and third, according to Nualart, Pardoux [6], p. 543, we have to show that $v^n \to 0, w^n \to 0$, both in $\mathbf{L}_{2,1}$, as $n \to \infty$. According to their definition and the boundedness of exp on the unit circle, convergence in $L^2(\Omega, \mathbf{F}, P)$ is no problem. Just apply proposition 3.1 and the mean value theorem as we already did. Let us concentrate on proving the more difficult

$$\int_0^1 \int_0^1 \| D_s v_t^n \|_2^2 \, dsdt \to 0,$$

$$\int_0^1 \int_0^1 \| D_s w_t^n \|_2^2 \, dsdt \to 0 \qquad as \qquad n \to \infty. \tag{6}$$

Using the abbreviations introduced above, for $n \in \mathbf{N}, s, t \in [0,1]$

$$D_s v_t^n = D_s u_t [Y_{b(t)}^n - Y_{b(t)}] + u_t D_s [Y_{b(t)}^n - Y_{b(t)}].$$

Hence

$$\int_0^1 \int_0^1 \| D_s v_t^n \|_2^2 \, dsdt$$

$$\leq 2\Big[\int_0^1 \int_0^1 \| D_s u_t [Y_{b(t)}^n - Y_{b(t)}] \|_2^2 \, dsdt$$

$$+ \int_0^1 \int_0^1 \| u_t D_s [Y_{b(t)}^n - Y_{b(t)}] \|_2^2 \, dsdt$$

$$\leq 2\Big[\Big(\int_0^1 \int_0^1 \| D_s u_t \|_4^4 \, dsdt \int_0^1 \| Y_{b(t)}^n - Y_{b(t)} \|_4^4 \, dt\Big)^{1/2}$$

$$+ \Big(\int_0^1 \| u_t \|_4^4 \, dt \int_0^1 \int_0^1 \| D_s [Y_{b(t)}^n - Y_{b(t)}] \|_4^4 \, dsdt\Big)^{1/2}\Big].$$

Now apply the mean value theorem and proposition 3.1 (for the first summand), and proposition 3.1 and corollary 3.2 (for the second summand) to prove the first half of (6). The second half follows similarly. For $n \in \mathbf{N}, s, t \in [0,1]$ we have $D_s w_t^n = D_s(u^n - u)_t Y_{b(t)}^n + (u^n - u)_t D_s Y_{b(t)}^n$. Hence

$$\int_0^1 \int_0^1 \| D_s w_t^n \|_2^2 \, dsdt$$
$$\leq 2[(\int_0^1 \int_0^1 \| D_s(u^n - u)_t \|_4^4 \, dsdt \int_0^1 \| Y_{b(t)} \|_4^4 \, dt)^{1/2}$$
$$+(\int_0^1 \| u_t^n - u_t \|_4^4 \, dt \int_0^1 \int_0^1 \| D_s[Y_{b(t)}^n - Y_{b(t)}] \|_4^4 \, dsdt)^{1/2}].$$

In this case we have to remember $D_s(u^n - u)_t = (f^n - f)(s,t)$ and that exp is bounded on the unit circle (for the first summand), and apply proposition 3.1 and corollary 3.1 to the second to establish the second half of (6). This completes the proof of the proposition.

We have been aiming at the following asymptotic result.

Proposition 3.6 *For $y \neq 0$ we have*

$$2y^{-2}(1 - \exp(iyU_1)) + 2iy^{-1} \lim_{n\to\infty} X_n = \int_0^1 \exp(iyU_s) u_s^2 ds$$

where $X_n = \sum_{J \in \mathbf{J}_n} \exp(iyU_{s^J}^n) u_{s^J}^n \Delta_J W, \qquad n \in \mathbf{N}$

Proof: From the proof of proposition 5.3 of Nualart [7] we obtain

$$1/2y^2 \int_0^1 \exp(iyU_s) u_s^2 ds$$
$$= 1 - \exp(iyU_1) + iy \lim_{n\to\infty} \sum_{J \in \mathbf{J}_n} \exp(iyU_{s^J}) \Delta_J U.$$

Now multliply by $\frac{2}{y^2}$ and apply proposition 3.5.

Remark: The aspect of proposition 3.6, which will be most useful in the sequel, is that X_n is a finite dimensional process. The importance of this property will become clear in section 5. In the following section we will use proposition 3.5 and some stochastic arguments to interpret (1).

4 Stochastic interpretation of Berman's condition

Recall (1). We actually can replace the integration ober **R** by an integration over $[-1,1]^c$ only, according to the following proposition.

Proposition 4.1 *U possesses a square integrable occupation density iff*

$$\int_{[-1,1]^c} \int_0^1 \int_0^1 E(\exp(iy(U_t - U_s)))u_s^2 u_t^2 dsdtdy < \infty. \quad (7)$$

Proof: Since exp is bounded by 1 on the unit circle, we have for $y \in \mathbf{R}$

$$\begin{aligned}
&\int_0^1 \int_0^1 \mid E(\exp(iy(U_t - U_s))u_s^2 u_t^2) \mid dsdt \\
&\quad \leq \int_0^1 \int_0^1 E(u_s^2 u_t^2) dsdt \\
&\quad \leq (\int_0^1 \parallel u_s \parallel_4^4 ds)^{1/2} \leq \sqrt{c_4} \parallel f \parallel_4^2 \qquad (proposition 3.1).
\end{aligned}$$

This immediately implies the equivalence of (1) and (7).

Now proposition 3.6 gives us the opportunity to describe (7) in a different way. For $y \in \mathbf{R}$ we have

$$\begin{aligned}
&\int_0^1 \int_0^1 \exp(iy(U_t - U_s))u_s^2 u_t^2 dsdt \\
&\quad = 4y^{-4}(1 - \exp(iyU_1))(1 - \exp(iy(-U_1))) \\
&\quad +4iy^{-3} \lim_{n\to\infty} [(1 - \exp(iy(-U_1)))X_n - (1 - \exp(iyU_1))\overline{X_n}] \\
&\quad +4y^{-2} \mid X_n \mid^2 . \qquad (8)
\end{aligned}$$

We proceed by simply considering the integrability properties of the terms on the right hand side of (8) separately. The first non-trivial term is the one containing $X_n, \overline{X_n}$ linearly. Due to the fact that it has y^3 in the denominator, a rather rough estimate, performed in the following proposition, is sufficient.

Proposition 4.2 *There is a constant c such that for all $y \in \mathbf{R}, n \in \mathbf{N}$*

$$\| X_n \|_1 \leq c[\| f^n \|_2 + \| f^n \|_4 + | y | (\| f^n \|_2^2 + \| f^n \|_4^2)].$$

Proof: As in the proof of proposition 3.5 we have

$$X_n = \sum_{J \in \mathbf{J}_n} \exp(iyU^n_{s_J})\delta(1_J u^n) = \delta(v^n) + \sum_{J \in \mathbf{J}_n} \int_J u^n_t D_t[\exp(iyU^n_{s_J})]dt$$

where $v^n = \sum_{J \in \mathbf{J}_n} 1_J u^n \exp(iyU^n_{s_J})$, $n \in \mathbf{N}$. Let us use the abbreviations introduced in the proof of proposition 3.5 again. We first estimate the second term, where c_1 is a universal constant.

$$\begin{aligned} &\| \int_0^1 u^n_t D_t Y^n_{b(t)} dt \|_1 \\ &\quad \leq (\int_0^1 \| u^n_t \|_2^2 \, dt \int_0^1 \| D_t Y^n_{b(t)} \|_2^2 \, dt)^{1/2} \\ &\quad \leq c_1 \| f^n \|_2 | y | \| f^n \|_2 \qquad (proposition\ 3.1, corollary\ 3.1)(9) \end{aligned}$$

For the first one, according to Nualart, Pardoux [6], p. 543,

$$\| \delta(v^n) \|_2 \leq (\int_0^1 \| v^n_t \|_2^2 \, dt)^{1/2} + (\int_0^1 \int_0^1 \| D_s v^n_t \|_2^2 \, dsdt)^{1/2}.$$

Now $D_s v^n_t = D_s u^n_t \quad Y^n_{b(t)} + u^n_t \quad D_s Y^n_{b(t)}$. Hence

$$\begin{aligned} &\| \delta(v^n) \|_2 \\ &\quad \leq (\int_0^1 \| u^n_t \|_2^2 \, dt)^{1/2} + (\int_0^1 \int_0^1 \| s \|^d_{D_s u^n_t} \quad Y^n_{b(t)} \, dt)^{1/2} \\ &\quad +(\int_0^1 \int_0^1 \| u^n_t \quad D_s Y^n_{b(t)} \|_2^2 \, dsdt)^{1/2} \\ &\quad \leq \| f^n \|_2 + [\int_0^1 \int_0^1 \| D_s u^n_t \|_4^4 \, dsdt \int_0^1 \| Y^n_{b(t)} \|_4^4 \, dt]^{1/4} \\ &\qquad\qquad (proposition\ 3.1, Cauchy - Schwarz) \\ &\quad +[\int_0^1 \| u^n_t \|_4^4 \, dt \int_0^1 \int_0^1 \| D_s Y^n_{b(t)} \|_4^4 \, dsdt]^{1/4} \\ &\quad \leq \| f^n \|_2 + \| f^n \|_4 + c_2 \| f^n \|_4 | y | \| f^n \|_4 \\ &\qquad\qquad (corollary\ 3.1) \qquad (10) \end{aligned}$$

with a universal constant c_2. Now combine (9) and (10).

By the preceding proposition, the crucial term for the existence of a square integrable occupation density is $\lim_{n\to\infty} E(|X_n|^2)$. To clarify the role of adaptedness as a special case, we will next split the "adapted contribution" off this expression. The "non-adapted contribution" can be given in a symmetric and a non-symmetric description. We prefer the non-symmetric one for its brevity. We could as well give a symmetric description, which might be more conclusive, since it exhibits underlying patterns more clearly.

Proposition 4.3 *For $n \in \mathbf{N}, y \in \mathbf{R}$ we have*

$$\begin{aligned}
E(|X_n|^2) &= \| f^n \|_2^2 \\
&+ 2 \sum_{s^J \le s^K} E(\cos(y(U^n_{s^K} - U^n_{s^J}))u^n_{s^K}\Delta_J W) f^n(s^K, s^J)\lambda(K) \\
&+ 2y \sum_{s^J \le s^K} E(\sin(y(U^n_{s^K} - U^n_{s^J})) \\
&\qquad \delta(1_{[s^J,s^K]} f^n(s^K,.))u^n_{s_j} u^n_{s^K}\Delta_J W)\lambda(K).
\end{aligned}$$

Proof: We have

$$E(|X_n|^2) = \sum_{J,K\in\mathbf{J}_n} E(\exp(iy(U^n_{s^K} - U^n_{s^J})))\delta(1_J u^n)\delta(1_K u^n).$$

Now fix $J, K \in \mathbf{J}_n$ such that $s^J \le s^K$. By partial integration (Nualart, Pardoux [6], p. 542) with the abbreviation $Y^n = \exp(iyU^n)$

$$E(Y^n_{s^K}\overline{Y^n_{s^J}}\delta(1_J u^n)\delta(1_K u^n)) = \int_K E(D_t[Y^n_{s^K}\overline{Y^n_{s^J}}\delta(1_J u^n)]1_K(t)u^n_{s^K}dt.$$

But for $t \in K$, proposition 3.2 gives

$$\begin{aligned}
&D_t[Y^n_{s^K}\overline{Y^n_{s^J}}\delta(1_J u^n)] \\
&= D_t[Y^n_{s^K}\overline{Y^n_{s^J}}]\delta[1_J u^n) + Y^n_{s^K}\overline{Y^n_{s^J}} D_t[\delta(1_J u^n)] \\
&= iyY^n_{s^K}\overline{Y^n_{s^J}}\delta(1_{[s^J,s^K]} f^n(s^K,.))\delta(1_J u^n) \\
&\qquad + Y^n_{s^K}\overline{Y^n_{s^J}}[1_J(t)u^n_{s^K} + f^n(s^K, s^J)\Delta_J W].
\end{aligned}$$

Therefore the balance we get is the following

$$\sum_{s^J \leq s^K} E(Y^n_{s^K} \overline{Y^n_{s^J}} \delta(1_J u^n)\delta(1_K)u^n))$$

$$= iy \sum_{s^J \leq s^K} Y^n_{s^K} \overline{Y^n_{s^J}} \delta(1_{[s^J,s^K]} f^n(s^K,.)) u^n_{s^J} u^n_{s^K} \Delta_J W \lambda(K)$$

$$+ \int_0^1 \| u^n_t \|_2^2 \, dt + \sum_{s^J \leq s^K} Y^n_{s^K} \overline{Y^n_{s^J}} u^n_{s^K} \Delta_J W f^n(s^K, s^J)\lambda(K).$$

Since by isometry $\int_0^1 \| u^n_t \|_2^2 \, dt = \| f^n \|_2^2$, and since the contributions for which $s^J \geq s^K$ are taken care of by reversing the sign in the exponent, the proposition follows from Euler's formula.

It now becomes clear that everything simplifies drastically, as expected, if u and U are adapted.

Remark: If $f(u,v) = 1_{[0,v]}(u) f(u,v)$, i.e. if u and U are adapted, U possesses a square integrable occupation density.

Proof: First, in this case for all $n \in \mathbf{N}, y \in \mathbf{R}$ $f^n(s^K, s^J) = 0$ and $\delta(1_{[s^J,s^K]} f^n(s^K,.)) = 0$ for $s^J \leq s^K$. Hence $E(| X_n |^2) = \| f^n \|_2^2$ by proposition 4.3 and thus $E(| X^n |) \leq \| f^n \|_2$. Consequently (8) gives

$$\int_0^1 \int_0^1 E(\exp(iy(U_t - U_s)) u_s^2 u_t^2) ds dt$$
$$\leq 16 y^{-4} + 32 | y |^{-3} \| f \|_2 + 4 y^{-2} \| f \|_2^2 .$$

Hence proposition 4.1 is applicable.

In case u and U are not adapted, the two additional terms of proposition 4.3 enter the scene and we obtain

Theorem 4.1 *For $y \in \mathbf{R}$ let*

$$A(y) = 2 \lim_{n \to \infty} \sum_{s^J \leq s^K} E(\cos(y(U^n_{s^K} - U^n_{s^J})) u^n_{s^K} \Delta_J W) f^n(s^K, s^J) \lambda(K),$$

$$B(y) = 2y \lim_{n \to \infty} \sum_{s^J \leq s^K} E(\sin(y(U^n_{s^K} - U^n_{s^J}))$$
$$\delta(1_{[s^J,s^K]} f^n(s^K,.)) u^n_{s^J} u^n_{s^K} \Delta_J W) \lambda(K).$$

Then U possesses a square integrable occupation density iff

$$\int_{[-1,1]^c} (A(y) + B(y))y^{-2}dy < \infty.$$

Proof: By (8) and proposition 4.2, (7) is equivalent with

$$\int_{[-1,1]^c} \lim_{n\to\infty} E(| X^n |^2)y^{-2}dy < \infty.$$

Now apply proposition 4.3.

Stochastically, A(y) and B(y) are not very transparent any more. Therefore, at this point, we will start the translation into purely analytical terms, using the device developed in [2]. This will be done in the following section.

5 The analytical integral criterion

Here it will finally turn out to be important to have finite dimensional processes such as u^n and U^n. This fact will enable us to use matrices in the first step of the translation procedure. In the second step those will be seen to converge to appropriate infinite dimensional operators. Let us first introduce the ones we need.

For $s, t \in [0,1], n \in \mathbf{N}$, let

$$H(s,t)(u,v) = f1_{[s,t]}(u,v) + 1_{[s,t]}f^*(u,v),$$

$$H^n(s,t)(u,v) = f^n 1_{[s^J,s^K]} + 1_{[s^J,s^K]}(f^n)^*(u,v),$$

where $s \in J \in \mathbf{J}_n, t \in K \in \mathbf{J}_n, (u,v) \in [0,1]$. Here $1_{[s,t]} = sgn(t-s)1_{[s\wedge t, s\vee t]}$. For $y \in \mathbf{R}, s,t \in [0,1], n \in \mathbf{N}$, let

$$F(s,t,y) = I + \frac{i}{2}yH(s,t), \quad F^n(s,t,y) = I + \frac{i}{2}yH^n(s,t).$$

H^n and F^n are finite dimensional operators. The following proposition on the translation from stochastics into analysis is general and will be specialized to our needs afterwards.

Proposition 5.1 *Let $n \in \mathbf{N}, g, h_1, h_2$ functions on the unit square which are constant on the rectangles $L \times M, L, M \in \mathbf{J}_n$. Suppose g is symmetric. Let G, H_1, H_2 be the H-S-operators induced by $g, h_1, h_2, F = I + iG$. Then*

$$E(\exp(i \sum_{L,M\in\mathbf{J}_n} g(s^L, s^M)\Delta_L W \Delta_M W \sum_{L,M\in\mathbf{J}_n} h_1(s^L, s^M)\Delta_L W \Delta_M W)$$
$$= det F^{-1/2} tr(F^{-1}H_1),$$
$$E(\exp(i \sum_{L,M\in\mathbf{J}_n} g(s^L, s^M)\Delta_L W \Delta_M W) \prod_{j=1,2} \sum_{L,M\in\mathbf{J}_n} h_j(s^L, s^M)\Delta_L W \Delta_M W)$$
$$= det F^{-1/2}[tr(F^{-1}H_1)tr(F^{-1}H_2) + tr(F^{-1}H_1F^{-1}H_2) + tr(F^{-1}H_1F^{-1}H_2^*)].$$

Proof: We will prove the more difficult second equation. To this end, we first introduce, besides the integral kernels given, their matrix descriptions corresponding to the unit vector $h = (h_L : L \in \mathbf{J}_n)$, where $h_L = \frac{1}{\sqrt{\lambda(L)}} 1_L, L \in \mathbf{J}_n$. Denote

$$\tilde{G} = (g(s^L, s^M)\sqrt{\lambda(L)\lambda(M)} : L, M \in \mathbf{J}_n),$$

and $\tilde{H}_1, \tilde{H}_2$ accordingly. Set $V = (\lambda(L)^{-1/2}\Delta_L W : L \in \mathbf{J}_n)$. V is a Gaussian unit vector. Now let O be an orthogonal matrix such that $O^*\tilde{G}O = D$ diagonal with diagonal elements $d_L, L \in \mathbf{J}_n$. This choice is possible since $\tilde{G}$ is symmetric. Finally, let $\tilde{F} = I + i\tilde{G}$. Then

$$E(\exp(i \sum_{L,M\in\mathbf{J}_n} g(s^L, s^M)\Delta_L W \Delta_M W) \prod_{j=1,2} \sum_{L,M\in\mathbf{J}_n} h_j(s^L, s^M)\Delta_L W \Delta_M W)$$
$$= E(\exp(iV^*\tilde{G}V) \quad V^*\tilde{H}_1V \quad V^*\tilde{H}_2V)$$
$$= E(\exp(iV^*ODO^*V) \quad V^*\tilde{H}_1V \quad V^*\tilde{H}_2V).$$

Now let $X = O^*V$. Then X is another Gaussian unit vector. Set, for abbreviation, $\tilde{K_1} = O^*\tilde{H}_1 O, \tilde{K_2} = O^*\tilde{H}_2 O$. Then, due to [2], p. 12, we have

$$
\begin{aligned}
&E(\exp(iV^*ODO^*V) \quad V^*\tilde{H}_1 V \quad V^*\tilde{H}_2 V) \\
&\quad = E(\exp(iX^*DX) \quad X^*\tilde{K_1}X \quad X^*\tilde{K_2}X) \\
&\quad = \sum_{L\neq M} E(exp(iX^*DX) \quad X_L^2 X_M^2) \\
&\qquad\qquad ((\tilde{K_1})_{LL}(\tilde{K_2})_{MM} + (\tilde{K_1})_{LM}(\tilde{K_2})_{ML} + (\tilde{K_1})_{LM}(\tilde{K_2})_{LM}) \\
&\quad + \sum_{L\in\mathbf{J}_n} E(\exp(iX^*DX)X_L^4)(\tilde{K_1})_{LL}(\tilde{K_2})_{LL} \\
&\quad = \sum_{L\neq M} \prod_{L,M\neq N\in\mathbf{J}_n} (1+id_N)^{-1/2}(1+id_L)^{-3/2}(1+id_M)^{-3/2} \\
&\qquad\qquad ((\tilde{K_1})_{LL}(\tilde{K_2})_{MM} + (\tilde{K_1})_{LM}(\tilde{K_2})_{ML} + (\tilde{K_1})_{LM}(\tilde{K_2})_{LM}) \\
&\quad +3 \sum_{L\in\mathbf{J}_n} \prod_{L\neq N\in\mathbf{J}_n} (1+id_N)^{-1/2}(1+id_L)^{-5/2}(\tilde{K_1})_{LL}(\tilde{K_2})_{LL} \\
&\quad = \sum_{L,M\in\mathbf{J}_n} \prod_{N\in\mathbf{J}_n} (1+id_N)^{-1/2}[(1+id_L)^{-1}(1+id_M)^{-1} \\
&\qquad\qquad ((\tilde{K_1})_{LL}(\tilde{K_2})_{MM} + (\tilde{K_1})_{LM}(\tilde{K_2})_{ML} + (\tilde{K_1})_{LM}(\tilde{K_2})_{LM})] \\
&\quad = \det(I+iD)^{-1/2}[tr((I+iD)^{-1}\tilde{K_1})tr((I+iD)^{-1}\tilde{K_2} \\
&\qquad\qquad +tr((I+iD)^{-1}\tilde{K_1}(I+id)^{-1}\tilde{K_2}) \\
&\qquad\qquad +tr((I+id)^{-1}\tilde{K_1}(I+id)^{-1}\tilde{K_2}^*)]. \qquad (11)
\end{aligned}
$$

We only have to translate the expressions in the last line of (11) into integral kernel form. First of all, in terms of the unit vector $k = O^*h$ we have $G = h^*\tilde{G}h = k^*Dk$. Hence, by Simon [9], p. 50

$$\det F = \det(I+iD). \qquad (12)$$

Moreover

$$
\begin{aligned}
&tr((I+iD)^{-1}\tilde{K_1}) && (13)\\
&\quad = tr(\tilde{F^{-1}}\tilde{H}_1) \qquad (Reed,\ Simon\ [8],\ p.\ 207) &&\\
&\quad = tr(F^{-1}H_1) && (14)
\end{aligned}
$$

Thus,

$$tr((I+iD)^{-1}\tilde{K}_1(I+iD)^{-1}\tilde{K}_2) = tr(F^{-1}H_1F^{-1}H_2). \tag{15}$$

Now use (12)–(15) in the last line of (11) to complete the proof.

Let us return to the crucial moments A(y), B(y) of theorem 4.1. Proposition 5.1 yields the two following results. To state them, let us abbreviate

$$A_n(y) = \\ 2\sum_{s^J \le s^K} E(\cos(y(U^n_{s^K} - U^n_{s^J}))u^n_{s^K}\Delta_J W) f^n(s^K, s^J)\lambda(K),$$

$$B_n(y) = \\ 2y\sum_{s^J \le s^K} E(\sin(y(U^n_{s^K} - U^n_{s^J}))\delta(1_{[s^J,s^K]} f^n(s^K,.))u^n_{s^J} n \in \mathbf{N}, y \in \mathbf{R}.$$

Proposition 5.2 *Let $n \in \mathbf{N}, y \in \mathbf{R}$. Then*

$$A_n(y) = 2\int_0^1\int_0^1 1_{[0,t]}(s) Re[\det F^n(s,t,y)^{-1/2} \\ F^n(s,t,y)^{-1}(s,.)f^n(t,s)]dsdt.$$

Proof: Fix $J, K \in \mathbf{J}_n$ such that $s^J \le s^K$. Now choose

$$g = \frac{1}{2}y[f^n 1_{[s^J,s^K]} + 1_{[s^J,s^K]}(f^n)^*],$$
$$h_1 = f^n(.,s^K)\lambda(K)f^n(s^K,.)1_J(.)$$

and apply the first part of proposition 5.1. Then by definition $F = F^n(s^J, s^K, y)$ and

$$tr(F^{-1}H_1) = F^n(s^J,s^K,y)^{-1}(s^J,.)f^n(.,s^K)\lambda(K)f^n(s^K,s^J)\lambda(J).$$

This gives the desired result.

Proposition 5.3 *Let $n \in \mathbf{N}, y \in \mathbf{R}$. Then, setting $b(t) = s^J$ for $t \in J \in \mathbf{J}_n$, we have*

$$B_n(y) \quad = 2y\int_0^1\int_0^1 1_{[0,t]}(s) Im[\det F^n(s,t,y)^{-1/2}$$

$$(f^n(t,.)1_{[b(s),b(t)]}F^n(s,t,y)^{-1}f^n(.,t)F^n(s,t,y)^{-1}(s,.)f^n(.,s)$$
$$+f^n(t,.)1_{[b(s),b(t)]}F^n(s,t,y)^{-1}f^n(.,s)F^n(s,t,y)^{-1}(s,.)f^n(.,t)$$
$$+f^n(t,.)1_{[b(s),b(t)]}F^n(s,t,y)^{-1}(.,s)(f^n)^*(s,.)F^n(s,t,y)^{-1}$$
$$f^n(.,t))]dsdt.$$

Proof: Again, fix $J, K \in \mathbf{J}_n$ such that $s^J \leq s^K$. This time, choose

$$\begin{aligned} g &= \frac{1}{2}y[f^n 1_{[s^J,s^K]} + 1_{[s^J,s^K]}(f^n)^*], \\ h_1 &= f^n(.,s^K)\lambda(K)f^n(s^K,.)1_{[s^J,s^K]}(.), \\ h_2 &= f^n(.,.)1_J(.). \end{aligned}$$

and apply the second part of proposition 5.1. Again $F = F^n(s^J, s^K, y)$ and

$$\begin{aligned} tr(F^{-1}H_1) &= f^n(s^K,.)1_{[s^J,s^K]}(.)F^n(s^J,s^K,y)^{-1}f^n(.,s^K)\lambda(K), \\ tr(F^{-1}H_2) &= F^n(s^J,s^K,y)^{-1}(s^J,.)f^n(.,s^J)\lambda(J). \end{aligned}$$

It is equally easy to identify the remaining two traces. The asserted formula follows readily.

The next question we have to answer is: What happens if $n \to \infty$ in the formulas of propositions 5.2 and 5.3? The answer will strongly rely on the convergence of H^n to H.

Proposition 5.4 *For all $p \geq 2$ we have*

$$\int_0^1 \cdots \int_0^1 \mid H^n(s,t)(u,v) - H(s,t)(u,v) \mid^p dudvdsdt \to 0 \quad as n \to \infty.$$

Proof: For $s, t \in [0,1], n \in \mathbf{N}$ such that $s \in J \in \mathbf{J}_n, t \in K \in \mathbf{J}_n, s \leq t$, we have

$$\begin{aligned} &\mid H^n(s,t)(u,v) - H(s,t)(u,v) \mid \\ &\quad \leq \mid f^n(u,v) - f(u,v) \mid + \mid f^n(v,u) - f(v,u) \mid \\ &+ \parallel f \parallel_\infty [\mid 1_{[s^J,s^K]}(u) - 1_{[s,t]}(u) \mid + \mid 1_{[s^J,s^K]}(v) - 1_{[s,t]}(v) \mid], \end{aligned}$$

$u, v \in [0,1]$.

Hence the proposition follows from continuity of f.

We now get the answer to the above question readily.

Proposition 5.5 *For all $p \geq 2, y \in \mathbf{R}$*

$$\det F^n(.,.,y)^{-1/2} \to \det_2 F(.,.,y)^{-1/2} in L^p([0,1]^2).$$

Proof: Due to the trace-free definition of H^n we have for all $n \in \mathbf{N}, J, K \in \mathbf{J}_n$

$$\begin{aligned} &\det F^n(s^J, s^K, y) \\ &\quad = \det_2 F^n(s^J, s^K, y) \exp(-\frac{i}{2} y tr(H^n(s^J, s^K))) \ (Simon\ [9],\ p.\ 107) \\ &\quad = \det_2 F^n(s^J, s^K, y). \end{aligned}$$

It therefore suffices to apply Simon [9], p. 107, according to which the mapping $A \mapsto \det_2(I + A)$ is continuous on the ideal of H-S-operators, and appeal to proposition 5.4.

As for the resolvents appearing in propositions 5.2 and 5.3, we note that according to Simon [9], pp. 106, 107, and Smithies [10], pp. 96-99,

$$F^n(s,t,y)^{-1} = I + D_2(\frac{i}{2} y H^n(s,t)) \det_2 F^n(s,t,y)^{-1}, \qquad s,t \in [0,1], y \in \mathbf{R}. \tag{16}$$

It is therefore sufficient to consider the limits of $D_2(i/2)$ as $n \to \infty$.

Proposition 5.6 *For all $p \geq 2, y \in \mathbf{R}$ we have as $n \to \infty$*

$$\int_0^1 \cdots \int_0^1 | D_2(\frac{i}{2} y H^n(s,t)(u,v) - D_2(\frac{i}{2} y H(s,t)(u,v) |^p \, du dv ds dt \to 0$$

Proof: See [3] for the following series representation of $D_2(\frac{i}{2}.)$. For a H-S-kernel G, define

$$G(u,v,u_1,\ldots,u_k) = \begin{pmatrix} G(u,v) & G(u,u_1) & G(u,u_2) & \ldots & G(u,u_k) \\ G(u_1,v) & 0 & G(u_1,u_2) & \ldots & G(u_1,u_k) \\ G(u_2,v) & G(u_2,u_1) & 0 & \ldots & \vdots \\ \vdots & \vdots & \vdots & \ddots & \vdots \\ \ldots & \ldots & \ldots & \ldots & G(u_{k-1},u_k) \\ G(u_k,v) & G(u_k,u_1) & \ldots & G(u_k,u_{k-1}) & 0 \end{pmatrix}$$

and

$$\tilde{\beta}_k(G)(u,v) = \int_0^1 \cdots \int_0^1 G(u,v,u_1,\ldots,u_k)du_1\ldots du_k, \; k \in \mathbf{N}_0.$$

Then $D_2(\frac{i}{2}yG) = \sum_{k=0}^{\infty} \frac{(\frac{i}{2}y)^k}{k!}\tilde{\beta}_k(G)$. It is important to note that this series converges uniformly, due to Hadamard's inequality for matrices (see Joergens [4], p. 124). Therefore it is sufficient to establish

$$\int_0^1 \cdots \int_0^1 \mid \tilde{\beta}_k(H^n(s,t))(u,v) - \tilde{\beta}_k(H(s,t))(u,v) \mid^p dudvdsdt \to 0$$

as $n \to \infty$ for all $k \in \mathbf{N}_0$. But, considering the representation of $\tilde{\beta}_k(H^n(s,t))$, $\tilde{\beta}_k(H(s,t))$, and using Hoelder's inequality, one easily sees that this reduces to proposition 5.4.

The preceding proposition now enables us to compute the analytical versions of A(y), B(y). In the following we will write $F(s,t,y)^{-1}(u,.)f(.,v)$ for the "more correct" expression

$$1_{\{u\neq v\}}f(u,v) + D_2(\frac{i}{2}yH(s,t))(u,.)f(.,v)\det{}_2F(s,t,y)^{-1},$$
$$u,v \in [0,1].$$

Proposition 5.7 *Let $y \in \mathbf{R}$. Then*

$$A(y) = 2\int_0^1 \int_0^1 1_{[0,t]}(s)Re[\det{}_2F(s,t,y)^{-1/2}$$
$$F(s,t,y)^{-1}(s,.)f(.,t)f(t,s)]dsdt.$$

Proof: Apply propositions 5.2,5.5 and 5.6 and remember that f is continuous.

Proposition 5.8 *Let $y \in \mathbf{R}$. Then*

$$B(y) =$$
$$2y\int_0^1 \int_0^1 1_{[0,t]}(s)Im[\det{}_2F(s,t,y)^{-1/2}$$
$$(f(t,.)1_{[s,t]}F(s,t,y)^{-1}f(.,t)F(s,t,y)^{-1}(s,.)f(.,s)$$
$$+f(t,.)1_{[s,t]}F(s,t,y)^{-1}f(.,s)F(s,t,y)^{-1}(s,.)f(.,t)$$
$$+f(t,.)1_{[s,t]}F(s,t,y)^{-1}(.,s)f^*(s,.)F(s,t,y)^{-1}f(.,t))]dsdt.$$

Proof: Apply propositions 5.3,5.4 and 5.5.

We summarize the results of this section in the following transcription of theorem 4.1.

Theorem 5.1 *U possesses a square integrable occupation density iff*

$$\int_{[-1,1]^c} (A(y) + B(y))y^{-2}dy < \infty,$$

where

$$A(y) = 2\int_0^1 \int_0^1 1_{[0,t]}(s) Re[\det_2 F(s,t,y)^{-1/2}$$
$$F(s,t,y)^{-1}(s,.)f(.,t)f(t,s)]dsdt,$$

$$B(y) =$$
$$2y\int_0^1 \int_0^1 1_{[0,t]}(s) Im[\det_2 F(s,t,y)^{-1/2}$$
$$(f(t,.)1_{[s,t]}F(s,t,y)^{-1}f(.,t) \quad F(s,t,y)^{-1}(s,.)f(.,s)$$
$$+f(t,.)1_{[s,t]}F(s,t,y)^{-1}f(.,s) \quad F(s,t,y)^{-1}(s,.)f(.,t)$$
$$+f(t,.)1_{[s,t]}F(s,t,y)^{-1}(.,s) \quad f^*(s,.)F(s,t,y)^{-1}f(.,t))]dsdt.$$

Proof: Apply propositions 5.7, 5.8 and theorem 4.1 to get the result for continuous f. It is an easy task to use continuity properties such as the ones expressed in propositions 5.5 and 5.6 to extend the result to square integrable f. See also Simon [9], pp. 67, 107.

This gives the following sufficient criterion.

Corollary 5.1 *Suppose that there exists a constant $\alpha > 0$ such that $y \mapsto (A(y) + B(y)) \mid y \mid^{-\alpha}$ is bounded on $[-1,1]^c$. Then U possesses a square integrable occupation density.*

Proof: Trivial from theorem 1, since for $\gamma > 1$ $\int_{[-1,1]^c} \mid y \mid^{-\gamma} < \infty$.

Remark: The theorems of this paper were clearly designed to exhibit in particular the adapted case. This is seen in the obvious equations A(y) = B(y) = 0. It is not clear whether the criteria given by them have more advantages to the more compact one of [2].

Bibliography

[1] Berman, S.M. *Local times and sample function properties of stationary Gaussian processes.* Trans. Amer. Math. Soc. 137 (1969), 277-300.

[2] Imkeller, P. *Occupation densities for stochastic integral processes in the second Wiener chaos.* Preprint, Univ. of B.C. (1990).

[3] Imkeller, P. *On the existence of occupation densities of stochastic integral processes via operator theory.* Preprint, Univ. of B.C. (1990).

[4] Joergens, K. *Lineare Integraloperatoren.* Teubner: Stuttgart (1970).

[5] Nualart, D., Zakai, M. *Generalized stochastic integrals and the Malliavin calculus.* Probab. Th. Rel. Fields 73 (1986), 255-280.

[6] Nualart, D., Pardoux, E. *Stochastic calculus with anticipating integrands.* Probab. Th. Rel. Fields 78 (1988), 535-581.

[7] Nualart, D. *Noncausal stochastic integrals and calculus.* LNM 1516. Springer: Berlin, Heidelberg, New York (1988).

[8] Reed, M., Simon, B. *Methods of modern Mathematical Physics. I: Functional Analysis.* Academic Press: New York (1972).

[9] Simon, B. *Trace ideals and their applications.* London Math. Soc. Lecture Notes Series 35. Cambridge Univ. Press: Cambridge, London (1979).

[10] Smithies, F. *Integral equations.* Cambridge Univ. Press: Cambridge, London (1965).

[11] Zakai, M. *The Malliavin calculus.* Acta Appl. Math. 3 (1985), 175-207.

A Skeletal Theory of Filtering

G. Kallianpur
Center for Stochastic Processes
Department of Statistics
University of North Carolina
Chapel Hill, NC 27599–3260 USA

Abstract

The article gives a brief survey of some recent developments in nonlinear filtering theory based on the use of finitely additive white noise in the observation model. The relationship with "skeletons" in other parts of stochastic analysis is clarified.

1 Introduction

In a series of papers beginning in 1983, R.L. Karandikar and I developed an approach to nonlinear filtering theory that would enable one to compute the optimal nonlinear filter without having to solve difficult infinite dimensional (measure valued) stochastic differential equations (SDE's) derived from the stochastic calculus theory. The point of view was not new with us, having previously been advocated by A.V. Balakrishnan and, prior to that in a different context, by I.E. Segal and L. Gross (see the references in [4]). At present the theory is restricted to filtering problems in which the system process and observation noise are independent. The extension to more general filtering models remains to be carried out and the difficulties standing in the way of such an extension have not yet been fully tested.

Apart from the paucity of algorithms to solve measure–valued Itô SDE's, other considerations have led to the development of the present theory. These have been adequately discussed elsewhere and it suffices to mention

Keywords Bayes formula, measure–valued equation, white noise, skeleton.

Acknowledgement Research supported by the Air Force Office of Scientific Research Contract No. F49620 85C 0144.

ISBN 0-12-481005-5

here that an important motivation has been the desire not to have to extend the basic space of observations in order to accommodate the mathematics [4]. It is pleasing to note that a similar (though by no means identical) trend is becoming perceptible in recent papers on infinite dimensional stochastic analysis in which skeletons are being discovered in stochastic closets (e.g. see [9]). We shall see later in this article that, in a specific technical sense, the nonlinear filter described here is a skeleton of the optimal filter obtained via the SDE's of the conventional theory. Indeed, the white noise theory provides a skeletal approach to the robust filtering discussed in the papers of J.M.C. Clark and M.H.A. Davis.

This article attempts to give only the briefest possible review of the theory to which the book by Karandikar and myself is devoted (see also the survey article [5]). Nevertheless, as explained above, I have written it with a slightly different perspective. In some recent work with G.W. Johnson and with A.S. Ustunel related to chaos expansions and Feynman integrals we have found the approach of the present paper helpful and illuminating [4,8]. Hence I have thought it more appropriate to call the finitely additive theory of nonlinear filtering, a skeletal approach to the problem.

2 Basic ingredients of the theory

Let H be a real, separable Hilbert space whose inner product and norm are denoted by (,) and $|\cdot|$ and let $\mathscr{C}$ be the field of finite dimensional Borel cylinder sets in H. A cylinder measure n on $\mathscr{C}$ is a finitely additive (f.a.) probability measure such that n is countably additive on the σ–field $\mathscr{C}_P$ for each $P \in \mathscr{P}(H)$, the class of orthogonal projections on H with finite dimensional range. $\mathscr{C}_P$ is defined to be the class of all sets $P^{-1}B$ where B is a Borel subset of PH.

We now introduce the basic ingredients of our theory. The probability space $(\Omega, \mathscr{A}, \Pi)$ is the space on which the unobserved H–valued random variable ξ is defined. The noise space is H itself, together with the canonical Gauss measure m on $\mathscr{C}$, i.e. m has characteristic functional $\exp(-\frac{1}{2}|h|^2)$. Let $E = \Omega \times H$, and $\mathscr{E}$ the field of subsets of E defined by $\mathscr{E} = \bigcup_{P \in \mathscr{P}} \mathscr{E}_P$, $\mathscr{E}_P$ being the σ–field $\mathscr{A} \times \mathscr{C}_P$.

Definition. A f.a. probability measure β on $(E, \mathscr{E})$ such that its restriction to each $\mathscr{E}_P$ is countably additive is called a quasi cylindrical probability (QCP). The QCP we will be most concerned with will be denoted by α and given by $\alpha = \Pi \otimes m$, the "product" of Π and the canonical Gauss measure m. More precisely, for $a \in \mathscr{A}$ and $C \in \mathscr{C}$, $\alpha(A \times C) = \Pi(A)m(C)$.

Associated with a QCP β is a triplet consisting of a probability space $(\tilde{\Omega}, \tilde{\mathscr{A}}, \tilde{\Pi})$, a mapping L from H into the space of r.v.'s on $\tilde{\Omega}$ and ρ: $\tilde{\Omega} \longrightarrow \Omega$, a measurable map with the following property: For $A \in \mathscr{A}$ and $C \in \mathscr{C}$,

$$C = \{h \in H: ((h,h_1),\ldots,(h,h_j)) \in B\} \text{ (B, Borel)},$$

$$\beta(A \times C) = \tilde{\Pi}\{\tilde{\omega}: \rho(\tilde{\omega}) \in A, (L(h_1)(\tilde{\omega}),\ldots,L(h_j)(\tilde{\omega})) \in B\}.$$

Definition $(\rho, L, \tilde{\Pi})$ is a space for β.

We are now in a position to introduce the concepts of lifting and accessible r.v.'s which are central to our work. Let $(S, \mathscr{B}(s))$ be a Polish space with its Borel sets. An S–valued function f on E is a cylinder function if it is of the form $\varphi(\omega,(h,h_1),\ldots,(h,h_k))$ where φ: $\Omega \times \mathbb{R}^k \longrightarrow S$ is a measurable function relative to the obvious σ–field. Let $R_\beta f$ be the r.v. $\varphi(\rho, L(h_1),\ldots,L(h_k))$ defined on the representation space of β. To extend this definition, let us consider the class of functions f: $E \longrightarrow S$ such that for all $P \in \mathscr{P}(H)$, $f_P(\omega,h) := f(\omega,Ph)$ is an $(\mathscr{E}_P, \mathscr{B}(S))$ measurable cylinder function. $R_\beta f_P$ is then well defined.

Definition $\mathscr{L}(E, \mathscr{E}, \beta; S)$ is the class of functions f defined above such that for all sequences $\{P_k\} \subseteq \mathscr{P}$ converging strongly to I ($P_k \xrightarrow{s} I$), the sequence $\{R_\beta f_{P_k}\}$ is Cauchy in $\tilde{\Pi}$–probability. Let

$$R_\beta f = \lim_{P_k \xrightarrow{s} I} \text{ in prob. } R_\beta f_{P_k}.$$

$R_\beta f$ is the β–lifting of f (or simply the lifting of f when the context makes it clear which β is being considered). Elements of $\mathscr{L}^0(E, \mathscr{E}, \beta; S)$ are called <u>accessible</u> r.v.'s. The symbol S will be suppressed when $S = \mathbb{R}$. It has been shown in [4,5] that the limit in probability in the above definition does not depend on the sequence $\{P_k\}$ chosen and indeed that an equivalent definition is given in terms of nets.

As is the case in many applications S is either $\mathbb{R}^d$ or a Banach space with norm $\|\cdot\|$. With this choice of S, it is useful to consider the following subclasses of $\mathscr{L}^0(E, \mathscr{E}, \beta; S)$. For $q \geq 1$, let

$$\mathscr{L}^q(E, \mathscr{E}, \beta; S) = \{f \in \mathscr{L}^0(E, \mathscr{E}, \beta; S) : \forall \{P_k\} \subseteq \mathscr{P}(H),$$

$$P_k \xrightarrow{s} I, \ (R_\beta(f_{P_k}))$$

is L^q–Cauchy in $(\tilde{\Omega}, \tilde{\mathscr{A}}, \tilde{P})\}$. For $f \in \mathscr{L}^q(E, \mathscr{E}, \beta; S)$ define

$$R_\beta(f) = L^q\text{–}\lim_{k \to \infty} R_\beta(f_{P_k}).$$

We then have

$$\int \|R_\beta(f)\|^q d\tilde{\Pi} < \infty \text{ and } \int \|R_\beta(f_{P_k}) - R_\beta(f)\|^q d\tilde{\Pi} \longrightarrow 0.$$

<u>Integration w.r.t. β.</u> Taking $S = \mathbb{R}$, for $f \in \mathscr{L}^1(E, \mathscr{E}, \beta)$ define the integral

$$\int_E f d\beta := \int_{\tilde{\Omega}} R_\beta(f) d\tilde{\Pi}.$$

The value of $\int_E f d\beta$ does not depend on the choice of representation.

From the standpoint of the present theory it is important to observe that the class $\mathscr{L}(E, \mathscr{E}, \beta, S)$ does not depend on the choice of the representation of β. (See [5]). What this means is that, given a QCP β, the class of accessible r.v.'s on H is independent of the particular $(\tilde{\Omega}, \tilde{\mathscr{A}}, \tilde{\Pi})$ chosen. As an example, consider an abstract Wiener space (H, B, μ) which can serve as a prototype for such examples.

Take $\beta = \alpha = \Pi \circledcirc m$. Let

$$\tilde{\Omega} = \Omega \times B, \qquad \tilde{A} = A \times \mathscr{B}(B), \qquad \tilde{\Pi} = \Pi \times \mu.$$

For $\tilde{\omega} = (\omega, x) \in \tilde{\Omega}$ and $h \in H$, define

$$\rho(\tilde{\omega}) = \omega \quad \text{and} \quad L(h)(\tilde{\omega}) = L_0(h)(x)$$

where L_0 is obtained in the following manner. Let γ: H $\longrightarrow$ B be the continuous injection and γ^*, the adjoint of γ which maps the dual B^* continuously into H^*, the dual of H (identified with H). Let $\{\phi_j\} \subseteq \gamma^*(B^*)$ be a complete orthonormal system (CONS) in H and let $f_j \in B^*$ such that $\phi_j = \gamma^*(f_j)$. For $h \in H$ and $x \in B$, define

$$
\begin{aligned}
L_0(h)(x) &= \sum_{j=1}^{\infty} (h, \varphi_j) f_j[x] \qquad \text{if the series converges,} \\
&= 0 \qquad \text{otherwise.}
\end{aligned}
$$

Then, taking $\bar{\Pi}$ to be the product measure $\Pi \times \mu$ on $\bar{\mathscr{A}}$ it can be shown that $(\rho, L, \bar{\Pi})$ as constructed above is a representation for α. In particular, the representation probability space is $(\Omega \times B, \mathscr{A} \times \mathscr{B}(B), \Pi \times \mu)$. The reader will find the pertinent details in [4]. Now if we have a measurable norm on H yielding a completion B′ different from B, the above procedure leads to another representation space $(\Omega \times B', \mathscr{A} \times \mathscr{B}(B'), \Pi \times \mu')$, μ' being the abstract Wiener measure determined by m on B′. The point of the remark, as illustrated by this particular example, is that the class of accessible r.v.'s on $(E, \mathscr{E}, \alpha)$ does not depend on the particular enlargement of H obtained by completing H with respect to different measurable norms on H. Without going into the details, it suffices to observe that in a stochastic calculus model of filtering, a specific choice of B is essential and the results obtained from such a theory involve B and cannot be stated only in terms of H.

The situation is similar in the case of the Malliavin–Watanabe calculus though it will take us too far afield to discuss it here.

The technical obstacles standing in the way of developing a skeletal theory of filtering have to do with providing suitable and rigorous counterparts to some of the basic concepts in (countably additive) probability theory. Of particular importance are absolute continuity and conditional expectation.

Absolute continuity The available definitions in the literature, including the $\epsilon - \delta$ definition are not suitable for our purpose since they do not lead to the R–N derivatives that are accessible r.v.'s. This point is discussed in [4]. We proceed in a different manner.

Definition Let β_1, β_2 be QCP's on $(E, \mathscr{E})$. β_1 is absolutely continuous w.r.t. β_2 $(\beta_1 << \beta_2)$ if there exists a function $f \in \mathscr{L}^1(E, \mathscr{E}, \beta_2)$, $f \geq 0$ such that for all $F \in \mathscr{E}$

$$\beta_1(F) = \int_F f d\beta_2.$$

f is called the Radon–Nikodym derivative.

Conditional expectation Here again, the conditional expectation in addition to possessing the usual properties should give an accessible r.v. Since σ–fields have no place in the theory what is needed is a definition of $E_\beta(g|\phi)$, the conditional expectation of an accessible r.v. $g \in \mathscr{L}^1(E, \mathscr{E}, \beta)$ with respect to an appropriate class of mappings ϕ.

Let $E_1 = \Omega_1 \times H_1$ where $(\Omega_1, \mathscr{A}_1)$ is a measurable space and H_1 is a separable Hilbert space with inner product $(\ ,\ )_1$. Let $\mathscr{E}_1$ have the same meaning as $\mathscr{E}$ defined earlier.

Definition A map $\phi: E \longrightarrow E_1$ is a quasi–cylindrical mapping (QCM) if it satisfies the following conditions: $\phi = (\phi_1, \phi_2)$, $\phi_1: E \longrightarrow \Omega_1$, $\phi_2: E \longrightarrow H_1$ where

(i) ϕ_1 is a cylinder function, or $\phi_1 \in \mathscr{L}^0(E, \mathscr{E}, \beta, \Omega_1)$ if Ω_1 is a Polish space with $\mathscr{A}_1 = \mathscr{B}(\Omega_1)$;

(ii) For each $h_1 \in H_1$, the function $(\omega, h) \longrightarrow (h_1, \phi_2(\omega, h))_1$ is an element of $\mathscr{L}(E, \mathscr{E}, \beta)$.

If ϕ is a QCM, the set function

$$\beta_1(F) := \beta[(\omega,h): \phi(\omega,h) \in F], \quad F \in \mathscr{E}_1$$

is a QCP on $(E_1, \mathscr{E}_1)$ and is denoted by $\beta \circ \phi^{-1}$, the induced QCP. Choose and fix $(\rho_1, L_1, \bar{\Pi})$, a representation of β_1 and let R_{β_1} denote the associated lifting map. Let S be a Polish space. For a QCM ϕ define

$\mathcal{U}(\phi) := \{f \in \mathcal{L}(E_1, \mathcal{E}_1, \mathcal{B}_1; S)$: $f \circ \phi \in \mathcal{L}(E, \mathcal{E}, \mathcal{B}, S)$ and $R_{\beta_1}(f) = R_\beta(f \circ \phi)\}$.

Definition Let $g \in \mathcal{L}^1(E, \mathcal{E}, \beta)$. Suppose there exists $g_1 \in \mathcal{U}(\phi)$ such that $g_1 \in \mathcal{L}^1(E_1, \mathcal{E}_1, \beta_1)$ and the following relation holds for all $f: E_1 \to \mathbb{R}$ of the form $f(\omega_1, h_1) = f_1(\omega_1)f_2(h_1)$ where f_1 is $\mathcal{A}_1$–measurable and f_2 is a cylinder function:

$$\int_E gf(\phi)d\beta = \int_E g_1(\phi)f(\phi)d\beta.$$

Then $g_1 \circ \phi$ is defined to be the conditional expectation of g given ϕ and denoted by $E_\beta(g|\phi)$. Besides the usual linearity property two other familiar properties are shared by $E_\beta(\cdot|\phi)$.

(a) Let $f \in \mathcal{U}(\phi)$ be such that g and $f(\phi)\cdot g \in \mathcal{L}^1(E, \mathcal{E}, \beta)$. Then $E_\beta(g\cdot f(\phi)|\phi)$ exists,

$$E_\beta(g|\phi)f(\phi) \in \mathcal{L}^1(E, \mathcal{E}, \beta)$$

and

$$E_\beta(gf(\phi)|\phi) = f(\phi)E_\beta(g|\phi).$$

(b) If $g^2 \in \mathcal{L}^1(E, \mathcal{E}, \beta)$ then $[E_\beta(g|\phi)]^2 \in \mathcal{L}^1(E, \mathcal{E}, \beta)$ provided $E_\beta(g|\phi)$ exists. Further,

$$\int[g-E_\beta(g|\phi)]^2 d\beta = \min_{f_0 \in \mathcal{U}(\phi)} \int[g-f_0(\phi)]^2 d\beta.$$

Since both g and $E_\beta(g|\phi)$ are accessible r.v.'s it is interesting to see whether $R_\beta[E_\beta(g|\phi)]$ is related to the conditional expectation of $R_\beta g$ with respect to some σ–field depending on ϕ. Let $\mathscr{D}_\phi$ be the minimal σ–field on $\tilde{\Omega}$ with respect to which the family $\{R_\beta(f\circ\phi): f \in \mathscr{F}\}$ is measurable together with $\tilde{\Pi}$–null sets.

Proposition 2.1 (i) If $g \in \mathscr{U}(\phi)$, then $R_\beta(g\circ\phi)$ is $\mathscr{D}_\phi$–measurable.

(ii) Suppose $g \in \mathscr{L}^1(E, \mathscr{E}, \beta)$ such that $E_\beta(g|\phi)$ exists. Then

$$R_\beta[E_\beta(g|\phi)] = E[R_\beta(g)|\mathscr{D}_\phi] \quad \text{a.s. } \tilde{\Pi}$$

(iii) Let $g \in \mathscr{L}^1(E, \mathscr{E}, \beta)$. If $g_1 \in \mathscr{U}(\phi)$ is such that

$$R_\beta(g_1\circ\phi) = E[R_\beta(g)|\mathscr{D}_\phi] \quad \text{a.s. } \tilde{\Pi}.$$

Then $E_\beta(g|\phi)$ exists and is equal to $g_1\circ\phi$.

3 Liftings and skeletons

Let $f \in \mathscr{L}^2(E, \mathscr{E}, \alpha; S)$. Choose an arbitrary CONS (φ_i) in H and let H_k be the subspace of H spanned by $\{\varphi_j, 1 \le j \le k\}$. If P_k is the orthogonal projection on H with range H_k, then $P_k \uparrow I$ and

$$f_{P_k}(\omega,h) = f(\omega, \sum_{j=1}^{k} (h,\varphi_j)\varphi_j).$$

Let $(\rho, L, \tilde{\Pi})$ be a representation of β. The lifting $R_\beta(f_{P_k})$ is given by

$$R_\beta f_{P_k}(\tilde{\omega}) = f(\rho(\tilde{\omega}), \sum_{j=1}^{k} L(\varphi_j)(\tilde{\omega})\varphi_j).$$

Hence

$$f(\tilde{\rho}(\tilde{\omega}), \sum_{j=1}^{k} L(\varphi_j)(\tilde{\omega})\varphi_j) \longrightarrow (R_\beta f)(\tilde{\omega})$$

in $\tilde{\Pi}$–probability.

We have defined $R_\beta f$ to be a lifting of f. On the other hand, if F is an S–valued r.v. on $(\tilde{\Omega}, \tilde{\mathscr{A}}, \tilde{\Pi})$ such that $F = R_\beta f$ for some $f \in \mathscr{L}^0(E, \mathscr{E}, \beta; S)$ then we define f to be a skeleton of F. If S is a Banach space with norm $\|\cdot\|$ and F is such that $F = R_\beta f$ for some $f \in \mathscr{L}^q(E, \mathscr{E}, \beta; S)$, $q \geq 1$, then

$$(3.1) \qquad E_{\tilde{\Pi}} \|f(\rho(\tilde{\omega}), \sum_{j=1}^{k} L(\varphi_j)(\tilde{\omega})\varphi_j) - F(\tilde{\omega})\|^q \longrightarrow 0.$$

The definition of skeleton (or equivalently, lifting) given above is in the spirit of the definition given by Zakai in his recent paper [9], but there are some differences. The most obvious one is that imposed by the requirements of filtering theory: our skeletons are defined on E and not on H. It is more important to note that $R_\beta f$ is the limit in probability (or in L^q as the case may be) along the directed set $\{\mathscr{P}(H), <\}$ where $<$ denotes partial ordering such that $P_1 < P_2$ means $P_1 H \subset P_2 H$. So, $R_\beta f$ does not depend on the sequence $\{P_k\}$ and hence is independent of the choice of CONS $\{\varphi_i\}$ in H.

To further illustrate the comparison let us consider the following special case. Let (H, B, μ) be an abstract Wiener space, $(\tilde{\Omega}, \tilde{\mathscr{A}}, \tilde{\Pi}) = (\Omega \times B, \mathscr{A} \times \mathscr{B}(B), \Pi \times \mu)$ and $\beta = \alpha = \Pi \circ m$. Let ρ and L be chosen as in the previous section.

Now suppose that F is a real valued r.v. on $\Omega \times B$ such that $F = R_\alpha f$ for some $f \in \mathscr{L}^2(E, \mathscr{E}, \alpha)$. Then f is a

φ–skeleton of F in the sense of Zakai for the specific CONS $\{\varphi_i\} \subseteq B^*$ appearing in the definition of L given in Section 1. (Our assumption about f makes it unnecessary to assume further that it is continuous. The latter condition ensures that f_P is a Borel cylinder function so that $R_\alpha f_P$ always exists).

From the definition of L, it follows that the $L_0(\varphi_i)$ are i.i.d. N(0,1) r.v.'s on B and we have from (3.1),

$$(3.2) \qquad f(\omega, \sum_{i=1}^{k} L_0(\varphi_i)(x)\varphi_i) \longrightarrow F(\omega,x)$$

in $L^2(\tilde{\Omega},\tilde{\Pi})$.

Earlier, in [2] and in the as yet unpublished work [3], G.W. Johnson and I have used the skeleton as defined in the present paper, in obtaining necessary and sufficient conditions for the existence of Stratonovich multiple Wiener integrals as well as to give a rigorous definition of Hu and Meyer's concept of the natural extension of a multiple Wiener integral (see [2] for the Hu–Meyer reference). It has been necessary in these papers to define σ–liftings or σ–skeletons for all $\sigma>0$ where σ is the variance parameter of the Wiener process. In a forthcoming paper with A.S. Ustunel, the skeleton plays a part in defining distributions on abstract Wiener spaces and, in particular, the Feynman distribution [8].

4 Bayes formula in the white noise theory

Let us first recall the familiar setup of the conventional or stochastic calculus theory.

A canonical form of the observation model in the conventional nonlinear filtering theory is given by

$$(4.1) \qquad Y_t = \int_0^t h_s(X_s)dB_s + W_t, \qquad 0\leq t\leq T$$

where $X = (X_t)$ is a Markov process, called the signal or

system process and $W = (W_t)$ is a standard Wiener process in $\mathbb{R}^n$. The state space of X_t is assumed to be $\mathbb{R}^d$ and $h: [0,T]\times\mathbb{R}^d \longrightarrow \mathbb{R}^n$ satisfies all the usual measurability conditions and

$$\int_0^T E|h_s(X_s)|^2 ds < \infty.$$

In our setup we also assume independence of signal and noise. The nonlinear filtering problem in its most general form is to find a recursive formula for the conditional distribution of X_t given Y_s, $0\leq s\leq t$, in other words, to find a stochastic differential equation (SDE) for the measure valued process $F_t^Y(\cdot) := P[X_t \epsilon (\cdot)|\mathcal{F}_t^Y]$. A crucial tool in deriving it is the Bayes formula (see [4] or [6] for details) given below.

Let g be $\mathcal{F}_T^X$–measurable and integrable on $(\Omega,\mathcal{A},\Pi)$ where $\mathcal{F}_T^X$ is the σ–field (with null sets added) generated by (X_t), $0\leq t\leq T$. We may assume the processes in (4.1) to be defined on the product space $(\Omega\times\Omega_0,\ \mathcal{A}\times\mathcal{A}_0,\ \Pi\times\Pi_0)$ where W is a Wiener process on $\Omega_0,\mathcal{A}_0,\Pi_0)$. Then

$$E_\Pi[g|\mathcal{F}_t^Y] = \frac{\sigma_t(g,Y)}{\sigma_t(1,Y)} \tag{4.2}$$

where

$$\sigma_t(g,Y) = \int_\Omega gq_t d\Pi$$

where

$$q_t = \exp\{\sum_{i=1}^n \int_0^t h_s^i(X_s)dY_s^i - \frac{1}{2}\int_0^t|h_s(X_s)|^2ds\}. \tag{4.3}$$

Taking $g = f\circ X_t$ where $f(X_t)$ is Π–integrable and writing $\sigma_t(f,Y)$ for $\sigma_t(g,Y)$ we have

$$\int f(x) F_t^Y(dx) = E[f(X_t) | \mathcal{F}_t^Y] = \frac{\langle f, \Gamma_t^Y \rangle}{\langle 1, \Gamma_t^Y \rangle}$$

where Γ_t^Y is the unnormalized conditional distribution

(4.4) $$\Gamma_t^Y(B) := \int_\Omega 1_{[X_t \in B]} q_t \, d\Pi$$

and

$$\langle f, \Gamma_t^Y \rangle = \int f(x) \Gamma_t^Y(dx).$$

Since

$$F_t^Y(\cdot) = \frac{\Gamma_t^y(\cdot)}{\Gamma_t^Y(\mathbb{R}^d)} .$$

It is sufficient and easier to handle the SDE for Γ_t^Y than the more complicated nonlinear SDE for F_t^Y. Suppose in addition that

(4.5) $$\Gamma_t^Y(B) = \int_B p_t(x, Y) dx, \quad B \in \mathcal{B}(\mathbb{R}^d).$$

Then the measure–valued SDE for Γ_t^Y or F_t^Y which we shall call the KunitaSzpirglas equation can be replaced by a stochastic partial differential equation for the unnormalized conditional density. The latter is the famous Zakai equation. The Bayes formula plays a key role in deriving this equation.

We turn now to the white noise or skeletal theory.

Bayes formula for the abstract filtering model

The white noise model of nonlinear filtering in an abstract setting is given as follows.

Let ξ be the process introduced in Section 2, i.e., ξ is an H valued r.v. defined on $(\Omega, \mathcal{A}, \Pi)$ and let e be the white noise on H, i.e., by definition, e is the identity map on H. The observation process y is defined on the quasi cylindrical probability space $(E, \mathscr{E}, \alpha)$,

(4.6) $$y = \xi + e$$

where $Y(\omega,h) = \xi(\omega,h) + e(\omega,h)$,

$$\xi(\omega,h) = \xi(\omega) \quad \text{and} \quad e(\omega,h) = e(h) = h.$$

Let Q be an arbitrary orthogonal projection on H and $H_1 = QH$. If g is an integrable r.v. on $(\Omega, \mathscr{A}, \Pi)$, the first task is to show that $E_\alpha(g|Qy)$ exists for which Qy has to be shown to be a QCM from $(E, \mathscr{E}, \alpha)$ to $(H_1, \mathscr{C}_1)$, $\mathscr{C}_1$ being the class of finite dimensional cylinder sets in H_1. All this has been shown in [4] and we have the following white noise version of the Kallianpur–Striebel Bayes formula.

Theorem 4.1 The conditional expectation of g given Qy exists and is given by

(4.7) $$E_\alpha(g|Qy) = \frac{\sigma_Q(g,Qy)}{\sigma_Q(1,Qy)}$$

where for $h \in H$ and $f \in \mathscr{L}^1(\Omega, \mathscr{A}, \Pi)$

(4.8) $$\sigma_Q(f,h) = \int_\Omega f(\omega)\exp\{(h,Q\xi(\omega)) - \tfrac{1}{2}|Q\xi(\omega)|^2\}d\Pi(\omega).$$

Before discussing applications of this formula it may be of some interest to note some properties of $\sigma_Q(g,Qy)$ and $E_\alpha(g|Qy)$ as functionals of y. We need, first of all, to define the spaces $L^r(H)$. For any Banach space B with norm $\|\cdot\|$ let $L(H,B)$ be the class of all bounded linear transformations from H to B. Let $L^0(H) = \mathbb{R}$, $L^1(H) = L(H,\mathbb{R})$ and for $r \geq 1$ let $L^{r+1}(H) = L(H,L^r(H))$. The Banach space $L^r(H)$ can be identified with the class of all bounded linear mappings from the r–fold product $H \times \ldots \times H$ into $\mathbb{R}$ with the norm $\|\cdot\|_r$ given by

$$\|f\|_r = \sup\{|f[h_1,\ldots,h_r]| : h_i \in H, \|h_i\| \le 1\}.$$

The map f: $H \longrightarrow \mathbb{R}$ will be said to be (r+1) times Fréchet differentiable if it is r times Fréchet differentiable and the rth Fréchet derivative $D^r f: H \longrightarrow L^r(H)$ is Fréchet differentiable. Let $L^r_{(2)}(H)$ be the subspace of $L^r(H)$ consisting of $g \in L^r(H)$ for which

$$\|g\|^2_{r,2} := \sum_{i_1,\ldots,i_r} |g[e_{i_1},\ldots,e_{i_r}]|^2 < \infty$$

for $\{e_i\}$ any CONS in H. Then $\|g\|_{r,2}$ does not depend on the choice of CONS. Furthermore, $L_2^{(r)}(H)$ is a Hilbert space with norm $\|\cdot\|_{r,2}$.

The above details as well as the results stated below are taken from [1].

Theorem 4.2 (i) $\sigma_Q(g,Qy)$ of Theorem 4.1 is r times Fréchét differentiable for all $r \ge 1$.

(ii) If $\int_\Omega \|Q\xi(\omega)\|^{2r} |g(\omega)| d\Pi(\omega) < \infty$

then

$$D^r\sigma_Q(g,Qy) \in \mathcal{L}(E, \mathcal{E}, \alpha; L^r_{(2)}(H)).$$

In other words, $D^r\sigma_Q(g,Qy)$ is an $L^r_{(2)}(H)$–valued accessible r.v.

Definition A real–valued functional $\theta(y)$ is an accessible, C^∞–functional of the observations if

(a) $\theta \in \mathcal{L}(E, \mathcal{E}, \alpha)$,

(b) $D^r\theta$ exists for all $r \ge 1$, and

(c) $D^r\theta \in \mathscr{L}(E, \mathscr{E}, \alpha; L^r_{(2)}(H))$.

Theorem 4.3 Let $E\|\xi\|^r < \infty$ for all $r \geq 1$ and let g be bounded. Then the conditonal expectation $E_\alpha(g|Qy)$ is an accessible, C^∞–functional of the observations.

The first application we make of Theorem 4.1 is to the white noise version of the conventional filtering model (4.1).

(4.9) $$y_s = h_s(X_s) + e_s, \quad 0 \leq s \leq T$$

where

(4.10) $$X = (X_s, \quad 0 \leq s \leq T)$$

is a Markov process with a Polish space S as its state space and defined on $(\Omega, \mathscr{A}, P)$;

(4.11) $$h: [0,T] \times S \longrightarrow \mathbb{R}^n$$

is such that

$$\int_0^T |h_s(X_s)|^2 ds < \infty \quad P\text{–a.s.}$$

In the abstract model (4.6) we have $\xi_s(\omega) = h_s(X_s(\omega))$, $0 \leq s \leq T$ so that $\xi(\omega) \in H$ where H is the Hilbert space $L^2([0,T];\mathbb{R}^n)$. The function h and the process X will be assumed to satisfy the usual measurability conditions.

(4.12) $$e = (e_s)$$

is Gaussian white noise on H independent of X.

Fix $Q = Q_t$, the orthogonal projection on H with range $H_t = \{\eta \in H: \int_t^T |\eta_s|^2 ds = 0\}$. Let $f: S \longrightarrow \mathbb{R}$ be such that $f \circ X_t \in L^1(\Omega, \mathscr{A}, \Pi)$. Then the Bayes formula of Theorem 1 takes the form

$$(4.13) \qquad E_\alpha[f(X_t)\,|\,Q_t y] = \frac{\sigma_t(f, Q_t y)}{\sigma_t(1, Q_t y)}$$

where we have written $\sigma_t(f,Q_t y)$ for $\sigma_{Q_t}(f,Q_t y)$ and $(Q_t y)(s) = y(s)$ for $0 \leq s \leq t$ and $= 0$ for $t < s \leq T$. σ_t is given by

$$(4.14) \quad \sigma_t(f,Q_t\eta) = \int_\Omega f(X_t)\exp\left\{\sum_{j=1}^n \int_0^t \eta_s^j h_s^j(X_s)ds - \frac{1}{2}\sum_{j=1}^n \int_0^t (h_s^j(X_s))^2 ds\right\} d\Pi = \int_\Omega f(X_t)q_t(\eta,X)d\Pi, \quad \text{say.}$$

It is convenient, in the later sections, to use the notation $\sigma_t(f,\eta)$ for the L.H.S. of (4.14).

The Markovian assumption on X is not needed for the Bayes formula and is used, together with the latter to derive the versions of the Zakai equation.

5 The Zakai equation in skeletal form

In the white noise theory the filtering model (4.9) differs from the conventional model (4.1) in the following sense. The SDE for the optimal filter and the Zakai stochastic partial differential equaiton (SPDE) for the unnormalized conditional density can now be replaced by "ordinary" differential equations in which the randomness occurs only in the coefficients. It will be assumed throughout this section that the process (X_t) in (4.9) is an $\mathbb{R}^d$–valued diffusion process with diffusion and drift coefficients denoted by $a = (a_{ij})$ and $b = (b_i)$ and whose weak generator is denoted by L. The general assumptions on the coefficients are that a_{ij} and b_i are measurable functions bounded on compact subsets of $[0,T]\times\mathbb{R}^d$ and that the matrix (a_{ij}) is nonnegative definite. The assumptions on (X_t) are such that when a and b are continuous, $C_0^{1,2}([0,T]\times\mathbb{R}^d) \subseteq \mathscr{D}$ where $\mathscr{D}$ is the domain of L. For $f \in C_0^{1,2}([0,T]\times\mathbb{R}^d)$,

$$(Lf)(t,x) = (\frac{\sigma}{\sigma t} + L_t)f(t,x)$$

where L_t is the differential operator

$$L_t = \frac{1}{2} \sum_{i,j} a_{ij}(t,x) \frac{\partial^2}{\partial x^i \partial x^j} + \sum_i b_i(t,x) \frac{\partial}{\partial x^i}.$$

Let $\mathscr{D}_0$ be the class of functions $f: \mathbb{R}^d \longrightarrow \mathbb{R}$ which, regarded as functions of (t,x) belong to $\mathscr{D}$. The first version of the Zakai equation pertains to the unnormalized conditional expectation which it is convenient here to present as an integral equation.

Theorem 5.1 In (4.9) suppose that h satisfies

$$E \int_0^T |h_s(X_s)|^2 ds < \infty. \tag{5.1}$$

Then for $f \in \mathscr{D}_0$ and for all $y \in H$ we have

$$\sigma_t(f,y) = E_\Pi[f(X_0)] + \int_0^t \sigma_s(L_s f,y)ds + \int_0^t [\sum_{i=1}^n \sigma_s(h_s^i f,y)y_s^i - \frac{1}{2}\sigma_s(|h_s|^2 f,y)]ds. \tag{5.2}$$

Under suitable conditions, the conditional distribution of X_t given $Q_t y$ (i.e., y_s, $0 \leq s \leq t$) admits a density $p_t(x,y)$ which can be characterized as the unique solution (generalized or classical in a sense to be made precise below) of a partial differential equation (PDE). This PDE is our skeletal version of the Zakai SPDE and is a perturbation of the Kolmogorov forward equation for the unconditional density of X_t.

First, some preliminary notation is essential.

Let M_t be a partial differential operator given by

$$M_t f(x) := \frac{1}{2} \sum_{i,j=1}^{d} a^0_{ij}(t,x) \frac{\partial^2 f}{\partial x^i \partial x^j}(x) + \sum_{i=1}^{d} b^0_i(t,x) \frac{\partial f}{\partial x^i}(x) + {}^0 \text{where } f \in$$

$C^2(\mathbb{R}^d)$, $(t,x) \in [0,T]\times\mathbb{R}^d$ and the coefficients satisfy the following conditions.

a^0_{ij}, b^0_i and c^0 are (t,x)–measurable functions and for some $\lambda > 0$,

$$\sum_{i,j} a^0_{ij}(t,x) z^i z^j \geq \lambda \sum_i (z^i)^2 \tag{5.3}$$

for all real $z^1,\ldots,z^d$ and for all $(t,x) \in [0,T]\times\mathbb{R}^d$.

(5.4) The coefficients a^0_{ij} are symmetric.

Let p_0 be a given real valued function on $\mathbb{R}^d$ and let M^*_t denote the formal adjoint of M_t. The problem of finding a function u such that

$$\frac{\partial u}{\partial t} = M^*_t u \tag{5.5}$$

for $0<t\leq T$ and

$$u(0,x) = p_0(x) \tag{5.6}$$

will be referred to as a Cauchy problem. The term is used in a slightly more general sense than is customary in the literature.

Definition (a). A function $u: [0,T]\times\mathbb{R}^d \to \mathbb{R}$ is a classical solution to the Cauchy problem (5.5)–(5.6) if $u \in C^{1,2}((0,T]\times\mathbb{R}^d) \cap C([0,T]\times\mathbb{R}^d)$ and the relations (5.5), (5.6) hold pointwise for $0<t\leq T$ and $x \in \mathbb{R}^d$. Further, it is assumed that c^0, p_0 are continuous.

(b) A measurable function $u:[0,T]\times\mathbb{R}^d \longrightarrow \mathbb{R}$ is said to be a generalized solution to (5.5), (5.6) if for all $\phi \in C_0^\infty(\mathbb{R}^d)$,

$$\int_{\mathbb{R}^d} u(t,x)\phi(x)dx = \int_{\mathbb{R}^d} p_0(x)\phi(x)dx + \int_0^t \int_{\mathbb{R}^d} u(s,x)M_s\phi(x)dxds$$

for all $t \in [0,T]$. It is assumed that all the integrals appearing above exist and are finite.

The above definitions are useful in establishing two variants of results on the existence of unique solution of the white noise Zakai equation.

Theorem 5.2 Suppose that the coefficients a,b in the generator L of (X_t) satisfy (5.3), (5.4) and (5.7).

For all i,j,

$$(5.7) \qquad a_{ij}, \ \frac{\partial a_{ij}}{\partial x^i}, \ \frac{\partial^2 a_{ij}}{\partial x^i \partial x^j}, \ b_i, \ \frac{\partial b_i}{\partial x^i}$$

are bounded, Hölder continuous functions on $[0,T]\times\mathbb{R}^d$. Further assume that X_0 has a density $p_0(x)$ which is continuous and satisfies condition

$$(5.8) \quad |p_0(x)| \leq K \exp(|x|^{2-\epsilon}) \quad \text{for some } \epsilon > 0,\ K, \text{ a positive constant.}$$

In the observation model (4.9) suppose that

$$(5.9) \qquad E \int_0^T |h_t(X_t)|^2 dt < \infty.$$

Then for all $y \in H$, the unnormalized conditional density $p_t(x,y)$ exists and is the unique generalized solution to the Cauchy problem

$$(5.10)\quad \frac{\partial p_t(x,y)}{\partial t} = L_t^* p_t(x,y) + \Big(\sum_{i=1}^n h_t^i(x) y_t^i - \tfrac{1}{2}|h_t(x)|^2\Big) p_t(x,y),\ 0<t\leq T,$$

$$(5.11)\quad p_0(x,y) = p_0(x).$$

The uniqueness is in the class of measurable functions f: $[0,T]\times\mathbb{R}^d \longrightarrow \mathbb{R}$ satisfying

$$(5.12)\quad \sup_{0\leq t\leq T} \int_{\mathbb{R}^d} |f(t,x)|dx < \infty$$

and

$$(5.13)\quad \int_0^T \int_{\mathbb{R}^d} |h_t(x)|^2\ |f(t,x)|dxdt < \infty.$$

The solution $p_t(x,y)$ has the following properties. If $y_k \longrightarrow y$ in H, then

$$(5.14)\quad \int_{\mathbb{R}^d} |p_t(x,y_k) - p_t(x,y)| \longrightarrow 0.$$

For all $y \in H$,

$$(5.15)\quad \lim_{t'\to t} \int |p_{t'}(x,y) - p_t(x,y)|dx = 0.$$

For the next result, we need to define

$$\mathscr{G}[0,T] = \{f \in C([0,T]\times\mathbb{R}^d): f \in C^{1,2}([0,T]\times\mathbb{R}^d)$$

and

$$\sup_{0\leq t\leq T} |f(t,x)| \leq \exp(K(1+|x|^2)^{\frac{1}{2}}) \text{ for some constant } K\}.$$

Let

$$H_0 = \{\phi \in H: t \longrightarrow \phi_t \text{ is Hölder continuous}\}.$$

Theorem 5.3 In addition to conditions (5.3) and (5.4) suppose that the functions given in (5.7) are locally Hölder continuous satisfying the growth condition

(5.16) $|g(t,x)| \leq K(1+|x|^2)^{\frac{1}{2}}$ for some $K < \infty$.

Let X_0 have a continuous density $p_0(x)$ which satisfies the growth condition

$$p_0(x) \leq \exp\{K(1+|x|^2)^{\frac{1}{2}-\epsilon}\} \tag{5.17}$$

for some constants K and ϵ, $(K<\infty,\ \epsilon>0)$. For the function h we assume local Hölder continuity in t and x.

Then, for all $y \in H_0$, the unnormalized conditional density $p_t(x,y)$ exists and is the unique classical solution to the Cauchy problem (5.10)–(5.11) in the class $\mathcal{G}[0,T]$.

If $y \in H$ is not continuous then $p_t(x,y)$ cannot be a classical solution of (5.10) – (5.11). However, when h is sufficiently smooth a result which gives a formula for $p_t(x,y)$ for every y in H has been given in [4] (Theorem VII. 4.5).

6. Measure–valued equations for the general optimal filter in the white noise theory.

It is useful to recast the Bayes formula in the abstract set up of Theorem 4.1 in the following general form.

Let $H = L^2([0,T],\mathcal{H})$ where $\mathcal{H}$ is a separable Hilbert space and $\xi_t = h_t(X_t)$ where X_t is an S–valued Markov process, defined on $(\Omega,\mathcal{A},\Pi)$, S being a Polish space with Borel σ–field $\mathcal{B}(S)$ and

$$h: [0,T] \times S \longrightarrow \mathcal{H}$$

is a measurable function such that

$$E_\Pi \int_0^T \|h_s(X_s)\|^2_{\mathcal{H}} ds < \infty. \tag{6.1}$$

Then

$$y_t = h_t(X_t) + e_t \tag{6.2}$$

where $e = (e_t)$ is $\mathcal{H}$–valued Gaussian white noise.

In deriving the Zakai equation it was assumed that $S = \mathbb{R}^d$, X_t, a d–dimensional diffusion and $\mathcal{H} = \mathbb{R}^n$. Then the Bayes formula (4.13) can be written in the form

$$E_\alpha(f(X_t)\,|\,Q_t y) = \int_S f(x)\ \hat{F}_t^y(dx) \tag{6.3}$$

where

$$\hat{F}_t^y(B) = \frac{\hat{\Gamma}_t^y(B)}{\hat{\Gamma}_t^y(S)}, \quad B \in \mathcal{B}(S). \tag{6.4}$$

In (6.4), $\hat{F}_t^y$ is the conditional distribution of X_t given Q_t^y and $\hat{\Gamma}_t^y$ is the unnormalized conditional distribution of X_t given $Q_t y$: For $\eta \in H$,

$$\hat{\Gamma}_t^\eta(B) = E_\Pi 1_B(X_t)\exp\{\int_0^t (h_s(X_s),\eta)_{\mathcal{H}} ds - \tfrac{1}{2}\int_0^t \|h_s(X_s)\|_{\mathcal{H}}^2 ds\}.$$

Let $\mathcal{M}(S)$ be the class of finite positive measures on $(S, \mathcal{B}(S))$.

The Bayes formula has been applied to derive the Zakai equation for the unnormalized conditional density and to investigate the existence and uniqueness of solution. In this case we assumed (X_t) to be a diffusion process with $S = \mathbb{R}^d$ and $\mathcal{H} = \mathbb{R}^n$. Equivalently, we could have derived a differential equation for $\hat{\Gamma}_t^y \in \mathcal{M}(\mathbb{R}^d)$ and the conditions under which it has a unique solution would give us a way of tackling the problem for the Zakai equation.

The solution of the filtering problem via the measure–valued equation is a more general approach and when $\mathcal{H}$ is not $\mathbb{R}^n$ but an infinite–dimensional Hilbert space, it is the only natural method available.

For the precise conditions imposed on (X_t), see [4].

They include

(6.5) (a) X_t is an S–valued Markov process whose paths are cadlag.

(b) (X_t) admits a transition probability function $P(\cdot,\cdot,\cdot,\cdot)$, satisfying

$$\Pi(X \in B | X_u, u \leq s) = P_{s,X_s}(B) \quad \text{a.s. } \Pi. \quad (B, \text{ a Borel set in } D).$$

We shall also denote by V_t^s $(s \leq t)$ the two–parameter semigroup associated with (X_t) and by L its generator.

For $\eta \in H$, let $c_s^\eta(x)$: $[0,T]\times S \longrightarrow \mathbb{R}$ be given by

$$c_s^\eta(x) = (h_s(x), \eta_s)_{\mathscr{H}} - \frac{1}{2}\|h_s(x)\|_{\mathscr{H}}^2$$

For $G \in \mathscr{M}(S)$, $\langle f,G \rangle$ stands for $\int f(x)G(dx)$ for a bounded, Borel measurable function f. Let $N_0 = \Pi \circ X_0^{-1}$ and let $\mathscr{K} \subseteq \mathscr{M}(S)$ be the class of measures K_t satisfying the following conditions: For $A \in \mathscr{B}(S)$, $K_0(A) = E_\Pi 1_A(X_0) = N_0(A)$. The map $t \longrightarrow K_t(A)$ is a bounded Borel measurable function and further, K_t is absolutely continuous with respect to $\Pi \circ X_t^{-1}$ with a uniformly bounded Radon–Nikodym derivative for all $t \in [0,T]$.

Theorem 6.1 Let (X_t) satisfy condition (6.5). Assume also that (6.1) holds. Then we have the following conclusions:

(a) $\hat{\Gamma}_t^y$ is the unique solution to the equation

$$\langle g,\hat{\Gamma}_t^y \rangle = \langle V_t^0 g, N_0 \rangle + \int_0^t \langle c_s^y(V_t^s g), \hat{\Gamma}_s^y \rangle ds;$$

(b) $\hat{F}^y_t$ is the unique solution to

$$<g,\hat{F}^y_t> = <V^0_t g, N_0> + \int_0^t <c^y_s(V^s_t g), \hat{F}^y_s>ds$$
$$- \int_0^t <c^y_s,\hat{F}^y_s> <V^s_t g,\hat{F}^y_s>ds.$$

(c) $\hat{\Gamma}^y_t$ is the unique solution to the $\mathcal{M}(S)$–valued equation:

$$<f(t,\cdot),\hat{\Gamma}^y_t> = <f(0,\cdot),N_0> + \int_0^t<(Lf)(s,\cdot)$$
$$+ c^y_s(\cdot)f(s,\cdot),\hat{\Gamma}^y_s>ds,$$

(d) $\hat{F}^y_t$ is the unique solution to the $\mathcal{M}(S)$–valued equation

$$<f(t,\cdot),\hat{F}^y_t> = <f(0,\cdot),N_0> + \int_0^t<(Lf)(s,\cdot)$$
$$+ c^y_s(\cdot)f(s,\cdot),\hat{F}^y_s>ds - \int_0^t<c^y_s,\hat{F}^y_s> <f(s,\cdot),\hat{F}^y_s>ds.$$

In (c) and (d) f is any function belonging to $\mathcal{D}$. The uniqueness in all the above assertions is in the class $\mathcal{K}$. The following result shows the unique solution $\hat{\Gamma}^y_t$ can be obtained by successive approximation. A similar result holds for $\hat{F}^y_t$.

Theorem 6.2 Assume that

$$\|h_s(x)\|_{\mathcal{H}} \leq a(s), \tag{6.6}$$

where a(s) is measurable and $\int_0^T a^2(s)ds < \infty$. Let $\{\hat{\Gamma}^y_{t,k}\}$ be defined inductively by (for $A \in \mathcal{B}(S)$)

$$\hat{\Gamma}^y_{t,0}(A) = E_\Pi 1_A(X_t), \quad \text{and for } k \geq 1$$

$$\hat{\Gamma}^y_{t,k+1}(A) = \hat{\Gamma}^y_{t,0}(A) + \int_0^t \langle c^y_s(V^s_t 1_A), \hat{\Gamma}^y_{s,k}\rangle ds.$$

Then

$$\sup_{0\le t\le T}\Big[\sup_{A\in\mathcal{B}(S)} |\hat{\Gamma}^y_{t,k}(A) - \hat{\Gamma}^y_t(A)|\Big] \longrightarrow 0 \quad \text{as } k \longrightarrow \infty.$$

Filtering problem for infinite dimensional processes

By an infinite dimensional filtering problem we mean one in which the model in the conventional theory is of the form

$$(6.7) \qquad Y_\ell = \int_0^t h_s(X_s) + W_t$$

where W_t is a Wiener process taking values in a separable infinite dimensional Banach space B and X_t is a sample continuous B–valued Markov process such that X and W are independent. Here we have taken S = B and $(\mathcal{H},B,P)$ is an abstract Wiener space. The function h: $[0,T] \times B \longrightarrow \mathcal{H}$ satisfies the condition (6.1), i.e., $E\int_0^T \|h_s(X_s)\|^2_{\mathcal{H}} ds < \infty$. The model can be used for certain problems of filtering of random fields. See [6] for concrete examples. The skeletal analog of this example is included in the general white noise model for such filtering problems described below.

Let $(\gamma,\mathcal{H},B)$ be an abstract Wiener space. Let $\mathcal{X} := C_0([0,T],B)$ and $H = L^2([0,T]),\mathcal{H})$. Then there exists a Gaussian measure μ on $(\mathcal{X},\mathcal{B}(\mathcal{X}))$ such that if W_t is the co–ordinate mapping on $\mathcal{X}$ then under μ, W is a B–valued Wiener process with

$$Ef_1[W_t] = 0, \quad Ef_1[W_t]f_2[W_s] = (s\wedge t)\,(\gamma^* f_1, \gamma^* f_2)_{\mathcal{H}}$$

where $f_1,f_2 \in B^*$, the dual of B and γ^* is the adjoint of the injection map γ. Let X_t be a process defined on

$(\Omega, \mathscr{A}, \Pi)$ taking values in B. The white noise filtering model is given by (6.2).

Theorems 6.1 and 6.2 can be used to solve the filtering problem. The unnormalized conditional measure $\hat{\Gamma}_t^y(\cdot) \in \mathscr{M}(B)$.

7 Consistency and robustness

By consistency of the white noise theory is meant the requirement that the optimal filter in the conventional approach can be obtained by a sequence of approximations of solutions to the white noise version of the problem. The theory given below is a typical representative of such results.

Theorem 7.1 $\hat{\Gamma}_t^y$ is an $\mathscr{M}(S)$–valued accessible r.v. on $(E, \mathscr{E}, \alpha)$ and

$$\Gamma_t^y = R_\alpha[\hat{\Gamma}_t^y]. \tag{7.1}$$

In other words, the unique solution of the $\mathscr{M}(S)$–valued differential eqaution of Theorem 6.1(c) is a skeleton (in the sense of Section 3) of Γ_t^y, the unnormalized conditional probability distribution of X_t given $\mathscr{F}_t^Y$.

Robustness There are two kinds of robustness results. Of these, statistical robustness is probably the more interesting and useful.

Theorem 7.2 Assume the conditions of Theorem 6.1. Then for y^1 and y^2 in H,

$$\sup_{A \in \mathscr{B}(S)} |\hat{\Gamma}_t^{y^1}(A) - \hat{\Gamma}_t^{y^2}(A)| \le K\|Q_t y^1 - Q_t y^2\| \exp[\tfrac{1}{2}|Q_t y^1|^2 + \tfrac{1}{2}|Q_t y^2|^2] \tag{7.2}$$

where $K = E_\Pi \int_0^T \|h_s(X_s)\|^2_{\mathscr{H}} ds < \infty$.

Theorem 7.3 (Statistical robustness). Let X and X^k

$(k \geq 1)$ be processes on $(\Omega, \mathscr{A}, \Pi)$ with paths in $D([0,T],S)$. Suppose that $\hat{\Gamma}_t^{y,k}$ is given by a formula analogous to $\hat{\Gamma}_t^y$ with X^k in place of X. Let the following conditions be satisfied.

(7.3) $\{X^k\}$ converges in distribution to X (as D–valued r.v.'s)

(7.4) $\Pi(X_t \neq X_{t-}) = 0$ for all t.

(7.5) $h: [0,T] \times S \to \mathscr{H}$ is continuous.

Then for all $y \in H$, $\hat{\Gamma}_t^{y,k} \to \hat{\Gamma}_t^y$ in the weak topology of $\mathscr{M}(S)$.

8 Nonlinear prediction and smoothing in the white noise theory.

Consider the problem of predicting the value of X_s, $s>t$ on the basis of the observations $\{y_u, 0 \leq u \leq t\}$. We have then to calculate the conditional expectation

$$E_\alpha[f(X_s) | Q_t y] := \pi_{st}(f,y).$$

Similarly, for the smoothing problem, $s < t$. In either case, the desired conditional expectation is given by the Bayes formula of Section 4.

$$(8.1) \qquad E_\alpha[f(X_s) | Q_t y] = \frac{1}{\hat{\Gamma}_{st}^y(\mathbb{R}^d)} \int f(x) \, \hat{\Gamma}_{s,t}^y(dx)$$

where

$\hat{\Gamma}_{s,t}^\phi \in \mathscr{M}(\mathbb{R}^d)$ for $0 \leq s, t \leq T$, $\phi \in H$ is given by

$$(8.2) \quad \hat{\Gamma}_{s,t}^\phi(B) = E_\Pi[1_B(X_s) q_t(\phi, \cdot)] \quad \text{for } B \in \mathscr{B}(\mathbb{R}^d).$$

In (8.1) f is any bounded, real valued measurable function

on $\mathbb{R}^d$ and (X_t) is a diffusion process as in Section 5. If the density $p_{st}(x,\phi)$ of $\hat{\Gamma}^{\phi}_{st}$ exists w.r.t. d–dimensional Lebesgue measure λ, we can obtain the corresponding Zakai equation for prediction and smoothing respectively and study the conditions under which $p_{st}(x,y)$ is the unique solution. We shall give the main result only for the case of prediction. The smoothing problem is treated similarly. Both problems are dealt with at length in [4]. First, an auxiliary result common to prediction and smoothing is of interest.

Theorem 8.1 Suppose that for all t, X_t has a density $p_t(x)$ w.r.t. λ.

(i) Then, for all $\phi \in H$, $\hat{\Gamma}^{\phi}_{s,t}$ is absolutely continuous w.r.t. λ. Further, we can choose a jointly measurable version $(s,t,x) \longrightarrow p_{st}(x,\phi)$ of the density $d\Gamma^{\phi}_{s,t}/d\lambda$.

(ii) If $y_k \longrightarrow y$ in H, then

$$\lim_{k\to\infty} \int_{\mathbb{R}^d} |p_{s,t}(x,y_k) - p_{s,t}(x,y)|\,\lambda(dx) = 0.$$

We conclude with the solution of the prediction problem via the Zakai equation contained in the following two results.

Theorem 8.2 Assume the conditions of Theorem 8.1. Further let the coefficients a_{ij}, b_i in the generator of X_t satisfy (5.3)–(5.5) and assume the growth condition (5.8) for p_0. Let h: $[0,T]\times\mathbb{R}^d \longrightarrow \mathbb{R}$ be a measurable function such that (5.9) is satisfied. Then the following conclusions hold.

(i) For each $y \in H$, the Cauchy problem

$$(8.3)\quad \frac{\partial v(s,x)}{\partial s} = L_s^* v(s,x) + \Big[\sum_{i=1}^{n} h_s^i(x) y_s^i - \tfrac{1}{2}|h_s(x)|^2\Big] v(s,x),$$

$$0<s\leq t,$$

$$(8.4)\quad v(0,x) = p_0(x)$$

has a unique generalized solution in the class of functions f satisfying (5.12) and

$$(8.5)\quad \int_0^t \int_{\mathbb{R}^d} |h_s(x)|^2 |f(s,x)| dx\, ds < \infty.$$

(ii) For each $y \in H$, the conditional density $p_{s,t}(x,y)$ satisfies

$$(8.6)\quad p_{s,t}(x,y) = v(s,x) \quad \text{for a.e. } x \text{ if } s \geq t,$$

where v is the unique generalized solution of (8.3)–(8.4).

Suppose (X_t) satisfies the conditions of Theorem 5.2 and 5.3. Let condition (5.18) hold. Denote by $p(t,z,s,x)$ the transition probability density of (X_t). Then we have

Theorem 8.3 For $s > t$, define $p_{s,t}(x,y)$ by

$$(8.7)\quad p_{s,t}(x,y) = \int_{\mathbb{R}^d} p_t(z,y) p(t,z,s,x) dx$$

for $y \in H$. (Here, the unnormalized conditional density exists by the preceding results).

(i) Then $p_{s,t}(x,y)$ is the unnormalized conditional density of X_s given $Q_t y$.

(ii) Suppose that for $y \in H$, $x \to p_t(x,y)$ is continuous. Then $p_{st}(x,y)$ is the unique classical solution to the Cauchy problem

$$(8.8)\quad \frac{\partial}{\partial s} p_{st}(x,y) = L_s^* p_{s,t}(\cdot,y)(x), \quad t<s\leq T,$$

$$p_{tt}(x,y) = p_t(x,y). \tag{8.9}$$

The uniqueness is in the class $\mathcal{G}[t,T]$ which is defined in the same way as $\mathcal{G}[0,T]$ replacing 0 by t.

Consistency and robustness results are also available for the prediction and smoothing problems.

References

[1] H.P. Hucke, G. Kallianpur and R.L. Karandikar, Smoothness properties of the conditional expectation in finitely additive white noise filtering, J. Mult. Analysis 27, (1988), 261–269.

[2] G.W. Johnson and G. Kallianpur, Some results on Hu and Meyer's paper and infinite dimensional calculus on finitely additive canonical Hilbert space, Theory of Probability and Its Applications, 34, (1989), 118–128.

[3] G.W. Johnson and G. Kallianpur, Homogeneous chaos, p–forms, scaling and the Feynman integral, University of North Carolina Center for Stochastic Processes Technical Report No. 274, (1989), submitted for publication.

[4] G. Kallianpur and R.L. Karandikar, *White Noise Theory of Prediction, Filtering and Smoothing*, Stochastic Monographs, vol. 3, Gordon and Breach Science Publishers, New York, London (1988).

[5] G. Kallianpur and R.L. Karandikar, White noise calculus and nonlinear filtering theory, Ann. Probab. 13, 1033–1107 (1985).

[6] G. Kallianpur, *Stochastic Filtering Theory*, Springer–Verlag, New York (1980).

[7] G. Kallianpur and R.L. Karandikar, The filtering problem for infinite dimensional stochastic processes, Stochastic differential systems, stochastic control theory and applications, IMA Vol. 10, Springer–Verlag, New York, (1988), 215–223.

[8] G. Kallianpur and A.S. Ustunel, Distributions, Feynman integrals and measures on abstract Wiener spaces (1990) submitted for publication.

[9] M. Zakai, Stochastic integration, trace and the skeleton of Wiener functionals, (1989), preprint.

EQUILIBRIUM IN A SIMPLIFIED DYNAMIC, STOCHASTIC ECONOMY WITH HETEROGENEOUS AGENTS

Ioannis Karatzas
Rutgers University

Peter Lakner
New York University

John P. Lehoczky
Carnegie Mellon University

Steven E. Shreve
Carnegie Mellon University

Abstract

We study a dynamic, stochastic economy with several agents, who may differ in their endowments (of a single commodity) and in their utilities. An equilibrium financial market is constructed, under the condition that all agents have infinite marginal utility at zero. If, in addition, the Arrow–Pratt indices of relative risk aversion for all agents are less than or equal to one, then uniqueness of equilibrium is also proved. When agents consume and invest in this equilibrium market so as to maximize their expected utility of consumption, their aggregate endowment is consumed as it enters the economy and all financial instruments are held in zero net supply. Explicit examples are provided.

1. Introduction

A fairly complete theory has been developed recently for the optimal consumption/investment problem of a small investor with a general utility function [3,4,13]. Using tools from stochastic calculus, explicit expressions for the optimal consumption policy and terminal wealth can be provided when stock prices are modelled by Itô processes. The present paper draws on the methodology of [3,4,13] to construct *equilibrium* in a multi–agent economy, and to establish uniqueness.

ISBN 0-12-481005-5

We suppose there is a finite number, N, of agents (small investors), each of whom receives an endowment stream denominated in units of a single, infinitely divisible commodity. The agents may have different endowment streams and utility functions. Each agent attempts to maximize his expected total utility from consumption of this commodity, over a finite horizon [0,T]. We shall construct a financial market, consisting of a bond and a finite number of stocks, which provides a vehicle for trading among the agents and thereby allows them to hedge the risk and smooth the nonuniformity associated with their respective endowments. The equilibrium problem is to construct this market in such a way that, when the stock and bond prices are accepted by the individual agents in the determination of their optimal policies, all the commodity is entirely consumed as it enters the economy and all the financial assets are held in zero net supply.

The present paper is quite similar to Duffie & Zame [9]. Both Duffie & Zame [9] and this work generalize the results of Cox, Ingersoll & Ross [5] in two important directions. First, heterogeneous agents are allowed, whereas in [5] all agents have the same endowments and the same utility functions. Secondly, endowment processes are adapted in a general way to an underlying d–dimensional Brownian motion, whereas in [5] this dependence on the underlying Brownian motion must be via a state process so that Markov methods could be employed. Duffie & Zame [9] and this paper both derive a formula for the endogenously determined equilibrium interest rate which agrees with that of [5] when specialized to their model. Both [9] and this paper derive formulas for the coefficients of the stock processes and the optimal consumption processes of the individual agents.

The Cox, Ingersoll & Ross interest rate formula is given in terms of an indirect utility function, J, derived from the single direct utility function, U, in their model. In our model, each agent has a utility function, U_n, and we construct a "representative agent" whose utility function will play the role of the Cox, Ingersoll & Ross function U. Roughly speaking, this representative agent acts as a proxy for the individual agents by receiving their aggregate endowment, solving his own optimization problem with utility function

$$(1.1) \qquad U(t,c;\Lambda) \triangleq \max_{\substack{c_1\geq 0,\ldots,c_N\geq 0\\ c_1+\ldots+c_N=c}} \Sigma_{n=1}^N \lambda_n U_n(t,c_n),$$

and then apportioning his optimal commodity consumption process to the agents, instead of actually consuming it. The search for equilibrium is reduced to a search for an appropriate vector $\Lambda \in (0,\infty)^N$ in (1.1); cf. Sections 9 and 12. At this point, our work differs from Duffie & Zame [9], who introduce the representative agent but construct equilibrium in an infinite–dimensional functional space. One advantage of posing the equilibrium problem in a finite–dimensional space is that in this context, one can develop arguments resolving the question of uniqueness, an issue not addressed by Duffie & Zame [9] and largely ignored in the finance literature.

We use the Knaster–Kuratowski–Mazurkiewicz lemma [2, p. 26] to give a very simple proof of the existence of equilibrium. A different proof under slightly different assumptions on the endowment processes can be obtained directly from Araujo and Monteiro [1]. Under the assumption that the agents' measure of relative risk aversion is less than or equal to one, we show by a separate simple argument that the agents' equilibrium optimal consumption processes, as well as the equilibrium interest rate, are unique. Furthermore, the coefficients of the equilibrium stock price processes are unique up to the formation of mutual funds.

Some generalizations of this model are possible. First, one could easily include capital assets which are owned by the N agents, pay dividends, and can be traded among the agents. The additional condition of equilibrium, i.e., that all such assets are exactly owned by the agents, can be easily met. A formula for the arbitrage–free price of such assets is given in Section 13. Secondly, throughout this paper we consider only individual agent utility functions satisfying the condition $U'_k(t,0) = \infty$. Generalization to the case in which $U'_k(t,0) < \infty$ for at least one of the agents is possible, but care is required.

To accommodate this case within our framework, one needs a more general model of the financial markets than we define in Section 2. For equilibrium to hold in general, both the stock and bond price processes must have singularly continuous components. One can describe the bond price process, but due to the singularly continuous component, there will be no interest rate process; see [14] and the earlier work contained in the appendix of [9]. There is an alternative model presented in [15], following the formulation of Duffie [7] and Duffie and Huang [8], which avoids requiring the financial assets to have singularly continuous components. We refer to this as the *moneyed model*; in it, prices are denominated in some currency, rather than in units of the commodity. There is also a commodity spot price process which gives the value of the commodity in that currency. In [15], the agents' commodity endowments and the prices of the financial assets are given exogenously, and the commodity spot price is determined endogenously by the equilibrium conditions. The existence and essential uniqueness of equilibrium are proved in [15] without any condition on $U_k'(t,0)$, $1 \le k \le N$. None of the financial assets will have singularly continuous parts in their price processes, but when those prices are divided by the commodity spot price to value them in commodity units, singularly continuous components can arise.

The present work is a self–contained companion to the more detailed and comprehensive article [15]. It is designed to be more accessible than [15] in that it deals exclusively with the moneyless model when all agents have infinite marginal utility at zero. These conditions obviate a number of complex technicalities; in particular, they permit a different proof of uniqueness for equilibrium, which is simpler than that appearing in [15]. Since this paper was first drafted, Dana & Pontier [6] have provided an equilibrium existence which does not require our assumption (3.2) and which accomodates a weakening of our assumption of a bounded aggregate endowment process. The existence proof of Dana & Pontier [6] is considerably simpler than our original proof, but similar to the proof we give here.

2. The Agents and their Endowments

We consider an economy consisting of N agents. Each agent, n, receives a nonnegative exogenous <u>endowment process</u> of a single commodity $\epsilon_n = \{\epsilon_n(t); 0 \le t \le T\}$, where T is the fixed, positive planning horizon. These endowment processes are uncertain, and we model them as Itô processes taking values in $[0,\infty)$. More precisely, let $W = (W_1,\ldots,W_d)^*$ be a d–dimensional Brownian motion on a complete probability space $(\Omega,\mathcal{F}, \mathbb{P})$, and let $\{\mathcal{F}_t\}$ denote the augmentation by null sets of the filtration generated by W. Assume that for $n = 1,\ldots,N$, there are bounded, $\{\mathcal{F}_t\}$–progressively measurable processes μ_n and ρ_n taking values in $\mathbb{R}$ and $\mathbb{R}^d$, respectively, such that

$$(2.1) \qquad \epsilon_n(t) = \epsilon_n(0) + \int_0^t \mu_n(s)ds + \int_0^t \rho_n^*(s)dW(s), \quad 0 \le t \le T,$$

where $\epsilon_n(0)$ is a deterministic, nonnegative constant.

We define the <u>aggregate endowment</u> $\epsilon(t) \triangleq \sum_{n=1}^N \epsilon_n(t)$, $0 \le t \le T$, and define also $\mu(t) \triangleq \sum_{n=1}^N \mu_n(t)$, $\rho(t) \triangleq \sum_{n=1}^N \rho_n(t)$, $0 \le t \le T$. Then

$$(2.2) \qquad \epsilon(t) = \epsilon(0) + \int_0^t \mu(s)ds + \int_0^t \rho^*(s)dW(s), \quad 0 \le t \le T.$$

We assume that for each n, ϵ_n is not identically zero, and that there exist positive constants k and K for which $k \le \epsilon(t) \le K$, $0 \le t \le T$, a.s.

3. The Agents' Utility Functions

We suppose that each agent, n, has a utility function $U_n : [0,T] \times (0,\infty) \to \mathbb{R}$ which is continuous and enjoys the following properties:

(i) for every $t \in [0,T]$, $U_n(t,\cdot)$ is strictly increasing and strictly concave;

(ii) the derivatives $\frac{\partial}{\partial t} U_n$, $\frac{\partial}{\partial c} U_n$, $\frac{\partial^2}{\partial t \partial c} U_n$, $\frac{\partial^2}{\partial c^2} U_n$ and $\frac{\partial^3}{\partial c^3} U_n$ exist and are continuous on $[0,T] \times (0,\infty)$;

(iii) for every $t \in [0,T]$, $U_n' \triangleq \frac{\partial}{\partial c} U_n$ satisfies

$$(3.1) \qquad U_n'(t,\infty) \triangleq \lim_{c \to \infty} U_n'(t,c) = 0, \quad U_n'(t,0) \triangleq \lim_{c \downarrow 0} U_n'(t,c) = \infty.$$

We define $U_n(t,0) \triangleq \lim_{c \downarrow 0} U_n(t,c)$, which may be $-\infty$.

In order to prove the uniqueness of equilibrium, we shall impose in Section 12 the additional condition

(iv) for every $t \in [0,T]$, the function $c \mapsto c\, U_n'(t,c)$ is nondecreasing.

Condition (iv) is equivalent to assuming that the Arrow–Pratt measure of relative risk aversion, $-cU_n''(t,c)/U_n'(t,c)$, is less than or equal to one [17, p. 69].

Examples of functions which satisfy conditions (i) – (iv) are $e^{-\alpha t} \log c$ and $\frac{1}{\gamma} e^{-\alpha t} c^\gamma$, where $\alpha \in \mathbb{R}$ and $0 < \gamma < 1$. When $\gamma < 0$, the function $\frac{1}{\gamma} e^{-\alpha t} c^\gamma$ violates condition (iv), but if all agents have this utility function, the uniqueness of equilibrium can be established by explicit computations; see Example 11.1.

4. The Financial Market

The agents in our model receive utility from consumption of the single commodity with which they are endowed. Because an individual agent's endowment process is typically random and non–uniform, he would find it advantageous to participate in a market which allows him both to hedge risk and to smooth his consumption. We shall create such a market endogenously by equilibrium considerations.

We introduce the financial market in this section; its coefficients will be specified in section 10, in terms of the endowment processes and utility functions of the individual agents. The market has $d+1$ assets. One of them is a pure discount bond, with price

$$P_0(t) = P_0(0)\exp\{\int_0^t r(s)ds\} \tag{4.1}$$

at time t. The remaining d assets are risky stocks, and the price per share $P_i(t)$ of the ith stock is modelled by the linear stochastic differential equation

$$dP_i(t) = P_i(t)[b_i(t)dt + \sum_{j=1}^{d} \sigma_{ij}(t)dW_j(t)]; \quad i = 1,\dots,d. \tag{4.2}$$

All these prices are denominated in units of the commodity with which the agents are endowed. The interest rate $r(\cdot)$ of the bond, the mean rate of return vector $b(\cdot) = (b_1(\cdot),\dots,b_d(\cdot))^*$ of the stocks, and the volatility matrix $\sigma(\cdot) = \{\sigma_{ij}(\cdot)\}_{1\le i,j\le d}$, will all be bounded, $\{\mathcal{F}_t\}$–progressively measurable processes. In addition, we shall impose the uniform nondegeneracy condition

$$\xi^*\sigma(t)\sigma^*(t)\xi \ge \delta\,\|\xi\|^2, \quad 0 \le t \le T, \text{ a.s.}, \tag{4.3}$$

for some $\delta > 0$. Under (4.3), the inverses of both $\sigma(\cdot)$ and $\sigma^*(\cdot)$ exist and are bounded. In particular, the relative risk process

$$\theta(t) \triangleq (\sigma(t))^{-1}[b(t) - r(t)\underset{\sim}{1}], \quad 0 \leq t \leq T, \tag{4.4}$$

is bounded and progressively measurable, where $\underset{\sim}{1}$ denotes the d–dimensional vector with every component equal to 1.

It follows then from the Girsanov theorem (e.g. [16, section 3.5]) that the exponential supermartingale

$$Z(t) \triangleq \exp\{-\int_0^t \theta^*(s)dW(s) - \frac{1}{2}\int_0^t \|\theta(s)\|^2 ds\}, \quad \mathscr{F}_t; \quad 0 \leq t \leq T, \tag{4.5}$$

is actually a martingale, and that $\tilde{W}(t) \triangleq W(t) + \int_0^t \theta(s)ds$; is Brownian motion under the probability measure $\tilde{\mathbb{P}}(A) \triangleq E(Z(T)1_A)$; $A \in \mathscr{F}_T$. Under this measure, the discounted stock price processes $\beta(t)P_i(t)$, with

$$\beta(t) \triangleq (P_0(t))^{-1} = \frac{1}{P_0(0)} \exp\{-\int_0^t r(s)ds\} \tag{4.6}$$

are martingales, a fact of great importance in the modern theory of continuous trading (cf. [10,11,18] for its connections with the notions of "absence of arbitrage opportunities" and "completeness" in the market model). We shall see in Remark 7.1 that the process

$$\zeta(t) \triangleq \beta(t)Z(t) \;;\; 0 \leq t \leq T, \tag{4.7}$$

acts as a "deflator", in the sense that multiplication by $\zeta(t)$ converts wealth held at time t to the equivalent amount of wealth at time zero.

We impose on ζ the condition

$$0 < k \le \zeta(t) \le K, \quad 0 \le t \le T, \text{ a.s.,} \tag{4.8}$$

for some constants k and K.

5. The Individual Agents' Optimization Problems

Once a financial market is specified, as it will be in Section 10, each agent, n, acts as a price–taker. He has at his disposal the choice of an $\mathbb{R}^d$–valued <u>portfolio process</u> $\pi_n(t) = (\pi_{n1}(t),\ldots,\pi_{nd}(t))^*$ and a nonnegative <u>consumption</u> rate <u>process</u> $c_n(t)$, $0 \le t \le T$. He must choose both these processes to be $\{\mathcal{F}_t\}$–progressively measurable and to satisfy $\int_0^T (c_n(t) + \|\pi_n(t)\|^2)dt < \infty$, almost surely. The interpretation here is that $\pi_{ni}(t)$ represents the amount of commodity invested at time t by the nth investor in the ith stock.

If we denote by $X_n(t)$ the wealth of the nth investor at time t, then $X_n(t) - \sum_{i=1}^d \pi_{ni}(t)$ is the amount invested in the bond. Neither this quantity nor the individual $\pi_{ni}(t)$'s are constrained to be nonnegative, i.e., borrowing at the interest rate $r(t)$ and short–selling of stocks are permitted.

The wealth X_n corresponding to a given portfolio/consumption pair (π_n, c_n) satisfies the equation

$$(5.1)\qquad dX_n(t) = [\epsilon_n(t) - c_n(t)]dt + \sum_{i=1}^{d} \pi_{ni}(t)[b_i(t) + \sum_{j=1}^{d} \sigma_{ij}(t)dW_j(t)]$$

$$+ [X_n(t) - \sum_{i=1}^{d} \pi_{ni}(t)]r(t)dt$$

$$= r(t)X_n(t)dt + [\epsilon_n(t) - c_n(t)]dt + \pi_n^*(t)\sigma(t)d\tilde{W}(t)$$

whose solution is

$$(5.2)\quad \beta(t)X_n(t) = \int_0^t \beta(s)[\epsilon_n(s) - c_n(s)]ds + \int_0^t \beta(s)\pi_n^*(s)\sigma(s)d\tilde{W}(s), \quad 0 \le t \le T.$$

5.1 Definition

A portfolio/consumption pair (π_n, c_n) is called admissible for agent n if the corresponding wealth process, X_n, is bounded from below and satisfies $X_n(T) \geq 0$, almost surely.

The nth agent's optimization problem is to maximize the expected total utility from consumption $E\int_0^T U_n(t, c_n(t))dt$ over all admissible pairs (π_n, c_n) that satisfy

$$(5.3)\qquad E\int_0^T \max\{0, -U_n(t, c_n(t))\}dt < \infty.$$

Condition (5.3) is imposed to ensure that $E\int_0^T U_n(t,c_n(t))dt$ is defined. We shall let $(\hat{\pi}_n,\hat{c}_n)$ denote an optimal pair for this problem, and let $\hat{X}_n$ denote the associated wealth process. The existence of $(\hat{\pi}_n, \hat{c}_n)$ is established in Section 7.

6. The Definition of Equilibrium

We are now in a position to define the notion of equilibrium.

6.1 Definition

We say that the financial market (more specifically, the processes $r(\cdot)$, $b(\cdot)$ and $\sigma(\cdot)$) introduced in Section 4 results in equilibrium if, in the notation of Section 5, we have almost surely

$$\sum_{n=1}^{N} \hat{c}_n(t) = \epsilon(t), \qquad 0 \le t \le T, \tag{6.1}$$

$$\sum_{n=1}^{N} \hat{\pi}_{ni}(t) = 0, \qquad 0 \le t \le T \text{ and } 1 \le i \le d, \tag{6.2}$$

$$\sum_{n=1}^{N} \hat{X}_n(t) = 0\,, \qquad 0 \le t \le T. \tag{6.3}$$

The above conditions enforce the clearing of the spot market in the commmodity, and the clearing of the stock and bond markets, respectively.

7. Solution of the nth Agent's Problem

In order to characterize an equilibrium financial market, we let a financial market be given and study individual agent behavior in its presence. Let us therefore consider an admissible pair (π_n, c_n) and evaluate the corresponding wealth process X_n at the stopping time $\tau_m \triangleq T \wedge \inf\{t \in [0,T];$ $\int_0^t \beta^2(s)\|\pi_n^*(s)\sigma(s)\|^2 ds \geq m\}$ for an arbitrary positive integer m. Taking expectation under $\tilde{\mathbb{P}}$ in (5.2) evaluated at $t = \tau_m$, we obtain $E\int_0^{\tau_m} \zeta(s)c_n(s)ds$ $= E\int_0^{\tau_m} \zeta(s)\epsilon_n(s)ds - E[\zeta(\tau_m)X_n(\tau_m)]$. Now we let $m \to \infty$. Admissibility and Fatou's lemma give $\underline{\lim}_{m\to\infty} E[\zeta(\tau_m)X_n(\tau_m)] \geq E[\zeta(T)X_n(T)] \geq 0$. This, coupled with the Monotone Convergence Theorem, yields in (7.1):

$$E\int_0^T \zeta(s)c_n(s)ds \leq E\int_0^T \zeta(s)\epsilon_n(s)ds. \tag{7.1}$$

7.1 Remark

Inequality (7.1) can be regarded as a <u>budget constraint</u>, and it justifies the terminology "deflator" for the process ζ of (4.7). It mandates that the expected total value of consumption, deflated back to the original time, does not exceed the expected total deflated value of endowment.

7.2 Proposition

Let a financial market be given. If (π_n, c_n) is an admissible pair for agent n, then (7.1) holds. Conversely, for any consumption process c_n satisfying (7.1), there exists a portfolio process π_n such that the pair (π_n, c_n) is admissible.

Proof:

It remains to justify the second claim; for any consumption process c_n satisfying (7.1), introduce the random variable $D_n \triangleq \int_0^T \beta(s)[\epsilon_n(s)-c_n(s)]ds$ and observe that (7.2) amounts to $\tilde{E}D_n \geq 0$. Now the $\tilde{\mathbb{P}}$–martingale $M_n(t) \triangleq \tilde{E}D_n - \tilde{E}(D_n | \mathcal{F}_t)$, can be written as a stochastic integral $M_n(t) = \int_0^t \beta(s)\pi_n^*(s)\sigma(s)d\tilde{W}(s)$ for a suitable portfolio process π_n, by virtue of the martingale representation theorem (cf. [16, Problem 3.4.16 and proof of Proposition 5.8.6]). Finally, the process

$$X_n(t) = \frac{1}{\beta(t)} \{\int_0^t \beta(s)[\epsilon_n(s)-c_n(s)]ds + M_n(t)\} \tag{7.2}$$

is obviously, from (5.2), the wealth associated with the pair (π_n, c_n) and satisfies

$$\zeta(t)X_n(t) = Z(t)\tilde{E}D_n - E\{\int_t^T \zeta(s)[\epsilon_n(s)-c_n(s)]ds | \mathcal{F}_t\}; \quad 0 \leq t \leq T, \text{ a.s.}$$

Both requirements of Definition 5.1 for admissibility follow easily from this representation, the boundedness of ϵ, and (4.8). □

We conclude from Proposition 7.2 that the nth agent's optimization problem can be cast thus: to maximize the expected utility from consumption $E\int_0^T U_n(t,c_n(t))dt$ over consumption processes c_n which satisfy (7.1) and (5.3).

In order to solve this problem, we introduce $I_n(t,\cdot)$, the inverse of the strictly decreasing mapping $U_n'(t,\cdot)$ from $(0,\infty)$ onto itself. It is a straightforward verification that

(7.3) $$U_n(t,I_n(t,y)) - yI_n(t,y) = \max_{c\geq 0}[U_n(t,c)-yc]; \ \forall\ (t,y) \in [0,T] \times (0,\infty).$$

Because I_n is jointly continuous (in fact, jointly C^1 because of condition (ii) of Section 3) and ζ satisfies (4.8), the function $\mathscr{X}_n(y) \triangleq E\int_0^T \zeta(t)I_n(t,y\zeta(t))dt$ maps $(0,\infty)$ onto itself and is continuous and strictly decreasing. Define y_n to be the unique positive number for which

(7.4) $$\mathscr{X}_n(y_n) = E\int_0^T \zeta(t)\epsilon_n(t)dt,$$

and set

(7.5) $$\hat{c}_n(t) \triangleq I_n(t,y_n\zeta(t)), \quad 0 \leq t \leq T.$$

Then $\hat{c}_n$ satisfies (7.1) with equality, and is bounded away from zero because ζ is bounded, so (5.3) holds. Let c_n be another consumption process satisfying (5.3) and (7.1). From (7.3) we have

$$
\begin{aligned}
&E\int_0^T U(t,\hat{c}_n(t))dt - E\int_0^T U(t,c_n(t))dt \\
&\geq E\int_0^T [U(t,I(t,y_n\zeta(t))) - y_n\zeta(t)I(t,y_n\zeta(t))]dt \\
&\quad - E\int_0^T [U(t,c_n(t)) - y_n\zeta(t)c_n(t)]dt \geq 0.
\end{aligned}
$$

Therefore, $\hat{c}_n$ is optimal. Proposition 7.2 guarantees the existence of $\hat{\pi}_n$.

8. Characterization of Equilibrium

The issue now is how to choose the market coefficients $r(\cdot)$, $b(\cdot)$ and $\sigma(\cdot)$ so that when, for each n, $\hat{c}_n$ is given by (7.5) and $\hat{\pi}_n$ is the corresponding portfolio process whose existence is guaranteed by Proposition 7.2, relations (6.1) – (6.3) are satisfied. It turns out that the only relevant aspect of $r(\cdot)$, $b(\cdot)$ and $\sigma(\cdot)$ is the process ζ they lead to, as shown by the following proposition.

8.1 Proposition

Let $r(\cdot)$, $b(\cdot)$ and $\sigma(\cdot)$, as described in Section 4, be given, and suppose that the equilibrium conditions (6.1) – (6.3) are satisfied. Then

$$
\epsilon(t) = \sum_{n=1}^{N} I_n(t,y_n\zeta(t)), \quad 0 \leq t \leq T, \tag{8.1}
$$

where y_n is defined by (7.4) and ζ is given by (4.7). Conversely, suppose there exist $r(\cdot)$, $b(\cdot)$ and $\sigma(\cdot)$ whose corresponding process ζ satisfies (8.1); then the equilibrium conditions (6.1) – (6.3) are also satisfied.

Proof:
For the first assertion, recall that for $n = 1,\dots,N$, the optimal consumption processes are given by (7.5). The spot market clearing condition (6.1) leads to (8.1).

For the converse assertion, note that for the ζ in question, the optimal consumption processes $\hat{c}_n$, $1 \le n \le N$, are again given by (7.5). Denote by $\hat{D}_n$, $\hat{M}_n$, $\hat{\pi}_n$ and $\hat{X}_n$ the corresponding processes constructed in Section 7, which now satisfy $\tilde{E}\hat{D}_n = 0$ and $\hat{X}_n(T) = 0$ a.s. From (8.1) we have $\sum_{n=1}^N \hat{D}_n = 0$, a.s. It follows then that $\sum_{n=1}^N \hat{M}_n(t) = \sum_{n=1}^N \hat{X}_n(t) = 0$, $0 \le t \le T$, a.s. Thus (6.1) and (6.3) are satisfied. Furthermore, the quadratic variation of $\sum_{n=1}^N \hat{M}_n$ on $[0,T]$, is equal to $\int_0^T \beta^2(s)\, \|\sigma^*(s) \sum_{n=1}^N \pi_n(s)\|^2 ds$, so this quantity is zero. Because σ^* is nonsingular, (6.2) must hold. □

9. The Representative Agent

For every $\Lambda = (\lambda_1,\dots,\lambda_N) \in (0,\infty)^N$, let us introduce the function

$$
(9.1)\quad U(t,c;\Lambda) = \max_{\substack{c_1 \ge 0,\dots,c_N \ge 0 \\ c_1+\dots+c_N = c}} \sum_{n=1}^N \lambda_n U_n(t,c_n); \quad (t,c) \in [0,T] \times (0,\infty),
$$

which inherits the basic properties of the individual utility functions U_n, as set out below. It is easily checked that the maximization in (9.1) is achieved by

$$
(9.2)\qquad c_n = I_n(t, \frac{1}{\lambda_n} H(t,c;\Lambda)),
$$

where $H(t,\cdot;\Lambda)$ is the inverse of the strictly decreasing function $I(t,\cdot;\Lambda)$ from $(0,\infty)$ onto itself, defined by

$$I(t,h;\Lambda) \triangleq \sum_{n=1}^{N} I_n(t, \frac{h}{\lambda_n}). \tag{9.3}$$

In order to examine the differentiability of $U(\cdot,\cdot;\Lambda)$, we first note that for each n, I_n is jointly C^1 because of condition (ii) of Section 3.1 and the Implicit Function Theorem. Differentiating the equation $U_n'(t, I_n(t,y)) = y$ twice with respect to y, one sees that $\frac{\partial^2}{\partial y^2} I_n$ exists and is continuous. Consequently, for each $\Lambda \in (0,\infty)^N$, $\frac{\partial}{\partial t} I(\cdot,\cdot;\Lambda)$, $\frac{\partial}{\partial y} I(\cdot,\cdot;\Lambda)$ and $\frac{\partial^2}{\partial y^2} I(\cdot,\cdot;\Lambda)$ exist and are continuous. Because $I(t,H(t,c;\Lambda);\Lambda) = c$ we can similarly conclude that $\frac{\partial}{\partial t} H$, $\frac{\partial}{\partial c} H$ and $\frac{\partial^2}{\partial c^2} H$ exist and are continuous. Finally

$$U(t,c;\Lambda) = \sum_{n=1}^{N} \lambda_n U_n(t, I_n(t, \frac{1}{\lambda_n} H(t,c;\Lambda))),$$

and differentiation with respect to c yields

$$U'(t,c;\Lambda) \triangleq \frac{\partial}{\partial c} U(t,c;\Lambda) = H(t,c;\Lambda) \frac{d}{dc} I(t,H(t,c;\Lambda);\Lambda) = H(t,c,\Lambda).$$

Therefore, $U_t'(t,c;\Lambda) \triangleq \frac{\partial^2}{\partial t \partial c} U(t,c;\Lambda)$, $U''(t,c;\Lambda) \triangleq \frac{\partial^2}{\partial c^2} U(t,c;\Lambda)$ and $U'''(t,c;\Lambda) \triangleq \frac{\partial^3}{\partial c^3} U(t,c;\Lambda)$ exist and are continuous on $[0,T] \times (0,\infty)$.

We have shown that $I(t,\cdot;\Lambda)$ defined by (9.3) is the inverse of $U'(t,\cdot;\Lambda)$, and so $U(\cdot,\cdot;\Lambda)$ satisfies conditions (i) – (iii) of Section 3. We call $U(\cdot,\cdot;\Lambda)$ the _utility function of a representative agent_ who assigns weights $\lambda_1,\dots,\lambda_N$ to the individual agents in the economy.

Making the identification $\Lambda = (\lambda_1,\dots,\lambda_N) = (\frac{1}{y_1},\dots,\frac{1}{y_N})$, equations (7.4) – (7.5), (8.1) may be rewritten as

$$\zeta(t) = U'(t,\epsilon(t);\Lambda), \quad 0 \le t \le T, \tag{9.4}$$

$$E\int_0^T U'(t,\epsilon(t);\Lambda)I_n(t,\frac{1}{\lambda_n}U'(t,\epsilon(t);\Lambda))dt = E\int_0^T U'(t,\epsilon(t);\Lambda)\epsilon_n(t)dt, \quad 1 \le n \le N, \tag{9.5}$$

and the search for equilibrium is equivalent to the search for a vector $\Lambda \in (0,\infty)^N$ _which satisfies (9.5)_. Once such a vector is found, the corresponding equilibrium ζ is given by (9.4), and the optimal consumption processes of the individual agents by

$$\hat{c}_n(t;\Lambda) \triangleq I_n(t,\frac{1}{\lambda_n}U'(t,\epsilon(t);\Lambda)), \quad 0 \le t \le T, \ 1 \le n \le N. \tag{9.6}$$

Note that ζ given by (9.4) satisfies (4.8) because of the assumption $k \le \epsilon(t) \le K$ and the continuity of $U'(\cdot,\cdot;\Lambda)$.

10. The Equilibrium Financial Market

In this section, we assume the existence of $\Lambda \in (0,\infty)^N$ satisfying (9.5), and we draw conclusions about the equilibrium financial market. The existence of such a Λ is established by explicit computation for certain special cases in Section 11 and in full generality by a fixed point argument in Section 12. It is apparent from (9.1) that for any $\Lambda \in (0,\infty)^N$ and $\eta > 0$,

$$(10.1)\qquad U(t,c;\ \eta\Lambda) = \eta\, U(t,c;\ \Lambda),\ \forall\ (t,c) \in [0,T] \times (0,\infty),$$

so a multiplicative constant on Λ cancels out of (9.5) and (9.6). Therefore, the existence of any solution Λ to (9.5) guarantees the existence of a one–parameter family of solutions. In Section 11 and under the additional assumption (iv) in Section 12, the solution to (9.5) is shown to be unique up to a positive multiplicative constant. It follows then from (9.6) and (10.1) that the equilibrium optimal consumption processes for the individual agents are uniquely determined.

10.1 Proposition

Assume that there exists $\Lambda \in (0,\infty)^N$ satisfying (9.5), and that this Λ is unique up to a positive multiplicative constant. Then an interest rate process $r(\cdot)$, a mean rate of return vector process $b(\cdot)$, and a volatility matrix process $\sigma(\cdot)$ lead to equilibrium if and only if

$$(10.2)\qquad r(t) = -\frac{1}{U'(t,\epsilon(t);\ \Lambda)}\,[U_t'(t,\epsilon(t);\ \Lambda) + \mu(t)\, U''\,(t,\epsilon(t);\ \Lambda)$$

$$+ \tfrac{1}{2}\, \|\rho(t)\|^2\, U'''\,(t,\epsilon(t);\ \Lambda)],$$

$$(10.3)\ \theta(t) \triangleq (\sigma(t))^{-1}[b(t) - r(t)\, \underset{\sim}{1}] = -\frac{U''\,(t,\epsilon(t);\ \Lambda)\ \rho(t)}{U'(t,\epsilon(t);\ \Lambda)},\quad 0 \le t \le T,$$

where Λ is determined by $P_0(0) \cdot U'(0,\epsilon(0);\ \Lambda) = 1$.

Proof:

From (4.5), (4.7), we have

$$(10.4)\qquad \zeta(t) = \frac{1}{P_0(0)} - \int_0^t r(s)\zeta(s)ds - \int_0^t \zeta(s)\ \theta^*(s)dW(s),\quad 0 \le t \le T.$$

Equilibrium occurs if and only if (9.4) holds, and recalling (2.2), we see that (9.4) is equivalent to

$$\zeta(t) = U'(0,\epsilon(0);\Lambda)+\int_0^t[\mu(s)U''(s,\epsilon(s);\Lambda) + \frac{1}{2}\|\rho(s)\|^2U'''(s,\epsilon(s);\Lambda)]ds + \int_0^t U''(s,\epsilon(s);\Lambda)\,\rho^*(s)dW(s), \quad 0 \le t \le T \tag{10.5}$$

Identifying coefficients in (10.4) and (10.5), we obtain $U'(0,\epsilon(0);\Lambda) = \frac{1}{P_0(0)}$, (10.2) and (10.3).

11. Examples

We cite a few special cases in which the equilibrium can be computed explicitly.

11.1 Example. $U_n(t,c) = \frac{1}{\gamma}e^{-\alpha t}c^\gamma$, $\forall\ (t,c) \in [0,T] \times (0,\infty)$, $n \in \{1,\ldots,N\}$, **where** $\alpha \in \mathbb{R}$ **and** $\gamma < 1$, $\gamma \neq 0$.

In this case, the vector $\Lambda = (\lambda_1,\ldots,\lambda_N) \in (0,\infty)^N$ with

$$\lambda_n^{\frac{1}{1-\gamma}} = [E\int_0^T e^{-\alpha t}\epsilon_n(t)\epsilon^{\gamma-1}(t)dt]\,[E\int_0^T e^{-\alpha t}\epsilon^\gamma(t)dt]^{-1}$$

is the unique solution to (9.5) subject to the normalizing condition $\sum_{n=1}^N \lambda_n^{\frac{1}{1-\gamma}} = 1$. The optimal consumption processes are $\hat{c}_n(t) = \lambda_n^{\frac{1}{1-\gamma}}\epsilon(t)$, and the equilibrium financial market satisfies

$$r(t) = \alpha + \frac{(1-\gamma)}{\epsilon(t)}\mu(t) - \frac{(1-\gamma)(2-\gamma)}{2\epsilon^2(t)}\|\rho(t)\|^2, \quad \theta(t) = \frac{1-\gamma}{\epsilon(t)}\rho(t).$$

The normalization of Λ we have adopted corresponds to $P_0(0) = \epsilon^{1-\gamma}(0)$.

11.2 Example. $U_n(t,c) = e^{-\alpha t}\log c \;\; \forall\, (t,c) \in [0,T] \times (0,\infty),\ n \in \{1,\dots,N\}$, **where** $\alpha \in \mathbb{R}$.

In this case, we obtain the formulas of Example 11.1 but with $\gamma = 0$. In particular,

$$\lambda_n = \begin{cases} \dfrac{\alpha}{1-e^{-\alpha T}}\, E\displaystyle\int_0^T e^{-\alpha t}\,\frac{\epsilon_n(t)}{\epsilon(t)}\,dt\,, & \alpha \neq 0, \\[2ex] \dfrac{1}{T}\, E\displaystyle\int_0^T \frac{\epsilon_n(t)}{\epsilon(t)}\,dt\,, & \alpha = 0, \end{cases}$$

provides the unique solution to (9.5) subject to the normalizing condition $\sum_{n=1}^{N} \lambda_n = 1$. The optimal consumption processes are $\hat{c}_n(t) = \lambda_n \epsilon(t)$, and the equilibrium financial market satisfies

$$r(t) = \alpha + \frac{1}{\epsilon(t)}\,\mu(t) - \frac{1}{\epsilon^2(t)}\,\|\rho(t)\|^2, \quad \theta(t) = \frac{1}{\epsilon(t)}\,\rho(t). \quad \square$$

If agents have different utility functions, it is not in general possible to compute the solution of the equilibrium problem in closed form. A special case in which such computations can be carried out arises when $N = 2$, $U_1(c) = \log c$ and $U_2(c) = \sqrt{c}$. Another special case is the following.

11.3 Example. Constant aggregate endowment $\epsilon(t) \equiv \epsilon > 0$ and time–independent utility functions.

In this case, the optimal consumption rates are constant:

$\hat{c}_n(t) \equiv \hat{c}_n \triangleq \frac{1}{T} E \int_0^T \epsilon_n(t)dt$, and every solution of (9.5) is a multiple of $\Lambda = (\frac{1}{U_1'(\hat{c}_1)}, \ldots, \frac{1}{U_N'(\hat{c}_N)})$. Constant aggregate endowment implies that $\mu \equiv 0$, $\rho \equiv 0$, so the equilibrium market must satisfy $r \equiv 0$ and $b \equiv 0$. The displayed Λ is normalized to correspond to $P_0(t) \equiv P_0(0) = 1$. Note, however, that in this model the individual agent endowments can be random and time–varying, in which case agents must trade with one another to finance their constant rates of consumption.

12. Existence and Uniqueness of Equilibrium

In this section we establish the major results of the paper: existence of an equilibrium financial market and its uniqueness in the sense of Proposition 10.1. The proof of existence is based on the Knaster–Kuratowski–Mazurkiewicz (KKM) Theorem [2, pg. 26] and requires only assumptions (i)–(iii) of Section 3, while our uniqueness proof requires the additional condition (iv). Example 11.1 shows, however, that condition (iv) is not necessary for uniqueness.

We begin with some notation adapted from [2]. Let $x^1, \ldots, x^{(n)}$ denote the elementary vectors of R^N, and let $\mathscr{N} = \{1, \ldots, N\}$. Suppose $A \subset \mathscr{N}$, then $\mathscr{S}_A$ denotes the convex hull of the elementary vectors $\{x^{(i)}; i \in A\}$, i.e., $\mathscr{S}_A = \{\sum_{i \in A} \lambda_i x^{(i)}; \lambda_i \geq 0 \; \forall \, i \text{ and } \sum_{i \in A} \lambda_i = 1\}$, and we define $\mathscr{S}_A^+ = \{\sum_{i \in A} \lambda_i x^{(i)}; \lambda_i > 0, \forall \, i \text{ and } \sum_{i \in A} \lambda_i = 1\}$. To set the stage for the next theorem, we define for $\Lambda \in \mathscr{S}_N$

$$R_n(\Lambda) \triangleq \begin{cases} E\int_0^T U'(t,\epsilon(t);\Lambda)[I_n(t,\frac{1}{\lambda_n}U'(t,\epsilon(t);\Lambda)) - \epsilon_n(t)]dt, & \text{if } \lambda_n > 0, \\ -E\int_0^T U'(t,\epsilon(t);\Lambda)\epsilon_n(t)dt, & \text{if } \lambda_n = 0, \end{cases}$$

and let $F_n = \{\Lambda \in \mathscr{S}_{\mathscr{N}}\,;\, R_n(\Lambda) \geq 0\}$.

12.1 Theorem

Under conditions (i) – (iii) of Section 3, there exists a vector $\Lambda \in \mathscr{S}_{\mathscr{N}}^{+}$ satisfying (9.5).

Proof:

With $\Lambda = (\lambda_1,\ldots,\lambda_N)$, we have from the dominated convergence theorem that $\lim_{\lambda_n \downarrow 0} R_n(\Lambda) = -E\int_0^T U'(t,\epsilon(t);\Lambda)\epsilon_n(t)dt < 0$. This, coupled with the smoothness conditions on U_n, proves that $R_n(\Lambda)$ is continuous on $\mathscr{S}_{\mathscr{N}}$ and F_n is closed. From (9.3) we have $\sum_{n=1}^{N} R_n(\Lambda) = 0$ for every $\Lambda \in \mathscr{S}_{\mathscr{N}}$. Suppose there were a Λ^* in $\mathscr{S}_{\mathscr{N}}$ which was not in $\bigcup_{n\in\mathscr{N}} F_n$. This would imply $\sum_{n=1}^{N} R_n(\Lambda^*) < 0$, a contradiction. Consequently, $\mathscr{S}_{\mathscr{N}} \subset \bigcup_{n\in\mathscr{N}} F_n$. More generally, if we let $A \subset \mathscr{N}$ and consider $\Lambda^* \in \mathscr{S}_A$, a similar argument shows that $\Lambda^* \in \bigcup_{n\in A} F_n$. Indeed, if $\Lambda^* \notin \bigcup_{n\in A} F_n$, then $R_n(\Lambda^*) < 0$ for all $n \in A$, again contradicting $\sum_{n=1}^{N} R_n(\Lambda^*) = 0$. By the KKM Theorem [2, page 26], $\bigcap_{n\in\mathscr{N}} F_n$ is nonempty. Choose $\hat{\Lambda} \in \bigcap_{n\in\mathscr{N}} F_n$. Then $R_n(\hat{\Lambda}) = 0$, $1 \leq n \leq N$, for otherwise we would have $\sum_{n=1}^{N} R_n(\hat{\Lambda}) > 0$, a contradiction. Thus (9.5) is satisfied by $\hat{\Lambda}$.

Finally, $\hat{\lambda}_n > 0$ or else $R_n(\hat{\Lambda})$ would be strictly negative. □

As observed following (10.1), once a vector in $\mathscr{S}_{\mathscr{N}}^{+}$ satisfying (9.5) is obtained, any positive multiple of this vector also satisfies (9.5). We next turn our attention to the question of uniqueness. Condition (iv) of Section 3 is equivalent to the assumption

$$\varphi_n(t,y) \triangleq y\, I_n(t,y) \quad \text{is nonincreasing in } y. \tag{12.1}$$

This leads to the following uniqueness result.

12.2 Theorem

Assume conditions (i)–(iv) of Section 3. Then the solution $\Lambda \in (0, \infty)^n$ of (9.5) is unique up to multiplication by a positive constant.

Proof:

We introduce the usual partial order in $(0,\infty)^N$: $\Lambda \leq M$ if and only if $\lambda_n \leq \mu_n$, $\forall\, n \in \{1, \ldots, N\}$. We write $\Lambda < M$ if $\Lambda \leq M$ and $\Lambda \neq M$. In particular, notice in (9.3) the implications

$$\Lambda \underset{(<)}{\leq} M \Longrightarrow I(t,h;\Lambda) \underset{(<)}{\leq} I(t,h;M) \quad \forall (t,h) \in [0,T] \times (0,\infty). \tag{12.2}$$

For $\Lambda \leq M$ we have from (12.2) that $U'(t,\epsilon(t);\Lambda) \leq U'(t,\epsilon(t);M)$. Let Λ and $\tilde{\Lambda}$ be two solutions of (9.5) and define $\eta \triangleq \max_{1 \leq n \leq N} \frac{\lambda_n}{\tilde{\lambda}_n}$ and $M = (\mu_1, \ldots, \mu_n)$

$= \eta\,\tilde{\Lambda}$, so M is a solution of (9.5) and $\Lambda \leq M$. If $\Lambda = M$, then $\tilde{\Lambda}$ is indeed a positive multiple of Λ. Therefore, it suffices to rule out the case $\Lambda < M$.

Suppose that $\Lambda < M$. From (12.2) we obtain $U'(t,\epsilon(t);\Lambda) < U'(t,\epsilon(t); M)$, $\forall\,(t,\omega) \in [0,T] \times \Omega$. Choose an integer $n \in \{1,\ldots,N\}$ satisfying $\lambda_n = \eta\tilde{\lambda}_n$ (and hence also $\lambda_n = \mu_n$). We have

$$E\int_0^T \frac{1}{\lambda_n} U'(t,\epsilon(t);\Lambda)\epsilon_n(t)dt < E\int_0^T \frac{1}{\mu_n} U'(t,\epsilon(t); M)\epsilon_n(t)dt$$

$$E\int_0^T \varphi_n(t, \frac{1}{\lambda_n} U'(t,\epsilon(t);\Lambda))dt \geq E\int_0^T \varphi_n(t, \frac{1}{\mu_n} U'(t,\epsilon(t); M))dt,$$

where φ_n is given by (12.1). Taking the difference of these two relations, we obtain $\frac{1}{\lambda_n} R_n(\Lambda) > \frac{1}{\mu_n} R_n(M)$. But Λ and M both solve (9.5), so $R_n(\Lambda) = R_n(M) = 0$, and a contradiction is obtained.

13. Variations of the Model

In addition to the financial assets of Section 4, one can allow the agents to trade in capital assets, and one can associate to each one of these assets a dividend process $\delta_m(\cdot)$, $1 \leq m \leq M$, denominated in units of the commodity. In contrast to financial assets, which are essentially contracts between the agents, capital assets have to maintain a positive net supply. One can show that the prices $S_m(\cdot)$ of these new assets have to be given as

$$\zeta(t)S_m(t) = E\,[\int_t^T \zeta(s)\delta_m(s)ds \mid \mathcal{F}_t]; \quad 0 \leq t \leq T, \tag{13.1}$$

in order to prevent "arbitrage opportunities". Once the deflator ζ has been determined by equilibrium considerations, relation (13.1) allows the endogenous computation of the capital asset prices $S_m(\cdot)$, $1 \le m \le M$. The details appear in [15].

Consider now an economy with <u>deterministic endowments</u> and <u>no financial market</u> except for a bond with deterministic interest rate. Agents can consume but cannot borrow or invest, are bound simply by the budget constraints

$$\int_0^T \beta(s)c_n(s)ds \le \int_0^T \beta(s)\epsilon_n(s)ds; \quad 1 \le n \le N,$$

(the deterministic analogue of (7.1)), and try to maximize their total utilities $\int_0^T U_n(t,c_n(t))dt$ from consumption. Equilibrium amounts to the requirements (6.1), (6.3) alone. In this simple model the results of sections 7–12 are valid, provided that one sets $\zeta(t) \equiv \beta(t)$, omits reference to θ, and drops the expectation signs in the formulas.

Acknowledgement: We are indebted to P. Dybvig for pointing out the relevence of the Knaster–Kuratowski–Mazurkiewicz Lemma. This permitted a simplification and strengthening of our original existence proof.

This research was supported by the National Science Foundation under grant DMS–87–23078 at Columbia University (Karatzas and Lakner) and under grants DMS–87–02537, DMS–90–02588 (Lehoczky and Shreve).

References

[1] A. Araujo and P.K. Monteiro, Equilibrium without uniform conditions, *J. Economic Theory* **48**, 1989, 416–427.

[2] K.C. Border, *Fixed Point Theorems with Applications to Economics and Game Theory*. Cambridge University Press, 1985.

[3] J.C. Cox and C.F. Huang, A variational problem arising in financial economics, Sloan School of Management, MIT Mimeo, 1987.

[4] J.C. Cox and C.F. Huang, Optimal consumption and portfolio policies when asset prices follow a diffusion process, *J. Economic Theory* **49**, 1989, 33–83.

[5] J.C. Cox, J.E. Ingersoll, and S.A. Ross, An intertemporal general equilibrium model of asset prices, *Econometrica* **53**, 1985, p. 363–384.

[6] R.–A. Dana, and M. Pontier, On the existence of a stochastic equilibrium. A remark, preprint, 1989.

[7] D. Duffie, Stochastic equilibria: existence, spanning number, and the "no expected financial gain from trade" hypothesis, *Econometrica* **54**, 1986, p. 1161–1383.

[8] D. Duffie and C.F. Huang, Implementing Arrow–Debreu equilibria by continuous trading of a few long–lived securities, *Econometrica* **53**, 1985, p. 1337–1356.

[9] D. Duffie and W. Zame, The consumption–based capital asset pricing model, *Econometrica* **57**, 1989, 1279–1297.

[10] J.M. Harrison, and S.R. Pliska, Martingales and stochastic integrals in the theory of continuous trading, *Stoch. Proc. Appl.* **11**, 1981, p. 215–260.

[11] J.M. Harrison and S.R. Pliska, A stochastic calculus model of continuous trading: complete markets, *Stoch. Proc. Appl.* **15**, 1983, p. 313–316.

[12] C.F. Huang, An intertemporal general equilibrium asset pricing model: the case of diffusion information, *Econometrica* **55**, 1987, p. 117–142.

[13] I. Karatzas. J.P. Lehoczky and S.E. Shreve, Optimal portfolio and consumption decisions for a "small investor" on a finite horizon, *SIAM J. Control & Optim.* **25**, 1987, p. 1557–1586.

[14] I. Karatzas, J.P. Lehoczky and S.E. Shreve, Equilibrium models with singular asset prices, preprint, 1990.

[15] I. Karatzas. J.P. Lehoczky and S.E. Shreve, Existence and uniqueness of multi–agent equilibrium in a stochastic,dynamic consumption/investment model, *Math. Operations Research* **15**, 1990, 80–128.

[16] I. Karatzas and S.E. Shreve, ***Brownian Motion and Stochastic Calculus.*** Springer–Verlag, 1987, New York.

[17] M. Rothschild and J.E. Stiglitz, Increasing risk II: its economic consequences, *J. Econ. Theory* **3**, 1971, p. 66–84.

[18] M. Taqqu and W. Willinger, The analysis of finite security markets using martingales, *Adv. Appl. Probab.* **19**, 1987, p. 1–25.

Feynman-Kac Formula for a Degenerate Planar Diffusion and an Application in Stochastic Control

Ioannis Karatzas
Department of Statistics
Columbia University
New York, NY 10027

Daniel L. Ocone
Department of Mathematics
Rutgers University
New Brunswick, NJ 08903

Abstract

A formula of the Feynman-Kac type is established for the degenerate, two-dimensional diffusion process introduced by Beneš, Karatzas & Rishel [2]. With its aid, a stochastic control problem with partial observations is solved explicitly. Our derivation combines probabilistic techniques with use of the so-called *principle of smooth fit*.

1 Introduction

The degenerate, two-dimensional diffusion process $(Y^{y,\xi}, Z^{y,\xi})$ given by the stochastic equation

$$(1.1) \qquad \begin{aligned} dY_t &= dW_t \ , \quad Y_0 = y \\ dZ_t &= -sgn(Y_t Z_t) dW_t \ , \quad Z_0 = \xi \end{aligned}$$

with W a standard, one-dimensional Brownian motion, was introduced and studied in [2]. It was shown there that (1.1) admits a

ISBN 0-12-481005-5

solution which is unique in the sense of the probability law for every $(y,\xi) \neq \underset{\sim}{0}$, and the paths of the resulting diffusion were studied in some detail.

We consider here functionals of this diffusion process, of the form

$$V(y,\xi) = E\int_0^\infty e^{-\lambda t} c(Y_t^{y,\xi}) \cosh(\theta Z_t^{y,\xi}) dt \ , \tag{1.2}$$

where $\theta > 0$, $\lambda > \theta^2/2$, and $c : \mathcal{R} \to [0,\infty)$ is an even, convex function, strictly increasing on $(0,\infty)$ and sufficiently smooth. Using probabilistic techniques and the "smooth-fit principle" (e.g. [3]), we compute the function V of (1.2) *explicitly* in Theorem 3.1, and establish some of its most interesting properties in Propositions 5.1, 5.4. In particular, we show that V is of class C^2 and satisfies the nonlinear partial differential equation

$$\frac{1}{2} V_{yy} + \min_{|u|\leq 1}\left[u.V_{y\xi} + \frac{1}{2}u^2(V_{\xi\xi} - \theta^2 V)\right] + c(y)\cosh(\theta\xi) = \alpha V \tag{1.3}$$

(with $\alpha = \lambda - \frac{\theta^2}{2}$) on $\mathcal{R}^2 \backslash \{\underset{\sim}{0}\}$, and that it is the unique such function subject to certain growth conditions.

These results are then employed in sections 6 and 7 to study a *partially observed stochastic control problem*, considered by [2] in the special case $c(x) = x^2$. We show that the solution found in [2] is still optimal for general even, convex and sufficiently smooth functions $c(\cdot)$.

2 A Degenerate, Planar Diffusion

The following two-dimensional diffusion process (Y, Z) was introduced and studied in [2].

2.1 PROBLEM: *To find a complete probability space $(\Omega, \mathcal{F}, P)$, a filtration $\{\mathcal{F}_t\}$ of sub-σ-fields of $\mathcal{F}$ which satisfies the usual conditions, as well as two continuous and $\{\mathcal{F}_t\}$-adapted processes Y, Z on this space, such that: (i) Y is a standard, one-dimensional Brownian motion process with $Y_0 = y \in \mathcal{R}$, and (ii) the equation*

$$Z_t = \xi - \int_0^t sgn(Y_s Z_s) dY_s \ ; \qquad 0 \leq t < \infty \tag{2.1}$$

is satisfied almost surely, for an arbitrary but fixed initial condition $(Y_0, Z_0) = (y, \xi) \in \mathcal{R}^2 \setminus \{\underset{\sim}{0}\}$. ◇

In other words, one seeks a *weak solution* for the degenerate, two-dimensional stochastic differential equation (1.1), for any given (y, ξ) in $\mathcal{R}^2 \setminus \{\underset{\sim}{0}\}$. It was shown in [2] that Problem 2.1 admits a solution, which is unique in the sense of probability law (cf. Theorem 2.3 below). The convention $sgn(x) := 1_{(0,\infty)}(x) - 1_{(-\infty,0]}(x)$ is employed throughout.

It should be observed that, for any solution of Problem 2.1, the process Z is an $\{\mathcal{F}_t\}$ - Brownian motion starting at $Z_0 \equiv \xi$ (because it is an $\{\mathcal{F}_t\}$ - local martingale with continuous paths and quadratic variation equal to $< Z >_t = t$, as is easily checked from (2.1)). It is an immediate consequence of (1.1) and the Tanaka formula for Brownian local time, that

$$|Y(t)| + |Z(t)| = |y| + |\xi| + L^Y(t) + L^Z(t)\ , \quad 0 \le t < \infty\ . \tag{2.2}$$

Here, $L^B(t)$ denotes the local time at zero of the Brownian motion B, up to time t.

The following result is proved in [2], [7].

2.3 THEOREM [2], [7]: *For any given* $(y, \xi) \neq \underset{\sim}{0}$, *there is a solution to Problem 2.1; this solution is unique in the sense of probability law. Introduce for this process the sequence of stopping times*

$$\begin{aligned} \sigma_k &\triangleq \inf\{t \ge 0;\ |Y(t)| + |Z(t)| \ge |y| + |\xi| + k\} \\ &= \inf\{t \ge 0;\ L^Y(t) + L^Z(t) \ge k\}\ , \quad k \in \mathrm{N}, \end{aligned} \tag{2.3}$$

(recall (2.2)); then for every $\lambda \in (0, \infty)$ *and* $0 < \epsilon < \sqrt{2\lambda}$, *there exists a positive constant* $M_\epsilon < \infty$ *such that*

$$Ee^{-\lambda\sigma_k} \le M_\epsilon e^{-(\sqrt{2\lambda}-\epsilon)k}\ , \quad \forall\ k \in \mathrm{N}\ . \tag{2.4}$$ ◇

It is also shown in [2] that with $y \neq 0$, $\xi = 0$ and $\tau = \inf\{t \ge 0;\ Y(t) = 0\}$,

$$sgn(Z_\tau) \ \text{ is independent of } \ \mathcal{F}^Y_\tau\ , \tag{2.5}$$

and that

$$\text{(2.6)} \qquad \left\{ \begin{array}{l} \text{the processes } Z, sgn(Z) \text{ are not adapted to} \\ \text{the filtration } \{\mathcal{F}_t^Y\},\ \mathcal{F}_t^Y = \sigma(Y_s; 0 \le s \le t) \end{array} \right\},$$

for any solution $(\Omega, \mathcal{F}, P), \{\mathcal{F}_t\}, (Y, Z)$ of Problem 2.1. In other words, *the equation (2.1) does not admit a strong solution.*

3 The Feynman - Kac Formula

We would like now to characterize and, if possible, compute explicitly, the resolvent function

$$\text{(3.1)} \qquad V(y,\xi) \triangleq E \int_0^\infty e^{-\lambda t} c(Y_t^{y,\xi}) \cosh(\theta Z_t^{y,\xi}) dt\ ,$$

for the unique (in the sense of probability law) solution $(Y^{y,\xi}, Z^{y,\xi})$ of Problem 2.1, corresponding to any given initial condition $(y,\xi) \neq (0,0)$. Here α, θ are positive real numbers, $\lambda := \alpha + \theta^2/2$, and $c : \mathcal{R} \to [0,\infty)$ is an even, strictly convex function of class C^4, strictly increasing on $(0,\infty)$, which satisfies $c(0) = 0$ and the polynomial growth condition

$$\text{(3.2)} \qquad |c^{(k)}(y)| \le K(1 + |y|^\beta)\ ; \qquad \forall\ y \in \mathcal{R}$$

for any $k = 0, 1, \ldots, 4$ and some positive real numbers β, K. In stating the main result it is convenient to use the function

$$G_b(\xi, u) = \left\{ \begin{array}{ll} \sqrt{\frac{2}{\lambda}}\, \frac{\sinh(\xi\sqrt{2\lambda})\sinh((b-u)\sqrt{2\lambda})}{\sinh(b\sqrt{2\lambda})} & ;\quad u \ge \xi \\ \sqrt{\frac{2}{\lambda}}\, \frac{\sinh(u\sqrt{2\lambda})\sinh((b-\xi)\sqrt{2\lambda})}{\sinh(b\sqrt{2\lambda})} & ;\quad u < \xi \end{array} \right\},$$

which is introduced in equation (4.5) and explained in equation (4.9).

3.1 THEOREM: *There is a unique function $Q : \mathcal{R}^2 \to \mathcal{R}$ of class C^2, which satisfies the equation*

$$\text{(3.3)} \quad \frac{1}{2}[Q_{yy} + Q_{\xi\xi}] - sgn(y\xi).Q_{y\xi} + c(y)\cosh(\theta\xi) = \lambda Q\ , \quad in\ \mathcal{R}^2$$

and a growth condition of the type

$$|Q(y,\xi)| \leq K e^{\theta(|y|+|\xi|)} \quad in \quad \mathcal{R}^2 , \tag{3.4}$$

for a suitable real constant $K > 0$. *This function agrees with the function of (3.1) away from the origin, i.e.,* $Q = V$, *in* $\mathcal{R}^2 \backslash \{\underset{\sim}{0}\}$, *and has the even symmetry* $Q(y,\xi) = Q(y,-\xi) = Q(-y,\xi) = Q(-y,-\xi)$, $\forall\ (y,\xi) \in \mathcal{R}^2$. *For* $y > 0$, $\xi > 0$, $Q(y,\xi)$ *is given explicitly as*

$$\begin{aligned} Q(y,\xi) &= \int_0^{y+\xi} G_{y+\xi}(\xi,u) c(y+\xi-u) \cosh(\theta u)\, du \\ &+ \frac{\sinh(\xi\sqrt{2\lambda})}{\sinh((y+\xi)\sqrt{2\lambda})} M(y+\xi) + \frac{\sinh(y\sqrt{2\lambda})}{\sinh((y+\xi)\sqrt{2\lambda})} N(y+\xi). \end{aligned} \tag{3.5}$$

At the origin

$$Q(0,0) = \int_0^\infty \frac{F(u)+G(u)}{\sinh^3(u\sqrt{2\lambda})} [\cosh(u\sqrt{2\lambda}) - 1] du\ . \qquad \diamond$$

The functions F, G, M, N appearing in (3.5) are given, respectively, by

$$F(s) \triangleq 2 \int_0^s \sinh(u\sqrt{2\lambda}) c(s-u) \cosh(\theta u) du \tag{3.6}$$

$$G(s) \triangleq 2 \int_0^s \sinh((s-u)\sqrt{2\lambda}) c(s-u) \cosh(\theta u) du \tag{3.7}$$

$$\begin{aligned} M(s) \triangleq \cosh(s\sqrt{2\lambda}) &\int_s^\infty \frac{F(u)\cosh(u\sqrt{2\lambda}) - G(u)}{\sinh^3(u\sqrt{2\lambda})} du \\ &+ \int_s^\infty \frac{G(u)\cosh(u\sqrt{2\lambda}) - F(u)}{\sinh^3(u\sqrt{2\lambda})} du \end{aligned} \tag{3.8}$$

$$\begin{aligned} N(s) \triangleq \cosh(s\sqrt{2\lambda}) &\int_s^\infty \frac{G(u)\cosh(u\sqrt{2\lambda}) - F(u)}{\sinh^3(u\sqrt{2\lambda})} du \\ &+ \int_s^\infty \frac{F(u)\cosh(u\sqrt{2\lambda}) - G(u)}{\sinh^3(u\sqrt{2\lambda})} du \end{aligned} \tag{3.9}$$

For properties of these functions, see Lemma 3.2 below; its proof is straightforward. The proof of Theorem 3.1 will be carried out in sections 4 and 5. In section 7 we shall use it as the tool for solving a stochastic control problem with partial observations.

The stochastic representation (3.1), and the explicit computation (3.5) which it produces, constitute a *Feynman-Kac formula* for the solution of the resolvent equation (3.3).

3.2 LEMMA: *The functions $F(\cdot), G(\cdot)$ of (3.6), (3.7) are positive, and bounded above by $\frac{2}{\theta}c(s)\sinh(\theta s)\sinh(s\sqrt{2\lambda})$. In particular, $M(s)$ and $N(s)$ are well-defined by (3.8) and (3.9), respectively, and satisfy*

$$(3.10) \qquad M(0+) = N(0+) = \int_0^\infty \frac{\cosh(u\sqrt{2\lambda}) - 1}{\sinh^3(u\sqrt{2\lambda})}[F(u) + G(u)]du$$

as well as the growth conditions,

(3.11)

$$G(s) \le Ks^\beta e^{s\sqrt{2\lambda}} \,, \quad M(s) \le K(1+s^{1+\beta})e^{\theta s} \,, \quad N(s) \le K(1+s^{1+\beta})$$

on $[0,\infty)$, for a suitable real constant $K > 0$. ◇

4 Analysis

From the even symmetry of the functions $c(\cdot), \cosh(\cdot)$, and from the uniqueness-in-law property of the stochastic equation (2.1), it is quite clear that the function $V : \mathcal{R}^2 \backslash \{\underset{\sim}{0}\} \to [0,\infty)$ of (3.1) has the *symmetry* property

$$(4.1) \qquad V(y,\xi) = V(-y,\xi) = V(y,-\xi) = V(-y,-\xi) \,.$$

On the other hand, it is easily seen (using the Hölder inequality and the growth assumption (3.2) for $k = 0$) that the *growth condition*

$$(4.2) \qquad 0 \le V(y,\xi) \le C(1 + |y|^\beta)e^{\theta|\xi|} \,, \quad \forall\ (y,\xi) \ne \underset{\sim}{0}$$

holds for a suitable real constant $C > 0$, where β, θ are as in the preceding section.

Let us start our analysis by studying the function V of (3.1) in the positive quadrant.

4.1 PROPOSITION: *For $y \geq 0$, $\xi \geq 0$ and $(y,\xi) \neq \underset{\sim}{0}$, we have*

$$
\text{(4.3)} \quad
\begin{aligned}
V(y,\xi) = & \int_0^{y+\xi} G_{y+\xi}(\xi,u) c(y+\xi-u)\cosh(\theta u)\,du \\
& + \frac{\sinh(\xi\sqrt{2\lambda})}{\sinh((y+\xi)\sqrt{2\lambda})}\tilde{M}(y+\xi) + \frac{\sinh(y\sqrt{2\lambda})}{\sinh((y+\xi)\sqrt{2\lambda})}\tilde{N}(y+\xi),
\end{aligned}
$$

where

$$
\text{(4.4)} \qquad \tilde{M}(s) \triangleq V(0,s), \quad \tilde{N}(s) \triangleq V(s,0)\,; \quad 0 < s < \infty\,,
$$

and G_b is the Green's function

$$
\text{(4.5)} \qquad G_b(\xi,u) \triangleq \left\{ \begin{array}{ll} \sqrt{\frac{2}{\lambda}}\,\frac{\sinh(\xi\sqrt{2\lambda})\sinh((b-u)\sqrt{2\lambda})}{\sinh(b\sqrt{2\lambda})} & ; \quad u \geq \xi \\ \sqrt{\frac{2}{\lambda}}\,\frac{\sinh(u\sqrt{2\lambda})\sinh((b-\xi)\sqrt{2\lambda})}{\sinh(b\sqrt{2\lambda})} & ; \quad u < \xi \end{array} \right\}.
$$

Proof: First, let us assume $y > 0, \xi > 0$, and consider the stopping time

$$
\text{(4.6)} \qquad \tau = \inf\{t \geq 0;\ Z_t^{y,\xi} \notin (0, y+\xi)\}\,, \qquad \text{so that}
$$

$$
\text{(4.7)} \qquad Y_t^{y,\xi} = y + \xi - Z_t^{y,\xi}\,, \quad \text{for} \quad 0 \leq t \leq \tau;
$$

putting $y + \xi = \beta$, we can write the function $V(y,\xi)$ of (3.1) as

$$
\text{(4.8)} \quad
\begin{aligned}
V(y,\xi) = & E\int_0^{\tau} e^{-\lambda t} c(\beta - Z_t^{y,\xi})\cosh(\theta Z_t^{y,\xi})dt \\
& + E\int_{\tau}^{\infty} e^{-\lambda t} c(Y_t^{y,\xi})\cosh(\theta Z_t^{y,\xi})dt\,.
\end{aligned}
$$

Now for any given $b > 0$, let us denote by $\mathcal{A}^b$ the generator of a Brownian motion, on the interval $[0,b]$, which is *killed* upon reaching the endpoints; that is, $\mathcal{A}^b = \frac{1}{2}\,\frac{d^2}{dx^2}$ in the domain $\{f \in C[0,b] \cap C^2(0,b);\ f(0) = f(b) = 0\}$. Then the first expectation in (4.8) is given precisely by

$$
\text{(4.9)} \qquad [\,(\lambda I - \mathcal{A}^{\beta})^{-1} f\,](\xi) = \int_0^{\beta} G_{\beta}(\xi,u) f(u)du\,,
$$

where $f(\xi) \triangleq c(\beta - \xi)\cosh(\theta\xi)$, $0 \le \xi \le \beta$. We obtain thus the first term on the right-hand side of the expression (4.3).

On the other hand, conditioning on $\mathcal{F}_\tau^{Y,Z}$, and using the fact that (Y, Z) of (2.1) is a strong Markov process (e.g. [6], p.322), we obtain for the second expectation of (4.8):

$$
\begin{aligned}
&E\int_\tau^\infty e^{-\lambda t} c(Y_t^{y,\xi})\cosh(\theta Z_t^{y,\xi})dt \\
&\quad = E[e^{-\lambda\tau}\; E\{\int_0^\infty e^{-\lambda t} c(Y_{\tau+t}^{y,\xi})\cosh(\theta Z_{\tau+t}^{y,\xi})dt \mid \mathcal{F}_\tau^{Y,Z}\}] \\
&\quad = E[e^{-\lambda\tau} V(Y_\tau^{y,\xi}, Z_\tau^{y,\xi})] \\
&\quad = E[e^{-\lambda\tau} 1_{\{Z_\tau^{y,\xi} = y+\xi\}}].V(0, y+\xi) \\
&\qquad\qquad + E[e^{-\lambda\tau} 1_{\{Z_\tau^{y,\xi} = 0\}}].V(y+\xi, 0)\,.
\end{aligned}
$$

The last two terms on the right-hand side of (4.3) follow now from the definitions of (4.4), and from the computations (e.g. [6], p.100):

$$
E[e^{-\lambda\tau} 1_{\{Z_\tau^{y,\xi} = y+\xi\}}] = \frac{\sinh(\xi\sqrt{2\lambda})}{\sinh((y+\xi)\sqrt{2\lambda})}\,,
$$

$$
E[e^{-\lambda\tau} 1_{\{Z_\tau^{y,\xi} = 0\}}] = \frac{\sinh(y\sqrt{2\lambda})}{\sinh((y+\xi)\sqrt{2\lambda})}\,.
$$

It is quite straightforward that the formula (4.3) remains valid if either one of y or ξ is equal to zero, as long as $(y, \xi) \neq \underset{\sim}{0}$. ◇

4.2 Remark: In the notation of the preceding proof, we have for $0 < x < b$,

$$
\begin{aligned}
\frac{1}{2}\frac{d^2}{dx^2}(\lambda I - \mathcal{A}^b)^{-1} f(x) &= \mathcal{A}^b(\lambda I - \mathcal{A}^b)^{-1} f(x) \\
&= -f(x) + \lambda(\lambda I - \mathcal{A}^b)^{-1} f(x)\,,
\end{aligned}
\tag{4.10}
$$

$$
\begin{aligned}
V(b-x, x) &= (\lambda I - \mathcal{A}^b)^{-1} f(x) \\
&\quad + \frac{\sinh(x\sqrt{2\lambda})}{\sinh(b\sqrt{2\lambda})}\tilde{M}(b) + \frac{\sinh((b-x)\sqrt{2\lambda})}{\sinh(b\sqrt{2\lambda})}\tilde{N}(b)\,,
\end{aligned}
\tag{4.11}
$$

for every $b \in (0,\infty)$. Here again, $f(x) = c(b-x)\cosh(\theta x)$, $0 \le x \le b$.

4.3 PROPOSITION: *The functions $\tilde{M}, \tilde{N}$ of (4.4) are continuous and four times continuously differentiable on $(0,\infty)$.*

Proof: A scaling argument shows that $(aY^{y,\xi}_{\cdot/a^2},\ aZ^{y,\xi}_{\cdot/a^2})$ is equivalent in law to the pair $(Y^{ay,a\xi},\ Z^{ay,a\xi})$ for any $a > 0$. Thus

$$
\begin{aligned}
\tilde{M}(a) &= E\int_0^\infty e^{-\lambda t} c(Y_t^{0,a})\cosh(\theta Z_t^{0,a})dt \\
&= E\int_0^\infty e^{-\lambda t} c(aY_{t/a^2}^{0,1})\cosh(\theta Z_{t/a^2}^{0,1})dt \qquad (4.12)\\
&= a^2.\ E\int_0^\infty e^{-\lambda t a^2} c(aY_t^{0,1})\cosh(\theta Z_t^{0,1})dt\ ,
\end{aligned}
$$

has as many continuous derivatives as the function c ; similarly for the function $\tilde{N}$.

4.4 PROPOSITION: *The function V is of class C^4 in the region $(y > 0, \xi > 0)$, and satisfies there the equation*

$$
(4.13)\quad \frac{1}{2}[V_{yy}(y,\xi)+V_{\xi\xi}(y,\xi)]-V_{y\xi}(y,\xi)+c(y)\cosh(\theta\xi) = \lambda V(y,\xi)\,.
$$

Proof: For any fixed $b \in (0,\infty)$, we have

$$
(4.14)\quad \frac{1}{2}\frac{d^2}{dx^2}V(b-x,x) = [\frac{1}{2}(V_{yy}+V_{\xi\xi}) - V_{y\xi}](b-x,x)\ , \quad \text{and}
$$

(4.15)

$$
\begin{aligned}
\frac{1}{2}\frac{d^2}{dx^2}V(b-x,x) &= -f(x) + \lambda\Big[(\lambda I - \mathcal{A}^b)^{-1}f(x) \\
&\quad + \frac{\sinh(x\sqrt{2\lambda})}{\sinh(b\sqrt{2\lambda})}\tilde{M}(b) + \frac{\sinh((b-x)\sqrt{2\lambda})}{\sinh(b\sqrt{2\lambda})}\tilde{N}(b)\Big] \\
&= -f(x) + \lambda\ V(b-x,x) = \lambda\ V(b-x,x) - c(b-x).\cosh(\theta x)\ ,
\end{aligned}
$$

from (4.10) and (4.11). The equation (4.13) follows, upon setting $x = \xi, b = y + \xi$ and equating terms in (4.14), (4.15). The required smoothness of the function V follows directly from its representation

(4.3) and from the corresponding smoothness of the functions $\tilde{N}, \tilde{M}$ (Proposition 4.3). ◇

The representation (4.3), and the symmetry properties (4.1), compute the function V on the entirety of $\mathcal{R}^2 \backslash \{\underset{\sim}{0}\}$ modulo *the determination of the functions* $\tilde{M}$ *and* $\tilde{N}$, a task which we now undertake.

We shall compute the functions $\tilde{M}$, $\tilde{N}$ by making the following assumption:

(4.16) *The function V is of class C^1 on $\mathcal{R}^2 \backslash \{\underset{\sim}{0}\}$.*

The Ansatz (4.16) will be vindicated in the "synthesis" section 5. Since we have guaranteed already that V is of class C^2 away from the axes (Proposition 4.4 and symmetry), the requirement (4.16) amounts to *the continuity of the gradient of V across the half-axes* $(y > 0, \xi = 0)$ *and* $(y = 0, \xi > 0)$. Now from the symmetry relation $V(y, \xi) = V(y, -\xi)$ we know that V_y, V_{yy} and $V_{\xi\xi}$ are continuous across the half-axis $(y > 0, \xi = 0)$; in order for V_ξ and $V_{y\xi}$ also to be continuous, we have to have

$$\text{(4.17)}(i) \qquad V_\xi(y, 0+) = 0, \quad 0 < y < \infty.$$

Similar considerations across the half-axis $(y = 0, \xi > 0)$ lead to

$$\text{(4.17)}(ii) \qquad V_y(0+, \xi) = 0, \quad 0 < \xi < \infty .$$

The requirements (4.17) are thus equivalent to the Ansatz (4.16).

Furthermore, if (4.17)(i), (ii) are true, then we must have that $V_{y\xi}(y, 0+) = 0$ for $y > 0$, and $V_{y\xi}(0+, \xi) = 0$ for $\xi > 0$. Hence $V_{y\xi}$ is also continuous across the axes. Since V_{yy} and $V_{\xi\xi}$ are automatically continuous across the axes because of the symmetry properties of V, it follows that the Anzatz (4.16) actually implies

(4.18) *V is of class C^2 on $\mathcal{R}^2 \backslash \{\underset{\sim}{0}\}$.*

Let us work in the quadrant $(y > 0, \xi > 0)$; differentating in (4.3) with respect to y and then letting $\xi \downarrow 0$, $y \downarrow 0$ we obtain, respectively, $V_y(y, 0+) = \tilde{N}'(y)$, $0 < y < \infty$ and, for $\xi > 0$,

(4.19)

$$V_y(0+, \xi) = \tilde{M}'(\xi) - \sqrt{2\lambda}\coth(\xi\sqrt{2\lambda}).\tilde{M}(\xi) + \frac{\tilde{N}(\xi)\sqrt{2\lambda}}{\sinh(\xi\sqrt{2\lambda})} + 2\int_0^\xi \frac{\sinh(u\sqrt{2\lambda})}{\sinh(\xi\sqrt{2\lambda})}\, c(\xi - u)\cosh(\theta u) du .$$

The requirement (4.17)(ii) thus amounts to the differential equation

$$\text{(4.20)}\quad \sinh(s\sqrt{2\lambda}).\tilde{M}'(s) = \sqrt{2\lambda}[\cosh(s\sqrt{2\lambda}).\tilde{M}(s) - \tilde{N}(s)] - F(s),$$

where $F(\cdot)$ is the function of (3.6). On the other hand, differentiating in (4.3) with respect to ξ and then letting $y \downarrow 0$, $\xi \downarrow 0$ we obtain, respectively, $V_\xi(0+, \xi) = \tilde{M}'(\xi)$, $0 < \xi < \infty$, and, for $y > 0$,
(4.21)

$$V_\xi(y+, 0) = \tilde{N}'(y) - \sqrt{2\lambda}\coth(y\sqrt{2\lambda}).\tilde{N}(y) + \frac{\sqrt{2\lambda}}{\sinh(y\sqrt{2\lambda})}\tilde{M}(y)$$
$$+ 2\int_0^y \frac{\sinh((y-u)\sqrt{2\lambda})}{\sinh(y\sqrt{2\lambda})} c(y-u)\cosh(\theta u)du \;.$$

Then the requirement (4.17)(i) amounts to the differential equation

$$\text{(4.22)}\quad \sinh(s\sqrt{2\lambda}).\tilde{N}'(s) = \sqrt{2\lambda}\,[\cosh(s\sqrt{2\lambda}).\tilde{N}(s) - \tilde{M}(s)] - G(s),$$

where $G(\cdot)$ is the function of (3.7).

4.5 PROPOSITION: *Recall the functions $\tilde{M}, \tilde{N}$ of (4.4) and M, N of (3.8), (3.9). The conditions (4.17)(i), (4.17)(ii) are satisfied if and only if*

$$\text{(4.23)}\qquad \tilde{M} \equiv M, \quad \tilde{N} \equiv N \;.$$

Proof: The conditions (4.17)(i),(ii) amount to the differential equations (4.20), (4.22). With $L_\pm(s) \triangleq [\tilde{M}(s) \pm \tilde{N}(s)]/\sinh(s\sqrt{2\lambda})$, these can be written equivalently as

$$\text{(4.24)}\quad \begin{aligned} (L_+(s).\tanh(\tfrac{s}{2}\sqrt{2\lambda}))' &= -\frac{F(s)+G(s)}{2\sinh(s\sqrt{2\lambda})\cosh^2(\frac{s}{2}\sqrt{2\lambda})} \\ (L_-(s).\coth(\tfrac{s}{2}\sqrt{2\lambda}))' &= -\frac{F(s)-G(s)}{2\sinh(s\sqrt{2\lambda})\sinh^2(\frac{s}{2}\sqrt{2\lambda})} \;. \end{aligned}$$

But from (4.2) and (4.4) we have $0 \le \tilde{M}(s) \le Ce^{\theta s}$, $0 \le \tilde{N}(s) \le C(1+s^\beta)$ and thus also $\lim_{s\to\infty} L_\pm(s) = 0$. We have from (4.24),

$$\frac{\tilde{M}(s)+\tilde{N}(s)}{2\cosh^2(\frac{s}{2}\sqrt{2\lambda})} = L_+(s).\tanh(\frac{s}{2}\sqrt{2\lambda})$$
$$= 2\int_s^\infty \frac{[F(u)+G(u)]\sinh^2(\frac{u}{2}\sqrt{2\lambda})}{\sinh^3(u\sqrt{2\lambda})}\,du,$$

$$\frac{\tilde{M}(s)-\tilde{N}(s)}{2\sinh^2(\frac{s}{2}\sqrt{2\lambda})} = L_-(s).\coth(\frac{s}{2}\sqrt{2\lambda})$$
$$= 2\int_s^\infty \frac{[F(u)-G(u)]\cosh^2(\frac{u}{2}\sqrt{2\lambda})}{\sinh^3(u\sqrt{2\lambda})}\,du\ ,$$

and (4.23) follows. ◇

4.6 PROPOSITION: *Suppose that the conditions (4.17)(i) and (4.17)(ii) are satisfied. Then we have*

$$\lim_{y\downarrow 0} V_y(y,0+) = \lim_{\xi\downarrow 0} V_\xi(0+,\xi) = 0\ , \quad \textit{and} \tag{4.25}$$

$$\begin{aligned}\lim_{y\downarrow 0} V_{yy}(y,0+) &= \lim_{\xi\downarrow 0} V_{yy}(0+,\xi)\\ &= 2\lambda\int_0^\infty \frac{\cosh(u\sqrt{2\lambda})G(u)-F(u)}{\sinh^3(u\sqrt{2\lambda})}\,du\end{aligned} \tag{4.26}$$

$$\begin{aligned}\lim_{\xi\downarrow 0} V_{\xi\xi}(0+,\xi) &= \lim_{y\downarrow 0} V_{\xi\xi}(y,0+)\\ &= 2\lambda\int_0^\infty \frac{\cosh(u\sqrt{2\lambda})F(u)-G(u)}{\sinh^3(u\sqrt{2\lambda})}\,du\ .\end{aligned} \tag{4.27}$$

Proof: Differentiating in (3.8), (3.9) we obtain for $0 < s < \infty$:

$$V_\xi(s,0+) = M'(s) =$$
$$\sqrt{2\lambda}\sinh(s\sqrt{2\lambda})\int_s^\infty \frac{\cosh(u\sqrt{2\lambda})F(u)-G(u)}{\sinh^3(u\sqrt{2\lambda})}\,du - \frac{F(s)}{\sinh(s\sqrt{2\lambda})}\ ,$$

$$V_y(0+,s) = N'(s) =$$
$$\sqrt{2\lambda}\sinh(s\sqrt{2\lambda})\int_s^\infty \frac{\cosh(u\sqrt{2\lambda})G(u)-F(u)}{\sinh^3(u\sqrt{2\lambda})}\,du - \frac{G(s)}{\sinh(s\sqrt{2\lambda})}\ .$$

The estimates of Lemma 3.2 guarantee that the indicated integrals are well-defined and finite for $0 \le s < \infty$. Letting $s \downarrow 0$ in these expressions, we obtain (4.25).

The straightforward proofs of (4.26), (4.27) are omitted.

4.7 Remark: The technique used here to compute $V(y,\xi)$ can be extended to compute the resolvent operator associated to the diffusion $(Y^{y,\xi}, Z^{y,\xi})$. We present this computation and its applications in the more complete work [7].

5 Synthesis

Consider now the function Q defined by (3.5) on $\mathcal{R}^2$. According to Propositions 4.5 and 4.6, this function is of class C^2 on $\mathcal{R}^2$ and satisfies the equation (4.13) in $(y > 0, \xi > 0)$, hence by symmetry the equation (3.3) on the entirety of $\mathcal{R}^2$. In particular, this function satisfies the analogues of (4.17), (4.25)-(4.27) (with V replaced by Q). It is also straightforward to see from (3.5) and (3.11) that the bound (3.4) holds, for a suitable real constant $K > 0$.

PROOF OF THEOREM 3.1: It remains to show that *any* function $Q : \mathcal{R}^2 \to \mathcal{R}$, which is of class C^2 and obeys the growth condition (3.4), has to agree with the function V of (3.1) on $\mathcal{R}^2 \backslash \{\underset{\sim}{0}\}$. Consequently, this will then be true for the function of (3.5).

In view of the representation (4.3), it obviously suffices to prove the identity $Q = V$ on each of the two axes $(y = 0), (\xi = 0)$. In particular, it will suffice to show that

$$Q(y,0) = V(y,0), \quad \forall\, y \neq 0 \tag{5.1}$$

holds (the other case being handled in exactly the same way).

Let us consider such a function Q, fix a point $(y,0) \neq \underset{\sim}{0}$, and recall the two-dimensional diffusion process (Y, Z) of Problem 2.1 with initial condition $(Y_0, Z_0) = (y, 0)$. Applying Itô's rule to the process $e^{-\lambda t}\, Q(Y_t, Z_t)$ in conjunction with the equations (2.1) and (3.3), we obtain for $0 \le t < \infty$,

(5.2)

$$\begin{aligned} &e^{-\lambda t} Q(Y_t, Z_t) + \int_0^t e^{-\lambda s} c(Y_s) \cosh(\theta Z_s) ds \\ &= Q(y,0) + \int_0^t e^{-\lambda s} Q_y(Y_s, Z_s) dY_s + \int_0^t e^{-\lambda s} Q_\xi(Y_s, Z_s) dZ_s \end{aligned}$$

almost surely. Now the sequence of stopping times $\{\sigma_k\}_{k=1}^{\infty}$ of (2.3) increases monotonically to infinity, and also satisfies (2.4) for every $0 < \epsilon < \sqrt{2\lambda} - \theta$. Substituting t by σ_k in (5.2) and taking expectations, we get

(5.3)

$$Q(y,0) = E\int_0^{\sigma_k} e^{-\lambda s} c(Y_s)\cosh(\theta Z_s)ds + E[e^{-\lambda\sigma_k} Q(Y_{\sigma_k}, Z_{\sigma_k})] ,$$

because the resulting stochastic integrals have expectation equal to zero; finally, letting $k \nearrow \infty$ in (5.3) we obtain (5.1), since (3.4) and (2.4) imply

$$\begin{aligned}|E(e^{-\lambda\sigma_k} Q(Y_{\sigma_k}, Z_{\sigma_k}))| &\leq K e^{\theta(|y|+k)} . E e^{-\lambda\sigma_k} \\ &\leq K M_\epsilon e^{\theta|y| - k(\sqrt{2\lambda}-\theta-\epsilon)} \to 0 .\end{aligned}$$

5.1 PROPOSITION: *The function* $U(y,\xi) \stackrel{\Delta}{=} Q_{y\xi}(y,\xi)$ *satisfies*

$$sgn U(y,\xi) = sgn(y\xi) , \quad \text{in} \quad \mathcal{R}^2 . \tag{5.4}$$

Proof: From the symmetry properties of Q, it clearly suffices to show that U *is positive in* $(y > 0, \xi > 0)$. From Proposition 4.4 we deduce that, in this region, U satisfies the equation

$$\frac{1}{2}(U_{yy} + U_{\xi\xi}) - U_{y\xi} + \theta c'(y)\sinh(\theta\xi) = \lambda U . \tag{5.5}$$

We also have the boundary conditions

(5.6) $U(y, 0+) = 0, \ \ 0 \leq y < \infty$ and $U(0+, \xi) = 0, \ \ 0 \leq \xi < \infty$.

By analogy with (4.3), a solution U to (5.5), (5.6) is given by

$$\begin{aligned}&\frac{1}{\theta}\sqrt{\frac{\lambda}{2}}\sinh((y+\xi)\sqrt{2\lambda})U(y,\xi) = \\ &\left[\sinh(\xi\sqrt{2\lambda})\int_0^y c'(u)\sinh(u\sqrt{2\lambda})\sinh(\theta(y+\xi-u))du + \right. \\ &\left. + \sinh(y\sqrt{2\lambda})\int_0^\xi c'(y+\xi-u)\sinh(u\sqrt{2\lambda})\sinh(\theta u)du\right],\end{aligned} \tag{5.7}$$

after a change of variables. On the other hand, *there is only one solution to (5.5), (5.6)* ; indeed, one easily sees that the change of variables $H(x_1, x_2) := U(x_1 + x_2, x_1 - x_2)$ transforms (5.5), (5.6) into the second-order ordinary differential equation

$$(5.8) \qquad \frac{1}{2}h''(x_2) - \lambda h(x_2) = -\theta c'(x_1 + x_2)\sinh(\theta(x_1 - x_2)) ,$$

$x_2 \in (-x_1, x_1)$, for the function $h(\cdot) = H(x_1, \cdot)$, $\forall\ x_1 > 0$. But (5.8) admits only one solution subject to $h(\pm x_1) = 0$. Clearly, the expression of (5.7) is positive for $y > 0, \xi > 0$. ◇

5.2 PROPOSITION: *The function*

$$(5.9) \qquad W(y,\xi) \triangleq U_{\xi\xi}(y,\xi) - \theta^2 U(y,\xi)$$

is positive in $(y > 0, \xi \geq 0)$.

Proof: We deduce easily from (5.7) that V is of class C^4 in the quadrant $(y > 0, \xi > 0)$. It follows from the equation (5.5), that the function $U_{\xi\xi}$ satisfies the equation

$$(5.10) \quad \frac{1}{2}\Big[(U_{\xi\xi})_{yy} + (U_{\xi\xi})_{\xi\xi}\Big] - (U_{\xi\xi})_{y\xi} + \theta^3 c'(y)\sinh(\theta\xi) = \lambda U_{\xi\xi}$$

in $(y > 0, \xi > 0)$. Therefore, in this region the function W of (5.9) satisfies the equation

$$(5.11) \qquad \frac{1}{2}(W_{yy} + W_{\xi\xi}) - W_{y\xi} = \lambda W$$

(from (5.5) and (5.10)) and, by analogy with (4.3) and the proof of Proposition 5.1, its unique solution is given by
(5.12)

$$W(y,\xi) = \frac{\sinh(\xi\sqrt{2\lambda}).W(0, y+\xi) + \sinh(y\sqrt{2\lambda}).W(y+\xi, 0)}{\sinh((y+\xi)\sqrt{2\lambda})} .$$

In order to determine the function W in the positive quadrant, we only have to compute its values on the axes. Now from (5.7) we obtain, by successively differentiating with respect to ξ, and by using the fact that $c'(0) = 0$, the formulae

$$\begin{aligned} &\frac{1}{\theta}\sqrt{\frac{\lambda}{2}}\sinh(y\sqrt{2\lambda}).U_\xi(y,0+) \\ &\quad = \sqrt{2\lambda}.\int_0^y c'(y-u)\sinh((y-u)\sqrt{2\lambda})\sinh(\theta u)du\ , \end{aligned} \tag{5.13}$$

$$\begin{aligned} &\frac{1}{\theta}\sqrt{\frac{\lambda}{2}}\sinh(y\sqrt{2\lambda}).U_{\xi\xi}(y,0+) \\ &\quad = \frac{4\lambda}{\sinh(y\sqrt{2\lambda})}\int_0^y c'(y-u)\sinh(u\sqrt{2\lambda})\sinh(\theta u)du \\ &\quad + 2\sqrt{2\lambda}.\int_0^y c''(y-u)\sinh((y-u)\sqrt{2\lambda})\sinh(\theta u)du\ . \end{aligned} \tag{5.14}$$

We obtain now from (5.12) - (5.14) that $W(y,\xi)$ equals

$$\begin{aligned} &\frac{4\theta\sqrt{2\lambda}.\sinh(y\sqrt{2\lambda})}{\sinh^3((y+\xi)\sqrt{2\lambda})}\int_0^{y+\xi} c'(y+\xi-u)\sinh(u\sqrt{2\lambda})\sinh(\theta u)du \\ &+ \frac{4\theta.\sinh(y\sqrt{2\lambda})}{\sinh^2((y+\xi)\sqrt{2\lambda})}\int_0^{y+\xi} c''(u)\sinh(u\sqrt{2\lambda})\sinh(\theta(y+\xi-u))du. \end{aligned} \tag{5.15}$$

This expression is positive for $(y > 0, \xi \geq 0)$, and vanishes for $y = 0$. ◇

5.3 Remark: From (4.2) and the identity $Q \equiv V$ of Theorem 3.1, one obtains the growth condition $0 \leq Q(y,\xi) \leq K(1+|y|^\beta)e^{\theta|\xi|}$, if $(y,\xi) \in \mathcal{R}^2$, for the function Q of (3.5), an obvious improvement over (3.4). On the other hand, it is quite easy to see (from (5.7) and (3.2)) that the function $U = Q_{y\xi}$ satisfies a growth condition of the type $|U(y,\xi)| \leq K\ e^{\theta(|y|+|\xi|)}\ ;\quad (y,\xi) \in \mathcal{R}^2$.

5.4 PROPOSITION: *The function Q of (3.5) satisfies*

$$\theta^2 Q > Q_{\xi\xi} \qquad \text{in} \quad \mathcal{R}^2\ . \tag{5.16}$$

Proof *(adapted from [2])*: Because of symmetry, it suffices to prove (5.16) in the quadrant $(y \geq 0, \xi \geq 0)$. From the equation (3.3) written in the form

$$\frac{1}{2}[Q_{yy} + Q_{\xi\xi}] + G = \lambda Q \quad \text{in} \quad (y > 0, \xi > 0) \tag{5.17}$$

where $G(y,\xi) \triangleq c(y)\cosh(\theta\xi) - U(y,\xi)$, and the boundary conditions

$$Q_\xi(y, 0+) = 0, \quad Q_y(0+, \xi) = 0 \tag{5.18}$$

(cf. (4.17)(i),(ii)), one can obtain the representation

$$\begin{aligned} Q(y,\xi) &= E\int_0^\infty e^{-\lambda t}\ G(|y+W_1(t)|,\ |\xi+W_2(t)|)\ dt \\ &= \iiint_{\mathcal{R}^3_+} e^{-\lambda t} q(t;y,u)q(t;\xi,\eta)G(u,\eta)du d\eta dt\ , \end{aligned} \tag{5.19}$$

where $\underset{\sim}{W} = (W_1, W_2)$ is a standard, two-dimensional Brownian motion process and, for $x > 0, y > 0$,

$$q(t;x,y) = \frac{1}{\sqrt{2\pi t}}[\exp\{-\frac{(x-y)^2}{2t}\} + \exp\{-\frac{(x+y)^2}{2t}\}]$$

is the transition probability density function for Brownian motion in the positive quadrant, with reflection on its sides. (The derivation of (5.19) is accomplished in a straightforward manner, by applying Itô's rule to the semimartingale $e^{-\lambda t}Q(|y+W_1(t)|,\ |\xi+W_2(t)|)$ and using (5.17), (5.18), as well as the growth conditions of Remark 5.3).

Integrating by parts twice in (5.19) we obtain, after some calculus:

$$\begin{aligned} &\theta^2 Q(y,\xi) - Q_{\xi\xi}(y,\xi) \\ &= \iiint_{\mathcal{R}^3_+} e^{-\lambda t} q(t;y,u)q(t;\xi,\eta)[\theta^2 G(u,\eta) - G_{\eta\eta}(u,\eta)]du d\eta dt \\ &\quad - 2\iint_{\mathcal{R}^2_+} e^{-\lambda t}\ q(t;y,u)\ \frac{e^{-\xi^2/2t}}{\sqrt{2\pi t}} G_\eta(u,0+)\ du dt\ . \end{aligned}$$

Now from the definition of G and Proposition 5.2: $\theta^2 G - G_{\xi\xi} = U_{\xi\xi} - \theta^2 U = W > 0$ in $(y > 0, \xi \geq 0)$, and from (5.13): $-G_\xi(y,0+) =$

$U_\xi(y, 0+) > 0$ for $y > 0$. It follows that the above expression is positive. ◇

6 A Control Problem With Partial Observations

Let us consider now a probability space $(\Omega, \mathcal{F}, P)$, $\{\mathcal{F}_t\}$, on which a solution of Problem 2.1 has been constructed with $Y_0 = y \neq 0$. We may always assume that this space is rich enough to support a random variable z *independent of* $\mathcal{F}_\infty$, with given distribution μ. We shall denote by $\{\mathcal{G}_t\}, \{\mathcal{F}_t^Y\}$ the P-augmentations of the filtrations $\{\sigma(z) \vee \mathcal{F}_t\}$, $\{\sigma(Y_s);\ 0 \le s \le t\}$, respectively.

6.1 DEFINITION: The class $\mathcal{U}$ of *wide-sense admissible control processes* consists of all $\{\mathcal{F}_t\}$ - progressively measurable processes $u = \{u_t\ ; 0 \le t < \infty\}$ with values in $[-1, 1]$. The class $\mathcal{U}_s$ of *strict-sense admissible control processes* consists of all processes $u \in \mathcal{U}$ which are adapted to $\{\mathcal{F}_t^Y\}$. ◇

For every $u \in \mathcal{U}$, we introduce the exponential $\{\mathcal{G}_t\}$ - martingale

$$(6.1) \qquad \Lambda_t^u \triangleq \exp\{z \int_0^t u_s dY_s - \frac{1}{2} z^2 \int_0^t u_s^2 ds\}\ , \quad 0 \le t < \infty$$

and the process

$$(6.2) \qquad W_t^u \triangleq Y_t - y - z \int_0^t u_s ds\ , \quad 0 \le t < \infty\ .$$

For every given $T \in (0, \infty)$, the process $\{W_t^u, \mathcal{G}_t\ ;\ \ 0 \le t \le T\}$ is Brownian motion on the interval $[0, T]$ and independent of the random variable z, under the probability measure

$$(6.3) \qquad P_T^u(A) \triangleq E[\Lambda_T^u . 1_A]\ , \qquad A \in \mathcal{G}_T$$

by virtue of the Girsanov theorem (cf. [6], §3.5).

We can formulate now our control problem with partial observations.

6.2 PROBLEM: *Minimize the expected discounted cost*

$$(6.4)\qquad J(u) \triangleq \lim_{T\to\infty} E_T^u \int_0^T e^{-\alpha t} c(Y_t)dt = \int_0^\infty e^{-\alpha t} E_t^u c(Y_t)dt$$

over $u \in \mathcal{U}$. Here $\alpha > 0$ is a given real constant, and the cost function $c(\cdot)$ is as in section 3. ◇

Put differently, one seeks to minimize the cost functional of (6.4), subject to the dynamics

$$(6.2)'\qquad dY_t = zu_t dt + dW_t^u\ , \qquad Y_0 = y\ ,$$

with W^u a Brownian motion independent of the random variable z. The minimization is to be over controls u which take values in the interval $[-1,1]$ and are adapted to the "observation filtration" $\{\mathcal{F}_t\}$. Because z is independent of $\{\mathcal{F}_t\}$, it is called an "unobservable" variable, and the stochastic control problem is one of *partial* (or incomplete) *observations*.

This independence of z and $\{\mathcal{F}_t\}$ allows us to cast (6.4) as

$$(6.5)\qquad \begin{aligned} J(u) &= \int_0^\infty e^{-\alpha t} E[c(Y_t)\Lambda_t^u]dt = E\int_0^\infty e^{-\alpha t} c(Y_t) E[\Lambda_t^u|\mathcal{F}_t]dt \\ &= E\int_0^\infty e^{-\alpha t} c(Y_t)\ F\Big(\int_0^t u_s^2 ds, \int_0^t u_s dY_s\Big)\ dt\ , \end{aligned}$$

where

$$(6.6)\quad F(t,x) \triangleq \int_{\mathcal{R}} \exp\{yx - \frac{1}{2}y^2 t\}\mu(dy)\ ; \quad (t,x) \in (0,\infty)\times\mathcal{R}\ .$$

On the other hand, the least-squares estimate $\hat{z}_t^u = E_t^u(z|\mathcal{F}_t)$ of the random variable z, given the observations $\mathcal{F}_t$ up to time t, is expressed by the Bayes rule ([6], p.193) as

$$(6.7)\qquad \hat{z}_t^u = \frac{E[z\Lambda_t^u|\mathcal{F}_t]}{E[\Lambda_t^u|\mathcal{F}_t]} = G\Big(\int_0^t u_s^2 ds, \int_0^t u_s dY_s\Big)\ ,$$

where

(6.8)

$$G(t,x) \triangleq \frac{1}{F(t,x)}\int_{\mathcal{R}} y\exp\{yx - \frac{1}{2}y^2 t\}\mu(dy)\ ; \quad (t,x) \in (0,\infty)\times\mathcal{R}\ .$$

6.3 Example: In the special case of a Bernoulli random variable z, with

$$P[z = \theta] = \rho \;, \quad P[z = -\theta] = 1 - \rho \tag{6.9}$$

for some $\theta \in (0, \infty)$ and $\rho \in (0, 1)$, the expressions (6.5) - (6.8) become

(6.10)

$$F(t, x; \theta) \triangleq \frac{\cosh(b + \theta x)}{\cosh b} \, e^{-\frac{t\theta^2}{2}} \;, \qquad G(t, x; \theta) \triangleq \theta . \tanh(b + \theta x)$$

(6.11)

$$J(u; \theta) \triangleq \frac{1}{\cosh(\theta\xi)} \, E \int_0^\infty e^{-\int_0^t (\alpha + \frac{\theta^2}{2} u_s^2) ds} \, c(Y_t) \cosh(\theta \xi_t^u) dt$$

$$\hat{z}_t^u = \theta \, \tanh(\theta \xi_t^u) \;, \qquad \text{where} \tag{6.12}$$

$$b = \tanh^{-1}(2\rho - 1) \;, \quad \xi = \frac{b}{\theta} \qquad \text{and} \tag{6.13}$$

$$\xi_t^u = \xi + \int_0^t u_s dY_s \;, \quad 0 \le t < \infty \;. \tag{6.14}$$

$\diamond$

In the *completely observable case*, where z is almost surely equal to a real constant, it is well-known that the optimal control law is of the form

$$u_t^{opt} = -sgn(zY_t) \;. \tag{6.15}$$

This result, first proved by Beneš [1] in 1974, was later established by different methods in [5], [4], [3] and [6], §6.5. In the partially observable setting of the present section, it is natural to guess that an optimal law can be obtained if one replaces in (6.15) z by its least-squares estimate $\hat{z}^* = E_t^*(z|\mathcal{F}_t)$, i.e.,

$$u_t^* = -sgn(\hat{z}_t^* Y_t) \;. \tag{6.16}$$

This was shown in [2] for a symmetric distribution on μ, and $c(y) = y^2$. It will be established in section 7 for a general cost function $c(\cdot)$ obeying the assumptions of section 3.

6.3 Example (Cont'd): For a Bernoulli random variable z of the type (6.9), we have from (6.12), (6.14) and (6.16): $u_t^* = -sgn(Y_t\xi_t^*)$ and $\xi_t^* = y - \int_0^t sgn(Y_s\xi_s^*)dY_s$, which is the equation (2.1). Thus, we can make the identification $\xi^* \equiv Z$, and try to show that the process

$$u_t^* = -sgn(Y_tZ_t)\ , \quad 0 \le t < \infty \tag{6.17}$$

is optimal for the Problem 6.2, in the case (6.9) of a Bernoulli random variable. This will be proved in Theorem 7.1 of the next section.

6.4 Remark: For any $u \in \mathcal{U}$, the *innovations process*

$$\nu_t^u \triangleq Y_t - y - \int_0^t u_s\hat{z}_s^u ds\ , \qquad 0 \le t < \infty \tag{6.18}$$

is adapted to $\{\mathcal{F}_t\}$ and, for every $T \in (0,\infty)$, we have that its restriction $\{\nu_t^u, \mathcal{F}_t;\ 0 \le t \le T\}$ is Brownian motion on $[0,T]$ under the probability measure P_T^u of (6.3). In the case of Example 6.3, it follows from (6.18), (6.12) that the processes Y and ξ^u (of (6.14)) satisfy the innovations-driven equations

$$dY_t = u_t\theta\tanh(\theta\xi_t^u)dt + d\nu_t^u\ , \quad Y_0 = y \neq 0 \tag{6.19}$$

$$d\xi_t^u = u_t^2\theta\tanh(\theta\xi_t^u)dt + u_td\nu_t^u\ , \quad \xi_0^u = \xi\ . \tag{6.20}$$

7 Solution to the Control Problem

Let us concentrate first on the Bernoulli case of (6.9). We introduce the function

$$\Phi(y,\xi) \triangleq \frac{Q(y,\xi)}{\cosh(\theta\xi)}\ , \quad (y,\xi) \in \mathcal{R}^2 \tag{7.1}$$

which is of class C^2 in $\mathcal{R}^2$ and satisfies the analogues

$$\begin{aligned} \alpha\Phi =& c(y) + \frac{1}{2}[\Phi_{yy} + \Phi_{\xi\xi}] \\ &- sgn(y\xi)[\Phi_{y\xi} + \theta\tanh(\theta\xi).\Phi_y] + \theta\tanh(\theta\xi).\Phi_y \end{aligned} \tag{7.2}$$

(7.3)
$$\frac{1}{2}\Phi_{\xi\xi} + \theta\tanh(\theta\xi)\Phi_\xi < 0\ , \quad sgn[\Phi_{y\xi} + \theta\tanh(\theta\xi)\Phi_y] = sgn(y\xi)$$

$$0 \le \Phi(y,\xi) \le K(1+|y|^\beta) \tag{7.4}$$

of (3.3), (5.16), (5.4). Thanks to (7.3), the equation (7.2) can be re-written as

(7.5)
$$\alpha\Phi = c(y) + \frac{1}{2}\Phi_{yy}$$
$$+ \min_{|u|\le 1}\left[u\{\Phi_{y\xi} + \theta\tanh(\theta\xi).\Phi_y\} + u^2\{\frac{1}{2}\Phi_{\xi\xi} + \theta\tanh(\theta\xi).\Phi_\xi\}\right]$$

the formal *Hamilton-Jacobi-Bellman (HJB) equation* corresponding to the problem of minimizing the expected discounted cost (6.4), subject to the dynamics (6.19), (6.20). The minimization in (7.5) is achieved by $u^* = -sgn[\Phi_{y\xi} + \theta\tanh(\theta\xi).\Phi_y] = -sgn(y\xi)$, again suggesting that the process (6.17) is optimal.

We can achieve now the following generalization of the main result in [2].

7.1 THEOREM: *For the Bernoulli case (6.9), we have*

$$J(u;\theta) \ge J(u^*;\theta) = \Phi(y,\xi)\ , \qquad \forall\ u\ \in\ \mathcal{U} \tag{7.6}$$

for any given $y \neq 0$, where ξ is given by (6.13) and u^* is the control process of (6.17).

Proof: From (6.11) we have $J(u^*;\theta) = \frac{V(y,\xi)}{\cosh(\theta\xi)}$, and so the equality in (7.6) is obvious from Theorem 3.1. Now for an arbitrary $u \in \mathcal{U}$ we apply Itô's rule to $e^{-\alpha t}\Phi(Y_t,\xi_t^u)$ and obtain, in conjunction with (6.19) and (6.20):

$$e^{-\alpha(T\wedge\tau_n)}\ \Phi(Y_{T\wedge\tau_n},\xi^u_{T\wedge\tau_n}) + \int_0^{T\wedge\tau_n} e^{-\alpha t}c(Y_t)dt$$
$$= \Phi(y,\xi) + \int_0^{T\wedge\tau_n} \beta_t^n dt \tag{7.7}$$
$$+ \int_0^{T\wedge\tau_n} e^{-\alpha t}[\Phi_y(Y_t,\xi_t^u) + u_t\Phi_\xi(Y_t,\xi_t^u)]d\nu_t^u,$$

where $T > 0$ is a constant, $\tau_n \triangleq \inf\{t \in [0,\infty)\,;\ |Y_t| \geq n \text{ or } |\xi_t^u| \geq n\}$. $n \in \mathbf{N}$ and the process

$$\begin{aligned}\beta_t^u \triangleq\ & \frac{1}{2}\Phi_{yy}(Y_t,\xi_t^u) + c(Y_t) - \alpha\Phi(Y_t,\xi_t^u)\\ &+ u_t[\Phi_{y\xi}(Y_t,\xi_t^u) + \theta\tanh(\theta\xi_t^u).\Phi_y(Y_t,\xi_t^u)]\\ &+ u_t^2[\frac{1}{2}\Phi_{\xi\xi}(Y_t,\xi_t^u) + \theta\tanh(\theta\xi_t^u)\Phi_\xi(Y_t,\xi_t^u)]\end{aligned}$$

is nonnegative, by virtue of the equation (7.5). If we take now expectations in (7.7) with respect to P_T^u, that of the stochastic integral is zero, and we can use condition (7.4) to show that

$$\begin{aligned}\lim_{n\to\infty} E_T^u[e^{-\alpha(T\wedge\tau_n)}\Phi(Y_{T\wedge\tau_n},\xi_{T\wedge\tau_n}^u)] &= e^{-\alpha T}E_T^u\Phi(Y_T,\xi_T^u)\\ &\leq C(1+T^{\beta/2})e^{-\alpha T}\ .\end{aligned}$$

Letting $T\to\infty$ in the resulting expression $E_T^u\int_0^T e^{-\alpha t}c(Y_t)dt + e^{-\alpha T}E_T^u\Phi(Y_t,\xi_T^u) \geq \Phi(y,\xi)$, we obtain the inequality in (7.6) and the optimality of u^*. ◇

The control u^* of (6.17) does *not* belong to $\mathcal{U}_s$, the class of strictly admissible control processes (recall (2.6)). It can be shown, as in [2], that $\inf_{u\epsilon\mathcal{U}_s} J(u;\theta) = \inf_{u\epsilon\mathcal{U}} J(u;\theta) = J(u^*;\theta)$, and that *no control process in $\mathcal{U}_s$ can be optimal.*

For a general *symmetric* distribution μ on the random variable z, the function $F(t,x)$ of (6.6) and the expected discounted cost $J(u)$ of (6.5) become, respectively

$$F(t,x) = 2\int_0^\infty e^{-\theta^2 t/2}\cosh(\theta x)\mu(d\theta),\quad J(u) = 2\int_0^\infty J(u;\theta)\mu(d\theta)\ .$$

It follows then from Theorem 7.1 that

$$J(u) \geq J(u^*) = 2\int_0^\infty J(u^*;\theta)\mu(d\theta)$$

holds for any $u \in \mathcal{U}$, thus proving the optimality of u^* in this class.

8 Acknowledgements

The research of I. Karatzas was supported in part by AFOSR Grant 86-0203. That of D. Ocone was supported in part by NSF grant DMS-89-03014.

9. References

[1] Beneš, V.E., Girsanov functionals and optimal bang-bang laws for final-value stochastic control. *Stoch. Processes & Appl.* **2** (1974), 127-140.

[2] Beneš, V.E., Karatzas, I. & Rishel, R.W., The separation principle for a Bayesian adaptive control problem with no strict-sense optimal law. *Stochastics Monographs*, (1989) to appear.

[3] Beneš, V.E., Shepp, L.A. & Witsenhausen, H.S., Some solvable stochastic control problems. *Stochastics* **4** (1980), 39-83.

[4] Davis, M.H.A. & Clark, J.M.C., On "predicted miss" stochastic control problems. *Stochastics* **2** (1979), 197-209.

[5] Ikeda, N. & Watanabe, S., A comparison theorem for solutions of stochastic differential equations, and its applications. *Osaka J. Math.* **14** (1977), 619-633.

[6] Karatzas, I. & Shreve, S.E., *Brownian Motion and Stochastic Calculus.* Springer-Verlag, New York (1988).

[7] Karatzas, I. & Ocone, D.L., The resolvent of a degenerate diffusion on the plane, with application to partially-observed stochastic control. Preprint.

On the Interior Smoothness of Harmonic Functions for Degenerate Diffusion Processes

N.V. Krylov
Department of Mechanics and Mathematics
Moscow State University, USSR

1 Introduction

There are many results in the theory of elliptic and parabolic partial differential equations of second order which can be interpreted and explained from the point of view of probability theory. Some times such an interpretation, revealing the genuine nature of the problem even provides new proofs and generalizations of the corresponding results in PDE theory. Let us mention only Gihman's proof of the solvability of degenerate parabolic equations based on an analysis of a probabilistic formula for solutions of these equations. He obtained a result which was new even in PDE theory. Let us mention also Malliavin's and Bismut's probabilistic proofs of Hörmander's hypoellipticity theorem explaining qualitatively a very deep analytical result. Nevertheless sometimes the probabilistic interpretation fails or seems to fail to prove the corresponding PDE theory result in its full generality or even under additional smoothness assumptions and we cannot even explain why this result can be true. The probabilist then feels that there is something inadequate in the probabilistic interpretation and tries to find it out.

We will discuss here two problems of this kind. Let E_d be an Euclidean space of dimension $d, (w_t, \mathcal{F}_t)$ a d_1 dimensional Wiener process defined on some probability space $(\Omega, \mathcal{F}, P), \sigma(x), b(x)$ smooth functions defined on E_d with values in the space of $d \times d_1$-matrices and in E_d respectively. Let us take a number $\lambda > 0$ and a function

ISBN 0-12-481005-5

f on E_d bounded with its derivatives and define

$$v(x) = E\int_0^\infty f(x_t)e^{-\lambda t}dt \tag{1}$$

where $x_t = x_t(x)$ is a (unique) solution of Ito's stochastic equation

$$dx_t = \sigma(x_t)dw_t + b(x_t)dt, \ x_0 = x. \tag{2}$$

It is well known that if v is twice continuously differentiable then $Lv - \lambda v + f = 0$ in E_d with

$$Lv := a^{ij}v_{x^i x^j} + b^i v_{x^i}, \ a = (a^{ij}) = \frac{1}{2}\sigma\sigma^*.$$

The investigation of the smoothness of v by probabilistic means follows Gihman's methods lately developed by Freidlin. We differentiate (1) formally with respect to x in the direction of $\xi \in E_d$ and get

$$v_{(\xi)}(x) = E\int_0^\infty f_{(\xi_t)}(x_t)e^{-\lambda t}dt \tag{3}$$

where ξ_t is a solution of the equation obtained from (2) by formal differentiation:

$$\begin{aligned} d\,\xi_t &= \sigma_{(\xi_t)}(x_t)dw_t + b_{(\xi_t)}(x_t)dt, \ \xi_0 = \xi, \\ u_{(\xi)} &:= \xi^i u_{x^i} = (\xi, u_x). \end{aligned} \tag{4}$$

One knows that if the right hand side of (3) is well defined then one really has the equality in (3). As $\left|f_{(\xi)}\right| \le \mathrm{N}\,|\xi|$, to justify (3) it suffices to prove that

$$E\,|\xi_t| \le \mathrm{N}e^{\epsilon t} \ \ \forall t \ge 0 \tag{5}$$

with a constant $\epsilon < \lambda$ and a constant N. Condition (5) is sometime necessary for v to have first derivatives. This is easily seen when $d = 1$, $\sigma = 0$, b is linear. But if a is non-degenerate and only continuous then from PDE theory we know that the solution of $Lv -$

$\lambda v + f = 0$ has bounded first derivatives regardless of whether or not (5) is fulfilled. Taking the difference in smoothness assumptions on a, b apart, we nevertheless see that basing on (3) with ξ_t defined by (4) we lose something and we cannot explain this effect even if $d = d_1 = 1, \sigma = 1, b$ is linear.

Another case is the following. Let $\mathcal{D}$ be a domain in E_d with smooth boundary $\partial\mathcal{D}, g$ a smooth function on E_d, $x_t = x_t(x)$,

$$\tau = \tau(x) = \tau_{\mathcal{D}}(x) = \inf\{t \geq 0, x_t \notin \mathcal{D}\}$$

$$v(x) = Eg(x_\tau) \tag{6}$$

and suppose that $\tau(x) < \infty$ (a.s.). Under natural assumptions one obtains

$$v_{(\xi)}(x) = Ev_{(\xi_\tau)}(x_\tau) \tag{7}$$

and to estimate $v_{(\xi)}(x)$ in $\mathcal{D}$ we have to estimate $E|\xi_\tau|$ and v_x on $\partial\mathcal{D}$. On the other hand if $\mathcal{D} = \{|x| < 1\}, d = d_1, \sigma^{ij} = \delta^{ij}, b = 0$ when $L = \Delta$, the function v is analytically expressed by the Poisson formula and $|v_x(0)|$ can be estimated in term of $\sup|g|$ only. Once more we see that we have to revise formula (7) which uses ξ_t from (4).

Here we want to explain that there is other processes ξ_t which can be used in formulas like (3) and (7) and which allow us to avoid the above mentioned shortcomings of solutions of (4). Solutions of (4) are derivatives of $x_t(x)$ with respect to x, the new ξ_t will be called quasi-derivatives. It is worth noting that the idea of their introduction is very close to some ideas used by Bismut in his proof of Hörmander's theorem. We will treat only the case of functions like (6) known as harmonic functions for $x_t(x)$ in $\mathcal{D}$ and here we will summarize some results of [1]. The function of type (1) are considered in [2].

2 Notion of Quasi-derivative

Let us write $u \in \mathcal{M}_1(= \mathcal{M}_1(\mathcal{D}, \sigma, b))$ if u is a real function defined and once continuously differentiable in $\mathcal{D}$ and such that $u(x_t(x))$

is a local martingale on $[0,\tau(x))$ for every $x \in \mathcal{D}$ with $x_t(x), \tau(x)$ defined by (2) and (6). Note that it is well known that if v in (6) is well defined then $v(x_t(x))$ is a local martingale on $[0,\tau(x))$ for every $x \in \mathcal{D}$. So if $v \in C^1(\mathcal{D}), v \in \mathcal{M}_1$.

Let $u \in \mathcal{M}_1, x \in \mathcal{D}, \xi \in E_d, \tau$ be a stopping time, $\tau \leq \tau(x), \xi_t, \xi_t^0$ be $\mathcal{F}_t$ - adapted continuous processes defined on $[0,\tau] \cap [0,\tau(x))$ with values in E_d, E_1 respectively and $\xi_0 = \xi$. We say that ξ_t is a u-quasi-derivative of $x_t(y)$ with respect to y along ξ at the point $y = x$ on $[0,\tau)$ if the process

$$u_{(\xi_t)}(x_t(x)) + \xi_t^0 u(x_t(x)) \tag{8}$$

is a local martingale on $[0,\tau)$. In this case ξ_t^0 is called u-accompanying process for ξ_t and we write

$$(\xi_\cdot, \xi_\cdot^0) \in \mathcal{D}_\tau(u,x,\xi)(= \mathcal{D}_\tau(u,\mathcal{D},\sigma,b,x,\xi))\,.$$

For $\tau = \tau(x)$ we drop here the index τ.

Now let the continuous functions $\sigma^1(x,\xi), \sigma_0(x,\xi), b^1(x,\xi)$ with values in the set of $d \times d_1$-matrices, in E_{d_1} and in E_d respectively be defined on $\mathcal{D} \times E_d$. Suppose that for every compact $\Gamma \subset \mathcal{D}$ there exists a constant N such that the inequality

$$\| \sigma^1(x,\xi) - \sigma^1(x,\eta) \| + |b^1(x,\xi) - b^1(x,\eta)| \leq N|\xi - \eta| \tag{9}$$

holds for all $x \in \Gamma, \xi, \eta \in E_d$ (with $\| A \|^2 := trAA^*$). Let $u \in \mathcal{M}_1$. We write

$$(\sigma^1, \sigma_0, b^1) \in \mathcal{L}(u,\mathcal{D})(= \mathcal{L}(u,\mathcal{D},\sigma,b))$$

if for every $x \in \mathcal{D}, \xi \in E_d$ we have $(\xi_\cdot, \xi_\cdot^0) \in \mathcal{D}(u,x,\xi)$ where ξ_t, ξ_t^0 are defined by

$$\xi_t = \xi + \int_0^t \sigma^1(x_s,\xi_s,)dw_s + \int_0^t b^1(x_s,\xi_s)ds, \tag{10}$$

$$\xi_t^0 = \int_0^t \sigma_0^*(x_s,\xi_s)dw_s \tag{11}$$

with $x_s = x_s(x)$. Note that condition (9) implies existence and uniqueness of solutions of (10) and (11) on $[0, \tau(x))$ anyway.

If $(\xi_\cdot, \xi_\cdot^0) \in \mathcal{D}_\tau(u, x, \xi)$ and a stopping time $\gamma \le \tau$, clearly $(\xi_\cdot, \xi_\cdot^0) \in \mathcal{D}_\gamma(u, x, \xi)$. The following theorem deals with expansion of the domain of definition $[0, \tau)$ of quasi-derivatives.

Theorem 1. a) Let $u \in \mathcal{M}_1, A$ be a set and for every $p \in A$ we are given functions $(\sigma^1, \sigma_0, b^1) = (\sigma^1, \sigma_0, b^1)(x, \xi, p)$ of the class $\mathcal{L}(u, \mathcal{D})$. b) Fix an integer $n \ge 1, p(1), \cdots, p(n) \in A, x_0 \in \mathcal{D}, \xi_0 \in E_d$ and stopping times $\tau_1, \cdots, \tau_n$ such that $\tau_i \le \tau(x_0)$ and the relations

$$\tau(x_0) > \tau_i = \tau_j = \inf_m \tau_m =: \tau$$

are impossible for any $w, i \neq j$. c) Let $(\xi_\cdot, \xi_\cdot^0) \in \mathcal{D}_\tau(u, x, \xi)$ and define η_t, η_t^0 so that $\eta_t = \xi_t, \eta_t^0 = \xi_t^0$ for $t \le \tau$ and on $[\tau, \tau(x_0))$ it holds

$$\eta_t = \xi_\tau + \sum_{i=1}^{n} I_{\tau=\tau_i} \left[\int_\tau^t \sigma^1 (x_s, \eta_s, p(i))\, dw_s + \int_\tau^t b^1 (x_s, \eta_s, p(i))\, ds \right],$$

$$\eta_t^0 = \xi_t^0 + \sum_{i=1}^{n} I_{\tau=\tau_i} \int_\tau^t \sigma_0^* (x_s, \eta_s, p(i))\, dw_s$$

where $x_s = x_s(x_0)$. Under these hypotheses we assert that

$$(\eta_\cdot, \eta_\cdot^0) \in \mathcal{D}(u, x, \xi).$$

This theorem is simply derived from the strong Markov property of $x_t(x)$. Using this theorem, replacing the process $p_t \in A = E_k$ by step processes p_t^n with finite sets of values and proceeding by recurrence from $p_{t\wedge t(n,i)}^n$ to $p_{t\wedge t(n,i+1)}^n$ we prove

Theorem 2. Assume condition a) of theorem 1 and let $A = E_k$, σ^1, σ_0, b^1 be linear functions with respect to p. Then for every $(x, \xi) \in \mathcal{D} \times E_d$ and every $\mathcal{F}_t$-adapted E_k-valued measurable process p_t on $[0, \tau(x))$ satisfying

$$\int_0^T |p_t|^2 dt < \infty \ \text{(a.s } \{T < \tau(x)\})$$

for every $T < \infty$ there exist processes ξ_t, ξ_t^0 uniquely defined on $[0, \tau(x))$ by

$$\begin{aligned} d\xi_t &= \sigma^1(x_t, \xi_t, p_t)dw_t + b^1(x_t, \xi_t, p_t)dt, \ \xi_0 = \xi, \\ d\xi_t^0 &= \sigma_0^*(x_t, \xi_t, p_t)dw_t, \ \xi_0^0 = 0 \end{aligned}$$

where $x_t = x_t(x)$. Moreover

$$(\xi_\cdot, \xi_\cdot^0) \in \mathcal{D}(u, x, \xi). \tag{12}$$

One more general result concerning quasi-derivatives in its proof makes use of the linearity of (8) with respect to ξ_t and to ξ_t^0.

Theorem 3. Under condition a) of theorem 1 for every $x \in \mathcal{D}, \xi \in E_d, n \geq 1, p(1), \ldots, p(n) \in A, \lambda(1), \ldots, \lambda(n) \in E_1$ with $\sum \lambda(i) = 1$ solutions of

$$\begin{aligned} d\xi_t &= \sum_{i=1}^n \lambda(i)\sigma^1(x_t, \xi_t, p(i))dw_t + \sum_{i=1}^n \lambda(i)b^1(x_t, \xi_t, p(i))dt, \ \xi_0 = \xi, \\ d\xi_t^0 &= \sum_{i=1}^n \lambda(i)\sigma_0^*(x_t, \xi_t, p(i))dw_t, \ \xi_0^0 = 0 \end{aligned}$$

(where $x_t = x_t(x)$) satisfy (12).

3 Examples of Quasi-derivatives

Lemma 1. Let $u, \epsilon\mathcal{M}_1, \rho \in E_1, \sigma_0 \in E_{d_1}, P$ be a $d_1 \times d_1$ skew symmetric matrix ($P^* = -P$), $x \in \mathcal{D}, x_t = x_t(x), \xi \in E_d$. Then every process ξ_t defined either by (4) or by one of the following equations:

$$d\xi_t = (\sigma_{(\xi_t)} + \rho\sigma)(x_t)dw_t + (b_{(\xi_t)} + 2\rho b)(x_t)dt, \ \xi_0 = \xi, \tag{13}$$

$$d\xi_t = (\sigma_{(\xi_t)} + \sigma P)(x_t)dw_t + b_{(\xi_t)}(x_t)dt, \ \xi_0 = \xi, \tag{14}$$

$$d\xi_t = \sigma_{(\xi_t)}(x_t)dw_t + (b_{(\xi_t)} - \sigma\sigma_0)(x_t)dt, \ \xi_0 = \xi \tag{15}$$

is a u-quasi-derivative of $x_t(y)$ at the point $y = x$ on $[0, \tau(x))$ along ξ with u-accompanying processes ξ_t^0 equal to zero in the cases of (4), (13) and (14) and defined by $\xi_t^0 = \sigma_0^* w_t$ in the case of (15).

Sketch of the Proof. Suppose first that $x_t(y)$ does not leave $\mathcal{D}$ if $y \in \mathcal{D}$. For real p sufficiently small let us define x_t^p as a solution of

$$x_t = x + p\xi + \int_0^t (1+p\rho)\sigma(x_s)dw_s + \int_0^t (1+p\rho)^2 b(x_s)ds. \quad (16)$$

The time change $t \to s = (1+p\rho)^2 t$ shows that $y_s^p := x_t^p$ satisfies (2) with $s, x+p\xi, \tilde{w}_s = (1+p\rho)w_t$ instead of t, x, w_t. Hence $u(x_t^p)(= u(y_s^p))$ is a local martingale on $[0,\infty)$. Its derivative with respect to p at the point $p=0$ is clearly a local martingale on $[0,\infty)$ too. This derivative coincides with $u_{(\xi_t)}(x_t(x))$ where ξ_t is defined by (13). This proves our assertion concerning (13). Equation (4) is a partial case of (13) when $\rho = 0$.

Let us pass to (14). Here instead of (16) we introduce x_t^p as a solution of

$$x_t = x + p\xi + \int_0^t \sigma(x_s)(\exp pP)dw_s + \int_0^t b(x_s)ds.$$

Since $\tilde{w}_t := (\exp pP)w_t$ is a Wiener process, $u(x_t^p)$ is a local martingale on $[0,\infty)$ and we finish the proof by the same argument as above.

In the case of equation (15) the process x_t^p is defind by

$$x_t = x + p\xi + \int_0^t \sigma(x_s)dw_s + \int_0^t [b(x_s) - p\sigma(x_s)\sigma_0]ds.$$

The Girsanov theorem shows that

$$u(x_t^p)\exp(p\sigma_0^* w_t - \frac{1}{2}p^2|\sigma_0|^2 t)$$

is a local martingale on $[0,\infty)$ and once more we achieve the proof by its differentiating at the point $p=0$.

To get rid of our extra hypothesis we take a compact domain $G \subset \bar{G} \subset \mathcal{D}$ and a C^∞-function ψ such that $\psi = 1$ on G, $\psi = 0$ on $E_d \backslash G$. Then in (2) we multiply σ by ψ and b by ψ^2. The random time change proves that $u \in \mathcal{M}_1(\mathcal{D}, \psi\sigma, \psi^2 b)$ and it is obvious that the process x_t corresponding to $\psi\sigma, \psi^2 b$ does not leave $\mathcal{D}$ starting in $\mathcal{D}$. Consequently, we can apply the above results and as $\psi\sigma = \sigma, \psi^2 b = b$

in G we see that solution of (4, 13-15) are u-quasi-derivatives of $x_t(x)$ on $[0, \tau_G(x))$. In fact it is exactly our assertion since G is arbitrary. □

It appears that equation (14) can be generalized. The proof of the following result makes use of the Euler polygonal method of solving the system (2),(14) with $\tilde{\sigma}$ instead of σP and on a direct calculation showing that

$$E(u_x(y + \sigma(y)w_s + b(y)s), \tilde{\sigma}(y)w_s) = 0.$$

Lemma 2. Let $\tilde{\sigma}(x)$ be a continuous $d \times d_1$-matrix function defined on $\mathcal{D}$. Suppose that the matrix $\sigma\tilde{\sigma}^*$ is skew symmetric in $\mathcal{D}$. Then $(\sigma_\xi + \tilde{\sigma}, 0, b_{(\xi)}) \in \mathcal{L}(u, \mathcal{D})$ for every $u \in \mathcal{M}_1$.

Taking $A = \{0, 1, 2, 3\}, \sigma^1(x, \xi, i), b^1(x, \xi, i)$ equal to $\sigma_{(\xi)}(x), b_{(\xi)}(x)$ if $i = 0$ and equal to coefficients of $(12 + i)$ for $i = 1, 2, 3$, $\lambda(0) = -2, \lambda(1) = \lambda(2) = \lambda(3) = 1$ and applying theorem 3 we get

Corollary 1. Under the hypotheses of lemma 1 if

$$\begin{aligned} d\xi_t &= [\sigma_{(\xi_t)} + \rho\sigma + \sigma P](x_t)dw_t + [b_{(\xi_t)} + 2\rho b - \sigma\sigma_0](x_t)dt, \ \xi_0 = \xi, \\ d\xi_t^0 &= \sigma_0^* dw_t, \ \xi_0^0 = 0 \end{aligned} \qquad (17)$$

then relation (12) holds.

There is a generalization of this fact.

Corollary 2. Let $(\sigma^1, \tilde{\sigma}_0, b^1) \in \mathcal{L}(u, \mathcal{D})$ and take ρ, σ_0, P from lemma 1. Then

$$(\sigma^1 + \rho\sigma + \sigma P, \tilde{\sigma}_0 + \sigma_0, b^1 + 2\rho b - \sigma\sigma_0) \in \mathcal{L}(u, \mathcal{D}).$$

To prove this we apply theorem 3 and corollary 1 to conclude that for every $\epsilon \neq 0$

$$\Big((1-\epsilon)\sigma^1 + \epsilon(\sigma_{(\xi)} + \frac{1}{\epsilon}\rho + \sigma\frac{1}{\epsilon}P), (1-\epsilon)\tilde{\sigma}_0 + \epsilon(\frac{1}{\epsilon}\sigma_0),$$
$$(1-\epsilon)b^1 + \epsilon(b_{(\xi)} + 2\frac{1}{\epsilon}\rho b - \sigma\frac{1}{\epsilon}\sigma_0)\Big) \in \mathcal{L}(u, \mathcal{D}) .$$

Now it suffices to let $\epsilon \to 0$.

As an immediate applicaiton of equation (13) let us consider the function v defined in (6) when $d = 2, d_1 = 1, \sigma^{11} = 1, \sigma^{21} = 0, b = 0, \mathcal{D} = \{|x| < 1\}$. Let us suppose that ($g$ and) v is smooth enough. Then by lemma 1 the process $v_{(\xi_t)}(x_t)$ is a local martingale on $[0, \tau(x))$ and since $\xi^1_\tau = \rho(x^1_\tau - x^1) + \xi^1, \xi^2_\tau = \xi^2$ this local martingale is bounded and we get (7). If in this formula we take $x = (0, \epsilon) \in \mathcal{D}, \rho = -\epsilon(1-\epsilon^2)^{-1}, \xi = (0,1)$, then $\xi_\tau = (\rho x^1_\tau, 1), x^2_\tau = \epsilon, |x_\tau| = 1$ and it follows that $\xi_\tau \perp x_\tau$, ξ_τ is tangent to $\partial\mathcal{D}$ at the point x_τ and

$$v_{(\xi_\tau)}(x_\tau) = g_{(\xi_\tau)}(x_\tau), \tag{18}$$

$$v_{x^2}(0,\epsilon) = Eg_{(\xi_\tau)}(x_\tau),\ E|\xi_\tau| = E(1+\rho^2(x^1_\tau)^2)^{\frac{1}{2}} = (1-\epsilon^2)^{-\frac{1}{2}},$$

$$|v_{x^2}(0,\varepsilon)| \leq \sup|g_x|(1-\epsilon^2)^{-\frac{1}{2}}\ . \tag{19}$$

We see that formula (7) with quasi-derivatives can explain why and how the smoothness of g implies the smoothness of v. Below we will give some other estimates of type (19). So let us mention that from derivation of (19) it is easily seen that sometimes there is equality instead of inequality (and $v \not\equiv 0$). It is also worth noting that even in the case of Wiener process the function v (defined by (6) with $x_t(x) = x + w_t$) does not have first derivatives bounded up to $\partial\mathcal{D}$ if g belongs only to C^1. Thus formula (7) with ordinary derivatives cannot be used at all for proving estimates like (19) since $\xi_t \equiv \xi$ and $v_{(\xi_\tau)}(x_\tau)$ cannot be estimated in terms of $\sup|g_x|$.

We used above a constant ρ to obtain (18) for a special choice of initial data x, ξ. In the general case we need random ρ, σ_0, p to turn ξ_t so that at the moment $t = \tau$ it becomes tangent to $\partial\mathcal{D}$ at the point x_τ.

Theorem 4. Let $x \in \mathcal{D}$ and suppose that for every $\xi \in E_d$ on $[0, \tau(x))$ we are given $\mathcal{F}_t$-adapted measurable with respect to (w, t, ξ) processes $\rho_t(\xi), \sigma_{0t}(\xi), P_t(\xi)$ with values in E_1, E_{d_1} and in the space of skew symmetric $d_1 \times d_1$-matrices respectively. Suppose further that equation (17) with $x_t = x_t(x)$ and $\rho_t(\xi_t), \sigma_{0t}(\xi_t)$ $P_t(\xi_t)$ instead of ρ, σ_0, P admit a solution $(\xi_\cdot, \xi^0_\cdot)$ on $[0, \tau(x))$ such that

$$\int_0^T \left[|\rho_t(\xi_t)|^2 + |\sigma_{0t}(\xi_t)|^2 + \|P_t(\xi_t)\|^2\right] dt < \infty\ (\text{a.s. } \{T < \tau(x)\})$$

for every $T < \infty$. Then $(\xi_\cdot, \xi^0_\cdot) \in \mathcal{D}(u, x, \xi)$ for every $u \in \mathcal{M}_1$.

Proof. Take $p = (\rho, \sigma_0, P)$

$$\begin{aligned} \sigma^1(x,\xi,p) &= \sigma_{(\xi)}(x) + \rho\sigma(x) + \sigma(x)P,\ \sigma_0(x,\xi,p) = \sigma_0, \\ b^1(x,\xi,p) &= b_{(\xi)}(x) + 2\rho b(x) - \sigma(x)\sigma_0, \end{aligned}$$

and for the above mentioned solution ξ_t define $p_t = (\rho_t(\xi_t), \sigma_0(\xi_t), P_t(\xi_t))$. After this it rests to apply corollary 1 and theorem 3. □

4 Applications to the Study of Interior Smoothness of Harmonic Functions

1: Let $\mathcal{D} = \{|x| < 1\}$, g be a C^∞-function and v be defined by (6) with $d_1 = d, \sigma^{ij} = \delta^{ij}, b = 0$ so that $x_t(x) = x + w_t$. Then it is well known that $v \in C^\infty(\bar{\mathcal{D}})$ and $\Delta v = 0$ in $\mathcal{D}, v = g$ on $\partial\mathcal{D}$. Using quasi-derivatives let us prove that

$$|v_x(0)| \le \mathcal{N} \max_{\partial\mathcal{D}} |g| \tag{20}$$

with a constant $\mathcal{N}$ independent of g.

To this end we take a nonnegative function h continuous in $\mathcal{D}$, $\xi \in E_d$ and define $(\xi_t,\ \xi^0_t)$ with the help of the following relations:

$$\xi_t = \xi - \int_0^t h(w_s)\xi_s ds,\ \xi^0_t = \int_0^t h(w_s)\xi^*_s dw_s.$$

Note that $x_t(0) = w_t$ so that by theorem 4 the process

$$\alpha_t = v_{(\xi_t)}(w_t) + \xi^0_t v(w_t)$$

is a local martingale on $[0, \tau(0))$. For $r \in (0,1)$ define

$$\mathcal{D}(r) = \{|x| < r\}, \gamma(r) = \inf\{t \ge 0 : |w_t| = r\}.$$

Obviously

$$\xi_t = \xi \exp\left(-\int_0^t h(w_s)ds\right),\ \xi^0_t = \int_0^t h(w_s)\exp\left(-\int_0^s h(w_r)dr\right)\xi^* dw_s,$$

$$|\xi_t| \le |\xi|,\quad E\int_0^{\gamma(r)} h^2(w_s)\,|\xi_s|^2\,ds \le \sup_{\mathcal{D}(r)} h^2\,|\xi|^2\,E\gamma(r) < \infty$$

It follows that $\alpha_{t\wedge\gamma(r)}$ is a martingale and

$$\begin{aligned} v_{(\xi)}(0) &= Ev_{(\xi_{\gamma(r)})}(w_{\gamma(r)}) + E\xi^0_{\gamma(r)}v(w_{\gamma(r)}) \\ &\le \sup_{\mathcal{D}} |v_x|\, E\left|\xi_{\gamma(r)}\right| + \sup_{\mathcal{D}} |v|\, E\left|\xi^0_{\gamma(r)}\right|. \end{aligned} \tag{21}$$

From (6) we see that $|v| \le \sup(g, \partial\mathcal{D})$ in $\mathcal{D}$. Moreover

$$(E|\xi^0_{\gamma(r)}|)^2 \le E|\xi^0_{\gamma(r)}|^2 = |\xi|^2 E \int_0^{\gamma(r)} h^2(w_s)\exp(-2\int_0^s h(w_u)\,du)ds.$$

Hence, to deduce (20) from (21) as $r \uparrow 1$ it suffices to find a function h such that

$$\int_0^{\tau(0)} h(w_s)\,ds = \infty,\ E\int_0^{\tau(0)} h^2(w_s)\exp\left(-2\int_0^s h(w_u)\,du\right)ds < \infty \tag{22}$$

a.s. Let

$$h(x) = c\exp\frac{1}{1-|x|^2},\quad u(x) = \exp\frac{1}{1-|x|^2}$$

and let us leave to the reader the verification that for a sufficiently large constant c we have $\Delta u - 4hu \le -2c^{-1}h^2$ in $\mathcal{D}$. Now by Ito's formula for $r < 1$

$$\begin{aligned} u(0) &= Eu\left(w_{\gamma(r)}\right)\exp\left(-2\int_0^{\gamma(r)} h(w_s)\,ds\right) \\ &\quad +\frac{1}{2}E\int_0^{\gamma(r)} (4hu - \Delta u)(w_t)\exp\left(-2\int_0^t h(w_u)\,du\right)dt \\ &\ge \exp\frac{1}{1-r^2}E\exp\left(-2\int_0^{\tau(0)} h(w_s)\,ds\right) \\ &\quad +c^{-1}E\int_0^{\gamma(r)} h^2(w_t)\exp\left(-2\int_0^t h(w_s)\,ds\right)dt \end{aligned}$$

Letting $r \uparrow 1$ we get immediately both relations in (22). Thus we have proved (20) constructing a quasi-derivative which vanishes when t approaches $\tau(0)$ so that the first expectation in (21) disappears and

$$v_{(\xi)}(0) = E\xi^0_{\tau(0)} g(w_{\tau(0)}) \ .$$

2: Let us use the possibility to make $\xi_{\tau(x)}$ tangent to $\partial\mathcal{D}$. Consider an example where

$$\mathcal{D} = \{x : x^d > 0\},\ b^d(x) \equiv -1,\ \sigma^{di} \equiv 0 \text{ for } i = 1, \ldots, d_1. \tag{23}$$

Theorem 5. Let $b, \sigma, g \in C^1_b(\bar{\mathcal{D}})$. Then the function v defined by (6) has in $\mathcal{D}$ first Sobolev derivatives and for every $\xi \in E_d$

$$|v_{(\xi)}(x)| \leq \mathcal{N}(|\xi| + (x^d)^{-\frac{1}{2}}|\xi|^d)e^{\mathcal{N}x^d} \sup_{x^d=0} |g_x| \text{ (a.e)} \tag{24}$$

with a constant $\mathcal{N}$ independent of g, x, ξ.

Sketch of the Proof. It is not hard to see that we can consider only the case when $v \in C^1_b(\bar{\mathcal{D}})$. In this case we want to use formula (7) with a quasi-derivative ξ_t such that $\xi_{\tau(x)}$ is tangent to $\partial\mathcal{D}$. As $\tau(x) = x^d$, we want to have $\xi^d_{x^d} = 0$. To assure this let us define

$$\psi(x) = x^d,\ B = (\epsilon|\xi|^2 + \psi^{-1}|\psi_{(\xi)}|^2)^{\frac{1}{2}} \exp \mathcal{N}\psi$$

where the constants $\mathcal{N}, \epsilon > 0$ will be chosen later. If we can construct a quasi-derivative ξ_t such that $B(x_t, \xi_t)$ is a supermartingale on $[0, \tau(x))$, then $B(x_t, \xi_t)$ will be bounded on $[0, \tau(x))$ (a.s) and it implies that $\psi_{(\xi_t)}(x_t) \to 0$ if $t \uparrow \tau(x)$ as we want.

Now let us define

$$\sigma^1 = \sigma_{(\xi)} + \psi^{-1}\psi_{(\xi)}\sigma,\ \sigma_0 = 0, b^1 = b_{(\xi)} + 2\psi^{-1}\psi_{(\xi)}b$$

and let ξ_t be a solution of (9) with $x_s = x_s(x)$ and fixed $x \in \mathcal{D}, \xi \in E_d$. By theorem (4) $(\xi_\cdot, 0) \in \mathcal{D}(v, x, \xi)$. Simple computation gives

$$dB^2(x_t, \xi_t) = \Gamma(x_t, \xi_t)dt + dm_t$$

where m_t is a local martingale on $[0, \tau(x))$,

$$\Gamma(y,\eta)\exp(-\mathcal{N}\psi(y)) = \psi^{-2}\psi_{(\eta)}^2\left[\epsilon\left|\sigma^k\right|^2 - 3\right] +$$

$$2\epsilon\psi^{-1}\psi_{(\eta)}\left[2(b,\eta) + \left(\sigma^k, \sigma_{(\eta)}^k\right)\right] + 2\epsilon\left(\eta, b_{(\eta)}\right) +$$

$$\epsilon\left|\sigma_{(\eta)}^k\right|^2 - 2\mathcal{N}\epsilon|\eta|^2 - 2\mathcal{N}\psi^{-1}\psi_{(\eta)}^2,$$

σ^k is the $k-th$ column of σ and we suppose summation over $k = 1, \ldots, d_1$. The estimates like

$$\epsilon\psi^{-1}|\psi_{(\eta)}| \cdot |\eta| \leq \epsilon\psi^{-2}\psi_{(\eta)}^2 + \epsilon|\eta|^2$$

show easily that we can first take a sufficiently small $\epsilon > 0$ and then a sufficiently large $\mathcal{N}$ so that $\Gamma(y,\eta) \leq 0$. These $\mathcal{N}, \epsilon$ being fixed, $B^2(x_t, \xi_t), B(x_t, \xi_t)$ are local super-martinagles on $[0, \tau(x))$. Now skipping some details from (7) we get

$$v_{(\xi)}(x) = EV_{(\xi_\tau)}(x_\tau) = Eg_{(\xi_\tau)}(x_\tau) \leq \mathcal{N}_1 EB(x_\tau, \xi_\tau) \leq \mathcal{N}_1 B(x, \xi)$$

where $\mathcal{N}_1 = \sup\{g_{(\xi)}(x)B^{-1}(x,\xi) : x^d = \xi^d = 0\}$. This obviously gives (24). □

Corollary 3. Let us drop condition (23) and suppose that the hypothesis of theorem 5 is fulfilled with $\bar{\mathcal{D}} = E_d$. Then the function $v(x,t) = Eg(x_t(x))$ in $E_d \times (0, \infty)$ has first Sobolev derivatives with respect to (x, t) and

$$|v_x| \leq \mathcal{N}e^{\mathcal{N}t}\sup|g_x|, \quad |v_t| \leq \frac{1}{\sqrt{t}}\mathcal{N}e^{\mathcal{N}t}\sup|g_x| \text{ (a.e.)} \tag{25}$$

with a constant $\mathcal{N}$ independent of g, x, t, ξ.

To prove this it suffice to apply theorem 5 to the process $y_t(y) = (x_t(x), x^{d+1} - t)$ where $y = (x, x^{d+1})$.

Remark 1. In (25) the term $\frac{1}{\sqrt{t}}$ is exact as is easily seen from explicit calculations for the simplest case when $d = d_1 = 1, x_t = x + w_t$, $g(x) \sim |x|$.

3: Here we will state a theorem treating the case of diffusion uniformly nondegenerate along the normal to $\partial\mathcal{D}$. Its proof also uses the technique and ideas described above.

Take a function $\psi \in C_b^3(E_d)$ and suppose that $\mathcal{D} = \{\psi > 0\}$, $|\psi_t| \geq 1$ on $\partial\mathcal{D}, L\psi \leq -1$ in $\mathcal{D}$, and

$$\| \sigma_{(\xi)} \|^2 - |\sigma^*_{(\xi)}\xi|^2 + 2(\xi, b_{(\xi)}) \leq -\delta + K(a\xi, \xi)$$

in $\mathcal{D}$ for all unit vectors $\xi \in E_d$ with some constants $K, \delta > 0$. Let $\sigma, b, g \in C_b^1(\bar{\mathcal{D}})$.

Theorem 6. Under the above hypotheses if $(a\psi_x, \psi_x) \geq 1$ on $\partial\mathcal{D}$, the function v is well defined by (6), continuous in $\bar{\mathcal{D}}$, has first Sobolev derivatives in $\mathcal{D}$ and for every $\xi \in E_d$

$$|v_{(\xi)}(x)| \leq \mathcal{N}(|\xi| + |\psi_{(\xi)} \ln \psi|) \sup(|g| + |g_x|) \text{ (a.e)} \tag{26}$$

with a constant $\mathcal{N}$ independent of g, x, ξ.

Remark 2. The term $\ln \psi$ in (26) cannot be avoided even in the simplest case when $d = d_1 = 2, \mathcal{D} = \{x^2 > 0\}, x_t = x + w_t$, as follows from direct calculations using the Poisson formula.

Remark 3. The author thinks that estimate (26) is true with $\frac{1}{\sqrt{t}}$ instead of $\ln \psi$ if we omit the condition $(a\psi_x, \psi_x) \geq 1$.

Bibliography

[1] N. V. Krylov, *On First Quasi-derivatives of Solutions of The Ito Stochastic Equations*, to appear in Izvestia An SSSR (in Russian).

[2] N. V. Krylov, *On Moment Estimates for Quasi-derivatives of Solutions of Stochastic Equations with Respect to the Initial Data, and their Application*, Math. USSR Sbornik, Vol 64 (1989), No. 2, p. 505-526.

The Stability and Approximation Problems in Nonlinear Filtering Theory

Hiroshi Kunita
Kyushu University

1 Introduction

Let $X_t, t \in [0,t]$ be a stochastic process with continuous time parameter, called a *system process.* Suppose that we wish to observe the sample $X_t, t \in [0,T]$, but what we can actually observe is a stochastic process perturbed by the noise W_t:

$$Y_t = \int_0^t h(X_s)ds + W_t,$$

where h is a continuous function and $W_t, t \in [0,T]$ is a Brownian motion independent of $\{X_t\}$. The *nonlinear filter* of X_t based on the observation data $\{Y_s; s \leq t\}$ is defined by the conditional distribution

$$\pi_t(dx) = P(X_t \in dx | \sigma(Y_s; s \leq t)).$$

It is known that under some conditions on the system process $\{X_t\}$, $\pi_t(dx)$ has a density function $\pi_t(x)$ $(= \pi_t(dx)/dx)$ and it satisfies a certain nonlinear stochastic partial differential equation, called a *Kushner-Stratonovich equation* ([6],[5]). Also, an unnormalized density function $\rho_t(x)$ (i.e., $\pi_t(x) = \rho(x)/\int \rho_t(x)dx$) satisfies a linear stochastic partial differential equation, called a *Zakai equation* ([7],[5]). Furthermore, *Kallianpur-Striebel formula* ([3],[5]) provides us another way of computing the nonlinear filter from the observation data Y_t.

One of the basic assumptions needed to obtain the above equations or formula is that the noise process W_t is a Brownian motion.

ISBN 0-12-481005-5

However in the real physical problem, the noise process W_t is not a Brownian motion. In fact, both Y_t and W_t have smooth derivatives $y_t = dY_t/dt$ and $w_t = dW_t/dt$. Then the physical observation process y_t can be written as

$$y_t = h(X_t) + w_t.$$

The noise process w_t is not a white noise though it may close the white noise. Unfortunately, however, in such a real physical system, nothing like Kushner-Stratonovich equation, Zakai equation or Kallianpur-Striebel formula is known. Hence no effective way of computing the filter is known. Thus it is an interesting problem whether the above mentioned equations or the formula approximate the filter of the real physical system or not, if the Brownian motion W_t approximates the physical noise process $\int_0^t w_s ds$ (or white noise dW_t/dt approximate w_t).

A similar problem arises in the filter of discrete time system. Let $X_k, k = 1, 2, \cdots$ be a system process and let

$$Y_k = h(X_k) + W_k, \qquad k = 1, 2, \cdots$$

be the observation process, where $\{W_k\}$ is a noise process independent of $\{X_k\}$. If the system process $\{X_k\}$ is a Gaussian process governed by a linear stochastic difference equation and the noise process $\{W_k\}$ is a sequence of independent Gaussian random variables, it is well known that the mean filter $\widehat{X}_k \equiv \int x\pi_k(dx)$ can be computed by celebrated *Kalman's algorithm* ([4]). On the other hand, if the noise process is not Gaussian, Kalman's algorithm is no longer valid, at least from the theoretical point of view. Then a practical question is whether the Kalman's algorithm approximates the real optimal filter if the Gaussian distribution approximates the distribution of the real noise $\{W_k\}$.

This paper provides a partial answer to these problems. In Section 2, we consider the approximation problem of the filter of discrete time. We will show that the last question concerning Kalman's algorithm can be solved positively. In Section 3, we consider the filter of continuous time system process based on the observation at discrete time. We discuss the asymptotic property of such filters when the frequency of the observation tends to infinity.

Approximation problem of continuous time filter will be discussed elsewhere.

2 Approximation of the filter with discrete parameter

Let $\{X_k, k \in \mathbf{N}\}$ be a stochastic process with values in $\mathbf{R}^d$ called a *system process*, where $\mathbf{N} = \{1, 2, \cdots\}$. Let $\{W_k, k \in \mathbf{N}\}$ be a stochastic process with values in $\mathbf{R}^e$, called a *noise process*. We assume that $\{W_k, k \in \mathbf{N}\}$ is independent of $\{X_k, k \in \mathbf{N}\}$. Let $h_k, k \in \mathbf{N}$ be continuous maps from $\mathbf{R}^d$ into $\mathbf{R}^e$. The *observation process* $\{Y_k, k \in \mathbf{N}\}$ is defined by

$$Y_k = h_k(X_k) + W_k, \qquad k \in \mathbf{N}. \tag{2.1}$$

The *filter* of X_k by the observation data $\{Y_1, \cdots, Y_k\}$ is defiend by the conditional distribution

$$\pi_k(E)(Y_1, \cdots, Y_k) = P(X_k \in E | \sigma(Y_1, \cdots, Y_k)), \quad k \in \mathbf{N}, \tag{2.2}$$

where E is a Borel set in $\mathbf{R}^d$. The sequence $\{\pi_k, k \in \mathbf{N}\}$ can be regarded as a stochastic process with values in the space of distributions. It is called a *filtering process.*

In the sequel we assume that the joint distribution of the ke-dimensional random variable $(W_1, \cdots, W_k)$ has a continuous density function, which we denote by $f_k(w_1, \cdots, w_k)$ ($w_i \in \mathbf{R}^e, i = 1, \cdots, k$). Hence

$$P((W_1, \cdots, W_k) \in A) = \int_A f_k(w_1, \cdots, w_k) dw_1 \cdots dw_k. \tag{2.3}$$

Let us denote by $G(x_1, \cdots, x_k)$ the joint distribution function of $(X_1, \cdots, X_k)$. Then the joint distribution of $(Y_1, \cdots, Y_k)$ has a density function and it is represented by

$$\int_{\mathbf{R}^{kd}} f_k(y_1 - h_1(x_1), \cdots, y_k - h_k(x_k)) dG_k(x_1, \cdots, x_k). \tag{2.4}$$

The following proposition is immediate from the definition of the conditional density function.

Proposition 2.1. The filter $\pi_k(E)(Y_1, \cdots, Y_k)$ is represented by

$$\pi_k(E)(Y_1, \cdots, Y_k) = \frac{\int_{\mathbf{R}^{(k-1)d} \times E} f_k(Y_1 - h_1(x_1), \cdots, Y_k - h_k(x_k)) dG_k(x_1, \cdots, x_k)}{\int_{\mathbf{R}^{kd}} f_k(Y_1 - h_1(x_1), \cdots, Y_k - h_k(x_k)) dG_k(x_1, \cdots, x_k)}. \tag{2.5}$$

Further if G_k has a continuous density function g_k, π_k has also a density function π_k and it is represented by

$$\pi_k(x_k)(Y_1, \cdots, Y_k) = \frac{\int f_k(Y_1 - h_1(x_1), \cdots, Y_k - h_k(x_k)) g_k(x_1, \cdots, x_k) dx_1 \cdots dx_{k-1}}{\int f_k(Y_1 - h_1(x_1), \cdots, Y_k - h_k(x_k)) g_k(x_1, \cdots, x_k) dx_1 \cdots dx_k}. \tag{2.6}$$

(In both expressions, we understand $0/0 = 0$). □

The above proposition shows that the filter depends crucially on the distributions of the system process and the noise process. However in physical problem, we may not know exactly the distributions of the system process and the noise processes. Hence it is important to know the stability or the sensitivety of the filter with respect to the change of the distributions of these processes. In the sequel we shall discuss the stability problem by approximating distribution functions of these processes.

For $\epsilon > 0$ we introduce hypothese (H_ϵ) and (H'_ϵ) to the distributions of $\{X_k\}$ and $\{w_k\}$.

Hypothesis (H_ϵ) The distribution of $(X_1, \cdots, X_k)$ is given by $G_k^{(\epsilon)}$.

Hypothesis (H'_ϵ) The distribution of $(W_1, \cdots, W_k)$ is given by

$$F_k^{(\epsilon)}(dw_1 \cdots dw_k) = f_k^{(\epsilon)}(w_1, \cdots, w_k) dw_1 \cdots dw_k$$

where $f_k^{(\epsilon)}$ is a continuous function.

Under the hypotheses (H_ϵ) and (H'_ϵ), the filter of X_k by the data $(Y_1, \cdots, Y_k)$ is represented by (2.5), replacing f_k and G_k by $f_k^{(\epsilon)}$ and $G_k^{(\epsilon)}$, respectively. We denote it by $\pi_k^{(\epsilon)}$.

Theorem 2.2. Assume the following (i) and (ii).

(i) The family $\{G_k^{(\epsilon)}\}_\epsilon$ converges weakly to a distribution function G_k as $\epsilon \to 0$ for any k.

(ii) The family $\{f_k^{(\epsilon)}\}_\epsilon$ converges boundedly and compact uniformly to a function f_k satisfying $\int f_k dx = 1$ as $\epsilon \to 0$ for any k.

Then the family $\{\pi_k^{(\epsilon)}\}_\epsilon$ converges weakly to π_k defined by (2.6) a.s. for any k.

Assume further that for any ϵ and k, $G_k^{(\epsilon)}$ has a density function $g_k^{(\epsilon)}$ converging to the density g_k of G_k compact uniformly, then density functions of the filters $\pi_k^{(\epsilon)}$ converge to the density function of π_k compact uniformly.

Proof. Let $x = (x_1, \cdots, x_k)$. For any $\delta > 0$ there exists $N > 0$ such that

$$G_k^{(\epsilon)}(|x| > N) < \delta \quad \text{and} \quad G_k(|x| > N) < \delta$$

holds for all ϵ. We may assume that the set $\{|x| > N\}$ is a continuity set of G_k. Since

$$f_k^{(\epsilon)}(Y_1 - h(x_1), \cdots, Y_k - h(x_k)) \to f_k(Y_1 - h(x_1), \cdots, Y_k - h(x_k))$$

uniformly on $\{|x| \leq N\}$, we obtain

$$\begin{aligned}\lim_{\epsilon \to 0} \int_{|x| \leq N} & g(x_k) f_k^{(\epsilon)}(Y_1 - h(x_1), \cdots, Y_k - h(x_k)) dG_k^{(\epsilon)}(x_1, \cdots, x_k) \\ &= \int_{|x| \leq N} g(x_k) f_k(Y_1 - h(x_1), \cdots, Y_k - h(x_k)) dG_k(x_1, \cdots, x_k)\end{aligned}$$

for any bounded continuous function g.

We have further

$$\left| \int_{|x| > N} g(x_k) f_k^{(\epsilon)}(Y_1 - h_1(x_1), \cdots, Y_k - h_k(x_k)) dG_k^{(\epsilon)}(x_1, \cdots, x_k) \right| < C\delta$$

where C is a positive constant such that $|gf_k^{(\epsilon)}| < C$. A similar inequality is valid for f_k and G_k. Since δ is arbitrary, we have

$$\lim_{\epsilon\to 0}\int g(x_k)f_k^{(\epsilon)}(Y_1-h(x_1),\cdots,Y_k-h(x_k))dG_k^{(\epsilon)}(x_1,\cdots,x_k)$$
$$=\int g(x_k)f_k(Y_1-h(x_1),\cdots,Y_k-h(x_k))dG_k(x_1,\cdots,x_k).$$

The above convergence yields

$$\lim_{\epsilon\to 0}\int g(x)\pi_k^{(\epsilon)}(dx)=\int g(x)\pi_k(dx).$$

This proves the first assertion of the theorem. The second assertion can be proved similarly. □

The mean filter $\widehat{X}_k$ is defined by the conditional expectation $E[X_k|\sigma(Y_1,\cdots,Y_k)]$. Under Hypothese (H_ϵ) and (H'_ϵ), it represented by $\int x\pi_k^{(\epsilon)}(dx)$, which we denote by $\widehat{X}_k^{(\epsilon)}$.

Corollary 2.3. Assume

$$\sup_{\epsilon}\int |x_k|^{1+\delta}G_k^{(\epsilon)}(dx_1\cdots dx_k)<\infty \tag{2.7}$$

holds for some $\delta > 0$. Then $\{\widehat{X}_k^{(\epsilon)}\}_\epsilon$ converges to $\widehat{X}_k = \int x\pi_k(dx)$ a.s. as $\epsilon \to 0$. □

Now suppose that the system process $\{X_k\}$ is governed by a linear stochastic difference equation

$$\begin{aligned} X_k &= \widetilde{A}_k X_{k-1} + \widetilde{B}_k V_k, \qquad k\in N\\ X_0 &= \widetilde{\xi} \end{aligned} \tag{2.8}$$

where $\widetilde{A}_k$, $\widetilde{B}_k$ are matrices, $\{V_k, k\in N\}$ is a noise process and $\widetilde{\xi}$ is an initial condition. Suppose

Hypothesis $(\widetilde{H}_\epsilon)$. (a) $\widetilde{A}_k = A_k^{(\epsilon)}$ and $\widetilde{B}_k = B_k^{(\epsilon)}$.

(b) The distribution of $(V_1,\cdots,V_k)$ is $\Phi_k^{(\epsilon)}$.

(c) ξ is independent of $\{V_k\}$ and is subject to the law μ_ϵ.

We denote the law of $(X_1, \cdots, X_k)$ under the hypothesis $(\widetilde{H}_\epsilon)$ by $G_k^{(\epsilon)}$. We assume the following condition (C)

Condition(C) (a) For any k, $\{A_k^{(\epsilon)}\}_\epsilon$ converges to A_k and $\{B_k^{(\epsilon)}\}_\epsilon$ converges to B_k.

(b) For any k, $\{\Phi_k^{(\epsilon)}\}_\epsilon$ converges weakly to Φ_k, which is equal to the distribution of k independent Gaussian random variables with mean 0 and covariance $Q_1, \cdots, Q_k$.

(c) $\{\mu_\epsilon\}_\epsilon$ converges to a Gaussian distribution μ.

Then for any k, $\{G_k^{(\epsilon)}\}_\epsilon$ converges weakly to a Gaussian distribution G_k, which is equal to the distribution of $(X_1, \cdots, X_k)$ under the hypothesis $(\widetilde{H}_0)$:

Hypothesis $(\widetilde{H}_0)$ (a) $\widetilde{A}_k = A_k$ and $\widetilde{B}_k = B_k$ for any $k \in N$.

(b) $V_1, \cdots, V_k$ are independent Gaussian random variables with means 0 and covariances $Q_1, \cdots, Q_k$.

(c) ξ is independent of $\{V_k\}$ with the law μ.

The mean and the covariance are computed by

$$E[X_k] = A_k \cdots A_1 m, \tag{2.9}$$

$$\begin{aligned} Cov(X_k) &= \sum_{i=1}^{k-1} A_k \cdots A_{i+1} B_i Q_i B_i' A_{i+1}' \cdots A_k' \\ &\quad + A_k \cdots A_1 \sigma A_1' \cdots A_k', \end{aligned} \tag{2.10}$$

where m and σ are the mean and the covariance of μ, respectively.

Suppose next that the observation process is given by

$$Y_k = H_k X_k + W_k, \qquad k \in N \tag{2.11}$$

where $H_k, k \in \mathbf{N}$ are matrices and $\{W_k; k \in \mathbf{N}\}$ is a noise process. For the noise process, we introduce Hypothesis (H_ϵ') and the following:

Hypothesis (H_0'): $W_k; k \in \mathbf{N}$ are independent Gaussian random variables with means 0 and nondegenerate covariances $R_k, k \in \mathbf{N}$.

Let $\pi_k^{(\epsilon)}$ be the filter of X_k under hypotheses $(\widetilde{H}_\epsilon)$ and (H_ϵ') and let π_k be the filter of X_k under hypotheses $(\widetilde{H}_0)$ and (H_0'). Then the previous theorem tells us that $\{\pi_k^{(\epsilon)}\}_\epsilon$ converges weakly to π_k a.s. We shall represent π_k explicitly. Since the distribution of W_l is written by

$$\phi_l(w) = \frac{1}{((2\pi)^l det R_l)^{\frac{1}{2}}} \exp^{-\frac{1}{2}(R_l^{-1}w,w)} \tag{2.12}$$

where (,) denotes the inner product, we have

$$f_k(w_1, \cdots, w_k) = \prod_{l=1}^{k} \phi_l(w_l).$$

Set

$$\begin{aligned} &\alpha(x_1, \cdots, x_k, y_1, \cdots, y_k) \\ &= \exp\left\{\sum_{l=1}^{k} \sum_{i,j=1}^{e} s_{ij}^{(l)} h_l^i(x_l) y_l^j - \frac{1}{2}\sum_{l=1}^{k} \sum_{i,j=1}^{e} s_{ij}^{(l)} h_l^i(x_l) h_l^j(x_l)\right\} \end{aligned} \tag{2.13}$$

where $R_l^{-1} = (s_{ij}^{(l)})$. Then π_k is represented by

$$\begin{aligned} &\pi_k(E) \\ &= \frac{\int_{\mathbf{R}^{(k-1)d}\times E} \alpha(x_1, \cdots, x_k, Y_1 \cdots, Y_k) dG_k(x_1, \cdots, x_k)}{\int_{\mathbf{R}^{kd}} \alpha(x_1, \cdots, x_k, Y_1, \cdots, Y_k) dG_k(x_1, \cdots, x_k)}, \end{aligned} \tag{2.14}$$

where G_k is the Gaussian distribution of $(X_1, \cdots, X_k)$.

For almost all $Y_1, \cdots, Y_k$, $\pi_k(\cdot)$ is a Gaussian distribution. Hence the filtering process $\{\pi_k; k \in \mathbf{N}\}$ can be regarded as a stochastic process with values in Gaussian distributions. It is well known that its means $\widehat{X}_k = \int x\pi_k(dx), k \in \mathbf{N}$ satisfy Kalman's linear stochastic difference equation

$$\begin{cases} \widehat{X}_k = A_k\widehat{X}_{k-1} + P_{k-1}^{-} H_k'(R_k + H_k P_k^{-} H_k')^{-1}(Y_k - H_k A_k \widehat{X}_{k-1}), \\ \qquad\qquad\qquad\qquad\qquad\qquad\qquad\qquad\qquad\qquad k \in \mathbf{N} \\ \widehat{X}_0 = \int x d\mu(x), \end{cases} \tag{2.15}$$

where $P_k^-, k = 0, 1, 2, \cdots$ are computed by

$$\begin{cases} P_k = P_k^- - P_k^- H_k'(R_k + H_k P_k^- H_k')^{-1} H_k P_k^-, & k \in \mathbf{N} \\ P_k^- = A_k P_{k-1} A_k' + B_k Q_k B_k', & k \in \mathbf{N} \\ P_0 = P_0^- = \sigma. \end{cases} \tag{2.16}$$

Theorem 2.4. Suppose that the system process $\{X_n\}$ is governed by a linear stochastic difference equation (2.8) and the observation process $\{Y_n\}$ is given by (2.11). We assume Hypotheses $(\widetilde{H}_\epsilon), (\widetilde{H}_0), (H_\epsilon'), (H_0')$ and Condition (C). Then the mean filter is approximated by Kalman filter computed by (2.15) and (2.16). □

3 Asymptotic property of discrete filter

Let $X_t, t \in [0, T]$ be a stochastic process with values in $\mathbf{R}^d$, whose sample paths are continuous in t a.s. Suppose that we observe the sample X_t perturbed by noises at discrete time $t = 0, \frac{1}{n}, \frac{2}{n}, \cdots$. For each n, the noise process $\{W_k^n; k \in \mathbf{N}\}$ is assumed to be a sequence of independent, identically distributed $\mathbf{R}^e$-valued random variables whose mean vectors are 0 and covariance matrices are R^n. The observation process is then defined by

$$Y_k^n = h(X_{\frac{k}{n}}) + W_k^n, \qquad k \in \mathbf{N}, \tag{3.1}$$

where h is a continuous map from $\mathbf{R}^d$ into $\mathbf{R}^e$. The filter of $X_{\frac{k}{n}}$ by the observation data $\{Y_1^n, \cdots, Y_k^n\}$ is denoted by π_k^n.

Now define sequences of stochastic processes with continuous time parameter $[0, T]$ by

$$W_t^{(n)} = \frac{1}{n} \sum_{k=1}^{[nt]} W_k^n, \tag{3.2}$$

$$Y_t^{(n)} = \frac{1}{n} \sum_{k=1}^{[nt]} Y_k^n, \tag{3.3}$$

$$X_t^{(n)} = X_{\frac{k}{n}} \quad \text{if} \quad \frac{k}{n} \le t < \frac{k+1}{n}, \tag{3.4}$$

$$\pi_t^{(n)} = \pi_k^n \quad \text{if} \quad \frac{k}{n} \le t < \frac{k+1}{n}, \tag{3.5}$$

where $[nt]$ is the integer part of nt. Then summing each term of (3.1) with respect to $k = 1, \cdots, [nt]$, we obtain

$$Y_t^{(n)} = \int_0^{\frac{[nt]}{n}} h(X_s^{(n)})ds + W_t^{(n)}. \tag{3.6}$$

Suppose that the sequence of noise processes $\{W_t^{(n)}\}_n$ converges uniformly to a Brownian motion W_t a.s., then the sequence of observation processes $\{Y_t^{(n)}\}_n$ converges uniformly a.s. and the limit process Y_t satisfies

$$Y_t = \int_0^t h(X_s)ds + W_t. \tag{3.7}$$

Let π_t be the filter of X_t based on the observation data $\{Y_s; s \le t\}$, i.e.,

$$\pi_t(E) = P(X_t \in E | \sigma(Y_s; s \le t)), \tag{3.8}$$

where E is a Borel set in $\mathbf{R}^d$. The purpose of this section is to discuss whether or not the sequence of filters $\{\pi_t^{(n)}\}_n$ converges to π_t a.s. for any t. To solve this problem, we introduce two assumptions. The first assumption is concerned with the sequence of noise processes. Note that the variance of $W_1^{(n)}$ is $\frac{1}{n}R^n$. Then the sequence $\{\frac{1}{n}R^n\}_n$ should converge. Let R be its limit. It is the variance of W_1, which we assume to be positive definite. Further for the convergence of $\{\pi_t^{(n)}\}_n$, the distributions of the normalized random variables $\widetilde{W}_k^n = \frac{1}{\sqrt{n}}W_k^n$ should converge to a Gaussian distribution fairly rapidly. Actually we assume the following condition on the distributions of $\{\widetilde{W}_k^n\}_n$, which is not necessary for the convergence of $\{W_t^{(n)}\}_n$.

Condition(A.1). For each n, the distribution of the random variable $\widetilde{W}_1^n = \frac{1}{\sqrt{n}}W_1^n$ has a density function $\tilde{f}^{(n)}(w)$ represented by

$$\tilde{f}^{(n)}(w) = \phi_{R^n/n}(w)\Big(1 + \frac{1}{\sqrt{n}}s^{(n)}(w)\Big), \tag{3.9}$$

where $\phi_{R^n/n}$ is the density function of the Gaussian distribution with mean 0 and covariance matrix R^n/n, and $s^{(n)}(w)$ is a C^1-function. Further the latter satisfies:

(1) The sequence $\{s^{(n)}(w)\}_n$ and $\{\partial_i s^{(n)}(w)\}_n, i = 1, \cdots, d$ converges to $s(w)$ and $\partial_i s(w), i = 1, \cdots, d$ uniformly on compact sets, respectively.

(2) There are positive constants c and $c^{(n)}, n = 1, 2, \cdots$ with $c^{(n)} \to 0$ and a continuous function $d(w)$ of the polynomial growth such that for any $i = 1, \cdots, d$,

$$|\partial_i s^{(n)}(w)| \leq c(1 + |w|)^2 + c^{(n)} d(w). \tag{3.10}$$

Next for the process $h(X_t)$, we assume the following.

Condition(A.2). $h(X_t)$ is a continous semimartingale. It is represented by $h(X_t) = M_t + \int_0^t C_1(X_s)ds$ and $\langle M \rangle_t = \int_0^t C_2(X_s)ds$, where C_1 and C_2 are bounded continuous functions on $\mathbf{R}^d$.

A way of computing the filter π_t of (3.8) is the Kallianpur-Striebel formula. In order to state the formula rigorously, we need to introduce the law of the systm process on the function space $\mathbf{C} = \mathbf{C}([0.T]; \mathbf{R}^d)$.

Let $\mathbf{C} = \mathbf{C}([0, T]; \mathbf{R}^d)$ be the set of all continuous maps from $[0, T]$ into $\mathbf{R}^d$. We denote elements of $\mathbf{C}$ by x and its value at $t \in [0, T]$ by x_t. Then the law of the system process X_t is defined as a probability measure P_X on $\mathbf{C}$ satisfying

$$P_X(\{x; (x_{t_1}, \cdots, x_{t_l}) \in E\}) = P(\{\omega; (X_{t_1}, \cdots, X_{t_l}) \in E\})$$

for any Borel sets E of $\mathbf{R}^{nl}$, where $l = 1, 2, \cdots$. We set

$$\alpha_t(x, Y) = \exp\left\{\int_0^t h(x_s)R^{-1}dY_s' - \frac{1}{2}\int_0^t h(x_s)R^{-1}h(x_s)'ds\right\}. \tag{3.11}$$

Here $Y(\omega) = Y = (Y_s; s \in [0, T])$ is regarded as an element of $\mathbf{C}([0, T]; \mathbf{R}^e)$, $h(x) = (h_1(x), \cdots, h_e(x))$ is a row vector function and $h(x)'$ is the transpose (column vector function) of $h(x)$. Then Kallianpur-Striebel formula states:

$$\pi_t(g)(Y) = \frac{\int \alpha_t(x, Y)g(x_t)dP_X}{\int \alpha_t(x, Y)dP_X}. \tag{3.12}$$

Theorem 3.1. Assume Conditions (A.1) and (A.2). Then for any t, the sequence $\{\pi_t^{(n)}\}$ converges to $\widehat{\pi}_t$ defined by

$$\widehat{\pi}_t(g)(Y) = \frac{\int \alpha_t(x,Y)\beta_t(x)g(x_t)dP_X}{\int \alpha_t(x,Y)\beta_t(x)dP_X}, \tag{3.13}$$

where

$$\beta_t(x) = \exp\Big\{-\sum_{i=1}^{d} \int \partial_i s(w)\phi_R(w)dw \int_0^t h_i(x_s)ds\Big\}. \tag{3.14}$$

In particular if

$$\int \partial_i s(w)\phi_R(w)dw = 0, \quad i = 1,\cdots,d, \tag{3.15}$$

then the sequence $\{\pi_t^{(n)}\}$ converges to the filter π_t defined by (3.8) a.s. □

We give the proof of the theorem in the case where $d = e = 1$, for simplicity of notations.

In the seqeul, we fix the number n for a moment. We first obtain a Bayes formula for the filter $\pi_t^{(n)}$. Since the noises $W_k^n, k = 1,2,\cdots$ are independent, the joint distribution of $(W_1^n,\cdots,W_l^n)$ has a density function $f_l^n(w_1,\cdots,w_l)$ which is represented by

$$f_l^n(w_1,\cdots,w_l) = \prod_{k=1}^{l} \phi_{R^n}(w_k)\left(1+\frac{1}{\sqrt{n}}s^{(n)}(\frac{1}{\sqrt{n}}w_k)\right). \tag{3.16}$$

Note that $Y_k^n = n\Delta Y_{\frac{k}{n}}^{(n)}$. Then by Proposition 2.1, the filter $\pi_t^{(n)}$ is represented by

$$\pi_t^n(g)(Y^{(n)})$$
$$= \frac{\int f_{[nt]}^n(n\Delta Y_{\frac{1}{n}}^{(n)} - h(x_{\frac{1}{n}}),\cdots,n\Delta Y_{\frac{[nt]}{n}}^{(n)} - h(x_{\frac{[nt]}{n}}))g(x_{\frac{[nt]}{n}})dP_X}{\int f_{[nt]}^n(n\Delta Y_{\frac{1}{n}}^{(n)} - h(x_{\frac{1}{n}}),\cdots,n\Delta Y_{\frac{[nt]}{n}}^{(n)} - h(x_{\frac{[nt]}{n}}))dP_X}. \tag{3.17}$$

Multiplying $f_{[nt]}^n(n\Delta Y_{\frac{1}{n}}^{(n)}, \cdots, n\Delta Y_{\frac{n}{[nt]}}^{(n)})$ to the numerator and the denominator of the right hand side. The term

$$f_{[nt]}^n(n\Delta Y_{\frac{1}{n}}^{(n)} - h(x_{\frac{1}{n}}), \cdots, n\Delta Y_{\frac{[nt]}{n}}^{(n)} - h(x_{\frac{[nt]}{n}})) \\ \times f_{[nt]}^{(n)}(n\Delta Y_{\frac{1}{n}}^{(n)}, \cdots, n\Delta Y_{\frac{[nt]}{n}}^{(n)})$$

is written by

$$\alpha_t^{(n)}(x^{(n)}, Y^{(n)})\beta_t^{(n)}(x^{(n)}, Y^{(n)})$$

where

$$\begin{aligned} &\alpha_t^{(n)}(x^{(n)}, Y^{(n)}) \\ &= \exp \frac{n}{R^n}\left\{\int_0^{\frac{[nt]}{n}} h(x_s^{(n)})dY_s^{(n)} - \frac{1}{2}\int_0^{\frac{[nt]}{n}} h(x_s^{(n)})^2 ds\right\} \end{aligned} \tag{3.18}$$

and

$$\begin{aligned} &\beta_t^{(n)}(x^{(n)}, Y^{(n)}) \\ &= \prod_{l=1}^{[nt]} \frac{1 + \frac{1}{\sqrt{n}}s^{(n)}(\sqrt{n}\Delta Y_{\frac{l}{n}}^{(n)} - \frac{1}{\sqrt{n}}h(x_{\frac{l}{n}}))}{1 + \frac{1}{\sqrt{n}}s^{(n)}(\sqrt{n}\Delta Y_{\frac{l}{n}}^{(n)})} \end{aligned} \tag{3.19}$$

and $x^{(n)}$ is defined by $x_t^{(n)} = x_{\frac{[nt]}{n}}$. Therefore (3.17) is represented by a Bayes type formula.

$$\begin{aligned} &\pi_t^{(n)}(g)(Y^{(n)}) \\ &= \frac{\int \alpha_t^{(n)}(x^{(n)}, Y^{(n)})\beta_t^{(n)}(x^{(n)}, Y^{(n)})g(x_t^{(n)})dP_X}{\int \alpha_t^{(n)}(x^{(n)}, Y^{(n)})\beta_t^{(n)}(x^{(n)}, Y^{(n)})dP_X}. \end{aligned} \tag{3.20}$$

Now in order to prove Theorem 3.1, we will show two facts. The first is that both $\{\alpha_t^{(n)}\}_n$ and $\{\beta_t^{(n)}\}_n$ converge to α_t and β_t of (3.11) and (3.13) a.s. P_X. The second is that both $\{\alpha_t^{(n)^2}\}_n$ and $\{\beta_t^{(n)^2}\}_n$ are uniformly P_X-integrable. These will be done in Lemma 3.2 and 3.3, respectively. The two lemmas will then establish Theorem 3.1.

Lemma 3.2. (i) For each t, the sequence $\{\alpha_t^{(n)}\}_n$ converges to α_t defined by (3.11), a.s.

(ii) For each t, the sequence $\{\beta_t^{(n)}\}_n$ converges to β_t defined by (3.13), a.s.

Proof. (i) We can rewrite (3.18) by

$$\begin{aligned}\alpha_t^{(n)} &= \exp\Big\{\frac{n}{R^n}\Big(h(x_t^{(n)})Y_t^{(n)} - \int_0^{\frac{[nt]}{n}} Y_s^{(n)} dh(x_s^{(n)})\Big) \\ &\quad - \frac{n}{2R^n}\int_0^{\frac{[nt]}{n}} h(x_s^{(n)})^2 ds\Big\}. \end{aligned} \tag{3.21}$$

Therefore the sequence $\{\alpha_t^{(n)}\}_t$ converges and the limit satisfies

$$\alpha_t = \exp\Big\{\frac{1}{R}\Big(h(x_t)Y_t - \int_0^t Y_s dh(x_s)\Big) - \frac{1}{2R}\int_0^t h(x_s)^2 ds\Big\}.$$

The right hand side of the above can be written as (3.11) by Itô's formula.

(ii) It is sufficient to prove

$$\exists \lim_{n\to\infty} \log \beta_t^{(n)} = -\int s'(w)\phi_R(w)dw \int_0^t h(x_s)ds. \tag{3.22}$$

Since $\log(1+z) = z - \frac{1}{2}z^2 + \frac{1}{3}\theta z^3$ $(|\theta| < 1)$, we have

$$\begin{aligned}\log \beta_t^{(n)} &= \frac{1}{\sqrt{n}}\Big\{\sum_{k=1}^{[nt]} s^{(n)}\big(\frac{1}{\sqrt{n}}(Y_k^n - h(x_{\frac{k}{n}}))\big) - \sum_{k=1}^{[nt]} s^{(n)}\big(\frac{1}{\sqrt{n}}Y_k^n\big)\Big\} \\ &- \frac{1}{2n}\Big\{\sum_{k=1}^{[nt]} s^{(n)}\big(\frac{1}{\sqrt{n}}(Y_k^n - h(x_{\frac{k}{n}}))\big)^2 - \sum_{k=1}^{[nt]} s^{(n)}\big(\frac{1}{\sqrt{n}}Y_k^n\big)^2\Big\} \\ &+ \frac{1}{3n\sqrt{n}}\Big\{\sum_{k=1}^{[nt]} \theta_k^{(n)} s^{(n)}\big(\frac{1}{\sqrt{n}}(Y_k^n - h(x_{\frac{k}{n}}))\big)^3 \\ &\qquad - \sum_{k=1}^{[nt]} \theta_k^{(n)} s^{(n)}\big(\frac{1}{\sqrt{n}}Y_k^n\big)^3\Big\} \\ &= I_1^{(n)} + I_2^{(n)} + I_3^{(n)}. \end{aligned} \tag{3.23}$$

We first consider $I_1^{(n)}$. By the mean value theorem,

$$I_1^{(n)} = -\frac{1}{n}\sum_{k=1}^{[nt]} h(x_{\frac{k}{n}})s^{(n)'}(\frac{1}{\sqrt{n}}(Y_k^n + \widetilde{\theta}_k^n h(x_{\frac{k}{n}}))).$$

Note that

$$\frac{1}{\sqrt{n}}(Y_k^n + \widetilde{\theta}_k^n h(x_{\frac{k}{n}})) = \frac{1}{\sqrt{n}}W_k^n + \frac{1}{\sqrt{n}}h(x_{\frac{k}{n}})(1 + \widetilde{\theta}_k^n).$$

The distributions of $\{\frac{1}{\sqrt{n}}W_k^n\}_n$ (independent of k) converge weakly to a Gaussian distribution with mean 0 and covariance R. For each k, the terms $\{\frac{1}{\sqrt{n}}h(x_{\frac{k}{n}})(1 + \widetilde{\theta}_k^n)\}_n$ converge to 0 as $n \to \infty$ a.s. Then using the law of the large numbers for $\{\frac{1}{\sqrt{n}}W_k^n\}_n$, we obtain

$$\begin{aligned}
&\lim_{n\to\infty} \frac{1}{n}\sum_{k=1}^{[nt]} s^{(n)'}(\frac{1}{\sqrt{n}}(Y_k^n + \widetilde{\theta}_k^n h(x_{\frac{k}{n}}))) \\
&= \lim_{n\to\infty} \frac{1}{n}\sum_{k=1}^{[nt]} s'(\frac{1}{\sqrt{n}}W_k^n) \\
&= t\int s'(w)\phi_R(w)dw \qquad \text{a.s.}
\end{aligned}$$

Then for any step process $x_t^{(n_0)}$ (n_0 is fixed), we have

$$\begin{aligned}
&\lim_{n\to\infty} \frac{1}{n}\sum_{k=1}^{[nt]} h(x_{\frac{k}{n}}^{(n_0)})s^{(n)'}(\frac{1}{\sqrt{n}}(Y_k^{(n)} + \widetilde{\theta}_k^{(n)} h(x_{\frac{k}{n}}))) \\
&= \int s'(w)\phi_R(w)dw \int_0^t h(x_s^{(n_0)})ds.
\end{aligned}$$

Now for any $\epsilon > 0$, choose n_0 such that $|h(x_t) - h(x_s)| < \epsilon$ holds if $|t-s| \le \frac{1}{n_0}$. Then for any $n \ge n_0$,

$$\begin{aligned}
&E[|\frac{1}{n}\sum_{k=1}^{[nt]}\Big(h(x_{\frac{k}{n}}) - h(x_{\frac{k}{n}}^{(n_0)})\Big)s^{(n)'}(\frac{1}{\sqrt{n}}(Y_k^n + \widetilde{\theta}_k^n h(x_{\frac{k}{n}})))|] \\
&\le \frac{\epsilon}{n}\sum_{k=1}^{[nt]} E[|s^{(n)'}(\frac{1}{\sqrt{n}}(Y_k^n + \widetilde{\theta}_k^n h(x_{\frac{k}{n}})))|] \\
&\le C\epsilon
\end{aligned}$$

holds with some positive constant C. Therefore we have

$$\lim_{n\to\infty} I_1^{(n)} = -\int s'(w)\phi_R(w)dw \int_0^t h(x_s)ds.$$

We shall next consider $I_2^{(n)}$. By the law of the large numbers, we have

$$\frac{1}{n}\sum_{k=1}^{[nt]} s^{(n)}(\frac{1}{\sqrt{n}}(Y_k^n - h(x_{\frac{k}{n}})))^2 \longrightarrow t\int s(w)^2\phi_R(w)dw,$$

$$\frac{1}{n}\sum_{k=1}^{[nt]} s^{(n)}(\frac{1}{\sqrt{n}}Y_k^n)^2 \longrightarrow t\int s(w)^2\phi_R(w)dw.$$

Therefore $I_2^{(n)}$ converges to 0 as $n\to\infty$. In the similar way, we can prove $\lim_{n\to\infty} I_3^{(n)} = 0$. Therefore (3.22) is proved. □

Lemma 3.3. Both of $\{\alpha_t^{(n)^2}\}_n$ and $\{\beta_t^{(n)^2}\}_n$ are P_X-uniformly integrable for any t and Y a.s.

Proof. Let $\delta > 0$. Since $\alpha_t^{(n)}$ is written by (3.18), we have

$$(\alpha_t^{(n)})^{2+\delta} = \exp\left\{\frac{n(2+\delta)}{R^n}Y_t^{(n)}h(x_t^{(n)}) + \int_0^t C_n(x_s)ds\right\} M_t^{(n)},$$

where $\{C_n(x)\}$ is a sequence of continuous functions such that there exists a positive constant C satisfying $|C_n(x)| \le C$ for all n and x. Further, $M_t^{(n)}$ is a P_X-martingale with P_X-mean 1 for any $Y^{(n)}$ a.s. Then,

$$\int(\alpha_t^{(n)})^{2+\delta}dP_X \le \exp\left\{\frac{n(2+\delta)}{R^n}|Y_t^n|\sup_x|h(x)| + tC\right\}.$$

Therefore $\sup_n \int(\alpha_t^{(n)})^{2+\delta}dP_X < \infty$ holds for any t and Y a.s. This proves the first assertion.

We next prove the P_X-uniform integrability of $\{\beta_t^{(n)^2}\}_n$ for any t

and Y a.s. We have the inequality

$$\beta_t^{(n)} \leq \frac{\exp\left\{\frac{1}{\sqrt{n}}\sum_{k=1}^{[nt]}\left(s^{(n)}(\frac{1}{\sqrt{n}}(Y_k^n - h(x_{\frac{k}{n}}))) - s^{(n)}(\frac{1}{\sqrt{n}}Y_k^n)\right)\right\}}{\prod_{k=1}^{[nt]}\left(1 + \frac{1}{\sqrt{n}}s^{(n)}(\frac{1}{\sqrt{n}}Y_k^n)\right)\exp\left\{-\frac{1}{\sqrt{n}}s^{(n)}(\frac{1}{\sqrt{n}}Y_k^n)\right\}} \tag{3.24}$$

The denominator of the right hand side converges to 1 as $n \to \infty$. For the numerator, we have by (A.1)(2)

$$\begin{aligned}
&\frac{1}{\sqrt{n}}\sum_{k=1}^{[nt]}|s^{(n)}(\frac{1}{\sqrt{n}}(Y_k^n - h(x_{\frac{k}{n}}))) - s^{(n)}(\frac{1}{\sqrt{n}}Y_k^n)| \\
&= \frac{1}{\sqrt{n}}\sum_{k=1}^{[nt]}|s^{(n)\prime}(\frac{1}{\sqrt{n}}(Y_k^n + \widetilde{\theta}_k^n h(x_{\frac{k}{n}})))||\frac{1}{\sqrt{n}}h(x_{\frac{k}{n}})| \\
&\leq \frac{1}{n}\sum_{k=1}^{[nt]}\Big\{c\Big(1 + \frac{1}{\sqrt{n}}(|Y_k^n| + c')\Big)^2 \\
&\qquad + c^{(n)}d(\frac{1}{\sqrt{n}}(Y_k^n + \widetilde{\theta}_k^n h(x_{\frac{k}{n}})))\Big\}c'.
\end{aligned} \tag{3.25}$$

where $|h| \leq c'$. We have

$$\lim_{n\to\infty}\frac{1}{n}\sum_{k=1}^{[nt]}\frac{1}{\sqrt{n}}Y_k^n = const. \quad \text{a.s.}$$

by the law of the large numbers, and

$$\lim_{n\to\infty} c^{(n)}\frac{1}{n}\sum_{k=1}^{[nt]} d(\frac{1}{\sqrt{n}}(Y_k^n + \widetilde{\theta}_k^n h(x_{\frac{k}{n}}))) = 0 \quad \text{a.s.}$$

Therefore, the supremum of (3.25) with respect to n is bounded (independent of $x \in \mathbf{C}$) for any t and Y a.s. Then the same property is valid to $\{\beta_t^{(n)}\}_n$. Consequently the sequence $\{\beta_t^{(n)^2}\}_n$ is P_X-uniformly integrable for any t and Y a.s. □

We can now complete the proof of Theorem 3.1. The sequence $\{\alpha_t^{(n)}\beta_t^{(n)}\}_n$ is P_X-uniformly integrable for any t and Y a.s., since $\alpha_t^{(n)}\beta_t^{(n)} \leq \frac{1}{2}(\alpha_t^{(n)^2} + \beta_t^{(n)^2})$ and both $\{\alpha_t^{(n)^2}\}$ and $\{\beta_t^{(n)^2}\}$ are P_X-uniformly integrable. Further, $\{\alpha_t^{(n)}\beta_t^{(n)}\}_n$ converges to $\alpha_t\beta_t$ a.s. P_X for any t and Y a.s. by Lemma 3.2. Therefore we have

$$\lim_{n\to\infty} \int \alpha_t^{(n)}\beta_t^{(n)} g(x_t^{(n)}) dP_X = \int \alpha_t\beta_t g(x_t) dP_X$$

for any t and Y a.s. The proof of Theorem 3.1 is complete.

Example 1. Let $\{\xi_j^n; n, j \in N\}$ be an array of 1-dimensional identically distributed random variables such that for each n, $\xi_j^n, j \in$ N are independent. We assume that their common distribution has a continuous density function $f(w)$ with mean 0 and variance 1. Suppose that for any n the noise process $W_k^n, k \in N$ is a sequence of independent random variables such that their common distribution is equal to that of $\sum_{j=1}^n \xi_j^n$. Let $\tilde{f}^{(n)}$ be the density function of the distribution of $\frac{1}{\sqrt{n}}W_k^n$. Then it is expanded by Edgeworth's series ([1]). Taking the first several series, we have

$$\begin{aligned} \tilde{f}^{(n)}(w) &= \phi_1(w)\Big\{1 + \frac{\beta_3}{6\sqrt{n}}h_3(w) + \frac{\beta_4}{24n}h_4(w) + \frac{\beta_3^2}{72n}h_6(w) \\ &\quad + \frac{1}{\sqrt{n}}r^{(n)}(w)\Big\}, \end{aligned}$$

where β_3 and β_4 are cumulants of the distribution density function $f(w)$ and are equal to $\beta_4 = \mu_3$ and $\beta_4 = \mu_4 - 3\mu_2^2$, where μ_2, μ_3 and μ_4 are the variance, the third and the fourth moments. Further h_3, h_4 and h_6 are Hermite polynomials of orders 3, 4 and 6, respectively. These are given by

$$\begin{aligned} h_3(w) &= w^3 - 3w, \qquad h_4(w) = w^4 - 6w^2 + 3, \\ h_6(w) &= w^6 - 15w^4 + 45w^2 - 15. \end{aligned}$$

Further $r^{(n)}(w) = O(\frac{1}{n})$. Set

$$s^{(n)}(w) = \frac{\beta_3}{6}h_3(w) + \frac{\beta_4}{24\sqrt{n}}h_4(w) + \frac{\beta_3^2}{72\sqrt{n}}h_6(w) + r^{(n)}(w).$$

Clearly it satisfies Condition (A.1). Therefore assuming (A.2) for the system process (X_t), Theorem 3.1 is valid. We have further $s(w) = \lim_{n\to\infty} s^{(n)}(w) = \frac{\beta_3}{6} h_3(w)$. Then $s'(w) = \frac{\beta_3}{2}(w^2-1)$, so that $\int s(w)\phi_1(w)dw = 0$ is satisfied. Therefore the sequence of filters $\{\pi_t^{(n)}\}$ converges to the filter π_t.

Example 2. We consider an example where $d = e = 1$. Let $\{W_k^n, k = 1, 2, \cdots\}$ be i.i.d. random variables for each n and suppose that the distribution of $\frac{1}{\sqrt{n}} W_1^n$ has a density function $\tilde{f}^{(n)}$ which is written by

$$\tilde{f}^{(n)}(w) = \phi_1(w)\Big(1 + \frac{1}{\sqrt{n}}(w^2 + w)\Big).$$

Then $\{W_k^n\}$ satisfies Condition (A.1) with $s^{(n)}(w) = w^2 + w$. Since $s'(w) = 2w + 1$, we have

$$\int s'(w)\phi_1(w)dx = 1.$$

Therefore the correction term β_t is needed for the limit of $\{\pi_t^{(n)}\}_n$. This means that it does not converge to the filter π_t.

Bibliography

[1] H. Cramer, *Mathematical method of statistics*, Princeton Univ. Press, 1946.

[2] M. Fujisaki, G. Kallianpur and H. Kunita, *Stochastic differential equations for the nonlinear filtering problem,* Osaka J. Math., 9(1972), 19-42.

[3] G. Kallianpur and C. Striebel, *Estimations of stochastic processes, Arbitrary system process with additive white noise observation errors,* Ann. Math. Statist., 39(1968), 785-801.

[4] R. E. Kalman, *A new approach to linear filtering and prediction problems,* J. Basic Eng. ASME, 82(1960), 33-45.

[5] H. Kunita, *Stochastic flows and stochastic differential equations,* Cambridge Univ. Press, 1990.

[6] H. J. Kushner, *Dynamical equations for optimal nonlinear filtering,* J. Differential Equations, 3(1967), 179-190.

[7] M. Zakai, *On the optimal filtering of diffusion processes,* Z. Wahrscheinlichkeitstheorie. Geb., 11(1969), 230-243.

Wong-Zakai Corrections, Random Evolutions, and Simulation Schemes for SDE's

Thomas G. Kurtz
Departments of Mathematics & Statistics
University of Wisconsin-Madison
Madison, WI 53706

Philip Protter
Departments of Mathematics & Statistics
Purdue University
W. Lafayette, IN 47907

Abstract

A general weak limit theorem for solutions of stochastic differential equations driven by arbitrary semimartingales is applied to give a unified treatment of limit theorems for random evolutions and consistency results for numerical schemes for stochastic differential equations. The asymptotic distribution of the error in an Euler scheme is studied. The Wong-Zakai correction in the random evolution limit arises through an integration by parts.

1 Introduction

For $n = 1, 2, ...$, let Y_n be an $\mathbb{R}^m$-valued semimartingale with respect to a filtration $\{\mathcal{F}_t^n\}$, that is a cadlag process such that Y_n can be written as $Y_n = M_n + A_n$ where M_n is an $\{\mathcal{F}_t^n\}$-local martingale and A_n has sample paths of finite variation. Suppose that $\{Y_n\}$ converges in distribution in the Skorohod topology to a process Y. We will say that the sequence is *good* if for every sequence $\{X_n\}$ of cadlag $k \times m$-matrix-valued processes such that X_n is $\{\mathcal{F}_t^n\}$-adapted and $(X_n, Y_n) \Rightarrow (X, Y)$, one has $\int X_n dY_n \Rightarrow \int X dY$. Note that the stochastic integrals are the usual semimartingale integrals given by

ISBN 0-12-481005-5

$$\begin{aligned} \int_0^t X dY &= \int_0^t X(s-)dY(s) \\ &= \lim \sum X(t_k)(Y(t_{k+1}) - Y(t_k)) \end{aligned} \tag{1.1}$$

where $\{t_k\}$ is a partition of $[0, t]$ and the limit is in probability and is taken as $\max_k |t_{k+1} - t_k| \to 0$. Jakubowski, Mémin, and Pages (1989) and Kurtz and Protter (1991a) give equivalent sets of sufficient conditions for "goodness". See also Mémin and Slomiński (1990). Kurtz and Protter (1991b) show that the conditions are in fact necessary. In particular, to state these conditions define $h_\delta : [0, \infty) \to [0, \infty)$ by $h_\delta(r) = (1 - \delta/r)^+$ and $J_\delta : D_{\mathbb{R}^m}[0, \infty) \to D_{\mathbb{R}^m}[0, \infty)$ by

$$J_\delta(x)(t) = \sum_{s \le t} h_\delta(|x(s) - x(s-)|)(x(s) - x(s-)) . \tag{1.2}$$

The mapping $x \to (x, J_\delta(x))$ is a continuous mapping of $D_{\mathbb{R}^m}[0, \infty) \to D_{\mathbb{R}^m \times \mathbb{R}^m}[0, \infty)$ (see Lemma 2.1 of Kurtz and Protter (1991a)), and hence $(X_n, Y_n) \Rightarrow (X, Y)$ in $D_{\mathbb{M}^{km} \times \mathbb{R}^m}[0, \infty)$ implies that $(X_n, Y_n, Y_n^\delta, J_\delta(Y_n)) \Rightarrow (X, Y, Y^\delta, J_\delta(Y))$ in $D_{\mathbb{M}^{km} \times \mathbb{R}^{3m}}[0, \infty)$ where $Y_n^\delta = Y_n - J_\delta(Y_n)$ and $Y^\delta = Y - J_\delta(Y)$. The following is Theorem 2.2 of Kurtz and Protter (1991a).

Theorem 1.1 *For each n, let (X_n, Y_n) be an $\{\mathcal{F}_t^n\}$-adapted process with sample paths in $D_{\mathbb{M}^{km} \times \mathbb{R}^m}[0, \infty)$, and let Y_n be an $\{\mathcal{F}_t^n\}$-semimartingale. Fix $\delta > 0$ (allowing $\delta = \infty$), and define $Y_n^\delta = Y_n - J_\delta(Y_n)$. (Note that Y_n^δ will also be a semimartingale.) Let $Y_n^\delta = M_n^\delta + A_n^\delta$ be a decomposition of Y_n^δ into an $\{\mathcal{F}_t^n\}$-local martingale and a process with finite variation. Suppose*

C1 For each $\alpha > 0$, there exist stopping times $\{\tau_n^\alpha\}$ such that $P\{\tau_n^\alpha \le \alpha\} \le \frac{1}{\alpha}$ and $\sup_n E[[M_n^\delta]_{t \wedge \tau_n^\alpha} + T_{t \wedge \tau_n^\alpha}(A_n^\delta)] < \infty$. ($T_t(A)$ denotes the total variation of A and $[M]$ denotes the quadratic variation of M.)

If $(X_n, Y_n) \Rightarrow (X, Y)$ in the Skorohod topology on $D_{\mathbb{M}^{km} \times \mathbb{R}^m}[0, \infty)$, then Y is a semimartingale with respect to a filtration to which X and Y are adapted, and $(X_n, Y_n, \int X_n dY_n) \Rightarrow (X, Y, \int X dY)$ in the Skorohod topology on $D_{\mathbb{M}^{km} \times \mathbb{R}^m \times \mathbb{R}^k}[0, \infty)$. If $(X_n Y_n) \to (X, Y)$ in probability, then the triple converges in probability.

In other words, if $\{Y_n\}$ is a sequence of semimartingales converging in distribution and satisfies C1, then $\{Y_n\}$ is good. This condition is shown to be necessary for goodness in Kurtz and Protter (1991b).

The following is a consequence of goodness.

Proposition 1.2 *Let $f : \mathbb{R}^k \to \mathbb{M}^{km}$ be bounded and continuous, for each n, let (U_n, Y_n) be an $\{\mathcal{F}_t^n\}$-adapted process in $D_{\mathbb{R}^k \times \mathbb{R}^m}[0,\infty)$, and let $\{Y_n\}$ be a good sequence of semimartingales with $(U_n, Y_n) \Rightarrow (U, Y)$. If for each n, X_n is a solution of*

$$X_n(t) = U_n(t) + \int_0^t f(X_n(s-))dY_n(s)\,, \tag{1.3}$$

then $\{(X_n, Y_n)\}$ is relatively compact (in the sense of convergence in distribution) and any limit point (U, X, Y) satisfies

$$X(t) = U(t) + \int_0^t f(X(s-))dY(s)\,. \tag{1.4}$$

(More general results can be found in Słomiński (1989) and Kurtz and Protter (1991a).)

It is well-known from the work of Wong and Zakai (1965) that the conclusion of Proposition 1.2 fails for many natural sequences $\{Y_n\}$ approximating Brownian motion.

Example 1.3 Let $\{\xi_k\}$ be independent and identically distributed with mean zero and variance σ^2. Define

$$W_n^a(t) = \frac{1}{\sqrt{n}} \sum_{k=1}^{[nt]} \xi_k\,. \tag{1.5}$$

Then it is easy to check that $\{W_n^a\}$ is a good sequence which, by the Donsker invariance principle, converges in distribution to σW where W is standard Brownian motion. □

Example 1.4 Let W be standard Brownian motion and let W_n^b satisfy $W_n^b(0) = 0$ and

$$\text{(1.6)} \qquad \frac{d}{dt}W_n^b(t) = n\left(W\left(\frac{[nt]+1}{n}\right) - W\left(\frac{[nt]}{n}\right)\right).$$

Then $W_n^b \Rightarrow W$, but $\{W_n^b\}$ is not good (e.g., $\int_0^t W_n^b dW_n^b \Rightarrow \int_0^t W dW + \frac{1}{2}t$). Since the sample paths of W_n^b have finite variation it is a semimartingale, but in the canonical decomposition of W_n^b into a local martingale plus a finite variation process, the local martingale is zero and $T_t(W_n^b) = O(\sqrt{n})$. (The decomposition used in Theorem 1.1 does not need to be the canonical decomposition, but, of course, the counter example ensures that no decomposition will satisfy C1.) □

Example 1.5 Let $\{\xi_k, k \geq 0\}$ be a finite, irreducible Markov chain with transition matrix $P = ((p_{ij}))$. Let $\pi = (\pi_1, \ldots, \pi_M)$ give the stationary distribution, and let f be a function satisfying

$$\text{(1.7)} \qquad \sum_m f(m)\pi_m = 0.$$

Define

$$\text{(1.8)} \qquad W_n^c(t) = \frac{1}{\sqrt{n}} \sum_{k=1}^{[nt]} f(\xi_k).$$

Letting $Pg(i) \equiv \sum_j g(j)p_{ij}$, by (1.7) there exists a function h such that $Ph - h = f$. Substituting in (1.8), we obtain

$$\begin{aligned} W_n^c(t) &= \frac{1}{\sqrt{n}} \sum_{k=1}^{[nt]} (Ph(\xi_k) - h(\xi_k)) \\ &= \frac{1}{\sqrt{n}} \sum_{k=1}^{[nt]} (Ph(\xi_{k-1}) - h(\xi_k)) \qquad \text{(1.9)} \\ &\quad + \frac{1}{\sqrt{n}} \left(Ph(\xi_{[nt]}) - Ph(\xi_0)\right) \\ &\equiv Y_n(t) + Z_n(t). \end{aligned}$$

Then $\{Y_n\}$ is a sequence of martingales satisfying C1 and converging in distribution to σW for some σ and $Z_n \Rightarrow 0$. (See Kurtz and Protter (1991a) Example 5.9 for details.) The sequence $\{W_n^c\}$, however, in general fails to satisfy C1. In particular, $\int_0^t Z_n dW_n^c \Rightarrow \alpha t$ where $\alpha = -2\Sigma\pi_i p_{ij}(Ph(j) - h(i))^2$. □

Note that W_n^b can also be written as the sum of a martingale and a process Z_n^b such that $Z_n^b \Rightarrow 0$. Specifically, $W_n^b(t) = W\left(\frac{[nt]+1}{n}\right) + \left(W_n^b(t) - W\left(\frac{[nt]+1}{n}\right)\right) \equiv Y_n^b + Z_n^b$. All these processes are adapted to the filtration given by $\mathcal{F}_t^n = \sigma\left(W(s) : s \le \frac{[nt]+1}{n}\right)$ and Y_n^b is an $\{\mathcal{F}_t^n\}$-martingale. With these last two examples in mind consider the equation

$$\begin{aligned} X_n(t) &= X_n(0) + \textstyle\int_0^t F(X_n(s-))dW_n(s) \\ &= X_n(0) + \textstyle\int_0^t F(X_n(s-))dY_n(s) \\ &\quad + \textstyle\int_0^t F(X_n(s-))dZ_n(s)\,. \end{aligned} \tag{1.10}$$

We have the following extension of the classical results of Wong and Zakai (1965).

Theorem 1.6 *Let Y_n and Z_n be $\{\mathcal{F}_t^n\}$-semimartingales, and let $X_n(0)$ be $\mathcal{F}_0^n$-measurable. Let $F : \mathbb{R}^k \to \mathbb{M}^{km}$ in (1.10) be bounded and have bounded and continuous first and second order derivatives. Define $H_n = ((H_n^{\beta\gamma}))$ and $K_n = ((K_n^{\beta\gamma}))$ by*

$$H_n^{\beta\gamma}(t) = \int_0^t Z_n^\beta(s-)dZ_n^\gamma(s) \tag{1.11}$$

and

$$K_n^{\beta\gamma}(t) = [Y_n^\beta, Z_n^\gamma]_t\,. \tag{1.12}$$

Suppose that $\{Y_n\}$ and $\{H_n\}$ satisfy C1 and that $(X_n(0), Y_n, Z_n, H_n, K_n) \Rightarrow (X(0), Y, 0, H, K)$. Then $\{(X_n(0), Y_n, Z_n, H_n, K_n, X_n)\}$ is relatively compact, and any limit point $(X(0), Y, 0, H, K, X)$ satisfies

$$X(t) = X(0) + \int_0^t F(X(s-))dY(s) + \sum_{\alpha,\beta,\gamma} \int_0^t \partial_\alpha F_\beta(X(s-))F_{\alpha\gamma}(X(s-))d(H^{\gamma\beta}(s) - K^{\gamma\beta}(s)) \tag{1.13}$$

where ∂_α denotes the partial derivative with respect to the αth variable and F_β denotes the βth column of F.

Proof The theorem follows by integrating the second term on the right of (1.10) by parts and then applying Theorem 1.2. See Kurtz and Protter (1991a), Theorem 5.10. □

2 Random evolutions

We now consider sequences of stochastic ordinary differential equations in $\mathbb{R}^k$ of the form

$$\dot{X}_n(t) = G(X_n(t), \xi(n^2t)) + nH(X_n(t), \xi(n^2t)) \tag{2.1}$$

where ξ is a stochastic process representing the random noise in the system. Models of this type were considered first by Stratonovich (1963, 1967) and Khas'minski (1966) and, in an abstract form, were dubbed random evolutions by Griego and Hersh (1969). Limit theorems for random evolutions are closely related to the results of the previous section.

Let ξ be a continuous time Markov chain with state space $E = \{1, \ldots, m\}$ and intensity matrix Q. We assume that Q is irreducible and hence that there is a unique stationary distribution π. Suppose $\Sigma_\beta H(x,\beta)\pi_\beta = 0$ and define V_n and W_n by

$$V_n^\beta(t) = \int_0^t I_{\{\beta\}}(\xi(n^2s))ds \tag{2.2}$$

and

$$W_n^\beta(t) = n(V_n^\beta(t) - \pi_\beta t) = n\int_0^t (I_{\{\beta\}}(\xi(n^2s)) - \pi_\beta)ds\,. \tag{2.3}$$

Then (2.1) becomes

$$X_n(t) = X_n(0) + \sum_{\beta=1}^{m} \int_0^t G(X_n(s),\beta)dV_n^\beta(s) \tag{2.4}$$
$$+ \sum_{\beta=1}^{m} \int_0^t H(X_n(s),\beta)dW_n^\beta(s)\,.$$

Let h_β satisfy

$$\sum_{k=1}^{m} q_{jk}h_\beta(k) = I_{\{\beta\}}(j) - \pi_\beta \tag{2.5}$$

(h_β exists by the uniqueness of π), and note that Y_n defined by

$$Y_n^\beta(t) = n \int_0^t (I_{\{\beta\}}(\xi(n^2 s)) - \pi_\beta)ds \tag{2.6}$$
$$-\tfrac{1}{n}h_\beta(\xi(n^2 t)) + \tfrac{1}{n}h_\beta(\xi(0))$$

is a martingale. Define Z_n by

$$Z_n^\beta(t) = \frac{1}{n}h_\beta(\xi(n^2 t)) - \frac{1}{n}h_\beta(\xi(0)) \tag{2.7}$$

so that $W_n = Y_n + Z_n$. Let $N_{ij}(t)$ denote the number of transitions of ξ from state i to state j up to time t. Then

$$[Y_n^\beta, Y_n^\gamma]_t = \sum_{ij=1}^{m} \frac{N_{ij}(n^2 t)}{n^2}(h_\beta(j) - h_\beta(i))(h_\gamma(j) - h_\gamma(i)) \tag{2.8}$$

$$[Y_n^\beta, Z_n^\gamma]_t = -[Y_n^\beta, Y_n^\gamma]_t \tag{2.9}$$

and

$$\int_0^t Z_n^\beta(s-)dZ_n^\gamma(s) = \tag{2.10}$$
$$\sum_{ij=1}^{m} \frac{N_{ij}(n^2 t)}{n^2}h_\beta(i)(h_\gamma(j) - h_\gamma(i)) - \frac{1}{n}h_\beta(\xi(0))Z_n^\gamma(t)\,.$$

As $n \to \infty$ we obtain

$$
\begin{aligned}
[Y_n^\beta, Y_n^\gamma]_t &\to C_{\beta\gamma} t \\
&\equiv \sum_{ij=1}^m \pi_i q_{ij}(h_\beta(j) - h_\beta(i))(h_\gamma(j) - h_\gamma(i))t
\end{aligned}
\tag{2.11}
$$

and

$$
\begin{aligned}
\textstyle\int_0^t Z_n^\beta(s-)dZ_n^\gamma(s) &\to D_{\beta\gamma} t \\
&\equiv \sum_{ij=1}^m \pi_i q_{ij} h_\beta(i)(h_\gamma(j) - h_\gamma(i))t\,.
\end{aligned}
\tag{2.12}
$$

The martingale central limit theorem (see, for example, Ethier and Kurtz (1986), Theorem 7.1.4) gives $Y_n \Rightarrow Y$ where Y is a Brownian motion with infinitesimal covariance $C = ((C_{\beta\gamma}))$. Theorem 1.6 gives the following

Theorem 2.1 *Let ξ in (2.1) be a finite Markov chain with state space $E = \{1, \ldots, m\}$ and intensity matrix Q, and let $X_n(0)$ be independent of ξ. Let G be bounded and continuous, and let H be bounded and have bounded and continuous first and second derivatives. Assume that Q is irreducible and that π, $((C_{\beta\gamma}))$ and $((D_{\beta\gamma}))$ are as above. Define $\bar{G} : \mathbb{R}^k \to \mathbb{R}^k$ by $\bar{G}(x) = \Sigma_\beta G(x,\beta)\pi_\beta$ and $H : \mathbb{R}^k \to \mathbb{M}^{km}$ by $H(x) = (H(x,1), \ldots, H(x,m))$. If $X_n(0) \Rightarrow X(0)$, then $\{X_n\}$ is relatively compact and any limit point satisfies*

$$
\begin{aligned}
X(t) &= X(0) + \textstyle\int_0^t \bar{G}(X(s))ds + \int_0^t H(X(s-))dY(s) \\
&\quad + \textstyle\sum_{\alpha,\beta,\gamma} \int_0^t \partial_\alpha H(X(s-),\beta) H_\alpha(X(s-),\gamma)(D_{\gamma\beta} + C_{\gamma\beta})ds\,.
\end{aligned}
\tag{2.13}
$$

Remark 2.2 Hersh and Papanicolaou (1972) and Kurtz (1973) prove the above result using functional analytic arguments.

3 Numerical schemes

Let Y be and $\{\mathcal{F}_t\}$-semimartingale in $\mathbb{R}^m$, and let $f : \mathbb{R}^k \to \mathbb{M}^{km}$ be continuous. A number of authors (for example, Mil'shtein (1974), Rumelin (1982), Pardoux and Talay (1985), Talay and Tubaro (1989), Wagner (1989)) have considered analogues of classical numerical schemes as means of simulating solutions of the stochastic differential equation

$$X(t) = X(0) + \int_0^t f(X(s-))dY(s) \,. \tag{3.1}$$

The results discussed above provide a natural approach to checking the consistency of these schemes, and we will see that they are also useful in a more careful study of the error in the scheme.

The simplest numerical scheme is, of course, the Euler scheme. Specifying a mesh $0 = t_0 < t_1 < \cdots$, define X_0 recursively by setting $X_0(0) = X(0)$ and

$$X_0(t_{k+1}) = X_0(t_k) + f(X_0(t_k))\Delta Y(t_k) \tag{3.2}$$

where $\Delta Y(t_k) = Y(t_{k+1}) - Y(t_k)$. If we extend the definition of X_0 to all t by setting $X_0(t) = X_0(t_k)$ for $t_k \le t < t_{k+1}$ and we define Y_0 by $Y_0(t) = Y(t_k)$ for $t_k \le t < t_{k+1}$, then

$$X_0(t) = X_0(0) + \int_0^t f(X_0(s-))dY_0(s) \,. \tag{3.3}$$

Note that if we define $\beta(t) = t_k$ for $t_k \le t < t_{k+1}$, then we can write $Y_0 = Y \circ \beta$. With this observation, consistency for the Euler scheme is a consequence of Proposition 1.2 and the following lemma. Note that the lemma would allow for a mesh determined by $\{\mathcal{F}_t\}$-stopping times.

Lemma 3.1 *Let Y be an $\{\mathcal{F}_t\}$-semimartingale. For each n, let β_n be a nonnegative, nondecreasing process such that for each $u \ge 0$, $\beta_n(u)$ is an $\{\mathcal{F}_t\}$-stopping time. If $\beta_n(u) \to u$ a.s. for each $u \ge 0$, then $Y \circ \beta_n \to Y$ a.s. in the Skorohod topology and $\{Y \circ \beta_n\}$ is a good sequence.*

Proof First observe that $\beta_n(u) \to u$ for each u implies $Y \circ \beta_n \to Y$ by Proposition 3.6.5 of Ethier and Kurtz (1986). To verify goodness for $\{Y \circ \beta_n\}$ it is enough to verify goodness for $\{Y^{\tau_m} \circ \beta_n\}$ ($Y^{\tau_m} \equiv Y(\cdot \wedge \tau_m)$) for some family of stopping times satisfying $P\{\tau_m \leq m\} \leq \frac{1}{m}$. In particular, let d be the metric for the Skorohod topology given in (3.5.2) of Ethier and Kurtz (1986), and suppose that $(X_n, Y \circ \beta_n) \Rightarrow (X, Y)$. Define X_n^m by setting $X_n^m(t) = X_n(t)$ for $t < \tau_m$ and $X_n^m(t) = X_n(\tau_m-)$ for $t \geq \tau_m$, and define X^m analogously. Then $(X_n^m, Y^{\tau_m} \circ \beta_n) \Rightarrow (X^m, Y^{\tau_m})$. Let

$$\begin{aligned} V_n^m(t) &= \textstyle\int_0^t X_n^m(s-)dY^{\tau_m} \circ \beta_n(s) \\ V^m(t) &= \textstyle\int_0^t X^m(s-)dY^{\tau_m}(s) \end{aligned} \tag{3.4}$$

and

$$V_n(t) = \int_0^t X_n(s-))dY \circ \beta_n(s) \quad V(t) = \int_0^t X(s-))dY(s) \tag{3.5}$$

and observe that

$$\begin{aligned} \overline{\lim}_{n\to\infty} E[d(V_n^m, V_n)] &\leq \overline{\lim}_{n\to\infty} E[e^{-\beta_n^{-1}(\tau_m)}] \\ &\leq e^{-m} + \tfrac{1}{m} \,. \end{aligned} \tag{3.6}$$

Consequently, if $V_n^m \Rightarrow V^m$ for each m, we have $V_n \Rightarrow V$ which would give the goodness for $\{Y \circ \beta_n\}$.

Fix $0 < \delta < \infty$, and define Y^δ as in Theorem 1.1. Let $Y^\delta = M^\delta + A^\delta$ be the canonical decomposition of Y^δ (so that the discontinuities of M^δ are bounded by 2δ and the discontinuities of A^δ are bounded by δ). Define $\tau_m = \inf\{t : [M^\delta]_t + T_t(A^\delta) > c_m\}$ where c_m is selected so that $P\{\tau_m \leq m\} \leq \frac{1}{m}$. Then

$$E[[M^\delta]_{\tau_m} + T_{\tau_m}(A^\delta)] < c_m + 3\delta \,. \tag{3.7}$$

We can write

$$\begin{aligned} Y^{\tau_m} \circ \beta_n &= M^\delta(\beta_n(\cdot) \wedge \tau_m) + A^\delta(\beta_n(\cdot) \wedge \tau_m) + J_\delta(Y^{\tau_m}) \circ \beta_n \\ &\equiv M_n + A_n + Z_n \end{aligned} \tag{3.8}$$

and since $E[[M_n]_t] \leq E[[M^\delta]_{\tau_m}]$ and $E[T_t(A_n)] \leq E[T_{\tau_m}(A^\delta)]$, we can apply Theorem 2.7 of Kurtz and Protter (1991a) to conclude that $V_n^m \Rightarrow V^m$. □

There is an alternative approach to representing the approximation given by the Euler scheme as a solution of a stochastic differential equation. Define $\eta(t) = t_k$ for $t_k \leq t < t_{k+1}$ (which is the same as β defined above, but the general assumptions that will be placed on the sequence $\{\eta_n\}$ below will be different from the assumptions placed on $\{\beta_n\}$). Let $\tilde{X}_0$ satisfy

$$\tilde{X}_0(t) = X(0) + \int_0^t f(\tilde{X}_0 \circ \eta(s-))dY(s)\,. \tag{3.9}$$

Then $\tilde{X}_0(t_k) = X_0(t_k)$. The consistency of the Euler scheme can also be obtained through the analysis of this equation.

Lemma 3.2 *For each n, let Y_n be an $\mathbb{R}^m$-valued $\{\mathcal{F}_t^n\}$-semimartingale, X_n a cadlag, $\mathbb{M}^{km}$-valued $\{\mathcal{F}_t^n\}$-adapted process, and η_n a right continuous, nondecreasing $\{\mathcal{F}_t^n\}$-adapted process. Suppose that $\eta_n(t) \leq t$ and $\eta_n(t) \to t$ for all $t \geq 0$. Assume that $\{Y_n\}$ is a good sequence and that $(X_n, Y_n) \Rightarrow (X, Y)$ in $D_{\mathbb{M}^{km}\times\mathbb{R}^m}[0,\infty)$. Then $\int X_n \circ \eta_n dY_n \Rightarrow \int X dY$.*

Proof First observe that for each fixed $\delta > 0$,

$$\int_0^\cdot \Big(J_\delta(X_n) \circ \eta_n(s-) - J_\delta(X_n)(s-)\Big) dY_n(s) \Rightarrow 0\,. \tag{3.10}$$

Consequently, there exist $\delta_n \to 0$ such that

$$\int_0^\cdot \Big(J_{\delta_n}(X_n) \circ \eta_n(s-) - J_{\delta_n}(X_n)(s-)\Big) dY_n(s) \Rightarrow 0\,. \tag{3.11}$$

But the asymptotic continuity of $X_n^{\delta_n}$ implies that
$X_n^{\delta_n} - X_n^{\delta_n} \circ \eta_n \Rightarrow 0$, so

$$\int_0^\cdot \Big(X_n^{\delta_n}(s-) - X_n^{\delta_n} \circ \eta_n(s-)\Big) dY_n(s) \Rightarrow 0 \tag{3.12}$$

and hence

$$\int_0^{\cdot} \Big(X_n(s-) - X_n \circ \eta_n(s-)\Big) dY_n(s) \Rightarrow 0 \tag{3.13}$$

which gives the lemma. □

Theorem 3.3 *For each n, let Y_n be an $\mathbb{R}^m$-valued $\{\mathcal{F}_t^n\}$-semimartingale and η_n a right continuous, nondecreasing $\{\mathcal{F}_t^n\}$-adapted process. Suppose that $\eta_n(t) \leq t$ and $\eta_n(t) \to t$ for all $t \geq 0$. Assume that $\{Y_n\}$ is a good sequence and that $Y_n \Rightarrow Y$. Let $f : \mathbb{R}^k \to \mathbb{M}^{km}$ be bounded and continuous, and let $\tilde{X}_n$ satisfy*

$$\tilde{X}_n(t) = X(0) + \int_0^t f(\tilde{X}_n \circ \eta_n(s-)) dY_n(s). \tag{3.14}$$

Then $\{(\tilde{X}_n, Y_n)\}$ is relatively compact and any limit point (X, Y) satisfies

$$X(t) = X(0) + \int_0^t f(X(s-)) dY(s). \tag{3.15}$$

If the Y_n are defined on the same sample space as Y, $\sup_{s \leq t} |Y_n(s) - Y(s)| \to 0$ in probability for each $t > 0$, and sample path uniqueness holds for the solution of (3.15), then $\sup_{s \leq t} |\tilde{X}_n(s) - X(s)| \to 0$ in probability for each $t > 0$.

Remark 3.4 For $Y_n = Y$, $n = 1, 2, \ldots$, and $\eta_n(t) = \tau_k^n$, $\tau_k^n \leq t < \tau_{k+1}^n$, for a sequence of stopping times $\{\tau_k^n\}$, this result is a special case of Theorem V.16 of Protter (1990).

Proof The relative compactness follows from Lemma 4.1 and Proposition 4.3 of Kurtz and Protter (1991a). The fact that any limit point satisfies (3.15) then follows from Lemma 3.2. Under the assumptions of the final assertion, we can treat $(\tilde{X}_n, X)$ as a solution of a single system. Using the uniqueness assumption, it follows that $(\tilde{X}_n, X) \Rightarrow (X, X)$ and hence that $\tilde{X}_n - X \Rightarrow 0$ which gives the desired conclusion. □

Note that if Y is a semimartingale and $Y_n = Y$ for all n, then $\{Y_n\}$ is good.

The next theorem gives an approach to the analysis of the error in the Euler scheme.

Theorem 3.5 *Let Y be an $\{\mathcal{F}_t\}$-semimartingale, and suppose that $f = (f_1, \ldots, f_m)$ is a bounded and continuously differentiable $k \times m$-matrix-valued function. For each n, let $0 = \tau_0^n < \tau_1^n < \cdots$ be $\{\mathcal{F}_t\}$-stopping times, define $\eta_n(t) = \tau_k^n$, $\tau_n^n \leq t < \tau_{k+1}^n$, and let $\tilde{X}_n$ satisfy (3.14). Let $\{\alpha_n\}$ be a positive sequence converging to infinity, set $U_n = \alpha_n(\tilde{X}_n - X)$, and define Z_n by*

$$Z_n^{ij}(t) = \alpha_n \int_0^t (Y_i(s-) - Y_i \circ \eta_n(s-))dY_j(s)\,. \tag{3.16}$$

Suppose that $\{Z_n\}$ is a good sequence with $(Y, Z_n) \Rightarrow (Y, Z)$. Then $U_n \Rightarrow U$ satisfying

$$\begin{aligned} U(t) &= \sum_i \textstyle\int_0^t \nabla f_i(X(s-))U(s-)dY_i(s) \\ &+ \sum_{ij} \textstyle\int_0^t \sum_k \partial_k f_i(X(s-)) f_{kj}(X(s-))dZ^{ij}(s)\,. \end{aligned} \tag{3.17}$$

Remark 3.6 a) For a discussion of linear stochastic differential equations, see Protter (1990), p271.
b) Rootzén (1980) gives the asymptotic distribution for the error in certain approximations for stochastic integrals.
c) Suppose

$$Y(t) = \begin{pmatrix} W(t) \\ t \end{pmatrix} \tag{3.18}$$

where W is an $(m-1)$-dimensional standard Brownian motion. Let $\eta_n(t) = \frac{[nt]}{n}$. Then, taking $\alpha_n = \sqrt{n}$, $(Y, Z_n) \Rightarrow (Y, Z)$ where Z is independent of Y, $Z^{im} = Z^{mi} = 0$, and for $1 \leq i, j \leq m-1$, Z^{ij} are independent mean zero Brownian motions with $E[(Z^{ij}(t))^2] = \frac{1}{2}t$.
d) With $m = 1$, let $1 < \beta < 3$, and let Y be the stable process with generator

$$(3.19)\quad Af(x) = \int_{-\infty}^{\infty} \Big(f(x+y) - f(x) - \frac{y}{1+y^2} f'(x)\Big) \frac{1}{|y|^\beta} dy\,.$$

Then, taking $\alpha_n = n^{\frac{1}{\beta-1}}$, Z is a process with stationary, independent increments and generator

$$A_Z f(x) =$$
$$(3.20)\quad \int_{-\infty}^{\infty} \int_{-\infty}^{\infty} \Big(f(x+zy) - f(x) - \frac{zy}{1+(zy)^2} f'(x)\Big) \frac{1}{|y|^\beta} dy \mu_\beta(dz)\,.$$

where μ_β is the distribution of $Y(1)$.

Proof Under the hypotheses of the theorem, the solution of (3.15) is unique, and $(\tilde{X}_n, X, Y, Z_n) \Rightarrow (X, X, Y, Z)$. For simplicity, assume $k = m = 1$. Then, noting that $\tilde{X}_n(s-) - \tilde{X}_n \circ \eta_n(s-) = f(\tilde{X}_n \circ \eta_n(s-))(Y(s-) - Y \circ \eta_n(s-))$

$$\begin{aligned} U_n(t) &= \textstyle\int_0^t \alpha_n \Big(f(\tilde{X}_n(s-)) - f(X(s-))\Big) dY \\ &\quad - \textstyle\int_0^t \alpha_n \Big(f(\tilde{X}_n(s-)) - f(\tilde{X}_n \circ \eta_n(s-))\Big) dY \\ &= \textstyle\int_0^t \frac{f(\tilde{X}_n(s-)) - f(X(s-))}{\tilde{X}_n(s-) - X(s-)} U_n(s-) dY(s) \\ &\quad - \textstyle\int_0^t \Big(f(\tilde{X}_n \circ \eta_n(s-) + f(\tilde{X}_n \circ \eta_n(s-))(Y(s-) - Y \circ \eta_n(s-))) \\ &\quad - f(\tilde{X}_n \circ \eta_n(s-))\Big)(Y(s-) - Y \circ \eta_n(s-))^{-1} dZ_n(s) \end{aligned}$$

(3.21)

where the integrands are defined in the obvious manner when the denominator vanishes. Let $\tau_n^a = \inf\{t : |U_n(t)| > a\}$. Then $\{U_n(\cdot \wedge \tau_n^a)\}$ is relatively compact, and any limit point will satisfy (3.17) on the time interval $[0, \tau^a]$ where $\tau^a = \inf\{t : |U(t)| > a\}$. But $\tau^a \to \infty$ as $a \to \infty$, so $U_n \Rightarrow U$. □

Bibliography

[1] Ethier, Stewart N. and Kurtz, Thomas G. (1986). *Markov Processes: Characterization and Convergence.* Wiley, New York.

[2] Griego, Richard J. and Hersh, Reuben (1969). Random evolutions, Markov chains, and systems of partial differential equations. *Proc. Nat. Acad. Sci.* USA 62, 305-308.

[3] Hersh, Reuben and Papanicolaou, George C. (1972). Non-commuting random evolutions, and an operator-valued Feynman-Kac formula. *Comm. Pure Appl. Math.* 25, 337-367.

[4] Jakubowski, A., Mémin, J., and Pagès, G. (1989). Convergence en loi des suites d'integrales stochastique sur l'espace D^1 de Skorokhod. *Probab. Th. Rel. Fields* 81, 111-137.

[5] Khas'minskii, R. Z. (1966). A limit theorem for the solutions of differential equations with random right-hand sides. *Theory Probab. Appl.* 11, 390-406.

[6] Kurtz, Thomas G. (1973). A limit theorem for perturbed operator semigroups with applications to random evolutions. *J. Funct. Anal.* 12, 55-67.

[7] Kurtz, Thomas G. and Protter, Philip (1991a). Weak limit theorems for stochastic integrals and stochastic differential equations. Ann. Probab. to appear

[8] Kurtz, Thomas G. and Protter, Philip (1991b). Characterizing weak convergence of stochastic integrals. (in preparation)

[9] Mémin, J. and Slomiński, L. (1990). Condition UT et stabilité en loi des solutions d'équations différentielles stochastiques. (preprint)

[10] Mil'shtein, G. N. (1974). Approximate integration of stochastic differential equations. *Theory Prob. App.* 19, 557-562.

[11] Pardoux, Etienne and Talay, Denis (1985). Discretization and simulation of stochastic differential equations. *Acta Applicandae Math.* 3, 23-47.

[12] Protter, Philip (1990). *Stochastic Integration and Differential Equations: A New Approach.* Springer-Verlag, New York.

[13] Rootzén, Holger (1980). Limit distributions for the error in approximation of stochastic integrals. *Ann. Probab.* 8, 241-251.

[14] Rumelin, Werner (1982). Numerical treatment of stochastic differential equations. *SIAM J. Numer. Anal.* 19, 604-613.

[15] Slomiński, Leszek (1989). Stability of strong solutions of stochastic differential equations. *Stochastic Process. Appl.* 31, 173-202.

[16] Stratonovich, R. L. (1963, 1967). *Topics in the Theory of Random Noise I, II.* Gordon Breach, New York.

[17] Talay, Denis, and Tubaro, L. (1989). Expansion of the global error for numerical schemes solving stochastic differential equations. (preprint).

[18] Wagner, W. (1989). Unbiased monte carlo estimators for functionals of weak solutions of stochastic differential equations, *Stochastics* 28, 1-20.

[19] Wong, Eugene, and Zakai, Moshe (1965). On the convergence of ordinary integrals to stochastic integrals. *Ann. Math. Statist.* 36, 1560-1564.

Nonlinear Filtering for Singularly Perturbed Systems

Harold J. Kushner
Division of Applied Mathematics
Brown University

Abstract

The nonlinear filtering problem for singularly perturbed or two time scale problems involves equations of high dimension and with high gains. It is desirable to simplify the filter by exploiting the multiple time scale property and do some sort of averaging of the "fast" dynamics. The problem is rather subtle. But it can be shown under broad conditions that the so-called averaged filter is nearly optimal for large scale separations.

1 Introduction

We are concerned with the use of an "averaged" filter for the so-called singularly perturbed jump-diffusion system:

$$dx^\epsilon = G(x^\epsilon, z^\epsilon)dt + \sigma(x^\epsilon, z^\epsilon)dw_1 + \int q_1(x^\epsilon, z^\epsilon, \gamma)N_1(dtd\gamma), \ x \in R^r, \tag{1.1}$$

where $x^\epsilon(0) = x(0)$, and

$$\epsilon dz^\epsilon = H(x^\epsilon, z^\epsilon)dt + \sqrt{\epsilon}v(x^\epsilon, z^\epsilon)dw_2 + \int q_2(x^\epsilon, z^\epsilon, \gamma)N_2^\epsilon(dtd\gamma), \ z \in R^{r'}. \tag{1.2}$$

The N_i are Poisson measures and the w_i are Wiener processes. Precise conditions will be given below. The observational process is defined by

$$dy^\epsilon = g(x^\epsilon, z^\epsilon)dt + dw_0, \quad y \in R^{r''} \tag{1.3}$$

ISBN 0-12-481005-5

Owing to the high dimension and high gains ($O(1/\epsilon)$) of the original system (1.1), (1.2), the construction of the optimal filter or of a good approximation to it can be much harder than ususal. Because of the differences in the "time scales" of the processes defined by (1.1) and (1.2), the possibility arises of using some sort of averaging method to get a simpler filter; say, one for a "slow" system which averaged out the z^ϵ−process. One would use the filter which is appropriate for the averaged system, but the input would be the actual physical measurements $y^\epsilon(\cdot)$.

Under appropriate conditions, the so-called averaged filter provides a very good approximation to the optimal filter for the singularly perturbed system and it can be substantially simpler. In Section 2, we state the conditions, and discuss the scalings. Some standard representation formulas for the nonlinear filter are reviewed in Section 3. The filtering problem and the averaged filter for the singularly perturbed problem are defined in Section 4. In Section 6, we prove that the average filter is "almost optimal" under conditions which seem to be quite reasonable. The conditions required to get the averaging result for the filtering problem are stronger than those needed to show that the system (1.1), (1.2), without filtering, can be well approximated by the averaged system for small ϵ mainly in the assumption concerning the convergence of the transition measure for the fixed-x "fast" process to the invariant measure of that process. If this condition does not hold, then the difference between the true filter and the averaged filter might not converge to zero, as $\epsilon \to 0$, as seen from the counterexample in Section 5. In any case, even when the cited convergence assumption does not hold, the averaged filter is nearly optimal in a specific and important sense: In particular, it can be shown to be nearly optimal with respect to a large and natural class of alternative filters. The reference [1] contains a comprehensive development of the singularly perturbed problem for both control and filtering, and efficient new averaging methods are introduced. There is also a discussion of the use of the averaged filters in control, and of the robustness (uniformly in the approximation parameter and ϵ) of a computational approximation to the averaged filter.

Similar results for the convergence to the averaged filter are in [2], where the methods are based on PDE techniques and require $\sigma = v =$identity, $q_i = 0$.

2 The Fixed-x Rescaled Fast Process and Assumptions

The fixed-x rescaled "fast" process. The basic averaging method exploits the "time scale" differences between $x^\epsilon(\cdot)$ and $z^\epsilon(\cdot)$. The averaged system will be similar to (1.1), (1.3), with the $z-$ process averaged out in some appropriate way. Let us examine (1.1) and (1.2) in a "stretched out" time scale. Define the processes $z_0^\epsilon(t) = z^\epsilon(\epsilon t)$ and $x_0^\epsilon(t) = x^\epsilon(\epsilon t)$. Then we have

$$dx_0^\epsilon = \epsilon G(x_0^\epsilon, z_0^\epsilon)dt + \sqrt{\epsilon}\sigma(x_0^\epsilon, z_0^\epsilon)d\tilde{w}_1 \tag{2.1}$$

$$dz_0^\epsilon = H(x_0^\epsilon, z_0^\epsilon)dt + v(x_0^\epsilon, z_0^\epsilon)d\tilde{w}_2, \tag{2.2}$$

where the $\tilde{w}_i(\cdot)$ are standard vector-valued Wiener processes and $\tilde{w}_i(t) = w_i(\epsilon t)/\sqrt{\epsilon}$. The "stretched out"process $x_0^\epsilon(\cdot)$ defined by (2.2) "varies slowly", and intuitively speaking, might be considered to be "nearly constant" over long time periods when ϵ is small. This suggests that the paths of $z_0^\epsilon(\cdot)$ are close to the paths that we would get if $x_0^\epsilon(\cdot)$ were "fixed" on appropriate large time intervals. It also suggests that if this "fixed-x" process were stationary ("modulo" an initial transient period), then the invariant measure of this stationary process could be used to average out the $z^\epsilon(\cdot)$ in (1.1) and (1.3). Now, define the *fixed-x process* $z_0(\cdot|x)$ (*written simply as* $z_0(\cdot)$ if the value of x is clear) by

$$dz_0 = H(x, z_0)dt + v(x, z_0)d\tilde{w}_2. \tag{2.3}$$

In (2.3), x is a *parameter*, which will sometimes be a random variable which is independent of $\tilde{w}_2(\cdot)$.

Notation and Assumptions. *Let A_x^0 denote the differential operator of the fixed-x process $z_0(\cdot|x)$, A^ϵ that of $(x^\epsilon(\cdot), z^\epsilon(\cdot))$, and $A^{0,\epsilon}$ that of $(x_0^\epsilon(\cdot), z_0^\epsilon(\cdot))$.*

We will use the following assumptions.

A2.1. *$G(\cdot)$ and $\sigma(\cdot)$ are continuous and are bounded by $K(1+|x|)$,* for some $K < \infty$. *$H(\cdot)$ and $v(\cdot)$ are continuous and bounded by $K(1+|z|)$. The $g(\cdot)$ and $q_i(\cdot)$ are bounded and continuous. w_0 is independent of $(w_1(\cdot), w_2(\cdot), x^\epsilon(\cdot), z^\epsilon(\cdot))$. The $N_1(\cdot)$ and $N_2^\epsilon(\cdot)$ are mutually independent Poisson measures which, together with the q_i- functions, satisfy the following conditions: $N_1(\cdot)$ has jump rate λ_1 and jump distribution $\Pi_1(\cdot)$; $N_2^\epsilon(\cdot)$ has jump rate λ_2/ϵ, and jump distribution $\Pi_2(\cdot)$; the $q_i(\cdot)$ are bounded and continuous in x, z for each value of γ and $q_i(x, z, 0) = 0$.*

A2.2. *For each initial condition and each x,* (2.3) *has a unique weak sense solution. For each x, $z_0(\cdot|x)$ has a unique invariant measure $\mu_x(\cdot)$. There is a continuous matrix valued function $\overline{\sigma}(\cdot)$ such that*

$$\overline{\sigma}(x)\overline{\sigma}'(x) = \int \sigma(x,z)\sigma'(x,z)\mu_x(dz) \equiv \overline{a}(x).$$

The factorization in (A2.2) is simply a convenience in the notation, so that we can represent the "averaged" process as a solution to a stochastic differential equation, instead of simply as a solution to a martingale problem.

Remarks on the Conditions. By the uniqueness of $\mu_x(\cdot)$ for each x, for any real valued, continuous and bounded $f(\cdot)$, $\int f(z)\mu_x(dz)$ is continuous in x. But $\overline{G}(x)$ and $\overline{\sigma}(x)$ are not necessarily Lipschitz continuous even if $G(x,z)$ and $\sigma(x)$ are, and we require (A2.3) below in order for the solution to (2.5) below to be well defined.

For notational convenience we suppose that $x^\epsilon(0) = x(0)$, not depending on ϵ. Define the averaged differential generator $\bar{A}$ by

$$\begin{aligned} \bar{A}f(x) &= f_x'(x)\bar{G}(x) + \tfrac{1}{2} \text{ trace } f_{xx}(x)\cdot \bar{a}(x) \\ &+\lambda_1 \textstyle\int [f(x+q_1(x,z,\gamma)) - f(x)]\Pi_1(d\gamma)\mu_x(dz). \end{aligned} \tag{2.4}$$

Define the averaged quantities

$$\overline{g}(x) = \int g(x,z)\mu_x(dz), \quad \overline{G}(x) = \int G(x,z)\mu_x(dz)$$

and the *averaged* system:

$$dx = \overline{G}(x)dt + \overline{\sigma}(x)dw + d\overline{J}(t), \tag{2.5}$$

where $\overline{J}(\cdot)$ is the jump process whose properties are defined by the operator (2.4). The initial value of (2.5) is that of (1.1). The observational data for the filter for the averaged system is

$$dy = \overline{g}(x)dt + dw_0, \tag{2.6}$$

where $w_0(\cdot)$ is a standard vector-valued Wiener process which is independent of $(w(\cdot), x(\cdot))$. We also need the following assumptions.

A2.3. (1.1), (1.2) *and* (2.5) *have unique weak sense solutions for each initial condition.*

A2.4. *The transition measure* $P_0(z, t, \cdot|x)$ *of* $z_0(\cdot|x)$ *satisfies: For* $f(\cdot)$ *bounded and continuous and* $z_0(\cdot)$ *denoting the fixed-*x *process with parameter* x,

$$E_{z_0} f(z_0(t)) \rightarrow \int f(z)\mu_x(dz)$$

as $t \rightarrow \infty$, *uniformly in* x *and* $z(0) = z_0$ *in any compact set.*

A2.5. $\{z^{\epsilon}(t), \epsilon > 0, t \leq T\}$ *is tight.*

The assumption (A2.5) is a "stability type" condition. Methods of stochastic stability can be used to prove that it holds in specific cases. In fact, if the fixed $x-$processes are appropriately stable, then (A2.5) holds under broad conditions. See the discussion in [1, Chapter 9].

A weak convergence result. When working with weak convergence of vector valued processes, we use the space $D^q[0,\infty)$ of R^q-valued paths which are right continuous and have left hand limits and with the Skorohod topology, for an appropriate value of q. The following result is a special case of [1, Theorem 4.1.2] (The cited theorem in [1] involves a sequence of controls also.)

Theorem 1. *Assume (A2.1)-(A2.3) and (A2.5). Let the fixed-*x *process* $z_0(\cdot|x)$ *have a unique invariant measure for each value of*

x. Then $\{x^\epsilon(\cdot), y^\epsilon(\cdot)\}$ is tight and converges weakly to a solution to (2.5),(2.6) for some mutually independent Wiener processes.

for any bounded and continuous function $f(\cdot)$, the sequence of processes with values

$$\int_0^t f(x^\epsilon(s), z^\epsilon(s))ds$$

converges weakly to the process with values

$$\int_0^t \bar{f}(x(s))ds,$$

where $\bar{f}(x) = \int f(x,z)\mu_x(dz)$.

3 The Representation Theorem

We are concerned with the relationships between the filtering problem associated with (1.1)-(1.3), and that associated with (2.5), (2.6). The filter for the averaged system is generally much simpler than that for the original physical system. This suggests that we construct the optimal non-linear filter for the averaged system (2.5), (2.6), but in lieu of using the "unobservable" observations $y(\cdot)$ as the input to that filter, we use the physical data $y^\epsilon(\cdot)$ instead. The resulting filter is referred to as the *averaged filter.* It turns out that the output process from this averaged filter is close to that from the true optimal filter for (1.1)-(1.3) under broad conditions. Even when it is not nearly optimal in a strict sense, it is nearly optimal with respect to a large class of alternative filters. We next define the representation for the optimal filter for the singularly perturbed problem as well as for the averaged filter.

We next state the well known representation for the non-linear filter for a signal process that is a jump-diffusion, and where the observation noise is "white Gaussian." The signal $x(\cdot)$ process is defined by

$$dx = b(x)dt + \sigma(x)dw + \int q(x,\gamma)N(dtd\gamma), \quad x \in R^r, \qquad (3.1)$$

and define the observation process by

$$dy = g(x)dt + dw_0, \quad y \in R^{r''}. \tag{3.2}$$

Let A denote the differential operator of the process defined by (3.1).
Let us assume:

A3.1. *$b(\cdot), \sigma(\cdot), q(\cdot), g(\cdot)$ are continuous. $q(\cdot)$ and $g(\cdot)$ are bounded and $b(\cdot)$ and $\sigma(\cdot)$ have at most a linear growth as $|x| \to \infty$. The standard vector-valued Wiener processes $w(\cdot)$ and $w_0(\cdot)$ are mutually independent. The Poisson measure $N(\cdot)$ has jump rate $\lambda < \infty$ and jump distribution $\Pi(\cdot)$, and is independent of the Wiener processes.*

For each $t \geq 0$, let E_t denote the expectation given the data $\{y(s), s \leq t\}$, and let $\phi(\cdot)$ be a bounded, continuous and real valued function of x. The representation theorem will be stated in the form in which it was initially derived in [3]. This form is equivalent to that obtained by measure transformation methods [4-5]. But it appears to be more convenient to use with the weak convergence type of calculations that we will use in Section 6.

Let $\hat{x}(\cdot)$ denote a process which has the same probability law that $x(\cdot)$ has, but which is independent of $(x(\cdot), y(\cdot), w_0(\cdot))$. Define $R(\cdot)$ by

$$R(t) = \exp\left[\int_0^t (g(\hat{x}(s)))'dy(s) - \frac{1}{2}\int_0^t |g(\hat{x}(s))|^2 ds\right].$$

We can write [3-5]:

$$E_t\phi(x(t)) = E_t\phi(\hat{x}(t))R(t)/E_tR(t). \tag{3.3}$$

Let $C_b(R^r)$ be the space of bounded and continuous functions on R^r. Equation (3.3) defines a measure valued process $P(\cdot)$ with values $P(t)$, the conditional distribution, by the relationship

$$(P(t), \phi) = E_t\phi(x(t)), \quad \text{all bounded, continuous } \phi(\cdot) \in C_b(R^r). \tag{3.4}$$

Define $C_0^2(R^r)$ to be the space of functions whose mixed partial derivatives up to second order are continuous, and which has compact support. Let $C_b^2(\bar{R}^r)$ denote the functions which are the sum of functions in $C_0^2(R^r)$ plus constants. Here, $\bar{R}^r$ denotes the one point compactification of R^r. For $\phi(\cdot)$ in $C_b^2(\bar{R}^r)$, the evolution equations for the conditional moments $E_t\phi(x(t))$ are defined by (3.5):

$$dE_t\phi(x(t)) = E_tA\phi(x(t))dt + [E_t\phi(x(t))g(x(t)) - E_t\phi(x(t))E_tg(x(t))] \cdot [dy - E_tg(x(t))dt]. \tag{3.5}$$

4 The Filter For The Singularly Perturbed System

We first write the filter for the singularly perturbed problem, and then the averaged filter will be formally defined. The proof of their closeness is in Section 6.

Definitions. The functions $\phi(\cdot)$ to be used in this section will be in $C_b^2(\overline{R}^r)$. Let E_t^ϵ denote the expectation conditioned on $\{y^\epsilon(s), s \leq t\}$. Let $(\hat{x}^\epsilon(\cdot), \hat{z}^\epsilon(\cdot))$ be processes with the same probability law that $(x^\epsilon(\cdot), z^\epsilon(\cdot))$ has, but which are independent of $(x^\epsilon(\cdot), z^\epsilon(\cdot), w_0(\cdot))$. Define the process $R^\epsilon(t)$ by

$$R^\epsilon(t) = \exp\left[\int_0^t g(\hat{x}^\epsilon(s), \hat{z}^\epsilon(s))'dy^\epsilon(s) - \frac{1}{2}\int_0^t |g(\hat{x}^\epsilon(s), \hat{z}^\epsilon(s))|^2 ds\right].$$

Then (3.3) implies that

$$E_t^\epsilon\phi(x^\epsilon(t)) = \frac{E_t^\epsilon\phi(\hat{x}^\epsilon(t))R^\epsilon(t)}{E_t^\epsilon R^\epsilon(t)}. \tag{4.1}$$

(4.1) defines the conditional probability process $P^\epsilon(\cdot)$ via:

$$(P^\epsilon(t), \phi) = E_t^\epsilon\phi(x^\epsilon(t)). \tag{4.2}$$

The filter for the averaged system (2.5),(2.6). Let $\hat{x}(\cdot)$ be a process with the same probability law that $x(\cdot)$ has, but which is independent of $(y(\cdot), w(\cdot), w_0(\cdot), x(\cdot))$. Define the process $R(t)$ by

$$R(t) = \exp\left[\int_0^t \bar{g}(\hat{x}(s))'dy(s) - \frac{1}{2}\int_0^t |\bar{g}(\hat{x}(s))|^2 ds\right].$$

By (3.3), we have

$$E_t\phi(x(t)) = \frac{E_t\phi(\hat{x}(t))R(t)}{E_tR(t)}. \tag{4.3}$$

The conditional moments of $x(\cdot)$ satisfy (3.4) with $\bar{A}$ replacing A. The conditional probability process $P(\cdot)$ is defined by

$$(P(t), \phi) = E_t\phi(x(t)). \tag{4.4}$$

Of course, the filter (4.3) cannot be used directly in applications since $y(\cdot)$ is not a physically available observation; $y^\epsilon(\cdot)$ is the only available data, and must be the input to any filter.

The averaged filter. Define $\overline{R}^\epsilon(\cdot)$ by

$$\overline{R}^\epsilon(t) = \exp\left[\int_0^t \overline{g}(\hat{x}(s))' dy^\epsilon(s) - \frac{1}{2}\int_0^t |\overline{g}(\hat{x}(s))|^2 ds\right].$$

Then the averaged filter (the filter for (2.5) and (2.6), but with input $y^\epsilon(\cdot)$) can be represented as

$$\overline{E}_t^\epsilon\phi(x^\epsilon(t)) = \frac{E_t^\epsilon\phi(\hat{x}(t))\overline{R}^\epsilon(t)}{E_t^\epsilon\overline{R}^\epsilon(t)}. \tag{4.5}$$

Equation (4.5) defines a probability-measure valued process $\overline{P}^\epsilon(\cdot)$:

$$(\overline{P}^\epsilon(t), \phi) = \overline{E}_t^\epsilon\phi(x^\epsilon(t)). \tag{4.6}$$

The expression (4.5) does represent the averaged filter, since it depends on ϵ only via $y^\epsilon(\cdot)$ which replaces $y(\cdot)$ in (4.3). The moment equations which are obtained from (4.5) satisfy

$$\begin{aligned} d\overline{E}_t^\epsilon\phi(x(t)) &= \overline{E}_t^\epsilon(A^\epsilon\phi)(x^\epsilon(t), z^\epsilon(t))dt \\ &+ [\overline{E}_t^\epsilon\phi(x^\epsilon(t))\overline{g}(x^\epsilon(t)) - \overline{E}_t^\epsilon\phi(x^\epsilon(t))\overline{E}_t^\epsilon\overline{g}(x^\epsilon(t))] \\ &\times (dy^\epsilon - \overline{E}_t^\epsilon\overline{g}(x^\epsilon(t))dt). \end{aligned} \tag{4.7}$$

The integral of the term in the last parentheses on the right is *not* generally an innovations process. We note that the moment equation

(4.7) is derived in the same way that (3.4) is, by using of Itô's Formula to "expand" the expression (4.5). The proof in either case does not require that the ratio in in (3.3) is a conditional expectation, but merely that it is a ratio of conditional expectations.

When doing calculations with E_t^ϵ acting on functions of $(\hat{x}^\epsilon(s), \hat{z}^\epsilon(s), \hat{x}(s), \hat{z}(s), y^\epsilon(\cdot))$ we can view it as an operator which averages out all the functions except the $y^\epsilon(\cdot)$, which is treated as a parameter when taking the expectation. The random variables $\overline{E}_t^\epsilon \phi(x^\epsilon(t))$ (resp. $\overline{P}^\epsilon(\cdot)$) are not usually conditional expectation (resp., conditional probabilities), although they "approximate" them (Section 6). It is close to the conditional expectations $E_t^\epsilon \phi(x^\epsilon(t))$ (resp., conditional probability $P^\epsilon(t)$) under appropriate conditions.

5 A Counterexample to the Averaged Filter

The convergence in (A2.4) seems to be essential to the proof of Theorem 2 in Section 6. Uniqueness of the invariant measure is not sufficient, although it is sufficient for the weak convergence result of Theorem 1. The filtering problem is rather subtle, since the "information" available even from functions which converge weakly to zero might not go to zero. We will give a simple example which shows the sort of problems which arise in the filtering problem, but not in the weak convergence problem of Theorem 1, and which seem to necessitate conditions such as (A2.4).

Define the process $z^\epsilon(\cdot) = (z^{1,\epsilon}(\cdot), z^{2,\epsilon}(\cdot))$ by

$$\epsilon \ddot{z}^{1,\epsilon} = -z^{1,\epsilon}$$

$$z^{2,\epsilon} = \dot{z}^{1,\epsilon}.$$

Choose the initial condition $z^\epsilon(0) = z(0)$ such that $z^{1,\epsilon}(t) = \sin(t/\sqrt{\epsilon}+\theta)$, where θ is a random variable which equals either 0 or π, each with probability 1/2. Define the stretched out system $z^\epsilon(\cdot)$ by

$$z_0^1(t) = \sin(t + \theta)$$

$$z_0^2(t) = \dot{z}_0^1(t).$$

The $z-$process just defined depends on the parameter θ, but we omit it from the notation for simplicity. In this example, the constant process $\theta(t) = \theta$ is the slow system. The $z-$process does not have a unique invariant measure. But if $|z^\epsilon(0)| = 1$, then there is a unique invariant measure $\mu(dz)$, and that measure is a uniform distribution on the unit circle. Condition (A2.4) does not hold, since the transition probability is periodic, and does not converge to the invariant measure. We next define the filtering problem.

Let the observational process be defined by $dy^\epsilon = g(z^\epsilon)dt + dw$, where $g(z) = \operatorname{sign} z^1$, and $w(\cdot)$ is a standard Wiener process and $(\theta, w(\cdot), z^\epsilon(0))$ are mutually independent and $|z^\epsilon(0)| = 1$. The same results are obtainable if $g(\cdot)$ is a smooth function, but the calculations are not so trivial. Since $\int g(z)\mu(dz) = 0$, the averaged system which corresponds to (2.5), (2.6) is

$$P(\theta = 0) = P(\theta = \pi) = \frac{1}{2}, \quad dy = dw_0. \tag{5.1}$$

It will next be shown that the true conditional probability $P(\theta = \pi \,|\, y^\epsilon(s),\, s \leq t) = P^\epsilon(t)$ converges to that given by the filter for the problem of estimating θ with the initial data

$$P(\theta = 0) = P(\theta = \pi) = \frac{1}{2} \tag{5.2}$$

and observations

$$dy = [I_{\{\theta=0\}} - I_{\{\theta=\pi\}}]dt + dw_0 \tag{5.3}$$

where $w_0(\cdot)$ is a standard Wiener process which is independent of θ. This is clearly different from the averaged filter. From an intuitive point of view, the "information" concerning the value of θ which is contained in the observations $y^\epsilon(\cdot)$ does not decrease to zero as $\epsilon \to 0$. If (A2.4) held, then the "information" would go to zero as $\epsilon \to 0$.

The triple $(P^\epsilon(\cdot), y^\epsilon(\cdot), \theta)$ does not converge weakly to the triple associated with (5.2), (5.3). But the $\{P^\epsilon(\cdot), \theta, \}$ does converge weakly to the pair which is given by (5.2), (5.3), for the observational data (5.3).

Proof of the Claim. To get the representation of the filter use (4.3). Let $\hat{\theta}$ have the same probability law that θ has, but be independent of $(\theta, w(\cdot))$, and let $\hat{z}^\epsilon(\cdot)$ denote the associated "fast" process.

Define the function $F(\hat{\theta}) = I_{\{\hat{\theta}=\pi\}}$. Then by (4.3), we can write the conditional expectation (actually a conditional probability here) as

$$P(\theta = \pi \mid y^\epsilon(s), s \le t)$$

$$= \frac{\int P(d\hat{\theta})F(\hat{\theta}) \exp\left[\int_0^t g(\hat{z}^\epsilon(s))dy^\epsilon(s) - \frac{1}{2}\int_0^t |g(\hat{z}^\epsilon(s))|^2 ds\right]}{\int P(d\hat{\theta}) \exp\left[\int_0^t g(\hat{z}^\epsilon(s))dy^\epsilon(s) - \frac{1}{2}\int_0^t |g(\hat{z}^\epsilon(s))|^2 ds\right]}. \quad (5.4)$$

The limit of the filters (5.4) will be obtained directly, via taking weakly convergent subsequences of the the terms as $\epsilon \to 0$. It is easy to see that

$$\lim_\epsilon \int_0^t g(\hat{z}^\epsilon(s))g(z^\epsilon(s))ds = \pm t,$$

according to whether

$$\{\theta = 0, \hat{\theta} = 0\}, \quad \{\theta = \pi, \hat{\theta} = \pi\} \quad (\text{ the limit is } +t) \quad (5.5a)$$

or

$$\{\theta = 0, \hat{\theta} = \pi\}, \quad \{\theta = \pi, \hat{\theta} = 0\} \quad (\text{the limit is } -t). \quad (5.5b)$$

Also, we have the identity

$$\int_0^t g^2(\hat{z}^\epsilon(s))ds = t.$$

Define the martingales

$$M_0^\epsilon(t) = \int_0^t g(\sin(s/\sqrt{\epsilon}))dw(s),$$

$$M^\epsilon(t) = \int_0^t g(\hat{z}^\epsilon(s))dw(s) = I_{\{\hat{\theta}=0\}}M_0^\epsilon(t) - I_{\{\hat{\theta}=\pi\}}M_0^\epsilon(t).$$

The sequence $\{M^\epsilon(\cdot),\ M_0^\epsilon(\cdot),\ \theta, \hat{\theta}\}$ is obviously tight since the integrands in the stochastic integrals are bounded . Let ϵ_n index a weakly convergent subsequence with limit denoted by $(M(\cdot), w_0(\cdot), \theta, \hat{\theta})$. Then we have the relationship

$$M(t) = I_{\{\theta=0\}} w_0(t) - I_{\{\hat{\theta}=\pi\}} w_0(t).$$

The process $w_0(\cdot)$ is a standard Wiener process, since it is a continuous martingale with quadratic variarion process t. It is independent of $(\theta, \hat{\theta})$. The asserted independence is a consequence of the mutual independence of $(M_0^\epsilon(\cdot), \theta, \hat{\theta})$ and the fact that independence is preserved under weak convergence.

Define the function $\overline{g}(\cdot)$ by $\overline{g}(0) = 1$, $\overline{g}(\pi) = -1$. Then the weak limit of the process with values $\int_0^t g(\hat{z}^\epsilon(s))dy^\epsilon(s)$ is the process with values

$$I_{\{\hat{\theta}=0\}}[t\overline{g}(\theta) + w_0(t)] - I_{\{\hat{\theta}=\pi\}}[t\overline{g}(\theta) + w_0(t)].$$

Now, we can combine all of the above limits and get that the limit of (5.4) (as $\epsilon_n \to 0$) equals

$$\frac{\int P(d\hat{\theta})F(\hat{\theta}) \exp[\overline{g}(\hat{\theta})(t\overline{g}(\hat{\theta}) + w_0(t)) - t/2]}{\int P(d\hat{\theta}) \exp[\overline{g}(\hat{\theta})(t\overline{g}(\hat{\theta}) + w_0(t)) - t/2]}. \tag{5.6}$$

Since (5.6) equals $P(\theta = \pi \mid s\overline{g}(\theta) + w_0(s),\ s \leq t)$, the assertion is proved.

6 The Almost Optimality of the Averaged Filter

Theorem 2 proves that the averaged filter defined by (4.5) or (4.7) is often a good approximation to the optimal filter for the system defined by (1.1)-(1.3). The theorem states that if $\bar{P}^\epsilon(t)$ is used as an "Erzatz" conditional probability, then the computed "conditional moments" of bounded and continuous functions of $x^\epsilon(t)$ are close to the exact conditional moments for small ϵ.

Remarks on the Proof of Theorem 2. In Section 5, we saw one aspect of the the subtlety of the convergence problem for the filtering problem. Now, we discuss why a direct weak convergence and averaging approach might not work very often. The main difficulty in the proof of Theorem 2 is caused by the fact that the simple averaging out of the $\hat{z}^\epsilon(\cdot)$ in the exponent of the $R^\epsilon(t)$ can yield the incorrect result. This is partly because the conditional expectations, given the $y^\epsilon(\cdot)$, do not necessarily converge to conditional expectations of the limit processes, given the limit $y(\cdot)$ process.

Let $x(t)$ and $y(t)$ be scalar valued. Define the "error" $\tilde{g}(x,z) = g(x,z) - \overline{g}(x)$ and define the process $\tilde{M}^\epsilon(\cdot)$ by

$$\tilde{M}^\epsilon(t) = \int_0^t \tilde{g}(\hat{x}^\epsilon(s), \hat{z}^\epsilon(s))dw_0(s).$$

The term which gives us the most trouble is

$$\exp\left[\tilde{M}^\epsilon(t) - \frac{1}{2}\int_0^t \tilde{g}^2(\hat{x}^\epsilon(s), \hat{z}^\epsilon(s))ds\right].$$

Note that $\tilde{M}^\epsilon(\cdot)$ is a martingale with quadratic variation $\int_0^t |\tilde{g}(\hat{x}^\epsilon(s), \hat{z}^\epsilon(s))|^2 ds$ and the set $\{\tilde{M}^\epsilon(\cdot)\}$ of processes is tight.

Suppose that $\{\tilde{M}^\epsilon(\cdot), x^\epsilon(\cdot), \hat{x}^\epsilon(\cdot), y^\epsilon(\cdot)\}$ is tight. For notational convenience, we let ϵ index a weakly convergent subsequence of $\{\tilde{M}^\epsilon(\cdot), x^\epsilon(\cdot), \hat{x}^\epsilon(\cdot), y^\epsilon(\cdot)\}$ and denote the limit by $(\tilde{M}(\cdot), x(\cdot), \hat{x}(\cdot), y(\cdot))$. Then $\tilde{M}(\cdot)$ is a martingale with quadratic variation $\int_0^t [g_0(\hat{x}(s))]^2 ds$, where

$$[g_0(x)]^2 = \int [\tilde{g}(x,z)]^2 \mu_x(dz).$$

It is well known that a continuous martingale whose quadratic variation is absolutely continuous can be represented as a stochastic integral. Thus, there is a standard Wiener process $\tilde{w}_0(\cdot)$ with respect to which $\hat{x}(\cdot)$ is non-anticipative and for which $\tilde{M}(\cdot)$ has the representation

$$\tilde{M}(t) = \int_0^t g_0(\hat{x}(s))d\tilde{w}_0(s).$$

Using an averaging result of the type stated in Theorem 1, it can be shown that $R^\epsilon(t)$ converges weakly to

$$\exp\int_0^t \left[\overline{g}(\hat{x}(s))dy - \frac{1}{2}\int_0^t |\overline{g}(\hat{x}(s))|^2 ds\right]$$

$$\times \exp\left[\tilde{M}(t) - \frac{1}{2}\int_0^t |g_0(\hat{x}(s))|^2 ds\right],$$

which is obviously not equal to $R(t)$. This implies that some method other than simply using a direct averaging of $R^\epsilon(t)$ and $\phi(x^\epsilon(t))$ and a direct weak convergence technique needs to be used to prove the approximation result. In fact, it will be necesary to exploit the way that the conditional expectation operator E_t^ϵ appears in (4.1) and (4.5).

Remark on weak convergence. Let $\mathcal{M}(\bar{R}^r)$ denote the space of measures with total finite mass on the (one point) compactified R^r, and with the Prohorov topology. On $\mathcal{M}(\overline{R}^r)$, the weak topology is equivalent to the Prohorov topology, and it is a complete and separable metric space under this topology. In the proof of Theorem 2, we introduce random variables $Z^\epsilon(t), \bar{Z}^\epsilon(t)$, with values in $\mathcal{M}(\overline{R}^r)$. The compactified space is used in order to avoid some details concerning the possibility of accumulation of mass at the point at infinity. This actually does not occur.

Define the space $D[\mathcal{M}(\bar{R}^r); 0, \infty)$, the space of $\mathcal{M}(\bar{R}^r)$−valued random processes whose paths are right continuous and have left hand limits. The following lemma (Theorem 1.6.2 in [1]) will be needed:

Lemma 1 *Let $\{Z_\alpha(\cdot)\}$ be a set of random processes with values in $D[\mathcal{M}(\overline{R}^r);\ 0, \infty)$. Suppose that $\{(Z_\alpha(\cdot), \phi)\}$ is tight in $D[0, \infty)$ for each $\phi(\cdot) \in C(\overline{R}^r)$. Then $\{Z_\alpha(\cdot)\}$ has a weakly convergent subsequence.*

The set $\{Z_\alpha(\cdot)\}$ is tight if for each $T < \infty$, $\sup_{\alpha, t \le T} EZ_\alpha(t, \overline{R}^r) < \infty$ holds, and for each $T < \infty$ and $\phi(\cdot) \in C(\overline{R}^r)$ we have

$$\lim_{\delta \to 0} \sup_{\alpha} \sup_{\tau \le T} E|(Z_\alpha(\tau + \delta) - Z_\alpha(\tau), \phi)| = 0. \tag{6.1}$$

In (6.1), the τ are stopping times.

Let $\Rightarrow$ denote weak convergence.

Theorem 2. (a) *Assume (A2.1)-(A2.5). Then the error process $\rho^\epsilon(\cdot)$ defined by*

$$\overline{E}_t^\epsilon \phi(x^\epsilon(t)) - E_t^\epsilon \phi(x^\epsilon(t)) = \rho^\epsilon(t),\ t \le T, \tag{6.2}$$

converges weakly to the "zero" process uniformly in $(x, z, P(0))$ in any compact set.

(b) *If (A2.4) is replaced by the existence and uniqueness of the invariant measure of the fixed x−process for each value of x, then we still have*

$$(\overline{E}_t^\epsilon \phi(x^\epsilon(t)), x^\epsilon(\cdot), y^\epsilon(\cdot)) \Rightarrow (E_t\phi(x(t)), x(\cdot), y(\cdot)), \tag{6.3}$$

uniformly in $(x, z, P(0))$ in any compact set.

Outline of Proof. Only Part (a) will be proved. It will be shown that the difference between the numerators of (4.1) and (4.5) goes to zero as $\epsilon \to 0$. Since the denomenators equal the numerators with $\phi(x) \equiv 1$, the differences between the respective denomenators will also go to zero. Stochastic differential equations for the numerators are derived and represented in the form of evolution equations by using an interchange of integration and conditional expectation. The solutions to these stochastic differential eqautions are represented in terms of processes $Z^\epsilon(\cdot)$ and $\overline{Z}^\epsilon(\cdot)$, the first process being an unnormalized conditional density for the filter for (1.1)-(1.3), and the second being an unnormalized density of the "approximating" measure given by the averaged filter. The $z^\epsilon(\cdot)$ in the coefficients in the stochastic differential equations are then "averaged" by use of (A2.4). It is then shown that the sequences $\{Z^\epsilon(\cdot)\}$, $\{\overline{Z}^\epsilon(\cdot)\}$ with paths in $D[\mathcal{M}(\bar{R}^r); 0, \infty)$, are tight, and that the limits satisfy the same stochastic differential equation. Since this equation has a unique solution, the limits are equal, and the desired result is obtained

Proof. In order to avoid unnecessary notation, let x, z and y be real valued and drop the jump components. The proof of the general case follows the same lines. The set $\{x^\epsilon(\cdot), y^\epsilon(\cdot), \hat{x}^\epsilon(\cdot)\}$ is tight. For notational simplicity, let the original sequence converge weakly and with limit denoted by $\{x(\cdot), y(\cdot), \hat{x}(\cdot)\}$. Part (a) follows if

$$\frac{E_t^\epsilon \phi(\hat{x}^\epsilon(t)) R^\epsilon(t)}{E_t^\epsilon R^\epsilon(t)} - \frac{E_t^\epsilon \phi(\hat{x}(t)) \overline{R}^\epsilon(t)}{E_t^\epsilon \overline{R}^\epsilon(t)} \Rightarrow \text{zero process} \tag{6.4}$$

for each continuous and bounded function $\phi(\cdot)$. The numerators in (6.4) are strictly positive with probability one, uniformly in ϵ, in the sense that

$$\lim_{\delta\to 0}\inf_{\epsilon>0} P\left(\inf_{t\leq T} E_t^\epsilon R^\epsilon(t) > \delta\right) = 1,$$

$$\lim_{\delta\to 0}\inf_{\epsilon>0} P\left(\inf_{t\leq T} E_t^\epsilon \overline{R}^\epsilon(t) > \delta\right) = 1.$$

Thus, it is only necessary to prove that

$$[E_t^\epsilon \phi(\hat{x}^\epsilon(t))R^\epsilon(t) - E_t^\epsilon \phi(\hat{x}(t))\overline{R}^\epsilon(t)] \Rightarrow \text{zero process.} \tag{6.5}$$

Since the $x^\epsilon(\cdot)$ are uniformly (in ϵ) bounded in probability on each bounded time interval, and the $\{\bar{R}^\epsilon(t), R^\epsilon(t), \epsilon > 0,\}$ are uniformly integrable, we need only show (6.5) for $\phi(\cdot) \in C_0(R^r)$, since

$$\lim_N \sup_\epsilon E\, E_t^\epsilon\, R^\epsilon(t)\, I_{\{|\hat{x}^\epsilon(t)|\geq N\}} = 0,$$

$$\lim_N \sup_\epsilon E\, E_t^\epsilon\, \overline{R}^\epsilon(t)\, I_{\{|\hat{x}^\epsilon(t)|\geq N\}} = 0.$$

Henceforth, let $\phi(\cdot) \in C_0^2(R^r)$.

Define

$$\tilde{R}^\epsilon(t) = \exp\left[\int_0^t \overline{g}(\hat{x}^\epsilon(s))dy^\epsilon(s) - \frac{1}{2}\int_0^t |\overline{g}(\hat{x}^\epsilon(s))|^2 ds\right].$$

Next, we want to average out the $z^\epsilon(\cdot)$ in

$$\int_0^t \overline{g}(\hat{x}^\epsilon(s))g(x^\epsilon(s), z^\epsilon(s)).$$

But the tightness of $\{x^\epsilon(\cdot), \hat{x}^\epsilon(\cdot)\}$ and the mutual independence of the "hatted" and the "unhatted" processes implies that we can average as though the $\{x^\epsilon(\cdot), \hat{x}^\epsilon(\cdot)\}$ were constants. Now applying Lemma 1, we have that

$$\int_0^t \overline{g}(\hat{x}^\epsilon(s))[g(x^\epsilon(s), z^\epsilon(s)) - \overline{g}(x^\epsilon(s))]ds \Rightarrow \text{zero process,}$$

$$\int_0^t \overline{g}(\hat{x}(s))[g(x^\epsilon(s), z^\epsilon(s)) - \overline{g}(x^\epsilon(s))]ds \Rightarrow \text{zero process.}$$

This result and the weak convergence $\hat{x}^\epsilon(\cdot) \Rightarrow \hat{x}(\cdot)$, implies that

$$\sup_{t \le T} E|E_t^\epsilon \phi(\hat{x}(t))\overline{R}^\epsilon(t) - E_t^\epsilon \phi(\hat{x}^\epsilon(t))\tilde{R}^\epsilon(t)| \xrightarrow{\epsilon} 0.$$

Consequently, we need only show that

$$E_t^\epsilon \phi(\hat{x}^\epsilon(t))\tilde{R}^\epsilon(t) - E_t^\epsilon \phi(\hat{x}^\epsilon(t))R^\epsilon(t) \stackrel{\epsilon}{\Rightarrow} \text{zero process.} \tag{6.6}$$

To facilitate this, we write the terms as solutions to stochastic differential equations. We can represent the process $\hat{x}^\epsilon(\cdot)$ as the solution to the stochastic differential equation $d\hat{x}^\epsilon = G(\hat{x}^\epsilon, \hat{z}^\epsilon)dt + \sigma(\hat{x}^\epsilon, \hat{z}^\epsilon)d\hat{w}_1^\epsilon$, where $\hat{w}_1^\epsilon(\cdot)$ is a standard vector-valued Wiener process which is independent of $(y^\epsilon(\cdot), x^\epsilon(\cdot), z^\epsilon(\cdot))$. By Itô's Formula and the mutual independence of $\hat{x}^\epsilon(\cdot)$ and $y^\epsilon(\cdot)$ (which implies that the second order differentials $d\hat{x}^\epsilon(t)dy^\epsilon(t)$ are zero), we can write

$$\begin{aligned} d(R^\epsilon(t)\phi(\hat{x}^\epsilon(t)) &= R^\epsilon(t)d\phi(\hat{x}^\epsilon(t)) + \phi(\hat{x}^\epsilon(t))dR^\epsilon(t) \\ dR^\epsilon(t) &= R^\epsilon(t)[\overline{g}(\hat{x}^\epsilon(t)) + \tilde{g}(\hat{x}^\epsilon(t), \hat{z}^\epsilon(t))]dy^\epsilon(t) \\ d\phi(\hat{x}^\epsilon(t)) &= (A^\epsilon\phi)(\hat{x}^\epsilon(t), \hat{z}^\epsilon(t))dt + \phi_x(\hat{x}^\epsilon(t))\sigma(\hat{x}^\epsilon(t), \hat{z}^\epsilon(t))d\hat{w}_1^\epsilon, \end{aligned} \tag{6.7}$$

where $\tilde{g}(x, z) = g(x, z) - \overline{g}(x)$ Also,

$$\begin{aligned} d\phi(\hat{x}^\epsilon(t))\tilde{R}^\epsilon(t) &= \tilde{R}^\epsilon(t)d\phi(\hat{x}^\epsilon(t)) + \phi(\hat{x}^\epsilon(t))d\tilde{R}^\epsilon(t), \\ d\tilde{R}^\epsilon(t) &= \tilde{R}^\epsilon(t)\overline{g}(\hat{x}^\epsilon(t))dy^\epsilon(t). \end{aligned}$$

Integrating the differential in (6.7) and taking conditional expectations of the result yields the expession

$$\begin{aligned} E_t^\epsilon R^\epsilon(t)\phi(\hat{x}^\epsilon(t)) = E_t^\epsilon \phi(\hat{x}^\epsilon(0)) &+ E_t^\epsilon \int_0^t R^\epsilon(s)(A^\epsilon\phi)(\hat{x}^\epsilon(s), \hat{z}^\epsilon(s))ds \\ &+ E_t^\epsilon \int_0^t R^\epsilon(s)\phi_x(\hat{x}^\epsilon(s))\sigma(\hat{x}^\epsilon(s), \hat{z}^\epsilon(s))d\hat{w}_1^\epsilon(s) \\ &+ E_t^\epsilon \int_0^t \phi(\hat{x}^\epsilon(s))R^\epsilon(s)[\overline{g}(\hat{x}^\epsilon(s)) + \tilde{g}(\hat{x}^\epsilon(s), \hat{z}^\epsilon(s))]dy^\epsilon(s). \end{aligned} \tag{6.8}$$

The next to last term on the right hand side of (6.8) is zero due to the independence of $y^\epsilon(\cdot)$ and $(\hat{x}^\epsilon(\cdot), \hat{z}^\epsilon(\cdot), \hat{w}_1^\epsilon(\cdot))$. Again using the mutual independence of the "hatted" and "unhatted" processes, (6.8) equals (with probability one)

$$E_t^\epsilon R^\epsilon(t)\phi(\hat{x}^\epsilon(t)) = E_0^\epsilon\phi(\hat{x}^\epsilon(0)) + \int_0^t [E_s^\epsilon R^\epsilon(s)(A^\epsilon\phi)(\hat{x}^\epsilon(s), \hat{z}^\epsilon(s))]ds$$
$$+ \int_0^t [E_s^\epsilon\phi(\hat{x}^\epsilon(s))R^\epsilon(s)\overline{g}(\hat{x}^\epsilon(s))]dy^\epsilon(s) \tag{6.9}$$
$$+ \int_0^t [E_s^\epsilon\phi(\hat{x}^\epsilon(s))R^\epsilon(s)\tilde{g}(\hat{x}^\epsilon(s), \hat{z}^\epsilon(s))]dy^\epsilon(s).$$

For use below, write the last term on the right of (6.9) as $\int_0^t L^\epsilon(s)dy^s$, where

$$L^\epsilon(s) = E_s^\epsilon\phi(\hat{x}^\epsilon(s))R^\epsilon(s)\tilde{g}(\hat{x}^\epsilon(s), \hat{z}^\epsilon(s)).$$

An analogous derivation yields the expression

$$E_t^\epsilon \tilde{R}^\epsilon(t)\phi(\hat{x}^\epsilon(t)) = E_0^\epsilon\phi(\hat{x}^\epsilon(0)) + \int_0^t [E_s^\epsilon \tilde{R}^\epsilon(s)(A^\epsilon\phi)(\hat{x}^\epsilon(s), \hat{z}^\epsilon(s))]ds$$
$$+ \int_0^t [E_s^\epsilon\phi(\hat{x}^\epsilon(s))\tilde{R}^\epsilon(s)\overline{g}(\hat{x}^\epsilon(s))]dy^\epsilon(s). \tag{6.10}$$

In order to simplify the notation used in the averaging of the second term on the right side of (6.9), define the function $p(\cdot)$ by

$$(A^\epsilon\phi)(x,z) = \phi_x(x)G(x,z) + \frac{1}{2}\phi_{xx}(x)\sigma^2(x,z) \equiv p(x,z).$$

Define the averaged function $\overline{p}(x) = \int p(x,z)\mu_x(dz)$.

The sequence of processes $\{R^\epsilon(\cdot)\}$ is tight. Thus, for small ϵ they can be approximated arbitrarily closely (in the mean) by the piecewise constant process with values $R^\epsilon(i\Delta)$ on $[i\Delta, i\Delta + \Delta)$ for small Δ. This and Lemma 1 yield that

$$\int_0^t E_s^\epsilon R^\epsilon(s)[p(\hat{x}^\epsilon(s), \hat{z}^\epsilon(s)) - \overline{p}(\hat{x}^\epsilon(s))]ds \overset{\epsilon}{\Rightarrow} \text{zero process.} \tag{6.11}$$

The limit (6.11) also holds if $\tilde{R}^\epsilon(\cdot)$ replaces $R^\epsilon(\cdot)$. By these calculations, the second terms on the right side of (6.9) and (6.10), resp., are asymptotically equivalent to, resp.,

$$\int_0^t E_s^\epsilon R^\epsilon(s)\overline{A}\phi(\hat{x}^\epsilon(s))ds, \quad \int_0^t E_s^\epsilon \tilde{R}^\epsilon(s)\overline{A}\phi(\hat{x}^\epsilon(s))ds.$$

We next show that the integrand $L^\epsilon(s)$ goes to zero as $\epsilon \to 0$, for almost all ω, s. This will be done by a sequence of approximations, as follows: Fix $\Delta > 0$. Then replace the terms $R^\epsilon(s)$ and $\hat{x}^\epsilon(s)$, resp., in $L^\epsilon(s)$ by the values $R^\epsilon(i\Delta)$ and $\hat{x}^\epsilon(i\Delta)$, resp., on each interval $[i\Delta, i\Delta+\Delta)$. The expression which results from these approximations is (6.12) below. After that, we show that the limit as $\epsilon \to 0$ of (6.12) is zero w.p.1. Finally, we let $\Delta \to 0$ to get the proof of the assertion.

Let $i \geq 0$, $s \in (i\Delta, i\Delta + \Delta)$ and let k be a positive real number. For $s - \epsilon k \geq 0$, we can write

$$E_s^\epsilon \phi(\hat{x}^\epsilon(i\Delta)) R^\epsilon(i\Delta) \tilde{g}(\hat{x}^\epsilon(i\Delta), \hat{z}^\epsilon(s)) \tag{6.12}$$

$$= E_s^\epsilon \{\phi(\hat{x}^\epsilon(i\Delta)) R^\epsilon(i\Delta) E_s^\epsilon[\tilde{g}(\hat{x}^\epsilon(i\Delta), \hat{z}^\epsilon(s)) \mid \hat{x}^\epsilon(u), \hat{z}^\epsilon(u), u \leq s - \epsilon k]\}.$$

Since $\tilde{g}(\hat{x}^\epsilon(i\Delta),\ \hat{z}^\epsilon(s))$ is independent of $y^\epsilon(\cdot)$, the conditioning on $\{y^\epsilon(u), u \leq s\}$ which is implicit in the definition of E_s^ϵ is irrelevant in the inner conditional expectation.

Let $\hat{z}_0^\epsilon(s - \epsilon k + \epsilon\, \cdot \mid \hat{x}^\epsilon(s - \epsilon k))$ denote the fixed-x process with parameter $x = \hat{x}^\epsilon(s - \epsilon k)$ and initial condition $\hat{z}^\epsilon(s - \epsilon k)$ at "starting" time $s - \epsilon k$. By the tightness of $\{\hat{x}^\epsilon(s),\ \hat{z}^\epsilon(s),\ s \leq T,\ \epsilon > 0\}$, and the linear bounds on the growth of the functions $G(\cdot)$ $H(\cdot)$, $\sigma(\cdot)$, $v(\cdot)$, the pair of sequences

$$\{\hat{z}^\epsilon(s - \epsilon k + \epsilon \cdot), \hat{z}_0^\epsilon(s - \epsilon k + \epsilon \cdot \mid \hat{x}^\epsilon(s - \epsilon k)), \epsilon > 0\} \tag{6.13}$$

is tight, and the difference between the two sequences in (6.13) converges weakly to the zero process as $\epsilon \to 0$. This latter fact implies that

$$E\Big| E[\tilde{g}(\hat{x}^\epsilon(i\Delta), \hat{z}^\epsilon(s)) \mid \hat{x}^\epsilon(u), \hat{z}^\epsilon(u), u \leq s - \epsilon k]$$

$$- \int \tilde{g}(\hat{x}^\epsilon(i\Delta), z) P_0(\hat{z}^\epsilon(s - \epsilon k), k, dz \mid \hat{x}^\epsilon(s - \epsilon k))\Big| \xrightarrow{\epsilon} 0, \tag{6.14}$$

where $P_0(\cdot \mid x)$ is the transition function for the fixed-x $z_0(\cdot \mid x)$ process. Now let $\epsilon \to 0$ and then $k \to \infty$ and use (6.14) and (A2.4) to get that

$$E\left|\int \tilde{g}(\hat{x}^\epsilon(i\Delta), z)[P_0(\hat{z}^\epsilon(s - \epsilon k), k, dz \mid \hat{x}^\epsilon(s - \epsilon k)) - \mu_{\hat{x}^\epsilon(s)}(dz)]\right| \to 0. \tag{6.15}$$

By (6.14) and (6.15), the limits in the mean (as $\epsilon \to 0$ and then $\Delta \to 0$) of the left side of (6.12) are the same as those of (as $\epsilon \to 0$ and then $\Delta \to 0$)

$$E_s^\epsilon \phi(\hat{x}^\epsilon(i\Delta))R^\epsilon(i\Delta) \int \tilde{g}(\hat{x}^\epsilon(i\Delta), z)\mu_{\hat{x}^\epsilon(s)}(dz). \tag{6.16}$$

Since $|\hat{x}^\epsilon(i\Delta) - \hat{x}^\epsilon(s)| \xrightarrow{P} 0$ (uniformly in ϵ) as $s - i\Delta \to 0$ and $\int \tilde{g}(x,z)\mu_x(dz) = 0$, we can conclude that (6.16) goes to zero in probability (uniformly in ϵ) as $s - i\Delta \to 0$. These results imply that $L^\epsilon(\cdot)$ goes to zero in (ω, s)-measure, as $\epsilon \to 0$. Using this last result and the fact that

$$\lim_N \sup_\epsilon E \int_0^t |L^\epsilon(s)|^2 I_{\{|L^\epsilon(s)| \geq N\}} ds = 0$$

implies the desired result

$$\int_0^t L^\epsilon(s) dy^\epsilon(s) \equiv \hat{p}^\epsilon(t) \Rightarrow \text{zero process.}$$

Define the measure valued processes $Z^\epsilon(\cdot)$ and $\tilde{Z}^\epsilon(\cdot)$, as follows (where $\phi(\cdot) \in C_b^2(\overline{R}^r)$):

$$(Z^\epsilon(t), \phi) = E_t^\epsilon R^\epsilon(t)\phi(\hat{x}^\epsilon(t)), \; (\tilde{Z}^\epsilon(t), \phi) = E_t^\epsilon \tilde{R}^\epsilon(t)\phi(\hat{x}^\epsilon(t)).$$

The $Z^\epsilon(t)$ and $\tilde{Z}^\epsilon(t)$ take values in $\mathcal{M}(\overline{R}^r)$.

The representations (6.9) and (6.10) imply that the processes $(Z^\epsilon(\cdot), \phi)$ and $(\tilde{Z}^\epsilon(\cdot), \phi)$ have continuous paths, w.p.1. Thus, the $Z^\epsilon(\cdot)$ have their paths in $D[\mathcal{M}(\overline{R}^r); 0, \infty)$. We now prove tightness of $\{Z^\epsilon(\cdot), \tilde{Z}^\epsilon(\cdot)\}$ by use of the criterion of Lemma 1. Given $T < \infty$, let τ be an arbitrary stopping time less than or equal to T. Then the representation (6.9) implies that for some $K < \infty$,

$$\lim_\delta \limsup_\epsilon \sup_\tau E|(Z^\epsilon(\tau + \delta) - Z^\epsilon(\tau), \phi)| = 0$$

$$\sup_{\epsilon, t \leq T} E|(Z^\epsilon(t), \phi)| \leq K \sup_x |\phi(x)|.$$

This and Lemma 1 imply that $\{Z^\epsilon(\cdot)\}$ is tight in $D[\mathcal{M}(\overline{R}^r); 0, T]$. The analogous result holds for $\{\tilde{Z}^\epsilon(\cdot)\}$.

Now to simplify the notation, let ϵ index a weakly convergent subsequence of $\{Z^\epsilon(\cdot),\ \tilde{Z}^\epsilon(\cdot),\ y^\epsilon(\cdot),\ \hat{x}^\epsilon(\cdot)\}$, and denote the limit by $(Z(\cdot), \tilde{Z}(\cdot), y(\cdot), \hat{x}(\cdot))$. Then using (6.9) to (6.11) and the approximation results proved above yields that $Z(\cdot)$ and $\tilde{Z}(\cdot)$ are both measure valued processes which satisfy the moment equation

$$(U(t),\phi) = (U(0),\phi) + \int_0^t (U(s),\overline{A}\phi)ds + \int_0^t (U(s),\overline{g}\phi)dy(s). \quad (6.17)$$

The initial condition in (6.17) is $(U(0),\phi) = (Z(0),\phi) = (\tilde{Z}(0),\phi) = E_0^\epsilon \phi(x^\epsilon(0))$. By the uniqueness result in [6, Theorem IV-1, V-1] any two solutions of (6.17) with the same initial condition and satisfying the bound

$$\sup_{t\leq T} E|(U(t),\phi)|^2 \leq K \sup_x |\phi(x)|^2, \quad (6.18)$$

for some constant $K < \infty$, must be the same w.p.1. The required inequality (6.18) follows from the weak convergence, Fatou's Lemma, and the fact that there is a $K_1 < \infty$ such that

$$\sup_{\epsilon,t\leq T} E|(\tilde{Z}^\epsilon(t),\phi)|^2 + \sup_{\epsilon,t\leq T} E|(Z^\epsilon(t),\phi)|^2 \leq K_1 \sup_x |\phi(x)|^2.$$

The proof of Part (a) is now complete. The proof of Part (b) of the theorem is easier, since we can directly average the $g(x^\epsilon(\cdot), z^\epsilon(\cdot))$ in the exponent in (4.5). Q.E.D.

Bibliography

[1] H.J. Kushner, *Weak Convergence Methods and Singularly Perturbed Stochastic Contol and Filtering Problems*, Birkhauser, Boston, 1990.

[2] A. Bensoussan, *Stochastic Control with Partial Information,* Cambridge University Press, 1989.

[3] H.J. Kushner, "Dynamical equations for non linear filtering," *J. Diff. Equations,* **3**, 1967, 179–190.

[4] R. Liptser and A.N. Shiryaev, *Statistics of Random Processes,* Springer, Berlin, 1977.

[5] M. Zakai, "On optimal filtering of diffusion processes," *Z. Wahrsch. verv. Gebeite,* **11**, 1969, 230–243.

[6] J. Szpirglas, "Sur l'equivalence d'equations differentielles stochastiques a valeurs mesures intervenant dans le filtrage markovien non lineare," *Ann. Inst. H. Poincaré,* **XIV**, 1978, 33–59.

Smooth σ-Fields

Paul Malliavin
Departement de Mathematique
Universite de Paris VI

In two beautiful papers [3], [5] Zakai and collaborators have recently developed the notion of a *faithful σ-field* on the Wiener space. On the other hand the notion of non-degenerate mappings [1] has led to a good theory of image measure and of desintegration. The drawback of the theory of non-degenerate mappings is its limitation to a finite dimensional range. The notion of a *differentiable σ-field* developed in this paper is an extension of non-degenerate mappings. We will construct a Riemannian structure on the image as the infinite dimensional parallel to H. Airault's finite dimensional case [1]. This abstract construction seems to be a first step to approach "hypoellipticity" in infinite dimension [2]. This paper is a preliminary work in these directions. Several theorems stated in full generality will be proved here only in special cases.

1. Smooth functional.

We will denote by X the Wiener space, by μ the Wiener measure, by H its Cameron Martin space. We denote $\mathcal{W}(X) = \bigcap L^p(X,\mu)$. More generally if G is an abstract Hilbert space $\mathcal{W}(X,G)$ is similarly defined. We will denote by $\mathcal{W}^\infty(X)$ the space of all indefinitely H-differentiable functions f which belong to $\mathcal{W}(X)$, such that $\nabla f \in \mathcal{W}(X;H), \nabla^2 f \in \mathcal{W}(X;H\otimes H)\ldots\nabla^n f \in \mathcal{W}(X, \bigotimes_{j=1}^{n} H_j)$ where $H\otimes H$ has been equipped with the Hilbert-Schmidt norm.

The family of semi-norms

$$\|f\|_{p,r} = \|f\|_{L^p(X)} + \|\nabla f\|_{L^p(X,H)} + \ldots + \|\nabla^r f\|_{L^p(X,H^{\otimes r})}$$

ISBN 0-12-481005-5

defines on $\mathcal{W}^\infty(X)$ a distance

$$d(f,f') = \sum_{p,r} \frac{\|f-f'\|_{p,r}}{1+\|f-f'\|_{p,r}} \frac{1}{(p^2+r^r)^2};$$

with this distance $\mathcal{W}^\infty$ is a *complete metric space.* Furthermore $\mathcal{W}^\infty(X)$ is an algebra with a continuous product. The derivation is a continuous map from $\mathcal{W}^\infty(X;G)$ to $\mathcal{W}^\infty(X,H\otimes G)$.

2. Smooth subalgebra.

We shall denote by A a *closed* vector subspace of $\mathcal{W}^\infty(X)$. We say that A is a *closed sub-algebra* if for any given polynomial P in n variables, $P(f_1,\ldots,f_n) \in A$, when $f_i \in$A.

Following Zakai and choosing a dense countable subset D of A, we shall denote

$$M_x = \bigcap_{f\in D} \{h \in H; (\nabla f(x) \mid h) = 0\}.$$

Up to a thin set M_x will not depend on the choice of D. Therefore M depends only upon A.

We will associate with M_x its orthogonal projection P_x upon M_x. We denote $P_x^\perp = 1 - P_x$. Given $h_1,h_2,h_3 \in H$, we introduce

$$B_x(h_1,h_2,h_3) = \left\{ \left(\frac{d}{d\epsilon} P_{x+\epsilon h_1} h_2 \mid h_3 \right) \right\}_{\epsilon=0}.$$

Assuming the existence of B, we state the following definition :

2.2. Definition. *We shall say that* A *is a smooth subalgebra if* $B \in \mathcal{W}^\infty(X : H\otimes H\otimes H)$.

We have the following example of a smooth algebra :

2.3. Theorem. *Let* $g \in \mathcal{W}^\infty(X; R^n)$. *Denote* $\sigma^{i,j} = (\nabla g^i \mid \nabla g^j)$. *Assume that* $(\det(\sigma))^{-1} \in \mathcal{W}(X)$. *Define*

$$A = \text{adherence}\{\tilde{u}; \tilde{u}(x) = u(g_1(x), \ldots, g_n(x))\}$$

where u *is a* C^∞ *function defined on* $\mathbf{R}^n$, *bounded with all its derivative. Then* A *is a smooth subalgebra.*

Proof. We have $M_x = \ker\ g'(x)$. Therefore $M_2^\perp$ has for basis $\{\nabla g^i(x)\}$. Denoting by γ the inverse of the covariance matrix σ, we have

$$P_x^\perp h = \Sigma \gamma_{i,k}(h \mid \nabla g^k) \nabla g^i(x)$$

$$P_x h = h - P_x^\perp h$$

$$B_x(h_1, h_2, h_3) = -\left(\frac{d}{d\epsilon} P_{x+\epsilon h_1}^\perp h_2 \mid h_3\right)$$

$$D_{h_1} P^\perp h_2 = \Sigma(D_{h_1}\gamma_{i,k})(h_2 \mid \nabla g^k)\nabla g^i$$
$$+\gamma_{i,k}\nabla^2 g^k(h_1, h_2)\nabla g^i + \gamma_{i,k} D_{h_2}(g^k) D_{h_1}(\nabla g^i).$$

As $\gamma_{i,k} \in \mathcal{W}^\infty$, we have only to show that the mappings of the following type belong to $\mathcal{W}^\infty(X, H \otimes H \otimes H)$

$$B_x^1(h_1, h_2, h_3) = (h_2 \mid \nabla g^k)(h_3 \mid \nabla g^s)(\nabla \gamma_{0,k} \mid h_1)$$

$$B^2(h_1, h_2, h_3) = (\nabla^2 g^k)(h_1, h_2)(\nabla g^i \mid \nabla g^s)(h_3 \mid \nabla g^\ell)$$

$$B^3(h_1, h_2, h_3) = (\nabla g^k \mid h_2)(\nabla^2 g^i)(h_1, \nabla g^s)(h_3 \mid \nabla g^\ell).$$

We have

$$\|B^1\|_{H \otimes H \otimes H} = \|\nabla g^k\|_H \|\nabla g^s\|_H \|\nabla \gamma_{u,k}\|_H$$

$$\|B^2\|_{H \otimes H \otimes H} \leq \|\nabla g^i\|_H \|\nabla g^s\|_H \|\nabla^2 g^k\|_{H \otimes H} \|\nabla g^\ell\|_H$$

$$\|B^3\|_{H \otimes H \otimes H} \leq \|\nabla g^k\|_H \|\nabla^2 g^i\|_{H \otimes H} \|\nabla g^s\|_H \|\nabla g^\ell\|_H,$$

which implies that $B^q \in \mathcal{W}(X; H \otimes H \otimes H)$. The B^q being algebraic expressions in $\nabla g^i, \nabla^2 g^s$, it is possible to differentiate those expressions an arbitrary number of times and the theorem is proved.

2.4. Proposition *(Frobenius identity). We have for $h_1, h_2 \in M_x, h_3 \in M_x^\perp$:*

$$B_x(h_1, h_2, h_3) - B_x(h_2, h_1, h_3) = 0.$$

Proof. Denote $Z_k(x) = P_x h_k \quad (k = 1, 2)$. We have that

$(Z_1 \mid \nabla u) = 0$, for all $u \in$A. Also

$D_{Z_2}(Z_1 \mid \nabla u) = (D_{Z_2} Z_1 \mid \nabla u) + (\nabla^2 u)(Z_1 \mid Z_2)$.

Therefore by the symmetry of the second derivative

$$(D_{Z_2} Z_1 - D_{Z_1} Z_2 \mid \nabla u) = 0 \text{ for all } u \in A \text{ or}$$

$$(D_{Z_2} Z_1 - D_{Z_1} Z_2) \in M_x.$$

If we now choose $h_1, h_2 \in M_{x_0}, h_3 \in M_{x_0}^\perp$, we will have $Z_1(x_0) = h_1, Z_2(x_0) = h_2$ and the proposition is proved.

3. Functional calculus.

A closed algebra A of smooth functionals will be said to be *functionally closed* if for every $u \in C^\infty(R^n)$ bounded with all its derivatives

$$u(f_1, \ldots f_n) \in A \text{ whenever } f_i \in \text{A}.$$

3.1. Proposition. *Assume that $A \cap L^\infty(X)$ is dense in A for the $\mathcal{W}^\infty$-topology and assume that* A *is closed, then* A *is functionally closed.*

Proof. Given u and given $f_1 \ldots f_n$ we approximate $f_1 \ldots f_n$ by $f_{1,\epsilon}, \ldots f_{n,\epsilon}$ which are bounded. Then their range defines a compact K of R^n. On K, by the Weierstrass approximation theorem, u can approximated in the $C^\infty(K)$ uniform topology by a sequence of polynomials P_q. Then $\lim_{q\to\infty} P_q(f_\epsilon)$ exists in $\mathcal{W}^\infty(X)$ and belongs to A. Therefore $u(f_{1,\epsilon},\ldots f_{n,\epsilon}) \in$A. Letting $\epsilon \to 0$ $\lim_{q\to\infty} u(f_{1,\epsilon},\ldots f_{n,\epsilon})$ exists in $\mathcal{W}^\infty$ and therefore $u(f_1,\ldots,f_n)$ belongs to A.

Remark. The density of $L^\infty \cap A$ in A can be relaxed by looking at the uniqueness method for the moment problem (see P. Malliavin, American Journal of Mathematics, 1959).

4. Smooth σ-field.

We shall call *smooth σ-field* a σ-field $\mathcal{A}$ on X which is generated by a smooth closed and functionally closed subalgebra A of $\mathcal{W}^\infty(X)$.

More precisely $\mathcal{A}$ is the smallest σ-field which contains all $\{u^{-1}(]r_1, r_2[)\}$ with $u \in$A.

From this definition we deduce

4.1. A *is dense in* $L^2(\mathcal{A})$.

This result implies that $\mathcal{W}^\infty(X) \cap L^2(\mathcal{A})$ is dense in $L^2(\mathcal{A})$. Therefore, using Zakai's terminology, a smooth σ-field is *faithful.*

4.2. Definition. *Denote by A* a *smooth algebra, let A be the smooth σ-field associated,* $E^{\mathcal{A}}$ *the conditional expectation. We shall say that* $E^{\mathcal{A}}$ *is smooth when*

$$E^{\mathcal{A}}(\mathcal{W}^\infty(X)) = A$$

and when $E^{\mathcal{A}}$ *is a continuous map from* $\mathcal{W}^\infty(X)$ *to* A.

The study of smoothness will depend upon the following concepts.

5. Basic vector field.

Given a vector field Z on $X, Z \in \mathcal{W}^\infty(X;H)$, we shall say that Z is *basic* when

$$(< Z, df >) \in A \text{ for all } f \in \text{A}.$$

5.1. Theorem. *Given a basic vector field Z, we have for all $u \in \mathcal{W}^\infty(X)$*

$$\begin{aligned} &< Z, d(E^{\mathcal{A}}(u)) > \\ &\quad = E^{\mathcal{A}}(< Z, du >) + E^{\mathcal{A}}(u \operatorname{div}_\mu(Z)) - E^{\mathcal{A}}(u)E^{\mathcal{A}}(\operatorname{div}_\mu(Z)). \end{aligned}$$

Proof. We shall prove this result when $\mathcal{A}$ is the σ-field associated to a non degenerate map $g \in \mathcal{W}^\infty(\text{X},\text{R}^n)$. Then we have the following desintegration formula (see H. Airault and P. Malliavin, Bulletin des Sciences Mathématiques, 1988).

$$(E^q(u))(x) = v(g(x)), \text{ where}$$

$$v(\xi) = \frac{1}{k(\xi)} \int_{g^{-1}(\xi)} u^*(x)\lambda_\xi(dx) \text{ and } k(\xi) = \lambda_\xi(1), \text{ with}$$

$$d\lambda = \frac{1}{\det[(g')^* g']} da^{\infty-n},$$

where $a^{\infty-n}$ is the Gaussian area of codimension n on the submanifold $g^{-1}(\xi)$.

Now we have that given $v \in C^\infty(R^n)$, there exists a $q \in C^\infty(R^n)$ such that

$$< Z, d(v \circ g) >= q \circ \text{g}.$$

The correspondence $v \to q$ is a derivation on the algebra $C^\infty(R^n)$. Therefore there exist a smooth vector field z on R^n such that

$$q =< z, dv > .$$

We have furthermore

$$(i) \qquad g'(x)Z_n = z_{g(x)};$$

conversely any vector field satisfying (i) is a basic vector field.

Denote by $U_\epsilon^Z, U_\epsilon^z$ the flows associated to the vector fields Z, z (we assume that the integrated flow of Z exists to get a transparent formulation of the following *differential* identity (iii) which can be verified by brute force derivation) ; then (i) can be integrated

$$(ii) \qquad g \circ U_\epsilon^Z = U_\epsilon^z \circ g;$$

therefore taking the inverse image of the differential form

$$(U_\epsilon^Z)^* g^* = g^* (U_\epsilon^z)^*;$$

and denoting by $\mathcal{L}_Z, \mathcal{L}_z$ the Lie derivatives on differential forms we get

$$(iii) \qquad \mathcal{L}_Z g^* = g^* \mathcal{L}_z;$$

now

$$(E^{\mathcal{A}} u)(x) = \frac{1}{k(g(x))} \int u^*(y) \lambda_{g(x)}(dy).$$

Denote by θ the differential form on R^n

$$\theta = k(\xi) d\xi^1 \wedge \ldots \wedge d\xi^n;$$

then

$$(iv) \qquad \left(\frac{1}{k(g(x))} \lambda_{g(x)} \right) \wedge g^* \theta = \mu.$$

The exterior product appearing here is of symbolic nature. When dim $X = N < +\infty$ it is the exterior product between a $(N-n)$-form and a n-form which gives the volume form of the space X.

Compute now the Lie derivative $\mathcal{L}_Z\theta$

$$\mathcal{L}_Z\theta = \left\{\frac{d}{d\epsilon}(U_\epsilon^z)^*\theta\right\}_{\epsilon=0}.$$

By the invariance of the integral of a n-form under a diffeomorphism, we have for every smooth form

$$\int (U_\epsilon^z)^* v (U_\epsilon^z)^*\theta = \int v\theta;$$

therefore by differentiation in ϵ

$$\int < z, dv > \theta + \int v\mathcal{L}(\theta) = 0$$

or

$$\mathcal{L}_z(\theta) = -(\mathrm{div}_\theta(z))\theta;$$

then applying $\mathcal{L}_Z$ to (iv) we get

$$\mathcal{L}_Z\left(\tfrac{1}{k}\lambda\right) \wedge g^*(\theta) + \frac{1}{k}\lambda \wedge \mathcal{L}_Z(g^*(\theta)) = \mathrm{div}_\mu(Z)\mu$$

$$\mathcal{L}_Z g^*\theta = g^*\mathcal{L}_z\theta = -\theta\ \mathrm{div}_\theta(z);$$

then

$$\mathcal{L}_Z\left(\tfrac{1}{k}\lambda\right) \wedge g^*\theta = (\mathrm{div}_\theta(z) - \mathrm{div}_\mu(Z))\mu.$$

(v) **Lemma.**

$$\mathrm{div}_\theta(z) = E^{\mathcal{A}}(\mathrm{div}_\mu(Z)).$$

Proof. We denote by $v \in C^\infty(R^n)$ with compact support and $\tilde{v} = v \circ g$; then

$$< Z, d\tilde{v} >_x = < z, dv >_{g(x)};$$

therefore

$$\int_{R^n} < z, dv >_\xi \theta(d\xi) = \int_X < Z, d\tilde{v} >_x \mu(dx)$$

or, by transposition

$$\int_{R^n} v \operatorname{div}_\theta(z) d\theta = \int \tilde{v} \ \operatorname{div}_\mu(Z) d\mu.$$

This identity, holding true for all v, implies the lemma.

We get then, denoting $\tilde{\lambda}(dx) = \frac{1}{k(x)}\lambda(dx)$

$$(vi) \qquad (\mathcal{L}_Z \tilde{\lambda}_{g(x)}) \wedge g^*\theta = (E^{\mathcal{A}}(\operatorname{div}_\mu(Z) - \operatorname{div}_\mu(Z))(\tilde{\lambda} \wedge g^*\theta).$$

We denote by V_ξ the submanifold $g^{-1}(\xi)$ of X ; then we have

$$(vii) \qquad U_\epsilon^Z(V_\xi) = V_{U_\epsilon^z(\xi)}$$

$$\int_{U_\epsilon^Z(V_\xi)} u(x)\lambda_{U_\epsilon^z(\xi)}(dx) = \int_{V_\xi} ((U_{-\epsilon}^Z)^* u) d((U_{-\epsilon}^Z)^*\lambda).$$

Differentiating in ϵ and taking $\epsilon = 0$ we get

$$< Z, d(E^A(u)) >= E^A(< Z, du >) - \int_{V_\xi} u\alpha_Z \tilde{\lambda}.$$

Going to (vi) we get the proof of the theorem.

We shall say that the σ-field $\mathcal{A}$ is *ample* if denoting by γ the set of basic vector fields, given p there exists p' and a constant c such that

$$\|f\|_{W^{1,p}} \leq c \sup \| < Z, df > \|_{L^p}; \quad \text{with } Z \text{ basic}, \|Z\|_{1,p'} \leq 1.$$

Corollary. *Assume that $\mathcal{A}$ is ample, then*

$$E^{\mathcal{A}}(\mathcal{W}^\infty) \subset \bigcap_p W^{1,p}.$$

Corollary. *Assume that g is a non-degenerate map, $g \in \mathcal{W}^\infty(X; R^n)$, then the associated σ-field is ample. Furthermore $E^{\mathcal{A}}(\mathcal{W}^\infty(X)) \subset \mathcal{W}^\infty(X)$.*

6. Basic differential form.

Differential forms on the Wiener space have been considered by I. Shigekawa [4].

We shall denote by H^* the space of continuous linear forms on H. Of course H^* can be identified to H but for our purpose we prefer to denote it for some time in this way. Given $M \subset H$, we denote by $M^0 = \{\ell \in H^*; \ell(z) = 0, \forall z \in M\}$.

Given $f \in \mathcal{W}^\infty(X)$ we denote by $df \in \mathcal{W}^\infty(X; H^*)$ defined by

$$< h, df >= (\nabla f \mid h).$$

We denote by $\Lambda^1(X) = \mathcal{W}^\infty(X; H^*)$; then $\Lambda^1(X)$ is a complete metric space.

Given a smooth subalgebra A we denote by $\Lambda^1(A)$ *the closure in* $\Lambda^1(X)$ of expression of the form

$$\sum_{k=1}^{n} f_k dg_k$$

where f_k and $g_k \in$A.

We shall call $\Lambda^1(A)$ the *basic differential form*

$$M_x^0 = \{\rho_x; \text{ with } \rho \in \Lambda^1(A)\}.$$

In the same way, introducing $H^* \wedge H^*$ the Hilbert space quotient of $H \otimes H$ by the vector space generated by $h_1 \otimes h_2 - h_2 \otimes h_1$, we define $\Lambda^2(X) = \mathcal{W}^\infty(X; H^* \otimes H^*), \Lambda^2$ the closure in $\Lambda^2(X)$ of expression of the form $\sum f_k dg_k \wedge dg'_k$. We denote by $\Lambda^1_{L^2}(X), \Lambda^1_{L^2}(A)$, the closure in L^2 of the corresponding spaces. Finally $E^{\Lambda A}$ will denote the orthogonal projection of $\Lambda^{\cdot}_{L^2}(X)$ onto $\Lambda^{\cdot}_{L^2}(A)$.

We shall define a metric on $\Lambda(A)$ by

$$\|\rho\|_\xi^2 = E^{\mathcal{A}}(\|\rho\|_{H^*}^2);$$

then this metric defines a "Riemannian structure" on the "quotient space" X/A.

Without using this "quotient structure" we want to recapture the construction of H. Airault [1].

We define the operator δ_A on $\Lambda^1(A)$ by

$$\delta_A \rho = E^{\mathcal{A}}(\delta\rho)$$

where $\delta = d^*$ in $\Lambda(X)$; then the Airault Laplacian on A is defined

$$\Delta_A = \delta_A d;$$

it is a mapping from A to A.

In higher degree δ_A is extended as an anti-derivation on forms of arbitrary degree ; then

$$\Box_A = \delta_A d + d\delta_A$$

is well defined ; then the Ricci tensor R on $X/\mathcal{A}$ is defined by the following Weitzenbock identity

$$E((\Box_A \rho \mid \rho)) = \|\rho\|^2_{W^{1,2}(A)} + \int_X (R\rho \mid \rho) d\mu \text{ where}$$

$$\|\rho\|^2_{W^{1,2}(A)} = E(\|P_{M^0 \otimes M^0} \nabla \rho\|^2_{H^*})$$

Define $W^{1,2}$ vector fields on $X/\mathcal{A}$ as the dual for the natural Riemannian metric of $W^{1,2}$ forms.

With these definitions the following theorem can be expected :
Assume that for all c

$$E\left(\exp\left(c\|\text{Ricci}\|_{\text{End}(M^0)}\right)\right) < +\infty \;;$$

then any vector field on $X/\mathcal{A}$ which is in $W^{1,2}$ *divergence relative to the image measure.*

BIBLIOGRAPHY

[1] H. Airault, *Projection of the infinitesimal generator of a diffusion.* Journal of Funct. Analysis, August 1989.

[2] P. Malliavin, *Hypoellipticity in infinite dimension.* Proceeding of the Northwesten conference (M. Pinsky editor), Birkhauser 1990.

[3] D. Nualart, A.S. Ustunel and M. Zakai, Some relations among classes of σ-fields on the Wiener space, Probability theory, 85, p.119-129, 1990.

[4] I. Shigekawa, The de Rham complex on the Wiener space, Osaka Mathematical Journal, 1984.

[5] A.S. Ustunel and M. Zakai, On the structure of independence on Wiener space, Journal of Functional Analysis, 1990.

-:-:-:-:-

10, rue Saint-Louis *en* l'Isle
75004 PARIS

Composition of Large Deviation Principles and Applications

Annie Millet
Université d'Angers, and
Université Paris 6 (UA 224)

David Nualart and Marta Sanz
Universitat de Barcelona

1 Introduction

The purpose of this paper is to present a general result on the composition of large deviation principles (Theorem 2.2) and to apply this theorem to obtain large deviations estimates for solutions of anticipating stochastic differential equations.

This problem was suggested to us by G. Benarous and we would like to thank him for his stimulating remarks.

The composition theorem is stated and proved in Section 2. Its proof is based on a recent result of P. Baldi and M. Sanz [2] about the equivalence between large deviations estimates and a continuity property. Section three contains the application of the composition result to deduce a large deviation principle for the solutions of anticipating stochastic differential equations. Two types of equations have been considered. First we deal with the Stratonovich stochastic differential equation studied by D. Ocone and E. Pardoux in [7]. In this case the large deviation principle we present is based on the large deviation principle for stochastic flows proved in [5], and it generalizes other large deviation results obtained there. The second type of anticipating equation is a quasilinear equation introduced by R. Buckdahn in [3]. For this equation we generalize the results of [4].

ISBN 0-12-481005-5

2 Composition of large deviation principles

Let $(\Omega, \mathcal{F}, P)$ be a probability space and (E, d) be a Polish space. Consider a family $(V^\varepsilon,\ \varepsilon > 0)$ of E–valued random variables which satisfies a large deviation principle (LDP) with rate function $\lambda : E \to [0, +\infty]$. That means, λ is lower semi-continuous, for every $a > 0$ the set $\{f \in E : \lambda(f) \le a\}$ is compact, and for every open (resp. closed) subset G (resp. C) of E we have

$$\begin{aligned}
\liminf_{\varepsilon \downarrow 0}\ \varepsilon \log P(V^\varepsilon \in G) &\ge -\inf\{\lambda(f) : f \in G\} \\
\limsup_{\varepsilon \downarrow 0}\ \varepsilon \log P(V^\varepsilon \in C) &\le -\inf\{\lambda(f) : f \in C\}.
\end{aligned}$$

In the applications presented in the next section, V^ε will be $\sqrt{\varepsilon}\, W$ where W is a standard Brownian motion.

Suppose that (E_2, d_2) is another Polish space and $(\xi^\varepsilon,\ \varepsilon > 0)$ a family of E_2–valued random variables. Assume also that there exists a map $\xi : \{\lambda < +\infty\} \to E_2$ such that the restriction of ξ to the compact sets $\{\lambda \le a\}$, $a \in [0, \infty)$, is continuous. The following result has been proved by P. Baldi and M. Sanz in [2, Theorem 3].

Proposition 2. 1 *The following properties are equivalent:*
(P1) The family $((V^\varepsilon, \xi^\varepsilon),\ \varepsilon > 0)$ *satisfies a* LDP *with rate function*

$$\tilde{\lambda}(f, e) = \begin{cases} \lambda(f) & \text{if } \lambda(f) < \infty \text{ and } e = \xi(f), \\ +\infty & \text{otherwise}, \end{cases}$$

for $f \in E$, $e \in E_2$.
(P2) For every $R > 0$, $\eta > 0$ *and* $f \in E$ *such that* $\lambda(f) < \infty$ *there exist* $\alpha > 0$ *and* $\varepsilon_0 > 0$ *such that for* $0 < \varepsilon \le \varepsilon_0$

$$P\Big(d_2(\xi^\varepsilon,\ \xi(f)) \ge \eta,\quad d(V^\varepsilon, f) \le \alpha\Big) \le \exp\left(-\frac{R}{\varepsilon}\right).$$

Let (F, ρ) be another Polish space and let K be a compact metric space. We denote by d_1 the Euclidean distance on $\mathbf{R}^d$ and by $\tilde{d}_1$ and $\tilde{\rho}$ the distances inducing the topology of uniform convergence on $\mathcal{C}(K, \mathbf{R}^d)$ and $\mathcal{C}(K, F)$, respectively. Let $\tilde{d}$ be a distance on $\mathcal{C}(\mathbf{R}^d, F)$ inducing the topology of uniform convergence on compact subsets of $\mathbf{R}^d$.

Given any $\varepsilon > 0$ we consider random variables $X^\varepsilon : \Omega \to \mathcal{C}(K, \mathbf{R}^d)$, $Y^\varepsilon : \Omega \to \mathcal{C}(\mathbf{R}^d, F)$ and set $\tilde{X}^\varepsilon = (V^\varepsilon, X^\varepsilon)$, $\tilde{Y}^\varepsilon = (V^\varepsilon, Y^\varepsilon)$. We introduce the following assumptions, which correspond to condition (P1) of Proposition 2.1.

(H1) There exists a map $X : \{\lambda < \infty\} \to \mathcal{C}(K, \mathbf{R}^d)$ such that its restriction to the compact sets $\{\lambda \le a\}$, $a \in [0, \infty)$, is continuous, and the family $(\tilde{X}^\varepsilon,\ \varepsilon > 0)$ satisfies a *LDP* with rate function

$$\tilde{\lambda}_1(f, g) = \begin{cases} \lambda(f) & \text{if } \lambda(f) < \infty \text{ and } g = X(f), \\ +\infty & \text{otherwise}, \end{cases}$$

for $f \in E$, $g \in \mathcal{C}(K, \mathbf{R}^d)$.

(H2) There exists a map $Y : \{\lambda < \infty\} \to \mathcal{C}(\mathbf{R}^d, F)$ such that its restriction to the compact sets $\{\lambda \le a\}$, $a \in [0, \infty)$, is continuous, and the family $(\tilde{Y}^\varepsilon,\ \varepsilon > 0)$ satisfies a *LDP* with rate function

$$\tilde{\lambda}_2(f, e) = \begin{cases} \lambda(f) & \text{if } \lambda(f) < \infty \text{ and } h = Y(f), \\ +\infty & \text{otherwise}, \end{cases}$$

for $f \in E$, $h \in \mathcal{C}(\mathbf{R}^d, F)$.

Then the following theorem states a *LDP* for the composition of Y^ε and X^ε.

Theorem 2. 2 *Suppose that the assumptions (H1) and (H2) are satisfied. For each* $\varepsilon > 0$, *let* $Z^\varepsilon : \Omega \to \mathcal{C}(K, F)$ *be defined by* $Z^\varepsilon = Y^\varepsilon \circ X^\varepsilon$. *Then* $(Z^\varepsilon, \varepsilon > 0)$ *satisfies a LDP with rate function defined on* $\mathcal{C}(K, F)$ *by*

$$\lambda_3(g) = \inf \{\lambda(f) : Y(f) \circ X(f) = g\} . \tag{2.1}$$

Proof : It suffices to show that the pair $(V^\varepsilon, Z^\varepsilon)$ satisfies a *LDP* with rate function

$$\tilde{\lambda}_3(f, g) = \begin{cases} \lambda(f) & \text{if } \lambda(f) < \infty \text{ and } g = Y(f) \circ X(f), \\ +\infty & \text{otherwise}, \end{cases}$$

where $f \in E, g \in C(K,F)$. Note that conditions (H1) and (H2) imply that the mapping $f \longmapsto Y(f) \circ X(f)$ is continuous on the level sets $\{\lambda \leq a\}$, $0 \leq a < \infty$. By Proposition 2.1 it is equivalent to verify the continuity property (P2) for $(Z^\varepsilon,\ \varepsilon > 0)$:
(C) : For every $R > 0$, $\eta > 0$ and $f \in E$ such that $\lambda(f) < \infty$ there exists $\alpha > 0$ and $\varepsilon_0 > 0$ such that for $0 < \varepsilon \leq \varepsilon_0$, if

$$A^\varepsilon = \left\{\tilde{\rho}(Y^\varepsilon \circ X^\varepsilon,\ Y(f) \circ X(f)) \geq \eta,\ d(V^\varepsilon, f) \leq \alpha\right\}$$

then

$$P(A^\varepsilon) \leq \exp\left(-\frac{R}{\varepsilon}\right). \tag{2.2}$$

This continuity property will be deduced from similar continuity properties for $(X^\varepsilon,\ \varepsilon > 0)$ and $(Y^\varepsilon,\ \varepsilon > 0)$, which are equivalent with our hypotheses (H1) and (H2), respectively, due again to Proposition 2.1. In order to complete the details of the proof of (C), fix $R > 0$, $\eta > 0$ and $f \in E$ such that $\lambda(f) < \infty$ and define, for any $\delta > 0$ and $\alpha > 0$

$$\begin{aligned}
B^\varepsilon &= \left\{\tilde{d}_1(X^\varepsilon, X(f)) \geq \delta,\ d(V^\varepsilon, f) \leq \alpha\right\} \\
C^\varepsilon &= \left\{\tilde{d}_1(X^\varepsilon, X(f)) < \delta, \quad d(V^\varepsilon, f) \leq \alpha,\ \tilde{\rho}(Z^\varepsilon, Y(f) \circ X^\varepsilon) \geq \frac{\eta}{2}\right\} \\
D^\varepsilon &= \left\{\tilde{d}_1(X^\varepsilon, X(f)) < \delta,\ \tilde{\rho}(Y(f) \circ X^\varepsilon,\ Y(f) \circ X(f)) \geq \frac{\eta}{2}\right\}.
\end{aligned}$$

Clearly $A^\varepsilon \subset B^\varepsilon \cup C^\varepsilon \cup D^\varepsilon$. By (H1) and Proposition 2.1 there exists $\alpha_1 > 0$, $\varepsilon_1 > 0$ such that for $0 < \alpha \leq \alpha_1$, $0 < \varepsilon \leq \varepsilon_1$, $P(B^\varepsilon) \leq \exp\left(-\frac{R}{\varepsilon}\right)$.

Suppose $0 < \delta \leq 1$, and set $\gamma = 1 + \sup_{k \in K} |X(f)(k)|$. Then $|X^\varepsilon(\omega)| \leq \gamma$ for each $\omega \in C^\varepsilon$, and there exists $\eta' > 0$ such that

$$\begin{aligned}
C^\varepsilon &\subset \left\{d(V^\varepsilon, f) \leq \alpha,\ \sup_{|x| \leq \gamma} \rho(Y^\varepsilon(x),\ Y(f)(x)) \geq \frac{\eta}{2}\right\} \\
&\subset \left\{d(V^\varepsilon, f) \leq \alpha,\ \tilde{d}(Y^\varepsilon,\ Y(f)) \geq \eta'\right\}.
\end{aligned}$$

By (H2) and Proposition 2.1 there exists $\varepsilon_2 > 0$, $\alpha_2 > 0$ such that for $0 < \varepsilon \le \varepsilon_2$, $0 < \alpha \le \alpha_2$, $P(C^\varepsilon) \le \exp\left(-\frac{R}{\varepsilon}\right)$. On the other hand, since $Y(f)$ is uniformly continuous on $\{x : |x| \le \gamma\}$ we can find $\delta \in (0,1]$ such that $D^\varepsilon = \emptyset$. Therefore for $\varepsilon \le \min(\varepsilon_1, \varepsilon_2)$ and $\alpha \le \min(\alpha_1, \alpha_2)$ we obtain $P(A^\varepsilon) \le 2\exp\left(-\frac{R}{\varepsilon}\right)$ which completes the proof of the continuity condition (C).

□

The following result shows that if two families of random variables are "close" and one of them satisfies a *LDP*, then the other one also satisfies a *LDP* with the same rate function. The proof, which uses well-known arguments (see e.g. [1]), is not given.

Proposition 2. 3 *Let (E,d) be a Polish space, and let $(\xi^\varepsilon,\ \varepsilon > 0)$ and $(\eta^\varepsilon,\ \varepsilon > 0)$ be E-valued random variables.*
Assume that:

(i) The family $(\xi^\varepsilon,\ \varepsilon > 0)$ satisfies a LDP with rate function $I : E \to [0, +\infty]$.

(ii) For any $\alpha > 0$

$$\limsup_{\varepsilon \downarrow 0}\ \varepsilon\ \log\ P\Big(d(\xi^\varepsilon,\ \eta^\varepsilon) \ge \alpha\Big) = -\infty\,. \tag{2.3}$$

Then the family $(\eta^\varepsilon,\ \varepsilon > 0)$ also satisfies a LDP with rate function I.

3 Large deviations for anticipating stochastic differential equations

In this section we will give two applications of Theorem 2.2 to the solution of anticipating stochastic differential equations. First we will consider the equations studied by Ocone and Pardoux in [7].

Let $(W_t,\ t \in [0,1])$ be a k-dimensional standard Brownian motion defined on the canonical probability space $(\Omega, \mathcal{F}, P)$. Suppose that $b, \sigma_i : \mathbf{R}^d \longrightarrow \mathbf{R}^d$, $1 \le i \le k$, and $m = \frac{1}{2}\sum_{i=1}^{k} \frac{\partial \sigma_i}{\partial x}\,\sigma_i : \mathbf{R}^d \longrightarrow$

$\mathbf{R}^d$ are functions of class C^2 with bounded partial derivatives up to order 2. Let $(\varphi^\varepsilon,\ \varepsilon > 0)$ denote the family of stochastic flows defined on $\mathbf{R}^d \times [0,1]$ by

$$\varphi_t^\varepsilon(x) = x + \int_0^t \sqrt{\varepsilon}\, \sigma_i\left(\varphi_s^\varepsilon(x)\right) \circ dW_s^i + \int_0^t b\left(\varphi_s^\varepsilon(x)\right) ds\,. \tag{3.1}$$

Here we made the convention of summation over repeated indices and the first stochastic integral is defined in the Stratonovich sense.

Define

$$\mathcal{H}_k = \Big\{ f : [0,1] \longrightarrow \mathbf{R}^k : f(t) = \textstyle\int_0^t \dot f(s)\, ds\ ,\quad \lambda(f) := \frac{1}{2} \int_0^1 |\dot f_s|^2 ds < \infty \Big\}\ ,$$

and set $\lambda(f) = +\infty$ if $f \notin \mathcal{H}_k$.

Given $f \in \mathcal{H}_k$ let $h = S(f)$ denote the solution of the ordinary differential equation

$$h_t(x) = x + \int_0^t \left[\sigma\left(h_s(x)\right) \dot f_s + b\left(h_s(x)\right)\right] ds\ , \tag{3.2}$$

called the skeleton of $\varphi_t^1(x)$.

Let $E := \Omega = \mathcal{C}_0([0,1], \mathbf{R}^k)$ be the set of continuous functions from [0,1] into $\mathbf{R}^k$ which vanish at 0, endowed with the distance d defined by the supremum norm on [0,1], and let $\mathcal{C}([0,1] \times \mathbf{R}^d, \mathbf{R}^d)$ be endowed with a distance $\tilde\rho$ inducing the topology of uniform convergence on compact sets.

Note that by Gronwall's lemma, the restriction of S to each level set $\{\lambda \le a\}$, $a \in [0,\infty)$, is continuous.

Then, given $f \in \mathcal{H}_k$, $\eta > 0$ and $R > 0$ there exists $\alpha > 0$ and $\varepsilon_0 > 0$ such that, for $0 < \varepsilon \le \varepsilon_0$

$$P\left(\tilde\rho(\varphi^\varepsilon,\ S(f)) \ge \eta,\ d(\sqrt{\varepsilon} W,\ f) \le \alpha\right) \le \exp\left(-\frac{R}{\varepsilon}\right). \tag{3.3}$$

This result is a kind of uniform Ventzell-Freidlin estimation and has been proved in [5, Theorem 2.1]. As a consequence, by Proposition 2.1, the pair $(\sqrt{\varepsilon}\, W,\ \varphi^\varepsilon)$ satisfies a LDP with rate function

$$I_2(f,h) = \begin{cases} \lambda(f) & \text{if } \lambda(f) < \infty \text{ and } h = S(f)\,, \\ +\infty & \text{otherwise}\,, \end{cases}$$

Furthermore, given any $\mathbf{R}^d$-valued random variable X_0^ε, the process $(Z_t^\varepsilon = \varphi_t^\varepsilon(X_0^\varepsilon),\ t \in [0,1])$ is a solution of the anticipating stochastic differential equation

$$Z_t^\varepsilon = X_0^\varepsilon + \int_0^t \sigma_i(Z_s^\varepsilon) \circ dW_s^i + \int_0^t b(Z_s^\varepsilon)\, ds \tag{3.4}$$

Then we have the following result:

Proposition 3. 1 *Let $(X_0^\varepsilon,\ \varepsilon > 0)$ be a family of $\mathbf{R}^d$–valued random variables verifying the following condition:*

(i) There exists a mapping $\zeta : \mathcal{H}_k \longrightarrow \mathbf{R}^d$ such that its restriction to the compact sets $\{\lambda \leq a\}$, $a \in [0,\infty)$, is continuous, and the pair $(\sqrt{\varepsilon}\, W,\ X_0^\varepsilon)$ satisfies a LDP on $\mathcal{C}_0([0,1], \mathbf{R}^k) \times \mathbf{R}^d$ with rate function

$$I_1(f,g) = \begin{cases} \lambda(f) & \text{if } \lambda(f) < \infty \text{ and } g = \zeta(f), \\ +\infty & \text{otherwise}, \end{cases}$$

Let $(\varphi^\varepsilon,\ \varepsilon > 0)$ be the stochastic flow solution of (3.1). Then $(Z_t^\varepsilon = \varphi_t^\varepsilon(X_0^\varepsilon),\ t \in [0,1])$ satisfies a LDP with rate function

$$I(g) = \inf\, \{I(f) : S(f)(\zeta(f)) = g\}, \tag{3.5}$$

for any $g \in \mathcal{C}([0,1], \mathbf{R}^d)$.

Proof: It suffices to apply Theorem 2.2 in the following context. Let $K = \{0\}$, $F = \mathcal{C}([0,1], \mathbf{R}^d)$, and as above $E = \mathcal{C}_0([0,1], \mathbf{R}^k)$. Then $\mathcal{C}(K, \mathbf{R}^d) \cong \mathbf{R}^d$ and $\mathcal{C}(\mathbf{R}^d, F) \cong \mathcal{C}([0,1] \times \mathbf{R}^d, \mathbf{R}^d)$. Condition (i) implies $(H1)$ for $X = \zeta$ and $X^\varepsilon = X_0^\varepsilon$.
Moreover we have seen that $Y^\varepsilon = \varphi^\varepsilon$ and $Y = S$ satisfy condition $(H2)$. This completes the proof of the proposition.

□

Remark 3. 2 Proposition 3.1 generalizes the large deviation estimates obtained in [4]. Indeed, the condition

$$\lim_{\varepsilon \downarrow 0} \varepsilon \log\, P(|X_0^\varepsilon - x_0| > \eta) = -\infty, \tag{3.6}$$

for any $\eta > 0$ and some $x_0 \in \mathbf{R}^d$, can be considered as a particular case of the assumption (i) in Proposition 3.1 . More precisely, let

$\zeta_0 : \mathcal{H}_k \longrightarrow \mathbf{R}^d$ be defined by $\zeta_0(f) = x_0$. Then (3.6) implies that the family $((\sqrt{\varepsilon}\, W,\ X_0^\varepsilon),\ \varepsilon > 0)$ satisfies a LDP on $\mathcal{C}_0([0,1], \mathbf{R}^k) \times \mathbf{R}^d$ with rate function

$$I_1(f,x) = \begin{cases} \lambda(f) & \text{if } x = x_0 \\ +\infty & \text{if } x \neq x_0, \end{cases}$$

In fact, by Proposition 2.3 it suffices to show that the family $((\sqrt{\varepsilon}\, W,\ x_0),\ \varepsilon > 0)$ satisfies a LDP with rate function I_1 and this is straightforward.

Now we proceed to deduce large deviation estimates for a different type of anticipating stochastic differential equation. Set $K = [0,1]$ and let $(\Omega, \mathcal{F}, P)$ be the canonical probability space associated with a standard one–dimensional Brownian motion. Fix a Lipschitz function $b : \mathbf{R} \longrightarrow \mathbf{R}$ and a constant $\sigma \neq 0$. For any $\varepsilon > 0$ and $t \in [0,1]$ set

$$\psi_t^\varepsilon(\omega) = \exp\left(\sqrt{\varepsilon}\ \sigma\ \omega_t - \frac{1}{2}\ \varepsilon\sigma^2 t\right) .$$

Consider the family of transformations $A_t^\varepsilon : \Omega \longrightarrow \Omega$ defined by

$$A_t^\varepsilon\,(\omega)\,(s) = \omega_s - \sqrt{\varepsilon}\ \sigma\,(t \wedge s)\ ,$$

and set $T_t^\varepsilon = (A_t^\varepsilon)^{-1}$, that means

$$T_t^\varepsilon\,(\omega)\,(s) = \omega_s + \sqrt{\varepsilon}\ \sigma\,(t \wedge s)\ .$$

We denote by $(z_t^\varepsilon\,(\omega, x)\,,\ t \in [0,1])$ the solution of the ordinary differential equation

$$z_t^\varepsilon\,(\omega, x) = x + \int_0^t\ [\psi_s^\varepsilon\ (T_t^\varepsilon\,(\omega))]^{-1}\ b\ [\psi_s^\varepsilon\ (T_t^\varepsilon\,(\omega))\ \ z_s^\varepsilon\,(\omega, x)]\ ds\ .$$

Then if $M^\varepsilon\,(t, x, \omega) = z_t^\varepsilon\ (A_t^\varepsilon\,(\omega),\ x)$ we have that

$$\begin{aligned} M^\varepsilon\,(t,x,\omega) = x\ + \int_0^t \exp\left(-\sqrt{\varepsilon}\,\sigma\,\omega_s + \frac{\varepsilon}{2}\,\sigma^2 s\right) \\ \times\ b\left[\exp\left(\sqrt{\varepsilon}\,\sigma\,\omega_s - \frac{\varepsilon}{2}\,\sigma^2 s\right)\ M^\varepsilon(s,x,\omega)\right] ds\,. \end{aligned} \qquad (3.7)$$

Furthermore, the stochastic flow (ζ_t^ε) solution of

$$\zeta_t^\varepsilon(x) = x + \sqrt{\varepsilon}\int_0^t \sigma\,\zeta_s^\varepsilon(x)\,dW_s + \int_0^t b(\zeta_s^\varepsilon(x))\,ds \tag{3.8}$$

is given by

$$\zeta_t^\varepsilon(x)(\omega) = \psi_t^\varepsilon(\omega)M^\varepsilon(t,x,\omega)\,. \tag{3.9}$$

Given a random variable $X_0^\varepsilon \in \mathbf{D}^{1,p} \cap L^q$ for some $p>1$, $q>2$, it has been proved in [3,4] that the process $Z_t^\varepsilon = \zeta_t^\varepsilon[X_0^\varepsilon(A_t^\varepsilon)]$ has a continuous version and it is the unique solution of the anticipating quasilinear stochastic differential equation

$$Z_t^\varepsilon = X_0^\varepsilon + \sqrt{\varepsilon}\int_0^t \sigma\,Z_s^\varepsilon\,dW_s + \int_0^t b(Z_s^\varepsilon)\,ds\,, \tag{3.10}$$

where the stochastic integral is defined in the Skorohod sense. We refer the reader to [6] for the definition and main properties of the Skorohod integral, the derivative operator D and the Sobolev spaces $\mathbf{D}^{1,p}$.

We at first prove that the pair $((\sqrt{\varepsilon}\,W, \zeta^\varepsilon),\ \varepsilon > 0)$ satisfies a LDP on $\mathcal{C}_0([0,1]) \times \mathcal{C}([0,1]\times\mathbf{R})$. The very particular nature of ζ^ε, with a constant diffusion coefficient in dimension one, allows to obtain this result under milder assumptions on b than those required for the general uniform Ventzell-Freidlin estimates. As before $\tilde{\rho}$ will denote the metric on $\mathcal{C}([0,1]\times\mathbf{R})$, which induces the topology of uniform convergence on compact sets.

Let $\eta^\varepsilon(t,x,\omega)$ be the solution of the ordinary differential equation (for each fixed $\omega \in \Omega$) :

$$\begin{aligned}\eta^\varepsilon(t,x,\omega) &= x + \int_0^t \exp\left(-\sigma\,\omega_s + \frac{\varepsilon}{2}\sigma^2 s\right)\\ &\qquad \times b\left[\exp\left(\sigma\,\omega_s - \frac{\varepsilon}{2}\,\sigma^2 s\right)\eta^\varepsilon(s,x,\omega)\right]ds\,,\end{aligned}$$

then $\eta^\varepsilon(t,x,\sqrt{\varepsilon}\,\omega) = M^\varepsilon(t,x,\omega)$. Similarly, let $\eta(t,x,\omega)$ be the solution of the differential equation

$$\eta(t,x,\omega) = x + \int_0^t \exp(-\sigma\,\omega_s) b\left[\exp(\sigma\,\omega_s)\,\eta(s,x,\omega)\right]ds\,.$$

Consider the mappings H^ε, $H : \mathcal{C}_0([0,1]) \longrightarrow \mathcal{C}([0,1] \times \mathbf{R})$ given by

$$H^\varepsilon(\omega)(t,x) = \exp\left(\sigma\,\omega_t - \frac{1}{2}\,\varepsilon\,\sigma^2\,t\right)\,\eta^\varepsilon(t,x,\omega) \tag{3.11}$$

$$H(\omega)(t,x) = \exp(\sigma\,\omega_t)\,\eta(t,x,\omega)\,. \tag{3.12}$$

Notice that for any $f \in \mathcal{H}_1 \subset \Omega$, $H(f)$ is the skeleton associated with f and the stochastic flow $(\zeta_t^\varepsilon(x))$ introduced in (3.8). That means, $H(f)$ is the solution of the ordinary differential equation

$$H(f)_t(x) = x + \int_0^t \left[\sigma\,H(f)_s(x)\dot{f}_s + b(H(f)_s(x))\right] ds\,.$$

Proposition 3. 3 *The pair $((\sqrt{\varepsilon}\,W, \zeta^\varepsilon),\ \varepsilon > 0)$ satisfies a LDP on $\mathcal{C}_0([0,1]) \times \mathcal{C}([0,1] \times \mathbf{R})$ with rate function*

$$I_2(f,g) = \begin{cases} \lambda(f) & \text{if } \lambda(f) < \infty \text{ and } g = H(f)\,, \\ +\infty & \text{otherwise,} \end{cases}$$

where

$$\lambda(f) = \begin{cases} \frac{1}{2}\int_0^1 |\dot{f}_s|^2\,ds & \text{if } f \in \mathcal{H}_1 \\ +\infty & \text{otherwise,} \end{cases}$$

denotes the rate function of the Brownian motion.

Proof : For any $\varepsilon > 0$ set $G^\varepsilon(\omega) = (\omega, H^\varepsilon(\omega))$ and $G(\omega) = (\omega, H(\omega))$. Then it is not difficult to check that G and G^ε are continuous functions on $\Omega = \mathcal{C}_0([0,1])$ and $\lim_{\varepsilon \downarrow 0} G^\varepsilon = G$ uniformly on compact subsets of Ω.
Let P^ε denote the law of $\sqrt{\varepsilon}\,W$ on Ω; then $(P^\varepsilon,\ \varepsilon > 0)$ satisfies a LDP on Ω with rate function λ. Let Q^ε denote the law of $G^\varepsilon(\sqrt{\varepsilon}\,W) = (\sqrt{\varepsilon}\,W,\ H^\varepsilon(\sqrt{\varepsilon}\,W))$. Then (cf. [8, Theorem 2.4]) $(Q^\varepsilon,\ \varepsilon > 0)$ satisfies a LDP with rate function

$$\begin{aligned} I_2(f,g) &= \inf\{\lambda(f') : G(f') = (f,g)\} \\ &= \begin{cases} \lambda(f) & \text{if } \lambda(f) < \infty \text{ and } g = H(f)\,, \\ +\infty & \text{otherwise,} \end{cases} \end{aligned}$$

Furthermore, by construction $H^\varepsilon(\sqrt{\varepsilon}\,\omega) = \exp\left(\sigma\sqrt{\varepsilon}\,\omega_t - \frac{1}{2}\,\varepsilon\,\sigma^2\,t\right)$ $\times\,\eta^\varepsilon(t,x,\sqrt{\varepsilon}\,\omega) = \psi_t^\varepsilon(\omega)\,M^\varepsilon(t,x,\omega) = \zeta_t^\varepsilon(x)(\omega)$.

□

Note that the restriction of H to the compact sets $\{\lambda \le a\}$ $a \in [0,\infty)$, is continuous.

The second ingredient in the proof of large deviation estimates for the solution of (3.10) will be a LDP for the pair $(\sqrt{\varepsilon}\,W,\ X_0^\varepsilon(A_0^\varepsilon))$.

Proposition 3. 4 *Let* $(X_0^\varepsilon,\ \varepsilon > 0)$ *be a family of real-valued random variables verifying the following conditions:*

(i) There exists a mapping $X : \mathcal{H}_1 \longrightarrow \mathbf{R}$ *such that its restriction to the compact sets* $\{\lambda \le a\}$, $a \in [0,\infty)$, *is continuous and the pair* $((\sqrt{\varepsilon}\ W,\ X_0^\varepsilon),\ \varepsilon > 0)$ *satisfies a LDP on* $\mathcal{C}_0([0,1]) \times \mathbf{R}$ *with rate function*

$$\lambda_0(f,g) = \begin{cases} \lambda(f) & \text{if } \lambda(f) < \infty \text{ and } g = X(f) \circ X(f), \\ +\infty & \text{otherwise}, \end{cases} \tag{3.13}$$

(ii) For each $\varepsilon > 0$, X_0^ε *belongs to* $\mathbf{D}^{1,p} \cap L^q$ *for some* $p > 1$ *and* $q > 2$, *and for any* $M > 0$ *there exists* $\varepsilon_0 > 0$ *such that*

$$\sup_{0<\varepsilon\le\varepsilon_0} E\int_0^1 \exp\left[M\,|D_s X_0^\varepsilon|^2\right] ds < \infty\,. \tag{3.14}$$

For any $\varepsilon > 0$ *and* $t \in [0,1]$, *set* $X_t^\varepsilon(\omega) = X_0^\varepsilon(A_t^\varepsilon(\omega))$. *Then* X^ε *has a version with continuous paths and* $((\sqrt{\varepsilon}\ W,\ X^\varepsilon),\ \varepsilon > 0)$ *satisfies a LDP with rate function*

$$I_1(f,g) = \begin{cases} \lambda(f) & \text{if } \lambda(f) < \infty \text{ and } g_t = X(f) \text{ for all } t \in [0,1], \\ +\infty & \text{otherwise}, \end{cases} \tag{3.15}$$

where $f \in \mathcal{C}_0([0,1])$, $g \in \mathcal{C}([0,1])$.

Proof : As it has been proved in [4, Proposition 1.3] the existence of a continuous version for (X_t^ε) follows from the formula

$$X_0^\varepsilon(A_t^\varepsilon) - X_0^\varepsilon(A_s^\varepsilon) = -\sigma\sqrt{\varepsilon}\int_s^t (D_r X_0^\varepsilon)(A_r^\varepsilon)\,dr\,,$$

for any $s \leq t$. Condition (i) implies that $((\sqrt{\varepsilon}\, W,\ X_0^\varepsilon),\ \varepsilon > 0)$ satisfies a LDP on $\mathcal{C}_0([0,1]) \times \mathcal{C}([0,1])$ with rate function (3.15). Here we have identified X_0^ε with a constant function. Then, using Proposition 2.3 it suffices to show that

$$\limsup_{\varepsilon \downarrow 0} \varepsilon \log P \left(\sup_{0 \leq t \leq 1} |X_0^\varepsilon(A_t^\varepsilon) - X_0^\varepsilon| \geq \alpha \right) = -\infty, \tag{3.16}$$

for any $\alpha > 0$; this has been proved in [4, Proof of Theorem 2.1].

□

Now we can state the LDP for the solution of (3.10).

Proposition 3. 5 *Let $(X_0^\varepsilon,\ \varepsilon > 0)$ be a family of real-valued random variables verifying the assumptions (i) and (ii) of Proposition 3.4. Then the family $(Z^\varepsilon,\ \varepsilon > 0)$ of solutions of (3.10) satisfies a LDP on $\mathcal{C}([0,1])$ with rate function*

$$\tilde{I}(g) = \inf \{\lambda(f) : H(f)(t,\ X(f)) = g_t\,,\ t \in [0,1]\}, \tag{3.17}$$

where H is defined in (3.12).

Proof : Set $d = 1$, $K = [0,1]$, $E = \mathbf{R}$, $V^\varepsilon = \sqrt{\varepsilon}\, W$ and $F = \mathcal{C}([0,1])$. We want to apply Theorem 2.2 to the random variables $X^\varepsilon = X_0^\varepsilon(A_\cdot^\varepsilon) : \Omega \longrightarrow \mathcal{C}(K, \mathbf{R})$ and $Y^\varepsilon = \zeta^\varepsilon : \Omega \longrightarrow \mathcal{C}(\mathbf{R},\ F)$. By Propositions 3.3 and 3.4 the hypotheses (H1) and (H2) hold. Consequently the family of random variables $\bar{Z}^\varepsilon : \Omega \longrightarrow \mathcal{C}([0,1],\ F) \cong \mathcal{C}([0,1]^2)$ defined by $\bar{Z}^\varepsilon(s,t) = \zeta_t^\varepsilon(X_s^\varepsilon)$ satisfies a LDP with rate function

$$\bar{I}(g) = \inf \left\{ \lambda(f) : H(f)(t,\ X(f)) = g(t,s)\,,\ (s,t) \in [0,1]^2 \right\}.$$

Let $\Pi : \mathcal{C}([0,1]^2) \longrightarrow \mathcal{C}([0,1])$ be the continuous mapping defined by $\Pi(g)_t = g(t,t)$. Then $Z_t^\varepsilon = \Pi(\bar{Z}^\varepsilon)_t$ and, therefore, the family $(Z^\varepsilon,\ \varepsilon > 0)$ satisfies a LDP with the rate function (3.17).

□

Just as in Remark 3.2, Proposition 3.5 generalizes the large deviations estimations results obtained in [4] for the anticipating quasi-linear equation.

Bibliography

[1] R. Azencott, *Grandes déviations et applications,* Ecole d'Été de Probabilités de Saint Flour. VIII-1978. Lecture Notes in Math. 774. Springer-Verlag. Berlin–Heidelberg–New York 1980.

[2] P Baldi, and M. Sanz, *Une remarque sur la théorie des grandes déviations*, Preprint.

[3] R. Buckdahn, *Quasilinear partial differential equations without nonanticipation requirement*, Sektion Mathematik der Humboldt–Universität. Preprint 176. Berlin, 1988.

[4] A. Millet, D. Nualart, and M. Sanz, *Small perturbations for quasilinear anticipating stochastic differential equations*, Preprint n° 102. Centre de Recerca Matemàtica. Institut d'Estudis Catalans.

[5] A. Millet, D. Nualart, and M. Sanz, *Large deviations for a class of anticipating stochastic differential equations*, Preprint.

[6] D. Nualart, and E. Pardoux, *Stochastic calculus with anticipating integrands*, Probab. Th. Rel. Fields 78,1988, p. 535–581.

[7] D. Ocone, and E. Pardoux, *A generalized Itô-Ventzell formula, Applications to a class of anticipating stochastic differential equations*, Ann. Inst. Henri Poincaré, 25,1989, p. 39-71.

[8] S. R. S . Varadhan, *Large deviations and Applications*, CBMS–NSF regional conference series in applied Mathematics, 46. SIAM. Philadelphia 1984.

Nonlinear Transformations of the Wiener Measure and Applications

David Nualart
Universitat de Barcelona

Abstract

The objective of this paper is to present some recent results on the absolute continuity of the image of the Wiener measure by anticipating Girsanov transformations. As an application we will discuss the Markov property of the solutions to second order stochastic differential equations.

Introduction

Suppose that $u = \{u_t,\ 0 \leq t \leq 1\}$ is a measurable process such that $\int_0^1 u_t^2\, dt < \infty$ almost surely, defined on the Wiener space $(\Omega, \mathcal{F}, P)$. That means, $\Omega = C_0([0,1])$ is the space of continuous functions which vanish at zero and P denotes the Wiener measure on Ω. A process of this type defines a transformation $T : \Omega \to \Omega$ given by $T(\omega)_t = \omega_t + \int_0^t u_s(\omega)ds$. Our purpose is to discuss the problem of the absolute continuity of the image measure $P \circ T^{-1}$ and the computation of its density functio with respect to the Wiener measure.

If the process u is adapted to the family of σ–fields generated by the Wiener process this problem was treated by I. V. Girsanov in [10]. In that case $P \circ T^{-1}$ is always absolutely continuous with respect to the Wiener measure. Moreover if the random variable $\xi = \exp(-\int_0^1 u_t d\omega_t - \frac{1}{2}\int_0^1 u_t^2\, dt)$ has expectation equal to one then the probability Q defined by $\frac{dQ}{dP} = \xi$ verifies $Q \circ T^{-1} = P$, in other words, $\{T(\omega)_t,\ 0 \leq t \leq 1\}$ is a Brownian motion under Q. These

ISBN 0-12-481005-5

results are well-known and can be obtained as an application of the stochastic calculus developed by Itô.

The case of a nonnecessarily adapted process u is much more delicate. The first to study this problem were R. H. Cameron and W. T. Martin in [6]. Under some strong smoothness conditions on the process u they proved the existence of a probability Q such that $Q \circ T^{-1} = P$ and they obtained an explicit formula for the density function of Q with respect to P. L. Gross [11] and H. H. Kuo [12] extended the work of R. H. Cameron and W. T. Martin for general abstract Wiener spaces. In [23] R. Ramer introduced an abstract version of the Itô integral for nonadapted processes and expressed the density function $\frac{dQ}{dP}$ (where Q verifies $Q \circ T^{-1} = P$) in terms of this nonadapted integral of the process u, under some assumptions on u, one of which being the continuity of the mapping $\omega \mapsto u(\omega)$ from Ω into $L^2(0, 1)$. This continuity condition was later removed in the work of S. Kusuoka [14].

The abstract stochastic integral defined by R. Ramer in [23] in order to compute the density function $\frac{dQ}{dP}$, coincides with the generalization of the Itô integral defined by A. V. Skorohod in [26]. On the other hand this nonadapted stochastic integral is precisely the divergence operator which is used in the stochastic calculus of variations. Recently a stochastic calculus for this anticipating integral has been developed. We refer to [15] and the references therein for an exposition of this theory. In connection with the recent progress in the noncausal stochastic calculus, the works of R. Ramer and S. Kusuoka provide a useful tool to deal with anticipating transformations and their results have been applied to some particular problems. For instance, in [17, 18] Kusuoka's theorem has been used to study the Markov property for the solutions of stochastic differential equations with boundary conditions. On the other hand, R. Buckdahn [2, 4, 5] (see also the work of A. S. Ustunel and M. Zakai in [28]) has obtained a new expression for the density function $\frac{dQ}{dP}$ which avoids the computation of a certain Carleman–Fredholm determinant.

Our first objective in these notes is to review some results concerning the problem mentioned above, that means, to find conditions for the absolute continuity of $P \circ T^{-1}$ with respect to P and to compute its density function. We will present in Section 1 the Ramer–Kusuoka's theorem and the results for a one-parameter family of transformations due to R. Buckdahn, A. S. Ustunel and M. Zakai. Section 2 will be devoted to describe the application of Ramer–Kusuoka's theorem to study the Markov property of the solution of stochastic differential equations with boundary conditions. In order to illustrate the methodology and the techniques used to study these problems, we will discuss in detail the case of a second order stochastic differential equation with Neumann boundary conditions.

1 Nonlinear Transformations of the Wiener Measure

Suppose that $(\Omega, \mathcal{F}, P)$ is the classical Wiener space. That means, $\Omega = C_0([0,1])$ is the space of continuous functions that vanish at zero, P is the Wiener measure and $\mathcal{F}$ is the Borel σ–field of Ω completed with respect to P. The coordinate process $W_t(\omega) = \omega(t)$ is the canonical Brownian motion.

Consider a measurable process $u = \{u_t,\ 0 \leq t \leq 1\}$ such that $\int_0^1 u_t^2\, dt < \infty$ a.s. and define the transformation $T : \Omega \to \Omega$ by

$$T(\omega)_t = \omega_t + \int_0^t u_s(\omega)\, ds\,. \tag{1.1}$$

We want to discuss the following problems:

(A) When is the image of the Wiener measure $P \circ T^{-1}$ absolutely continuous with respect to P, and which is the value of the density function $\frac{d[P \circ T^{-1}]}{dP}$ in terms of the process u.

(B) Find a new probability Q on $(\Omega, \mathcal{F})$, absolutely continuous with respect to P, such that $\{\omega_t + \int_0^t u_s(\omega)\, ds, 0 \leq t \leq 1\}$ has the law of a Brownian motion on the probability space $(\Omega, \mathcal{F}, Q)$. In other words, Q must satisfy $Q \circ T^{-1} = P$.

These two problems are related in the following way.

Proposition 1.1 *The probability $P \circ T^{-1}$ is absolutely continuous with respect to P if and only if there exists a probability Q absolutely* *continuous with respect to P such that $Q \circ T^{-1} = P$. Under these assumptions, if $X = \frac{d[P \circ T^{-1}]}{dP}$ and $Y = \frac{dQ}{dP}$ we have $E(Y \mid T) = \frac{1}{X(T)}$.*

Proof: Suppose first that $P \circ T^{-1}$ is absolutely continuous with respect to P and set $X = \frac{d[P \circ T^{-1}]}{dP}$. Define the measure Q by $dQ = \frac{1}{X(T)} dP$, and observe that $P\{X(T) = 0\} = (P \circ T^{-1})\{X = 0\} = 0$. Then we have

$$Q(T^{-1}(B)) = \int_{T^{-1}(B)} \frac{1}{X(T)} dP = \int_B \frac{1}{X} d[P \circ T^{-1}] = P(B),$$

for any Borel subset B of Ω.

Conversely let Q be a probability absolutely continuous with respect to P such that $Q \circ T^{-1} = P$. Then $P(B) = Q(T^{-1}(B)) = 0$ implies $P(T^{-1}(B)) = 0$ and, therefore, $P \circ T^{-1}$ is absolutely continuous with respect to P. Finally, if $Y = \frac{dQ}{dP}$ and $X = \frac{d[P \circ T^{-1}]}{dP}$, then it is easy to check that $E(Y \mid T) = \frac{1}{X(T)}$. □

Suppose that u_t is $\mathcal{F}_t$–measurable for all $t \in [0, 1]$, where $\mathcal{F}_t$ is the σ–field generated by $\{W_s,\ 0 \leq s \leq t\}$ and the P–null sets. In this case it holds that $P \circ T^{-1} << P$. Moreover, if $E(\xi) = 1$, where $\xi = \exp(-\int_0^1 u_t dW_t - \frac{1}{2} \int_0^1 u_t^2\, dt)$, the probability Q given by $\frac{dQ}{dP} = \xi$ verifies $Q \circ T^{-1} = P$.

In order to describe the results obtained in the nonnecessarily adapted case we have to introduce some preliminary notations and definitions of the theory of noncausal stochastic calculus.

1.1 Some Elements of Noncausal Stochastic Calculus

We refer the reader to the references [15, 16] for a more detailed exposition of the basic results of the noncausal stochastic calculus. Let us first recall briefly the notions of derivation on Wiener space and Skorohod integral.

Let us denote by H the Hilbert space $L^2(0,1)$. For any $h \in H$ we will denote by $W(h)$ the Wiener integral $\int_0^1 h(t)dW_t$. We denote by $\mathcal{S}$ the dense subset of $L^2(\Omega)$ consisting of those random variables of the form

$$F = f(W(h_1), \ldots, W(h_n)), \tag{1.2}$$

where $n \geq 1$, $h_1, \ldots, h_n \in H$ and $f \in C_b^\infty(\mathbb{R}^n)$ (that means, f and all its partial derivatives are bounded). The random variables of the form (1.2) are called *smooth functionals.* For a smooth functional $F \in \mathcal{S}$ of the form (1.2) we define its derivative DF as the stochastic process $\{D_t F,\ 0 \leq t \leq 1\}$ given by

$$D_t F = \sum_{i=1}^{n} \frac{\partial f}{\partial x_i} (W(h_1), \ldots, W(h_n)) h_i(t) . \tag{1.3}$$

Then D is a closable unbounded operator from $L^2(\Omega)$ into $L^2(\Omega \times [0,1])$. We will denote by $\mathbb{D}^{1,2}$ the completion of $\mathcal{S}$ with respect to the norm $\| \cdot \|_{1,2}$ defined by

$$\| F \|_{1,2} = \| F \|_2 + \| DF \|_{L^2(\Omega \times [0,1])}, \quad F \in \mathcal{S} .$$

We will denote by δ the adjoint of the derivation operator D. That means, δ is a closed and unbounded operator from $L^2(\Omega \times [0,1])$ into $L^2(\Omega)$ defined as follows: The domain of δ, Dom δ, is the set of processes $u \in L^2(\Omega \times [0,1])$ such that there exists a positive constant c_u verifying

$$\left| E \int_0^1 D_t F u_t dt \right| \leq c_u \| F \|_2 , \tag{1.4}$$

for all $F \in \mathcal{S}$. If u belongs to the domain of δ then $\delta(u)$ is the square integrable random variable determined by the duality relation

$$E\left(\int_0^1 D_t F u_t dt \right) = E(\delta(u) F), \quad F \in \mathbb{D}^{1,2} . \tag{1.5}$$

The operator δ is an extension of the Itô integral in the sense that the class L_a^2 of processes u in $L^2(\Omega \times [0,1])$ which are adapted to the Brownian filtration is included in Dom δ and $\delta(u)$ is equal to the Itô integral if $u \in L_a^2$. The operator δ is called the Skorohod stochastic integral. Define $\mathbb{L}^{1,2} = L^2([0,1]; \mathbb{D}^{1,2})$. Then the space $\mathbb{L}^{1,2}$ is included into the domain of δ. The operators D and δ are *local* in the following sense:

(a) $1_{\{F=0\}} DF = 0$, for all $F \in \mathbb{D}^{1,2}$,

(b) $1_{\{\int_0^1 u_t^2 dt = 0\}} \delta(u) = 0$, for all $u \in \mathbb{L}^{1,2}$.

Using these local properties one can define the spaces $\mathbb{D}^{1,2}_{\text{loc}}$ and $\mathbb{L}^{1,2}_{\text{loc}}$ by a standard localization procedure. For instance $\mathbb{D}^{1,2}_{\text{loc}}$ is the space of random variables F such that there exists a sequence $\{(\Omega_n, F_n),\ n \geq 1\}$ such that $\Omega_n \in \mathcal{F}$, $\Omega_n \uparrow \Omega$ a.s., $F_n \in \mathbb{D}^{1,2}$, and $F_n = F$ on Ω_n for each $n \geq 1$. By property (a) the derivation operator D can be extended to random variables of the space $\mathbb{D}^{1,2}_{\text{loc}}$. In a similar way we can extend the Skorohod integral δ to the space $\mathbb{L}^{1,2}_{\text{loc}}$.

Let us recall the definition of the Stratonovich integral and its relation with the Skorohod integral. Let $\{u_t, 0 \leq t \leq 1\}$ be a measurable process such that $\int_0^1 u_t^2\, dt < \infty$ a.s. Then u is said to be *Stratonovich integrable* if

$$\sum_{j=1}^{n-1} \left(\frac{1}{t_{j+1} - t_j} \int_{t_j}^{t_{j+1}} u_s ds \right) (W(t_{j+1}) - W(t_j))$$

converges in probability as $\pi \to 0$, where $\pi = \{0 = t_1 < \cdots < t_n = 1\}$ runs over all finite partitions of $[0,1]$ and $\pi = \max_j (t_{j+1} - t_j)$. The limit will be called the *Stratonovich integral* of u and it will be denoted by $\int_0^1 u_t \circ dW_t$.

Let $\mathbb{L}^{1,2}_C$ denote the set of processes $u \in \mathbb{L}^{1,2}$ such that:

(i) The set of functions $\{s \to D_t u_s, 0 \leq s \leq t\}$, $t \in [0,1]$, with values in $L^2(\Omega)$ is equicontinuous for some version of Du, and similarly, the set of functions $\{s \to D_t u_s, t \leq s \leq 1\}$, $t \in [0,1]$, is also equicontinuous for a version (possibly different) of Du.

(ii) $\text{ess sup}_{s,t}\, E(D_s u_t^2) < \infty$.

For a process u in the class $\mathbb{L}^{1,2}_C$ we define

$$D_t^+ u = \lim_{\epsilon \downarrow 0} D_t u_{t+\epsilon} \qquad D_t^- u = \lim_{\epsilon \downarrow 0} D_t u_{t-\epsilon}$$

Then we have (cf. Theorem 7.3 of [16]):

Proposition 1.2 *If u belongs to $\mathbb{L}^{1,2}_C$ then u is Stratonovich integrable and*

$$\int_0^1 u_t \circ dW_t = \delta(u) + \frac{1}{2} \int_0^1 (D_t^+ u + D_t^- u) dt.$$

From the point of view of the stochastic calculus, the Stratonovich integral behaves as an ordinary pathwise integral. We refer to [16] for a detailed discussion of this fact.

The next proposition provides a different kind of sufficient conditions for the existence of the Stratonovich integral.

Proposition 1.3 *Let u be a process in $\mathbb{L}^{1,2}_{\text{loc}}$. Suppose that the integral operator from H into H associated with the kernel $Du(\omega)$ is a nuclear (or a trace class) operator, for all ω a.s. Then u is Stratonovich integrable and we have*

$$\int_0^1 u_t \circ dW_t = \delta(u) + \mathrm{Tr} Du\,. \tag{1.6}$$

Proof: From Proposition 6.1 of [20] we know that for any complete orthonormal system $\{e_i, i \geq 1\}$ in H the series

$$\sum_{i=1}^{\infty} \left(\int_0^1 u_t e_i(t) dt \right) \int_0^1 e_i(s) dW_s$$

converges in probability to $\delta(u) + \mathrm{Tr}\, Du$. Then the Proposition follows from the results of [21]. □

1.2 Extended Versions of Girsanov Theorem

Let $u = \{u_t, 0 \leq t \leq 1\}$ be a measurable process such that $\int_0^1 u_t^2\, dt < \infty$ a.s., and consider the nonlinear transformation given by (1.1). In order to explain the main ideas of Girsanov theorem we will study in detail the particular case of an elementary process. Denote by $\{e_i,\ i \geq 1\}$ a complete orthonormal system in H. We will say that the process u is an *elementary process* if it is of the form

$$u_t = \sum_{j=1}^{N} \psi_j \left(W(e_1), \ldots, W(e_N) \right) e_j(t)\,, \tag{1.7}$$

where $\psi_j \in C_b^\infty(\mathbb{R}^N)$. For a process u of the form (1.7) the transformation $T : \Omega \to \Omega$ is given by

$$T(\omega)_t = \omega_t + \sum_{j=1}^{N} \psi_j \left(W(e_1), \ldots, W(e_N) \right) \int_0^t e_j(s) ds\,.$$

Let us denote by Δ the $N \times N$ random matrix equal to the Jacobian matrix of ψ evaluated at the point $(W(e_1), \ldots, W(e_N))$, namely $\Delta = \frac{\partial \psi}{\partial x}(W(e_1), \ldots, W(e_N))$. We define the following random variable

$$\eta = \det(I + \Delta) \exp\left(- \sum_{j=1}^{N} < u, e_j > W(e_j) - \frac{1}{2} \sum_{j=1}^{N} < u, e_j >^2 \right). \tag{1.8}$$

Consider a nonnegative random variable of the form

$$F = f(W(e_1), \ldots, W(e_N), G) \tag{1.9}$$

where G denotes the vector $(W(e_{N+1}), \ldots, W(e_M))$, $M > N$ and f belongs to the space $C_b^\infty(\mathbb{R}^M)$. The composition $F(T)$ is given by

$$F(T) = f(W(e_1) + < u, e_1 >, \ldots, W(e_N) + < u, e_N >, G).$$

Let us compute the expectation

$$\begin{aligned} E[\eta\, F(T)] &= E[\det(I + \Delta) \\ &\quad \cdot \exp\left(- \sum_{j=1}^{N} < u, e_j > W(e_j) - \frac{1}{2} \sum_{j=1}^{N} < u, e_j >^2 \right) \\ &\quad \cdot f(W(e_1) + < u, e_1 >, \ldots, W(e_N) + < u, e_N >, G)] \\ &= E \int_{\mathbb{R}^N} \left| \det\left(I + \frac{\partial \psi}{\partial x} \right) \right| \exp\left(- \frac{1}{2} \sum_{j=1}^{N} (x_j + \psi_j(x))^2 \right) \\ &\quad \cdot f(x_1 + \psi_1(x), \ldots, x_N + \psi_N(x), G)(2\pi)^{-N/2} dx_1 \cdots dx_N. \end{aligned}$$

Suppose that the mapping $\varphi(x) = x + \psi(x) : \mathbb{R}^N \to \mathbb{R}^N$ is *one-to-one* (in this case, we will say that u is a *one-to-one elementary process*). If we make the change of variables $x_j + \psi_j(x) = y_j$ in the integral over $\mathbb{R}^N$ we obtain

$$\begin{aligned} E[\eta F(T)] &= E \int_{\varphi(\mathbb{R}^n)} \exp\left(- \frac{1}{2} \sum_{j=1}^{N} y_j^2 \right) \\ &\quad \cdot f(y_1, \ldots, y_N, G)\, (2\pi)^{-N/2} dy_1 \cdots dy_N \end{aligned}$$

$$
\begin{aligned}
&\leq E \int_{\mathbb{R}^n} \exp\left(-\frac{1}{2}\sum_{j=1}^{N} y_j^2\right) f(y_1, \ldots, y_N, G)(2\pi)^{-N/2} dy_1 \cdots dy_N \\
&= E(F). \qquad (1.10)
\end{aligned}
$$

We claim that the random variable η defined in (1.8) can also be written as

$$
\eta = d_c(-Du) \ \exp(-\delta(u) - \frac{1}{2}\int_0^1 u_t^2 \, dt), \qquad (1.11)
$$

where $d_c(-Du)$ denotes the Carleman–Fredholm determinant of the square integrable kernel $-Du$ (we refer to [25, 27] for the properties of this determinant). In order to deduce the above equality, let us first remark that for an elementary process u of the form (1.7) the Skorohod integral $\delta(u)$ can be evaluated as follows (see [15])

$$
\begin{aligned}
\delta(u) &= \sum_{j=1}^{N} \psi_j(W(e_1), \ldots, W(e_N)) W(e_j) \\
&\quad - \sum_{j=1}^{N}\sum_{i=1}^{N} \frac{\partial \psi_j}{\partial x_i}(W(e_1), \ldots, W(e_N)) < e_i, e_j > \\
&= \sum_{j=1}^{N} < u, e_j > W(e_j) - \mathrm{Tr}\Delta. \qquad (1.12)
\end{aligned}
$$

On the other hand, for a process u given by (1.7), the derivative $D_s u_t$ is equal to

$$
\sum_{j=1}^{N}\sum_{i=1}^{N} \frac{\partial \psi_j}{\partial x_i} (W(e_1), \ldots, W(e_N))\, e_j(t)\, e_i(s), \qquad (1.13)
$$

that means,

$$
Du = \sum_{j=1}^{N}\sum_{i=1}^{N} \Delta_{ij}\ e_i \otimes e_j, \qquad (1.14)
$$

and the Carleman–Fredholm determinant of the kernel $-Du$ is equal to $\det(I + \Delta) \exp(-\mathrm{Tr}\Delta)$, which completes the proof of (1.11).

This random variable η will play the role of the Girsanov exponential $\exp(-\int_0^1 u_t dW_t - \frac{1}{2}\int_0^1 u_t^2 dt)$ in the adapted case. If we compare both expressions we see that the Itô integral $\int_0^1 u_t\, dW_t$ has been replaced by the Skorohod integral $\delta(u)$ and a new factor appears which is equal to $d_c(-Du)$.

Using these ideas we can obtain the following results for the problems (A) and (B) introduced at the beginning of this section (see D. Nualart and M. Zakai [20, Section 7]).

Proposition 1.4 *Let u be an elementary process of the form (1.7). Then, if $d_c(-Du) \neq 0$ a.s., the probability $P \circ T^{-1}$ is absolutely continuous with respect to P.*

Proof: Notice that the inequality (1.10) holds for any nonnegative and bounded random variable F, by a monotone class argument. Therefore, if $E(F) = 0$, then $E(\eta F(T)) = 0$, which implies $E(F(T)) = 0$, because $\eta > 0$ a.s., by hypothesis. □

In Proposition 7.3 of [20] the above result is extended to processes of the form $u_t = \sum_{i=1}^N F_i e_i(t)$ where the random variables F_i belong to the space $\mathbb{D}^{2,2}$. We do not know if the result is true for an arbitrary process u in the space $\mathbb{L}^{1,2}$.

Concerning problem (B) one can deduce the following result (see [20, Theorem 7.4]).

Theorem 1.1 *Let u be a one-to-one elementary process of the form (1.7). Suppose that $E(\eta) = 1$, where η is the random variable given by (1.11). Then the probability Q defined by $\frac{dQ}{dP} = \eta$ verifies $Q \circ T^{-1} = P$, that means, the process $\{T(\omega)_t,\ 0 \leq t \leq 1\}$ is a Wiener process under Q (which implies $P \circ T^{-1} << P$ by Proposition 1.1).*

Furthermore, this is still true if u is a process of the space $\mathbb{L}^{1,2}$ such that there exists a sequence $\{u^n,\ n \geq 1\}$ of one-to-one elementary processes which converge to u in the norm of $\mathbb{L}^{1,2}$.

Proof: The inequality (1.10) implies that

$$\int_\Omega F d(Q \circ T^{-1}) \leq \int_\Omega F dP \tag{1.15}$$

for any bounded nonnegative random variable F and this implies $Q \circ T^{-1} = P$, because $Q \circ T^{-1}$ is a probability. The last part of the theorem follows by Fatou's lemma. $\square$

Theorem 1.1 can be regarded as a nonadapted version of the classical Girsanov theorem. In general the condition $E(\eta) = 1$ is not easy to check. This difficulty can be avoided using the results obtained by Kusuoka in [14]. The hypotheses imposed by Kusuoka require a smoothness property on the mapping $\omega \mapsto u(\omega)$ which is stronger than the fact that $u \in \mathbb{L}^{1,2}$. Let us introduce the following definition.

Definition 1.1 *Let u be a measurable process such that $\int_0^1 u_t^2(\omega)dt < \infty$, for all $\omega \in \Omega$. We will say that u is $\mathcal{H}$–C^1 (namely, the mapping $\omega \mapsto u(\omega)$ from Ω into H is $\mathcal{H}$–C^1) if there exists a random kernel $Du(\omega) \in L^2([0,1]^2)$ such that:*

(i) $\| u(\omega + \int_0^\cdot h_s\, ds) - u(\omega) - Du(\omega)(h) \|_H = o(\| h \|_H)$ *for all* $\omega \in \Omega$.

(ii) *The mapping $h \mapsto Du(\omega + \int_0^\cdot h_s\, ds)$ is continuous from H into $L^2([0,1]^2)$ for all $\omega \in \Omega$.*

It holds that (see the proof of Theorem 5.2 in Kusuoka [14]) if u is $\mathcal{H}$–C^1 then u belongs to $\mathbb{L}^{1,2}_{\text{loc}}$ and the kernel Du verifying the conditions (i) and (ii) of Definition 1.1 is precisely the derivation operator D applied to u. Furthermore if $u : \Omega \to H$ is continuously Fréchet differentiable then u is $\mathcal{H}$–C^1, and for every $\omega \in \Omega$, $Du(\omega) \in H \otimes H$ is the derivative of u.

Using this definition we can state the following result proved by Kusuoka ([14, Theorem 6.4]).

Theorem 1.2 *Let $u = \{u_t, 0 \le t \le 1\}$ be a measurable process which defines an $\mathcal{H}$–C^1 map from Ω into $H = L^2(0,1)$. Suppose that:*

(i) *The transformation $T : \Omega \to \Omega$ given by $T(\omega)_t = \omega_t + \int_0^t u_s(\omega)ds$ is bijective.*

(ii) *$I + Du : H \to H$ is invertible, for all $\omega \in \Omega$ a.s.*

Then there exists a probability Q *given by*

$$\frac{dQ}{dP} = d_c(-Du)\exp(-\delta(u) - \frac{1}{2}\int_0^1 u_t^2\,dt)\,, \qquad (1.16)$$

such that $Q \circ T^{-1} = P$.

If the integral operator associated with the kernel $-Du$ is nuclear, then the Carleman–Fredholm determinant of $-Du$ can be evaluated as follows:

$$d_c(-Du) = \det(I + Du)\ \exp(-\mathrm{Tr}Du)\,, \qquad (1.17)$$

and using Proposition 1.3 the density function (1.16) can be expressed as

$$\frac{dQ}{dP} = \det\,(I + Du)\exp\left(-\int_0^1 u_t \circ dW_t - \frac{1}{2}\int_0^1 u_t^2\,dt\right)\,. \qquad (1.18)$$

The method used in [14] to show Theorem 1.2 consists in first to deal with the case where the mapping $\omega \mapsto \int_0^{\cdot} u_s(\omega)\,ds$ is a contraction from Ω into H. The general case is then treated by decomposing the space Ω into measurable subsets. The $\mathcal{H}$–C^1 property plays a basic role in the proof, and we do not know if the conclusion is true if we replace it by some weaker smoothness assumption like the fact that $u \in \mathbb{L}^{1,2}_{\mathrm{loc}}$. In a recent paper [4], R. Buckdahn has succeeded in proving Theorem 1.2 for a process $u \in \mathbb{L}^{1,2}$ which verifies a Novikov-type condition and such that the norm of the derivative Du is strictly less than one. More precisely, the result proved by Buckdahn in [4] is as follows.

Theorem 1.3 *Let* u *be a process of the space* $\mathbb{L}^{1,2}$ *such that*
(i) $\| Du \|_{H\otimes H} < 1$,
(ii) $E\left(\exp\left(\frac{q}{2}\int_0^1 u_t^2 dt\right)\right) < \infty$ *for some* $q > 1$.
Then the transformation $T(\omega)_t = \omega_t + \int_0^t u_s(\omega)ds$ *verifies* $P \circ T^{-1} << P$ *and there exists another transformation* $A : \Omega \to \Omega$ *verifying* $P \circ A^{-1} << P$ *and* $T(A\omega) = A(T\omega) = \omega$, P*–almost surely. Furthermore*

the probability $Q = P \circ A^{-1}$ *(which clearly satisfies* $Q \circ T^{-1} = P$*) has a density function given by*

$$\begin{aligned}\frac{dQ}{dP} &= \exp\Big(- \delta(u) - \tfrac{1}{2} \textstyle\int_0^1 u_t^2 dt \\ &- \textstyle\int_0^1 \int_0^t D_s u_t (D_t(u_s(A_t)))(T_t) ds dt \Big),\end{aligned} \tag{1.19}$$

where $\{T_t, \, 0 \leq t \leq 1\}$ *is the one–parameter family of transformations of* Ω *defined by*

$$(T_t \omega)_s = \omega_s + \int_0^{s \wedge t} u_r(\omega) dr \tag{1.20}$$

and $\{A_t, \, 0 \leq t \leq 1\}$ *is the corresponding family of inverse transformations, in the sense that* $T_t(A_t \omega) = A_t(T_t \omega)$ *a.s.,* $P \circ T_t^{-1} << P$, *and* $P \circ A_t^{-1} << P$ *for all* $t \in [0,1]$.

What is interesting in the above theorem is the introduction of the one–parameter family of transformation (1.20) which allows to obtain a formula for the density function without using the Carleman–Fredholm determinant. In the next section we will explain in more detail the role played by these families of transformations.

If we compare the expressions (1.16) and (1.19), which were obtained under different hypotheses, we deduce the following formula for the Carleman–Fredholm determinant of $-Du$:

$$d_c(-Du) = \exp\left(- \int_0^1 \int_0^t D_s u_t \, (D_t(u_s(A_t)))(T_t) \, ds \, dt \right) . \tag{1.21}$$

1.3 One-parameter Families of Transformations

Suppose that $u = \{u_t, \, 0 \leq t \leq 1\}$ is an adapted and measurable process such that $\int_0^1 u_t^2 \, dt < \infty$ a.s. Define

$$M_t = \exp\left(- \int_0^t u_s dW_s - \frac{1}{2} \int_0^t u_s^2 \, ds \right) . \tag{1.22}$$

The process $M = \{M_t, \, 0 \leq t \leq 1\}$ is a continuous local martingale which verifies the linear stochastic differential equation

$$M_t = 1 - \int_0^t u_s \, M_s \, dW_s \, . \tag{1.23}$$

Under the additional hypothesis $E(M_1) = 1$, the process M is a martingale and there exists a probability Q defined by $\frac{dQ}{dP} = M_1$ such that $Q \circ T^{-1} = P$, where T is the nonlinear transformation defined in terms of u, that means, $T(\omega)_t = \omega_t + \int_0^t u_s(\omega)\, ds$.

A natural problem is to try to extend these results to the non-adapted case. That means we would like to relate the density function $\frac{dQ}{dP}$ corresponding to a nonlinear anticipating transformation T, with the solution to a linear equation like (1.23) where now the process u_s is not necessarily adapted and the stochastic integral is taken in the Skorohod sense. In some recent papers (see [2, 4, 5]), R. Buckdahn has developed a new approach to deal with this problem. The basic ideas of this approach are the following. Instead of studying the transformation T associated with u let us introduce a one-parameter family of transformations $\{T_t,\ 0 \leq t \leq 1\}$ defined by the integral equations:

$$(T_t\omega)(s) = \omega_s + \int_0^{t\wedge s} u_r(T_r\omega)\, dr\,. \tag{1.24}$$

Denote by $\mathbb{L}^{1,\infty}$ the space of processes $u \in \mathbb{L}^{1,2}$ such that $\int_0^1 \| u_t \|_\infty^2\, dt + \int_0^1 \| \| Du_t \|_H \|_\infty^2\, dt < \infty$. Then, Buckdahn has shown (cf. [5]) the following result.

Theorem 1.4 *Let u be a process belonging to the space $\mathbb{L}^{1,\infty}$. Then there exists a unique family of transformations $\{T_t, 0 \leq t \leq 1\}$ which satisfy the equation (1.24). Furthermore this family verifies the following properties:*

(i) *For each $t \in [0,1]$ there exists a transformation $A_t : \Omega \to \Omega$ such that $T_t \circ A_t = A_t \circ T_t = Id$ a.s., $P \circ T_t^{-1} << P$, $P \circ A_t^{-1} << P$, and the density functions of $P \circ T_t^{-1}$ and $P \circ A_t^{-1}$ are given by*

$$M_t = \frac{d[P\circ A_t^{-1}]}{dP} = \exp\Big\{-\int_0^t u_s(T_s)dW_s - \tfrac{1}{2}\int_0^t u_s(T_s)^2\, ds - \int_0^t \int_0^s (D_r u_s)(T_s)\, D_s[u_r(T_r)]drds\Big\} \tag{1.25}$$

$$L_t = \frac{d[P \circ T_t^{-1}]}{dP} = \exp\Big\{ \int_0^t u_s(T_s A_t) dW_s - \frac{1}{2}\int_0^t u_s(T_s A_t)^2\, ds$$
$$- \int_0^t \int_0^s (D_s u_r)(T_r A_t) D_r[u_s(T_s A_t)]\, dr ds \Big\} \tag{1.26}$$

(ii) The density process $\{L_t,\, 0 \le t \le 1\}$ *satisfies the linear stochastic differential equation* $L_t = 1 + \int_0^t u_s L_s dW_s$*, in the sense that* $uL\mathbf{1}_{[0,t]} \in \operatorname{Dom}\delta$ *for each* $t \in [0,1]$.

Remarks:

(1) Suppose, in particular, that $u \in \mathbb{L}^{1,\infty}$ is an adapted process. In this case we claim that $(A_t\omega)(s) = \omega_s - \int_0^{t\wedge s} u_r(\omega)dr$. In fact, applying (1.24) to $A_t\omega$ we obtain

$$\omega = A_t\omega + \int_0^{t\wedge\cdot} u_r(\omega)dr, \quad \text{a.s.}$$

Then, from (1.26) we have $L_t = \exp\left(\int_0^t u_s dW_s - \frac{1}{2}\int_0^t u_s^2 ds\right)$, which coincides with the classical solution of the linear differential equation $L_t = 1 + \int_0^t u_s L_s dW_s$.

(2) Suppose that for some $t \in [0,1]$ the transformation T_t (namely, the associated process $\overline{u}_t = u_t(T_t)$) verifies the assumptions of Theorem 1.2; then $Q = P \circ A_t^{-1} = P \circ T_t$ and comparing (1.16) with (1.25) we deduce the following alternative expression for the Carleman–Fredholm determinant of $D\overline{u}_t = D[u_t(T_t)]$:

$$d_c(-D\overline{u}) = \exp\left(-\int_0^1 \int_0^t (D_s u_t)(T_t)\, D_t[u_s(T_s)]\, dsdt\right).$$

The method used by Buckdahn to show the preceding theorem consists in first proving the result for a smooth process of the form $u_t = f(t, W_{t_1}, \ldots, W_{t_n})$, where $0 < t_1 < \cdots < t_n < 1$ and $f \in C_b^\infty(\mathbb{R}^{n+1})$. The general case is obtained by means of a limit argument. A. S. Ustunel and M. Zakai (cf. [28]) have extended the above theorem to processes u such that:

$$E \int_0^1 \exp(\lambda u_r^2) dr < \infty, \quad \text{and} \quad \int_0^1 \| \| \| Du_t \|_H \|_\infty^4 \, dt < \infty,$$

for some $\lambda > 0$. In their work the conclusions of Theorem 1.3 are deduced from a general expression of the Radon-Nikodym derivative associated with smooth flows of transformations of Ω.

2 Stochastic Differential Equations with Boundary Conditions

In this section we will describe some recent applications of the anticipating Girsanov theorems discussed in Section 1, to investigate the Markov property of the solutions of some stochastic differential equations with boundary conditions. We will discuss first the case of first order stochastic differential equations studied by D. Nualart and E. Pardoux in [17] . For the sake of simplicity we will restrict the study to the one dimensional case.

2.1 Nonlinear Stochastic Differential Equations with Boundary Conditions

Consider the following equation

$$\left.\begin{aligned} dX_t + f(X_t)\, dt &= dW_t\,, \quad 0 \le t \le 1 \\ h(X_0, X_1) &= 0 \end{aligned}\right\} \tag{2.1}$$

where W is a one dimensional Wiener process and $f : \mathbb{R} \to \mathbb{R}$ and $h : \mathbb{R}^2 \to \mathbb{R}$ are measurable and locally bounded functions. That means, instead of giving the initial value X_0 as it is customary, we impose a nonlinear functional relation between X_0 and X_1.

The first step will be to find the explicit solution of the equation (2.1) in the linear case, namely, when $f(x) = \lambda x$ for some real number λ. In this case we have

$$\left.\begin{aligned} dY_t + \lambda Y_t\, dt &= dW_t\,, \quad 0 \le t \le 1 \\ h(Y_0, Y_1) &= 0 \end{aligned}\right\} \tag{2.2}$$

The boundary value problem (2.2) is equivalent to the following system

$$\left.\begin{array}{l} Y_t = e^{-\lambda t}(Y_0 + \int_0^t e^{\lambda s} dW_s) \\ h(Y_0,\, e^{-\lambda}(Y_0 + \int_0^1 e^{\lambda s} dW_s)) = 0 \end{array}\right\} \tag{2.3}$$

In order to solve the equation (2.3) we will impose the following assumption:

(H.1) For all $z \in \mathbb{R}$ the equation $h(y, e^{-\lambda}(y+z)) = 0$ has a unique solution $y = g(z)$.

Under this assumption we obtain

$$Y_t = e^{-\lambda t}(g(\xi_1) + \xi_t)\,, \tag{2.4}$$

where $\xi_t = \int_0^t e^{\lambda s} dW_s$. As in Section 1, Ω will denote the space $C_0([0,1])$ of continuous functions on $[0,1]$ which vanish at zero. On the other hand we denote by Σ the space of continuous functions $x : [0, 1] \to \mathbb{R}$ such that $h(x_0, x_1) = 0$. Then, the hypothesis (H.1) implies that there is a bijection $\Psi : \Omega \to \Sigma$ such that $Y = \Psi(W)$.

In order to study the case of a nonlinear function f, we introduce the decomposition $f(x) = \lambda x + \overline{f}(x)$ where λ is some fixed real number, such that (H.1) holds. Notice that in the particular case of periodic boundary conditions (that means, $h(x, y) = x - y$) the condition (H.1) holds only for $\lambda \neq 0$. We define the process $u_t = \overline{f}(Y_t)$, where Y_t is given by (2.4), and the corresponding transformation $T : \Omega \to \Omega$ given by

$$T(\omega)_t = \omega_t + \int_0^t \overline{f}(Y_s(\omega))\, ds\,. \tag{2.5}$$

The anticipating Girsanov formulas discussed in Section 1 will be used in the following way. Suppose that there exists a probability Q on Ω, absolutely continuous with respect to P, such that $Q \circ T^{-1} = P$. Then, the process $\overline{W}_t = W_t + \int_0^t \overline{f}(Y_s)\, ds$ will be a Brownian motion under Q, and we have

$$dY_t + \lambda Y_t dt \;= dW_t = d\overline{W}_t - \overline{f}(Y_t)\, dt\,. \tag{2.6}$$

Therefore, the law of $\{Y_t\}$ under Q coincides with the law of $\{X_t\}$ under P. Consequently the probabilistic properties of the process

$\{X_t\}$, like the Markov property, will be translated into the corresponding properties for the process $\{Y_t\}$ given by (2.4), under a new probability law Q.

To carry out this program we have to show that the process $u_t = \overline{f}(Y_t)$ satisfies the assumptions of the extended Girsanov theorem presented in Section 1.2. More precisely we will make use of Theorem 1.2 due to Kusuoka [14]. In particular we have to find sufficient conditions for the transformation $T : \Omega \to \Omega$ to be bijective. We remark that if T is bijective, then there is a unique solution to the equation (2.1) given by $X(\omega) = \Psi^{-1}(T(\omega))$. The following proposition provides some sufficient conditions for T to be bijective. It can be proved in the same way as Proposition 2.3 below. We refer to [17] for other types of conditions.

Proposition 2.1 *Suppose that $\overline{f}$ and g are nondecreasing, locally Lipschitz functions with linear growth. Then T is bijective.*

The application of Theorem 1.2 leads to the following result that we state without proof (see [17] for a complete proof in higher dimensions).

Theorem 2.1 *Suppose that $\overline{f}$ and g are continuously differentiable functions such that the mapping T given by (2.5) is bijective and*

$$g'(\xi_1)\left[1 - \exp\left(-\int_0^1 \overline{f}'(Y_s)\,ds\right)\right] \neq -1\,. \tag{2.7}$$

Then there exists a probability Q given by

$$\frac{dQ}{dP} = \left|1 + g'(\xi_1)\left[1 - \exp\left(-\int_0^1 \overline{f}'(Y_t)\,dt\right)\right]\right| \times \exp\left\{\frac{1}{2}\int_0^1 \overline{f}'(Y_t)dt - \int_0^1 \overline{f}(Y_t)\circ dW_t - \frac{1}{2}\int_0^1 \overline{f}(Y_t)^2 dt\right\}, \tag{2.8}$$

such that $Q \circ T^{-1} = P$.

The assumptions of Theorem 2.1 are satisfied in the following cases (see [17]):

(i) $\overline{f}$ and g are of class C^1, $\overline{f}' \geq 0, g' \geq 0$ and $\overline{f}$ and g have linear growth (see Proposition 2.1).

(ii) $\overline{f}$ and g are of class C^1, $g' > -1$, g has linear growth, $\lim_{a\to\infty} \frac{1}{a} \sup_{x\le a} \overline{f}(x) = 0$, and for some $\lambda \in \mathbb{R}$, it holds that $f' > \lambda$ and $h(x,y) = h(\overline{x}, \overline{y}) = 0$ implies $e^{-\lambda}x - \overline{x} \le y - \overline{y}$ (see [17]).

Theorem 2.1 can be used to investigate the Markov property of the process $\{X_t\}$. Let us first recall the following types of Markov property:
(1) We say that a (d-dimensional) stochastic process $\{Z_t,\ 0 \le t \le 1\}$ is a *Markov process* if for any $t \in [0,1]$ the past and the future of $\{Z_s\}$ are conditionally independent, given the present state Z_t.
(2) We say that $\{Z_t,\ 0 \le t \le 1\}$ is a *Markov field* if for any $0 \le s < t \le 1$, the values of the process inside and outside the interval $[s,\ t]$ are conditionally independent, given Z_s and Z_t.

It is not difficult to show that the process $\{Y_t\}$ given by (2.4) is a Markov field. Under the hypotheses of Theorem 2.1 the process $\{X_t\}$ will be a Markov field if and only if $\{Y_t\}$ is a Markov field under the probability Q. Heuristically, for the Markov field property of $\{Y_t\}$ to hold under Q one needs a factorization of the density function $\frac{dQ}{dP}$ into a product like $J_i J_e$, where J_i is measurable with respect to the σ–field $\sigma\{Y_u, u \in [s,\ t]\}$ and J_e is $\sigma\{Y_u, u \in (s,\ t)^c\}$–measurable. Then, if $\overline{f}'$ is not constant, the factor $\left|1 + g'(\xi_1)\left[1 - \exp(-\int_0^1 \overline{f}'(Y_t)\,dt)\right]\right|$ prevents this kind of factorization from existing. Using these ideas and some technical arguments, one can show the following result (see [17]).

Theorem 2.2 *Suppose that f and g are functions of class C^2 such that $g' > -1$, g' is not identically zero, $\overline{f}' \ge 0$ and T is bijective. Then $\{X_t\}$ is a Markov field if and only if $\overline{f}'' = 0$.*

These results have been extended by C. Donati-Martin [9] to the case of an equation of the type

$$\left.\begin{array}{rcl} dX_t + f(X_t)\,dt & = & \sigma X_t \circ dW_t\,, \quad 0 \le t \le 1 \\ \alpha X_0 + \beta X_1 & = & \gamma \end{array}\right\}$$

2.2 Second Order Stochastic Differential Equations with Boundary Conditions

Consider the second order differential equation

$$\frac{d^2X_t}{dt^2} + f(X_t) = \frac{dW_t}{dt}, \quad 0 \le t \le 1, \tag{2.9}$$

where $f : \mathbb{R} \to \mathbb{R}$ is a measurable and locally bounded function and $W = \{W_t, \ 0 \le t \le 1\}$ is a one dimensional Wiener process. The existence and uniqueness of a solution for this equation and its Markov properties have been investigated in [18] assuming that the solution $X = \{X_t, 0 \le t \le 1\}$ satisfies the Dirichlet boundary conditions $X_0 = a$ and $X_1 = b$, and assuming that the function f may also depend on $\frac{dX_t}{dt}$. Concerning the Markov property the main result of [18] says that the process $\{(X_t, \dot{X}_t), 0 \le t \le 1\}$ is a Markov field if and only if f is an affine function, assuming some additional hypotheses on f. If f only depends on X_t we require f to be of class C^2 with linear growth and $f' \le 0$. Our purpose is to obtain a similar result in the case of Neumann boundary condition, namely

$$\dot{X}_0 = \dot{X}_1 = 0, \tag{2.10}$$

where $\dot{X}_t$ denotes $\frac{dX_t}{dt}$.

The corresponding integral equation can be written as

$$\left.\begin{aligned} \dot{X}_t + \int_0^t f(X_s)\,ds &= W_t, \quad 0 \le t \le 1 \\ \dot{X}_1 &= 0 \end{aligned}\right\} \tag{2.11}$$

Notice that in the particular case $f \equiv 0$ the equation (2.11) has no solution unless $W_1 = 0$, which has probability zero. For this reason the treatment of this equation will be slightly different form the approach used in [18] for the case of Dirichlet boundary conditions. That means, instead of finding the explicit solution when f equals zero we will first consider the particular case $f(x) = -\lambda^2 x$ where $\lambda > 0$ is some fixed constant. Let $Y = \{Y_t, \ 0 \le t \le 1\}$ be the solution of

$$\left.\begin{aligned} \dot{Y}_t - \lambda^2 \int_0^t Y_s\,ds &= W_t, \quad 0 \le t \le 1 \\ \dot{Y}_1 &= 0 \end{aligned}\right\} \tag{2.12}$$

We can write in matrical form

$$\begin{bmatrix} Y_t \\ \dot{Y}_t \end{bmatrix} = \begin{bmatrix} Y_0 \\ 0 \end{bmatrix} + \int_0^t A \begin{bmatrix} Y_s \\ \dot{Y}_s \end{bmatrix} ds + \begin{bmatrix} 0 \\ W_t \end{bmatrix}$$

where A denotes the matrix $\begin{bmatrix} 0 & 1 \\ \lambda^2 & 0 \end{bmatrix}$. Let $\{B_t\}$ be the fundamental solution of the linear equation

$$\begin{aligned} \dot{B}_t &= AB_t\,, \quad 0 \le t \le 1 \\ B_0 &= I\,, \end{aligned}$$

that means,

$$B_t = \begin{bmatrix} \frac{1}{2}(e^{\lambda t} + e^{-\lambda t}) & \frac{1}{2\lambda}(e^{\lambda t} - e^{-\lambda t}) \\ \frac{\lambda}{2}(e^{\lambda t} - e^{-\lambda t}) & \frac{1}{2}(e^{\lambda t} + e^{-\lambda t}) \end{bmatrix}. \tag{2.13}$$

Then we obtain

$$\begin{bmatrix} Y_t \\ \dot{Y}_t \end{bmatrix} = B_t \begin{bmatrix} Y_0 \\ 0 \end{bmatrix} + \int_0^t B_{t-s} \begin{bmatrix} 0 \\ dW_s \end{bmatrix},$$

that means

$$Y_t = \tfrac{1}{2}(e^{\lambda t} + e^{-\lambda t})Y_0 + \tfrac{1}{2\lambda}\textstyle\int_0^t (e^{\lambda(t-s)} - e^{-\lambda(t-s)})\, dW_s$$

$$\dot{Y}_t = \tfrac{\lambda}{2}(e^{\lambda t} - e^{-\lambda t})Y_0 + \tfrac{1}{2}\textstyle\int_0^t (e^{\lambda(t-s)} + e^{-\lambda(t-s)})\, dW_s\,.$$

Using the condition $\dot{Y}_1 = 0$ we deduce

$$Y_0 = (\lambda(e^{-\lambda} - e^{\lambda}))^{-1} \int_0^1 (e^{\lambda(1-s)} + e^{-\lambda(1-s)})\, dW_s$$

and

$$\begin{aligned} Y_t = \; & \tfrac{e^{\lambda t} + e^{-\lambda t}}{2\lambda(e^{-\lambda} - e^{\lambda})} \textstyle\int_0^1 (e^{\lambda(1-s)} + e^{-\lambda(1-s)})\, dW_s \\ & + \tfrac{1}{2\lambda} \textstyle\int_0^t (e^{\lambda(t-s)} - e^{-\lambda(t-s)})\, dW_s\,. \end{aligned} \tag{2.14}$$

We remark that the right hand side of (2.14) is well defined for any continuous function $W \in \Omega = C_0([0,1])$. Indeed, integrating by parts we get

$$\begin{aligned} Y_t = \; & \beta_t(2W_1 + \lambda \textstyle\int_0^1 W_s(e^{\lambda(1-s)} - e^{-\lambda(1-s)})ds) + \\ & + \tfrac{1}{2} \textstyle\int_0^t W_s(e^{\lambda(t-s)} + e^{-\lambda(t-s)})\, ds\,, \end{aligned}$$

where $\beta_t = \left(e^{\lambda t} + e^{-\lambda t}\right)\left(2\lambda(e^{-\lambda} - e^{\lambda})\right)^{-1}$. For the derivative of Y_t we obtain

$$\begin{aligned} \dot{Y}_t = & \tfrac{e^{\lambda t} - e^{-\lambda t}}{2(e^{-\lambda} - e^{\lambda})} \int_0^1 \left(e^{\lambda(1-s)} + e^{-\lambda(1-s)}\right) dW_s \\ + & \tfrac{1}{2} \int_0^t \left(e^{\lambda(t-s)} + e^{-\lambda(t-s)}\right) dW_s . \end{aligned} \tag{2.15}$$

Notice that the mapping $W \mapsto Y(\omega)$ from Ω into the set Σ of continuously differentiable functions y on [0,1] such that $\dot{y}_0 = \dot{y}_1 = 0$ is bijective.

Now we turn to the case of a general function f. Consider the decomposition $f(x) = \overline{f}(x) - \lambda^2 x$, where $\lambda > 0$, and define the transformation $T : \Omega \to \Omega$ by

$$T(\omega)_t = \omega_t + \int_0^t \overline{f}(Y_s(\omega))\, ds . \tag{2.16}$$

We first have the following result.

Proposition 2.2 *If the transformation T given by (2.16) is bijective then for any $W \in C_0([0,1])$, the equation (2.11) has a unique solution given by $X = Y(T^{-1}(W))$.*

Proof: Suppose first that $T(\eta) = W$. Then we claim that $X_t = Y_t(\eta)$ is a solution of the equation (2.11). In fact, we have

$$\begin{aligned} \dot{X}_t &= \dot{Y}_t(\eta) = \lambda^2 \int_0^t Y_s(\eta) ds + \eta_t \\ &= \lambda^2 \int_0^t X_s\, ds + W_t - \int_0^t \overline{f}(X_s)\, ds = W_t - \int_0^t f(X_s)\, ds . \end{aligned}$$

Conversely, let $\{X_t\}$ be a solution of the equation (2.11) such that $\dot{X}_0 = \dot{X}_1 = 0$. Then $T(Y^{-1}(X)) = W$. In fact, if we set $\eta = Y^{-1}(X)$ we have

$$\begin{aligned} T(\eta)_t &= \eta_t + \int_0^t \overline{f}(Y_s(\eta))\, ds \\ &= \eta_t + W_t - \dot{X}_t + \lambda^2 \int_0^t X_s\, ds = W_t . \end{aligned}$$

□

The following proposition will provide some sufficient conditions on the function f for the transformation T given by (2.16) to be bijective.

Proposition 2.3 *Suppose that $\overline{f}$ is nonincreasing, locally Lipschitz and with linear growth. Then T is bijective.*

Proof: Given $\eta \in C_0([0,1])$ we have to show that there exists a unique function $W \in C_0([0,1])$ such that $T(W) = \eta$. Set $V = \eta - W$. Then V satisfies the differential equation

$$\begin{aligned} \dot{V}_t = \; & \overline{f}(\xi_t - \beta_t(2V_1 + \lambda \int_0^1 V_s\,(e^{\lambda(1-s)} - e^{-\lambda(1-s)})\,ds) \\ & -\tfrac{1}{2}\int_0^t V_s\,(e^{\lambda(t-s)} + e^{-\lambda(t-s)})\,ds), \end{aligned}$$

where

$$\begin{aligned} \xi_t = \; & \beta_t(2\eta_1 + \lambda \int_0^1 \eta_s(e^{\lambda(1-s)} - e^{-\lambda(1-s)})\,ds \\ & +\tfrac{1}{2}\int_0^t \eta_s(e^{\lambda(t-s)} + e^{-\lambda(t-s)})\,ds). \end{aligned}$$

For any $y \in \mathbb{R}$ we consider the differential equation

$$\begin{cases} \dot{V}_t(y) = \overline{f}(\xi_t - \beta_t y - \tfrac{1}{2}\int_0^t V_s(y)\,(e^{\lambda(t-s)} + e^{-\lambda(t-s)})\,ds, \\ V_0(y) = 0. \end{cases}$$

By a comparison theorem for ordinary differential equations and using the monotonicity properties of f we get that the mapping $y \mapsto V_t(y)$ is continuous and nonincreasing for each $t \in [0,1]$. Therefore, $2V_1(y) + \lambda \int_0^1 V_s(y)(e^{\lambda(1-s)} - e^{-\lambda(1-s)})\,ds$ is a nonincreasing and continuous function of y, and this implies the existence of a unique real number y such that $y = 2V_1(y) + \lambda \int_0^1 V_s(y)\,(e^{\lambda(1-s)} - e^{-\lambda(1-s)})\,ds$. This completes the proof of the proposition. □

As in [18], it is possible to show that T is bijective assuming that $\overline{f}$ is globally Lipschitz and the Lipschitz constant of $\overline{f}$ is small enough. On the other hand, if $f(x) = \alpha x + \beta$ is an affine function then, as we have seen when $\alpha = -\lambda^2$, there exists a unique solution to the equation (2.11) provided $\alpha \neq 0$.

Once we have some sufficient conditions for the existence and uniqueness of the solution of equation (2.11) we can proceed to investigate the Markov property of the solution. To study the Markov property we will make use of the extended Girsanov theorem presented in Section 1. More precisely we are going to apply Theorem 1.2 to the transformation T defined by (2.16), and we will find a new probability Q such that $Q \circ T^{-1} = P$, and Q is absolutely continuous with respect to P. The law of the process $\{X_t\}$ under P will be equal to the law of $\{Y_t\}$ under Q, because $X = Y \circ T^{-1}$, and in this way we will translate the problem of the Markov property of $\{X_t\}$ into the problem of the Markov property of $\{Y_t\}$ under Q.

Proposition 2.4 *Let $\overline{f}$ be a function of class C^2, with linear growth and such that $\overline{f}' \leq 0$, and let Y_t be the process defined by (2.14). Then the process $u_t = \overline{f}(Y_t)$ verifies the conditions of Theorem 1.2 and we have*

$$\frac{dQ}{dP} = \dot{Z}_1 \exp\left(-\int_0^1 \overline{f}(Y_t) \circ dW_t - \frac{1}{2}\int_0^1 \overline{f}(Y_t)^2\, dt\right), \tag{2.17}$$

Where $\dot{Z}_1$ is given by the second order differential equation

$$\ddot{Z}_t + \alpha_t Z_t - \lambda^2 Z_t = 0 \tag{2.18}$$

$$Z_0 = 2(\lambda(e^{\lambda} - e^{-\lambda}))^{-1}, \quad \dot{Z}_0 = 0,$$

where $\alpha_t = \overline{f}'(Y_t)$.

Proof: From Proposition 2.3 we already know that the transformation $T(\omega)_t = \omega_t + \int_0^t u_s(\omega)\, ds$ is bijective. Furthermore, the process u is $\mathcal{H}$–C^1 (see Definition 1.1) because the mapping $\omega \mapsto u(\omega)$ is continuously differentiable from Ω into H. So, it remains to show that $I + Du : H \to H$ is invertible a.s. From the properties of the operator D we deduce

$$D_s Y_t = \beta_t(e^{\lambda(1-s)} + e^{-\lambda(1-s)}) + \frac{1}{2\lambda}(e^{\lambda(t-s)} - e^{-\lambda(t-s)})\mathbf{1}_{\{s\leq t\}}, \tag{2.19}$$

and, therefore

$$D_s u_t = \alpha_t \left[\beta_t(e^{\lambda(1-s)} + e^{-\lambda(1-s)}) + \frac{1}{2\lambda}(e^{\lambda(t-s)} - e^{-\lambda(t-s)})\mathbf{1}_{\{s\leq t\}}\right]. \tag{2.20}$$

From the Fredholm alternative, in order to show that $I + Du$ is invertible, it suffices to check that -1 is not an eigenvalue of Du. Let $h \in L^2(0,1)$ be such that $(I + Du)h = 0$. Then

$$\begin{aligned} h_t \; &+ \alpha_t\beta_t \int_0^1 (e^{\lambda(1-s)} + e^{-\lambda(1-s)})h_s\, ds \\ &+ \tfrac{\alpha_t}{2\lambda} \int_0^t (e^{\lambda(t-s)} - e^{\lambda(t-s)})\, h_s ds = 0. \end{aligned}$$

Set $g_t = \frac{1}{2\lambda} \int_0^t (e^{\lambda(t-s)} - e^{\lambda(t-s)}) h_s \, ds$. Then we have
$\dot{g}_t = \frac{1}{2} \int_0^t (e^{\lambda(t-s)} + e^{-\lambda(t-s)}) h_s \, ds$, and $\ddot{g}_t = h_t + \lambda^2 g_t$.
Thus we obtain

$$\ddot{g}_t - \lambda^2 g_t + \alpha_t g_t + 2\alpha_t \beta_t \dot{g}_1 = 0 . \tag{2.21}$$

We denote by M_t the matrix $\begin{bmatrix} 0 & 1 \\ \lambda^2 - \alpha_t & 0 \end{bmatrix}$. Let Φ_t be the solution of the linear differential equation

$$\begin{aligned} d\Phi_t &= M_t \Phi_t \, dt \\ \Phi_0 &= I , \end{aligned} \tag{2.22}$$

and set $\Phi(t,s) = \Phi_t \Phi_s^{-1}$. Then the solution of the second order differential equation (2.21) is given by

$$g_t = -2\dot{g}_1 \int_0^t \Phi_{12}(t,s) \alpha_s \beta_s \, ds ,$$

which implies

$$\dot{g}_t = -2\dot{g}_1 \int_0^t \Phi_{22}(t,s) \, \alpha_s \beta_s \, ds ,$$

and we obtain $\dot{g}_1 = 0$ because $\alpha_s \le 0$, $\beta_s \le 0$ and $\Phi_{22} \ge 0$. To finish the proof of the proposition we have to compute the Carleman–Fredholm determinant of the kernel Du given by (2.20). It is more convenient to write this kernel in the following way:

$$D_s u_t = \alpha_t (2\lambda(e^{-\lambda} - e^{\lambda}))^{-1} \psi(s \wedge t) \, \varphi(s \vee t) , \tag{2.23}$$

where $\psi(t) = e^{\lambda t} + e^{-\lambda t}$ and $\varphi(t) = e^{\lambda(1-t)} + e^{-\lambda(1-t)}$.

From the expression (2.23) we deduce that integral operator associated with the kernel $D_s u_t$ is nuclear and, therefore, we can use the formula (1.18) in order to compute the density function. Actually, it suffices to show that $\det(I + Du) = \dot{Z}_1$, where $\dot{Z}_1$ is given by (2.18). Notice that $\dot{Z}_1 \ge 0$. We can write

$$\begin{aligned} \det(I + Du) &= \sum_{n=0}^{\infty} \frac{1}{n} \int_{[0,1]^n} \det(D_{t_i} u_{t_j}) \, dt_1 \cdots dt_n \\ &= \sum_{n=0}^{\infty} \int_{\{0 < t_1 < \cdots < t_n < 1\}} \det(D_{t_i} u_{t_j}) \, dt_1 \cdots dt_n . \end{aligned}$$

Define $\Lambda = (\prod_{i=1}^n \alpha_{t_i})(2\lambda(e^{-\lambda} - e^{\lambda}))^{-n}$. The determinant of the matrix $(D_{t_i}u_{t_j})$ for $0 < t_1 < t_2 < \cdots < t_n < 1$ is equal to

$$\Lambda \det \begin{bmatrix} \psi(t_1)\varphi(t_1) & \psi(t_1)\varphi(t_2) & \cdots & \psi(t_1)\varphi(t_n) \\ \psi(t_1)\varphi(t_2) & \psi(t_2)\varphi(t_2) & \cdots & \psi(t_2)\varphi(t_n) \\ \cdots & \cdots & & \cdots \\ \psi(t_1)\varphi(t_n) & \psi(t_2)\varphi(t_n) & \cdots & \psi(t_n)\varphi(t_n) \end{bmatrix}$$

$$= \Lambda \left(\prod_{i=1}^n \psi(t_i) \right)^2 \det \begin{bmatrix} \frac{\varphi(t_1)}{\psi(t_1)} & \cdots & \frac{\varphi(t_n)}{\psi(t_n)} \\ \cdots & & \cdots \\ \frac{\varphi(t_n)}{\psi(t_n)} & \cdots & \frac{\varphi(t_n)}{\psi(t_n)} \end{bmatrix}$$

$$= \Lambda \quad (\textstyle\prod_{i=1}^n \psi(t_i))^2 \left(\frac{\varphi(t_1)}{\psi(t_1)} - \frac{\varphi(t_2)}{\psi(t_2)} \right) \left(\frac{\varphi(t_2)}{\psi(t_2)} - \frac{\varphi(t_3)}{\psi(t_3)} \right) \cdots$$
$$\cdots \left(\frac{\varphi(t_{n-1})}{\psi(t_{n-1})} - \frac{\varphi(t_n)}{\psi(t_n)} \right) \frac{\varphi(t_n)}{\psi(t_n)}$$

$$= \Lambda \quad (\varphi(t_1)\psi(t_2) - \psi(t_1)\varphi(t_2)) (\varphi(t_2)\psi(t_3) - \psi(t_2)\varphi(t_3)) \cdots$$
$$\cdots (\varphi(t_{n-1})\psi(t_n) - \varphi(t_n)\psi(t_{n-1})) \, \varphi(t_n)\psi(t_1).$$

Using the identity

$$\varphi(t_i)\psi(t_{i+1}) - \psi(t_i)\varphi(t_{i+1}) = (e^{\lambda(t_{i+1}-t_i)} - e^{-\lambda(t_{i+1}-t_i)})(e^{\lambda} - e^{-\lambda})$$

we obtain

$$\det(D_{t_i}u_{t_j}) = \quad (2\lambda)^{-n}(-1)^n (\textstyle\prod_{i=1}^n \alpha_{t_i})$$
$$\times \left(\textstyle\prod_{i=1}^{n-1} (e^{\lambda(t_{i+1}-t_i)} - e^{-\lambda(t_{i+1}-t_i)}) \right) \frac{\varphi(t_n)\psi(t_1)}{(e^{\lambda}-e^{-\lambda})}.$$

Therefore, we have

$$\det(I + Du) = \textstyle\sum_{n=0}^{\infty} (-\frac{1}{2\lambda})^n \frac{1}{e^{\lambda}-e^{-\lambda}} \int_{\{0<t_1<\cdots<t_n<1\}} (e^{\lambda t_1} + e^{-\lambda t_1})$$
$$\times (e^{\lambda(t_2-t_1)} - e^{-\lambda(t_2-t_1)}) \cdots (e^{\lambda(t_n-t_{n-1})} - e^{-\lambda(t_n-t_n-1)})$$
$$\times (e^{\lambda(1-t_n)} + e^{-\lambda(1-t_n)}) \alpha_{t_1} \cdots \alpha_{t_n} \, dt_1 \cdots dt_n \, . \tag{2.24}$$

On the other hand, the solution of the second order differential equation (2.18) can be expressed as

$$\begin{bmatrix} Z_t \\ \dot{Z}_t \end{bmatrix} = B_t \begin{bmatrix} Z_0 \\ \dot{Z}_0 \end{bmatrix} + \int_0^t B_{t-s} \begin{bmatrix} 0 \\ -\alpha_s Z_s \end{bmatrix} ds\,,$$

where B_t is the matrix given by (2.13). If we impose the initial conditions $Z_0 = 2(\lambda(e^\lambda - e^{-\lambda}))^{-1}$, $\dot{Z}_0 = 0$, we obtain

$$\begin{aligned} \dot{Z}_t = \ & (e^{\lambda t} - e^{-\lambda t})(e^\lambda - e^{-\lambda})^{-1} \\ & - \textstyle\int_0^t \frac{1}{2}(e^{\lambda(t-s)} + e^{\lambda(t-s)})\alpha_s Z_s\, ds \end{aligned} \tag{2.25}$$

and

$$\begin{aligned} Z_t = \ & \tfrac{1}{\lambda}(e^\lambda - e^{-\lambda})^{-1}(e^{\lambda t} + e^{-\lambda t}) \\ & - \textstyle\int_0^t \frac{1}{2\lambda}(e^{\lambda(t-s)} - e^{-\lambda(t-s)})\alpha_s Z_s\, ds\,. \end{aligned} \tag{2.26}$$

Therefore, we deduce

$$\begin{aligned} Z_t = \ & \tfrac{e^{\lambda t} + e^{\lambda t}}{\lambda(e^\lambda - e^{-\lambda})} + \textstyle\sum_{n=1}^\infty \int_{\{0<t_1<\cdots<t_n<t\}} (-1)^n \frac{1}{\lambda(2\lambda)^n} \\ & (e^\lambda - e^{-\lambda})^{-1}(e^{\lambda t_1} + e^{-\lambda t_1})(e^{\lambda(t_2-t_1)} - e^{-\lambda(t_2-t_1)}) \cdots \\ & \cdots (e^{\lambda(t-t_n)} - e^{\lambda(t-t_n)})\alpha_{t_1} \cdots \alpha_{t_n}\, dt_1 \cdots dt_n\,, \end{aligned} \tag{2.27}$$

and by substitution of the value of Z_t given by (2.27) into (2.25) we deduce that $\dot{Z}_1$ equals to the right hand side of (2.24), which completes the proof of the theorem. □

Now we are going to use Proposition 2.4 to study the Markov property of the process $\{X_t\}$. In the case $\overline{f} \equiv 0$ (that means, $f(x) = -\lambda^2 x$) we have the following result.

Proposition 2.5 *The process $\{(Y_t, \dot{Y}_t),\ 0 \le t \le 1\}$ defined by the equations (2.14) and (2.15) is a Markov process.*

Proof: Let $\psi(x, y)$ be a real valued bounded and measurable function.

Fix $s < t$ and set $\rho = \int_0^1 (e^{\lambda(1-s)} + e^{-\lambda(1-s)})dW_s$. Then we have

$$E\left(\psi(Y_t, \dot{Y}_t) \mid (Y_r, \dot{Y}_r),\ 0 \le r \le s\right)$$

$$= E\left(\psi\left(\beta_t\rho + \tfrac{1}{2\lambda}\int_0^t (e^{\lambda(t-\theta)} - e^{\lambda(t-\theta)})dW_\theta,\ \dot{\beta}_t\rho\right.\right.$$

$$\left.\left.+\tfrac{1}{2}\int_0^t (e^{\lambda(t-\theta)} + e^{-\lambda(t-\theta)})dW_\theta\right) \quad \mid \rho,\ W_r,\ 0 \le r \le s\right)$$

$$= \int_{\mathbb{R}^2} \psi\left(\beta_t\rho + \tfrac{1}{2\lambda}\int_0^s (e^{\lambda(t-\theta)} - e^{-\lambda(t-\theta)})dW_\theta + \tfrac{x}{2\lambda},\right.$$

$$\left.\dot{\beta}_t\rho + \tfrac{1}{2}\int_0^s (e^{\lambda(t-\theta)} + e^{-\lambda(t+\theta)})dW_\theta + \tfrac{y}{2}\right)\ \mu(dx, dy),$$

where μ is the conditional law of the vector

$$\left(\int_s^t (e^{\lambda(t-\theta)} - e^{\lambda(t-\theta)})dW_\theta,\ \int_s^t (e^{\lambda(t-\theta)} + e^{-\lambda(t-\theta)})dW_\theta\right)$$

given $\int_s^1 (e^{\lambda(1-\theta)} + e^{-\lambda(1-\theta)})dW_\theta$. Consequently, the above conditional expectation will be a function of the random variables

$$\beta_t\rho + \frac{1}{2\lambda}\int_0^s (e^{\lambda(t-\theta)} - e^{\lambda(t-\theta)})dW_\theta,$$

$$\dot{\beta}_t\rho + \frac{1}{2}\int_0^s (e^{\lambda(t-\theta)} + e^{-\lambda(t+\theta)})dW_\theta \quad \text{and}$$

$$\int_s^1 (e^{\lambda(1-\theta)} + e^{-\lambda(1-\theta)})dW_\theta,$$

which are functions of Y_s and $\dot{Y}_s$. □

One can show by a direct argument that Proposition 2.5 still holds when f is a general affine function $\alpha x + \beta$ with $\alpha \neq 0$. If f is not affine the following theorem shows that the Markov property (and even the Markov field property) cannot be satisfied for the process $\{(X_t, \dot{X}_t)\}$ solution of (2.11).

Theorem 2.3 *Let f be a function of class C^2, with linear growth and such that $f' + \lambda^2 \le 0$ for some $\lambda > 0$. Then, if the process $\{(X_t, \dot{X}_t),\ 0 \le t \le 1\}$ is a Markov field we have $f'' = 0$.*

Proof: Let Q be the probability measure on $C_0([0,1])$ given by (2.17). The law of the process $\{X_t\}$ under P is the same as the law of $\{Y_t\}$ under Q. Therefore we assume that $\{(Y_t, \dot{Y}_t),\ 0 \le t \le 1\}$ is a Markov field under Q. We can factorize the density function $J = \frac{dQ}{dP}$ given by (2.17) as follows

$$J = \dot{Z}_1 L_t L^t ,$$

where

$$L_t = \exp(-\int_0^t \overline{f}(Y_s) \circ dW_s - \frac{1}{2}\int_0^t \overline{f}(Y_s)^2\, ds),$$

and

$$L^t = \exp(-\int_t^1 \overline{f}(Y_s) \circ dW_s - \frac{1}{2}\int_t^1 \overline{f}(Y_s)^2\, ds),$$

for any fixed $t \in (0,1)$. We define the σ–algebras

$$\begin{aligned} \mathcal{F}_t &= \sigma\{(Y_s, \dot{Y}_s),\ 0 \le s \le t\}, \\ \mathcal{F}^t &= \sigma\{(Y_s, \dot{Y}_s),\ t \le s \le 1\}, \\ \mathcal{F}_0^t &= \mathcal{F}^t \vee \sigma\{Y_0, \dot{Y}_0\} = \mathcal{F}^t \vee \sigma(\rho), \quad \text{and} \\ \mathcal{G}_t &= \sigma\{Y_t, \dot{Y}_t, \rho\}, \end{aligned}$$

where $\rho = \int_0^1 (e^{\lambda(1-s)} + e^{-\lambda(1-s)})\, dW_s$.

The Markov field property of $\{(Y_t, \dot{Y}_t)\}$ with respect to Q implies that for any $t \in (0,1)$ and for any $\mathcal{F}_0^t$–measurable and nonnegative random variable ξ, the conditional expectation

$$\Gamma_\xi = E_Q(\xi \mid \mathcal{F}_t)$$

is $\mathcal{G}_t$–measurable.

Notice that L_t is $\mathcal{F}_t$–measurable and L^t is $\mathcal{F}_0^t$–measurable. Thus, applying the Markov field property of $\{(Y_t, \dot{Y}_t)\}$ under P (Proposition 2.5) we deduce that

$$\Gamma_\xi = \frac{E_P(\xi J \mid \mathcal{F}_t)}{E_P(J \mid \mathcal{F}_t)} = \frac{E_P(\xi \dot{Z}_1 L^t \mid \mathcal{F}_t)}{E_P(\dot{Z}_1 L^t \mid \mathcal{F}_t)}$$

is $\mathcal{G}_t$–measurable. By chosing $\xi = (L^t)^{-1}\gamma$ and $\xi = (L^t)^{-1}$ we obtain that the quotient

$$\Lambda_\gamma = \frac{E_P(\gamma \dot{Z}_1 \mid \mathcal{F}_t)}{E_P(\dot{Z}_1 \mid \mathcal{F}_t)} \tag{2.28}$$

for $\mathcal{G}_t$–measurable for any $\mathcal{F}_0^t$–measurable and nonnegative random variable γ. Now we decompose the random variable $\dot{Z}_1$ as follows:

$$\dot{Z}_1 = \Phi_{21}(1,t)Z_t + \Phi_{22}(1,t)\,\dot{Z}_t \tag{2.29}$$

where $\Phi(t,s) = \Phi_t\Phi_s^{-1}$ and Φ_t is the solution of the linear differential equation (2.22). Define

$$\varphi_t = \frac{\Phi_{21}(1,t)}{\Phi_{22}(1,t)} \quad \text{and} \quad \psi_t = \frac{\dot{Z}_t}{Z_t} = \frac{\Phi_{21}(t)}{\Phi_{11}(t)}. \tag{2.30}$$

Notice that φ_t and ψ_t are continuously differentiable processes on $[0,1]$ such that $\varphi_1 = 0, \psi_0 = 0$, $\varphi_t > 0$ for $t \in [0,1)$ and $\psi_t > 0$ for $t \in (0,1]$. Then we can write from (2.29) and (2.30)

$$\dot{Z}_1 = \Phi_{22}(1,t)Z_t\,(\varphi_t + \psi_t)\,.$$

In the sequel we will denote by $\overline{F}$ the conditional expectation of a random variable F with respect to the σ–algebra $\mathcal{G}_t$. The random variables $\dot{Z}_t$ and ψ_t are $\mathcal{F}_t$–measurable and, on the other hand, $\Phi_{22}(1,t)$ and φ_t are $\mathcal{F}_0^t$–measurable. Therefore, from (2.28) and using the Markov field property of $\{(Y_t,\dot{Y}_t)\}$ under P, we deduce that

$$\Lambda_\gamma = \frac{\overline{\gamma\Phi_{21}(1,t)} + \overline{\gamma\Phi_{22}(1,t)}\,\psi_t}{\overline{\Phi_{21}(1,t)} + \overline{\Phi_{22}(1,t)}\,\psi_t}$$

is $\mathcal{G}_t$–measurable, and we obtain the equality

$$(\Lambda_\gamma\overline{\Phi_{22}(1,t)} - \overline{\gamma\Phi_{22}(1,t)})\psi_t + (\Lambda_\gamma\overline{\Phi_{21}(1,t)} - \overline{\gamma\Phi_{21}(1,t)}) = 0, \tag{2.31}$$

which is valid for any $\mathcal{F}_0^t$–measurable and bounded random variable γ. We are going to choose two particular variables γ:

$$\gamma_1 = \Phi_{22}(1,t)^{-1} \quad \text{and} \quad \gamma_2 = \Phi_{21}(1,t)^{-1}. \tag{2.32}$$

Notice that for $t \in (0,1)$ we have $\Phi_{22}(1,t) \geq 1$ and $\Phi_{21}(1,t) \geq \lambda^2(1-t)$. Define the set

$$G_t = \left\{\Lambda_{\gamma_1}\overline{\Phi_{21}(1,t)} = \overline{\gamma_1\Phi_{21}(1,t)}\right\}\bigcap\left\{\Lambda_{\gamma_2}\overline{\Phi_{21}(1,t)} = \overline{\gamma_2\Phi_{21}(1,t)}\right\}.$$

Notice that $G_t \in \mathcal{G}_t$. On the set G_t we have

$$\Lambda_{\gamma_1} = \frac{\overline{\gamma_1 \Phi_{22}(1,t)}}{\overline{\Phi_{22}(1,t)}} = \frac{\overline{\gamma_1 \Phi_{21}(1,t)}}{\overline{\Phi_{21}(1,t)}}, \tag{2.33}$$

$$\Lambda_{\gamma_2} = \frac{\overline{\gamma_2 \Phi_{22}(1,t)}}{\overline{\Phi_{22}(1,t)}} = \frac{\overline{\gamma_2 \Phi_{21}(1,t)}}{\overline{\Phi_{21}(1,t)}}. \tag{2.34}$$

Consequently, from (2.32), (2.33) and (2.34) we obtain

$$(\overline{\varphi_t^{-1}}) = \frac{\overline{\Phi_{21}(1,t)}}{\overline{\Phi_{22}(1,t)}} = \overline{\varphi_t}$$

By the strict Jensen inequality on the measure space $(G_t, \mathcal{F}_{G_t}, P)$ we deduce that the random variable $\mathbf{1}_{G_t}\varphi_t$ is $\mathcal{G}_t$–measurable. From the equation (2.31) we get that $\mathbf{1}_{G_t^c}\psi_t$ is $\mathcal{G}_t$–measurable.

We are going to translate the measurability with respect to $\mathcal{G}_t$ into an analytical condition. To do this let us first compute the derivatives of the generators of $\mathcal{G}_t$,

$$D_s Y_t = \beta_t(e^{\lambda(1-s)} + e^{-\lambda(1-s)}) + \frac{1}{2\lambda}(e^{\lambda(t-s)} - e^{-\lambda(t-s)})\mathbf{1}_{\{s \le t\}}, \tag{2.35}$$

$$D_s \dot{Y}_t = \frac{e^{\lambda t} - e^{-\lambda t}}{2(e^{-\lambda} - e^{\lambda})}(e^{\lambda(1-s)} + e^{-\lambda(1-s)}) + \frac{1}{2}(e^{\lambda(t-s)} + e^{-\lambda(t-s)})\mathbf{1}_{\{s \le t\}}, \tag{2.36}$$

$$D_s \rho = e^{\lambda(1-s)} + e^{-\lambda(1-s)}. \tag{2.37}$$

The following lemma has been proved in [17, Lemma 4.5]:

Lemma 2.1 *Let $\mathcal{G}$ be the σ–algebra generated by the random variables $\int_0^1 h_i(s)dW_s$, $1 \le i \le M$, where $h_i \in H$. Let $F \in \mathbb{D}^{1,2}_{\mathrm{loc}}$ and $G \in \mathcal{G}$ be such that $F\mathbf{1}_G$ is $\mathcal{G}$–measurable. Then DF belongs a.s. to the span of $\{h_1, \ldots, h_M\}$.*

Applying this lemma to the random variables Y_t, $\dot{Y}_t$ and ρ which generate $\mathcal{G}_t$, and to the pairs (φ_t, G_t) and (ψ_t, G_t^c), respectively (note that φ_t and ψ_t belong to $\mathbb{D}^{1,2}_{\mathrm{loc}}$), and using the relations (2.35), (2.36),

(2.37) we deduce the existence of random variables $\Gamma_1(t)$, $\Gamma_2(t)$ and $\Gamma_3(t)$ such that

$$D_s\varphi_t = (e^{\lambda(1-s)} + e^{-\lambda(1-s)})\,\Gamma_1(t)\,, \tag{2.38}$$

for all $s \in [t,1]$, $\omega \in G_t$, a.e., and

$$D_s\phi_t = e^{\lambda s}\Gamma_2(t) + e^{-\lambda s}\Gamma_3(t)\,, \tag{2.39}$$

for all $s \in [0,t]$, $\omega \in G_t^c$, a.e.

On the other hand, from the linear differential equations satisfied by $\Phi(t)$ and $\Phi(1,t)$ we can derive Ricatti type differential equations for φ_t and ψ_t. In fact, differentiating with respect to t the equations

$$\begin{aligned}\Phi_{21}(1,t) &= \varphi_t\Phi_{22}(1,t)\\ \Phi_{21}(t) &= \psi_t\Phi_{11}(t)\end{aligned}$$

we obtain

$$\dot{\varphi}_t - \varphi_t^2 + \lambda^2 - \alpha_t = 0 \quad , \quad \varphi_1 = 0 \tag{2.40}$$

$$\dot{\psi}_t + \psi_t^2 - \lambda^2 + \alpha_t = 0 \quad , \quad \psi_0 = 0 \tag{2.41}$$

Applying the operator D, which commutes with the derivative with respect to the time variable, to the equations (2.40) and (2.41) yields

$$\frac{d}{dt}D_s\varphi_t - 2\varphi_t D_s\varphi_t - D_s\alpha_t = 0 \quad , \quad D_s\varphi_1 = 0 \tag{2.42}$$

$$\frac{d}{dt}D_s\psi_t + 2\psi D_s\psi_t + D_s\alpha_t = 0 \quad , \quad D_s\psi = 0. \tag{2.43}$$

Set $\gamma_{uv} = \exp(\int_u^v 2\varphi_\theta d\theta)$ for any $u, v \in [0,1]$. Notice that $D_\theta\alpha_t = f''(Y_t)D_\theta Y_t$. From (2.42) and (2.35) we obtain

$$\begin{aligned}D_s\varphi_t = {} & -\textstyle\int_t^1 \gamma_{tu} D_s\alpha_u\, du\\ = {} & (e^{\lambda(1-s)} + e^{-\lambda(1-s)}) \textstyle\int_t^s \gamma_{tu}\, f''(Y_u)\frac{e^{\lambda u}+e^{-\lambda u}}{2\lambda(e^{\lambda}-e^{-\lambda})}\, du\\ & +(e^{\lambda s} + e^{-\lambda s}) \textstyle\int_s^1 \gamma_{tu}\, f''(Y_u)\frac{e^{\lambda(1-u)}+e^{-\lambda(1-u)}}{2\lambda(e^{\lambda}-e^{-\lambda})}\, du\\ = {} & (e^{\lambda(1-s)} + e^{\lambda(1-s)}) \textstyle\int_t^1 \gamma_{tu}\, f''(Y_u)\frac{e^{\lambda u}+e^{-\lambda u}}{2\lambda(e^{\lambda}-e^{-\lambda})}\, du\\ & +\frac{1}{2\lambda}\textstyle\int_s^1 \gamma_{tu}\, f''(Y_u)(e^{\lambda(s-u)} - e^{-\lambda(s-u)})\, du\,.\end{aligned} \tag{2.44}$$

According to (2.38), on the set G_t and for $s \in [t, 1]$, $D_s \varphi_t$ must be a multiple of $(e^{\lambda(1-s)} + e^{-\lambda(1-s)})$. Taking into account the equation (2.44), this is only possible if $f''(Y_u) = 0$ for all $u \in [t, 1]$ and $\omega \in G_t$ a.e. A similar argument, using the process ψ_t yields $f''(Y_u) = 0$ for all $u \in [0, t]$ and $\omega \in G_t^c$ a.e. Therefore $f''(Y_u) = 0$ a.e. and the proof of the theorem is complete. □

References

[1] D.R. Bell, *The Malliavin Calculus.* Pitman Monographs and Surveys in Pure and Applied Math., 34, Longman Scientific and Technical, New York, 1987.

[2] R. Buckdahn, *Girsanov transformation and linear stochastic differential equations without nonanticipation requirement.* Preprint 180, Sekt. Mak. der Humbold-Univ., 1988.

[3] R. Buckdahn, *Transformations on the Wiener space and Skorohod-Type stochastic differential equations.* Seminarbericht Nr. 105, Seck. Mat. der Humbold-Univ, 1989.

[4] R. Buckdahn, *Anticipative Girsanov transformations.* Preprint.

[5] R. Buckdahn, *Linear Skorohod stochastic differential equations.* Preprint.

[6] R.H. Cameron and W.T. Martin, *The transformation of Wiener integrals by nonlinear transformations.* Trans. Amer. Math. Soc. 66, 253–283, 1949.

[7] A.B. Cruzeiro, *Equations différentielles ordinaires: non explosion et mesures quasi-invariantes.* J. Functional Anal. 54, 193–205, 1983.

[8] A.B. Cruzeiro, *Equations différentielles sur l'espace de Wiener et formules de Cameron Martin non linéaires.* J. Functional Anal. 54, 206–227, 1983.

[9] C. Donati-Martin, *Equations différentielles stochastiques dans* $\mathbb{R}$ *avec conditions au bord.* Preprint.

[10] I.V. Girsanov, *On transforming a certain class of stochastic processes by absolutely continuous substitution of measures.* Theory Prob. Appl. 5, 285–301, 1960.

[11] L. Gross, *Integration and nonlinear transformations in Hilbert space.* Trans. Amer. Math. Soc. 94, 404–440, 1960.

[12] H.H. Kuo, *Integration theory in infinite dimensional manifolds.* Trans. Amer. Math. Soc. 159, 57–78, 1971.

[13] H.H. Kuo, *Gaussian measures on Banach space*, Lecture Notes in Math. 463, Springer–Verlag, 1975.

[14] S. Kusuoka, *The nonlinear transformation of Gaussian measure on Banach space and its absolute continuity.* J. Fac. Sci. Tokyo Univ. Sec. I.A., 567–597, 1985.

[15] D. Nualart, *Noncausal stochastic integrals and calculus.* Lecture Notes in Math. 1316, 80–126, 1988.

[16] D. Nualart and E. Pardoux, *Stochastic calculus with anticipating integrands*, Probab. Theory Rel. Fields 78, 535–581, 1988.

[17] D. Nualart and E. Pardoux, *Boundary value problems for stochastic differential equations*, Annals of Probability, to appear.

[18] D. Nualart and E. Pardoux, *Second order stochastic differential equations with Dirichlet boundary conditions.* Stoch. Proc. and Their Appl., to appear.

[19] D. Nualart and E. Pardoux, *Stochastic differential equations with boundary conditions.* Preprint.

[20] D. Nualart and M. Zakai, *Generalized stochastic integrals and the Malliavin Calculus.* Probab. Theory Rel. Fields, 73, 255-280, 1986.

[21] D. Nualart and M. Zakai, *On the relation between the Stratonovich and the Ogawa integrals.* Ann. Probab. 17, 1536–1540.

[22] E. Pardoux, *Applications of anticipating stochastic calculus to stochastic differential equations.* Lecture Notes in Math. 1444, 63–105, 1990.

[23] R. Ramer, *On nonlinear transformations of Gaussian measures.* J. Functional Anal. 15, 166–187, 1974.

[24] A. Russek, *Gaussian n–Markovian processes and stochastic boundary value problems.* Z. Wahrscheinlichkeitstheorie verw. Gebiete, 53, 117–122, 1980.

[25] B. Simon, *Trace ideals and their applications.* London Math. Soc., Lecture Note Series 35, Cambridge University Press 1979.

[26] A.V. Skorohod, *On a generalization of a stochastic integral.* Theory Probab. Appl. 20, 219–233, 1975.

[27] F. Smithies, *Integral equations.* Cambridge University Press, 1958.

[28] A.S. Ustunel and M. Zakai, *Transformation of Wiener measure under anticipative flows.* Preprint.

[29] S. Watanabe, *Lectures on stochastic differential equations and the Malliavin Calculus.* Tata Institute of Fundamental Research, Springer–Verlag, 1984.

Finite Dimensional Approximate Filters in the case of High Signal–to–Noise Ratio

E. Pardoux, M.C. Roubaud
Université de Provence and INRIA

Abstract

We present some recent results on nonlinear filtering problems with high signal–to–noise ratio. We concentrate mainly on the scalar case, where the observation function is not one to one. We describe two situations where a good suboptimal filter is provided by a collection of one dimensional filters, together with statistical tests for choosing which filter should be followed.

1 Introduction

It is by now well known (see e.g. Pardoux [9]) that the nonlinear filtering problem is a difficult one, whose optimal solution is in most cases given only by the solution of an infinite dimensional equation, the Zakai equation.

On the other hand, up to now all practical filtering problems are solved by approximate linear filters, in particular the well–known extended Kalman filter (see e.g. [9]). However, the extended Kalman filter does not rely on any mathematical foundation, it is known both from theory (see [9], Picard [14] and section 2 below) and experience that it sometimes behaves very poorly, while it gives very satisfactory results in many situations.

A good framework for a mathematical analysis of approximate filters, including the extended Kalman filter, is the situation of a

ISBN 0-12-481005-5

high signal–to–noise ratio, i.e. we consider the following non linear filtering problem :

$$(1) \begin{cases} dX_t &= f(X_t)\,dt + g(X_t)dV_t \\ dY_t &= h(X_t)\,dt + \varepsilon dW_t \end{cases}$$

where $\{X_t\}$ is the unobserved process to be filtered, $\{Y_t\}$ is the observation process, $\{V_t\}$ and $\{W_t\}$ are mutually independent standard Wiener processes, all processes being scalar for simplicity. The goal is to obtain asymptotic results as $\varepsilon \to 0$ (with $\varepsilon > 0$).

In the case where h is one to one, this problem has first been analysed by Bobrovsky, Zakai [2] and Katzur, Bobrovsky, Schuss [6]. Jean Picard has then given a very complete mathematical analysis of this problem, see [10], [11], [12], [13], and Bensoussan [1] has given another proof of most of Picard's results. Those results can be very roughly summarized as follows : for small $\varepsilon > 0$, the variance of the conditional law is of order ε, the optimal and suboptimal filters have a short memory (old observations are quickly "forgotten" by the filter), and there exist various finite dimensional suboptimal filters whose output is close to the conditional expectation of X_t given $\{Y_s, 0 \leq s \leq t\}$ (including the extended Kalman filter and a one dimensional filter whose "error" is of the order of ε). Let us note that analogous results have been established for discrete–time problems by Milheiro de Oliveira [7].

A second class of problems concerns the case where h is *not* one to one. Two cases of such an h are as follows. Case A is where h is locally one to one ; in the situation $\dim X = \dim Y = 1$ which we shall consider below, this means that h is piecewise monotone. Case B is where $\dim X > \dim Y$ (say $\dim X = 2, \dim Y = 1$); and h is a function of say X_t^1 only, h being either monotone or piecewise monotone.

The aim of this paper is to present some recent results for the two above problems. We shall mainly be concerned with case A, and give some hints concerning case B.

The organisation of the paper is as follows. In section 2, we shall present case A, discuss the problem in case $\varepsilon = 0$ (no observation noise), introduce two "detectability assumptions" and compare them.

In section 3, we shall treat in some detail case A under one of the two detectability assumptions. In section 4, we shall comment on some recent results concerning case B.

Let us insist upon the fact that we shall not try in this paper to formulate the most general known results, but rather to present some of the main ideas on simple examples. More general results can be found in the references which we shall give below.

2 Case A : Piecewise Monotone Observation Function.

In this section, we want to formulate the nonlinear filtering problem (1) with small observation noise where all processes are scalar and h is piecewise monotone.

To be more specific, let us assume that $g \equiv 1$; $f, h \in C^1(\mathbb{R})$ with bounded derivatives ; h has a unique minimum at $x = 0$, such that $h(0) = h'(0) = 0$ and $h'(x) < 0$ for $x < 0, h'(x) > 0$ for $x > 0$.

Remark 2.1 : *Inefficiency of the extended Kalman filter.* The extended Kalman filter for the above situation is :

$$\begin{cases} d\hat{X}_t &= f(\hat{X}_t)\,dt + \varepsilon^{-2}R_t h'(\hat{X}_t)(dY_t - h(\hat{X}_t)\,dt) \\ dR_t/dt &= 2f'(\hat{X}_t)R_t + 1 - \varepsilon^{-2}h'(\hat{X}_t)^2 R_t^2 \end{cases}$$

Note that (except possibly near $t = 0$) R_t is of the order of ε, hence $\varepsilon^{-2}R_t h'(\hat{X}_t)$ is of the order of ε^{-1}. Replacing dY_t by its expression in the first equation above yields :

$$d\hat{X}_t = [f(\hat{X}_t) + \varepsilon^{-2}R_t h'(\hat{X}_t)(h(X_t) - h(\hat{X}_t))]\,dt + \varepsilon^{-2}R_t h'(\hat{X}_t)\,dW_t$$

We note that, thanks to the leading term in the drift, the extended Kalman filter is such that $h(\hat{X}_t)$ tends to follow closely $h(X_t)$. However this effect of the drift is counterbalanced by the noise term, which is also of the order of ε^{-1}. But the main point is that the extended Kalman filter has no tendency whatsoever to correct a wrong sign : if $\hat{X}_t$ and X_t have the same sign, the drift tends to get them closer, and if $\hat{X}_t$ and X_t have opposite signs, $\hat{X}_t$ and X_t tend to stay far away one from the other (while the drift tends to get $h(\hat{X}_t)$ and $h(X_t)$ closer). In fact, if $f(0) = 0$, then $\hat{X}_t$ never changes sign, since $h'(0) = 0$, while X_s changes sign after arbitrarily large times with positive probability.◇

Our aim is to present an efficient finite dimensional filter for the above problem in two particular cases. In order to simplify the sequel, we shall from now on assume that h is piecewise linear, i.e. :

$$h(x) = \begin{cases} h_+ x & , x \geq 0 \\ h_- x & , x \leq 0 \end{cases}$$

with $h_+ > 0, h_- < 0$. Of course, h is no longer C^1.

We want to consider cases where the variance of the conditional law of X_t, given $\{Y_s; 0 \leq s \leq t\}$ is small (at least "most" of the time). In order to see what kind of condition is needed, let us now consider the (simpler) case where $\varepsilon = 0$:

$$\begin{cases} dX_t & = f(X_t)\,dt + dV_t \\ dY_t/dt & = h(X_t) \end{cases}$$

Since $h(X_t)$ is completely observed, it suffices, in order to recover X_t, to recover its sign. We first note that we know exactly when X_t reaches 0, and when it does not change sign. Hence the problem is : given a time interval $[a,b]$ such that $X_t \neq 0$, $t \in [a,b]$, can one recover the sign of X_t, from the observation of $h(X_t)$, $t \in [a,b]$?

This is clearly impossible in the case $f \equiv 0$ and $h_- = -h_+$, since there is no way to recover the sign of a Wiener process from its absolute value. Therefore we need to introduce what we call a "detectability assumption". There are two such possible assumptions. The first one is :

$$(DA1) \quad |h_+| \neq |h_-|$$

In this case, we have that :

$$\frac{d}{dt} < h(X_.) >_t = h_+^2, \; t \in [a,b] \text{ if } X_t > 0, \; t \in [a,b]$$

and

$$\frac{d}{dt} < h(X_\cdot) >_t = h_-^2, \ t \in [a,b] \text{ if } X_t < 0, \ t \in [a,b]$$

In other words, under $(DA1)$, the quadratic variation of the process $\{h(X_t)\}$ tells us instantaneously the sign of X_t.

If $(DA1)$ does not hold (say $h_+ = -h_- = 1$, i.e. $h(x) = |x|$), we can still do something, provided the drift helps us. We now formulate the second detectability assumption, assuming for simplicity that $h(x) = |x|$.

$$(DA2) \quad f(x) + f(-x) \neq 0, x \in \mathbb{R}$$

Let $Z_t = h(X_t)$. If $X_t > 0$, $t \in [a,b]$, then

$$dZ_t = f(Z_t)dt + dV_t, \ t \in [a,b].$$

If $X_t < 0$, $t \in [a,b]$, then

$$dZ_t = -f(-Z_t)dt - dV_t, \ t \in [a,b]$$

i.e. we observe a Wiener process plus a drift, which differs depending on the sign of X_t, thanks to $(DA2)$. The log–likelihood ratio is given to us in this case by the Girsanov theorem : for $a \leq t \leq b$,

$$L(a,t) = \int_a^t [f(Z_s) + f(-Z_s)]dZ_s - \frac{1}{2}\int_a^t [f^2(Z_s) - f^2(-Z_s)]ds$$

Note that if $X_t > 0$, $t \in [a,b]$,

$$L(a,t) = \frac{1}{2}\int_a^t [f(Z_s) + f(-Z_s)]^2 ds + \int_a^t [f(Z_s) + f(-Z_s)]dV_s$$

and if $X_t < 0$, $t \in [a,b]$,

$$L(a,t) = -\frac{1}{2}\int_a^t [f(Z_s) + f(-Z_s)]^2 ds - \int_a^t [f(Z_s) + f(-Z_s)]dV_s$$

Hence $L(a,t)$ is likely to be positive in case $X_t > 0$, $a \leq t \leq b$ and to be negative in case $X_t < 0$. The quality of the test (i.e. the probability of making the right decision) depends on the value of $U_t = \int_a^t [f(Z_s) + f(-Z_s)]^2 ds$, which of course depends on $t - a$. The larger U_t is, the smaller the probability of making a wrong decision is, since from the strong law of large numbers :

$$\frac{L(a,t)}{U_t} \to \frac{1}{2} \text{ as } t \to \infty, \text{ a.s. on } \{U_\infty = +\infty\}$$

if $X_t > 0$, $t \geq a$, and the limit is $-\frac{1}{2}$ if $X_t < 0$, $t \geq a$. However, in most cases X_t changes sign after some time, and we don't want to wait too long before making a decision.

Clearly, the situation is very different under $(DA1)$ and under $(DA2)$. Under $(DA1)$, the sign of X_t is detected instantaneously, while under $(DA2)$ some time is needed for the probability of a wrong decision to be small.

Let us now describe the strategy for a finite dimensional suboptimal filter in the case $\varepsilon > 0$. We expect that most of the time the conditional law of X_t, given $\{Y_s, 0 \leq s \leq t\}$ will be almost completely concentrated on one side of 0. Hence a good estimate of X_t should be given by an approximate filter for problem (2.1) with $h(x)$ replaced by h_+x (resp. h_-x) if $X_t > 0$ (resp. < 0). Therefore we consider the two auxiliary filtering problems :

$$(2.1_+) \quad \begin{cases} dX_t &= f(X_t)\,dt + g(X_t)\,dV_t \\ dY_t &= h_+X_t\,dt + \varepsilon dW_t \end{cases}$$

and

$$(2.1_-) \quad \begin{cases} dX_t &= f(X_t)\,dt + g(X_t)\,dV_t \\ dY_t &= h_-X_t\,dt + \varepsilon dW_t \end{cases}$$

to which we associate, following Picard [13], the two filters :

$$(2.2_+) \quad d\hat{X}_t^+ = f(\hat{X}_t^+)\,dt + \varepsilon^{-1}(dY_t - h_+\hat{X}_t^+ dt)$$

$$(2.2_-) \quad d\hat{X}_t^- = f(\hat{X}_t^-)\,dt - \varepsilon^{-1}(dY_t - h_-\hat{X}_t^- dt)$$

The filtering procedure which we propose consists in following alternatively the filter (2.2_+) and the filter (2.2_-). Note that since these two filters are given by stiff equations, the way they are initialized at the time where we start to follow them is irrelevant. In order to choose which filter to follow, we need :

a) to isolate time–intervals on which $\{X_t\}$ is very unlikely to change sign.

b) to decide which is the sign of $\{X_t\}$ on a time interval on which we believe that this sign is fixed.

We then follow the corresponding filter until a possible zero crossing by X_t is detected.

When X_t is close to zero and/or we cannot decide its sign, we estimate it by 0.

This program has been rigorously developped under $(DA1)$ by Fleming, Ji, Pardoux [3] and Roubaud [15] in the piecewise linear case ([15] allows noise correlation and a piecewise constant diffusion coefficient) and by Fleming, Pardoux [5] in the nonlinear case with a piecewise monotone observation function. Numerical experiments are reported in Fleming, Ji, Salame, Zhang [4].

The same program has been developped under $(DA2)$ by Roubaud [15] in the piecewise linear case, and numerical experiments are reported in Milheiro de Oliveira, Roubaud [8]. In the next section, we shall present some of the ideas in Roubaud [15], on the above example.

3 Case A : Piecewise Monotone Observation Function under the Detectability Assumption (DA2)

We consider again the filtering problem (2.1) under the condition $(DA2)$:

$$(3.1) \quad \left\{ \begin{array}{lcl} dX_t & = & f(X_t)\,dt + dV_t \\ dY_t & = & |X_t|\,dt + \varepsilon dW_t \end{array} \right.$$

with the assumption

$$(3.2) \quad \exists k \text{ s.t. } |f(x)| \le k(1+|x|),\ x \in \mathbb{R},$$

and the initial condition $X_0 = x_0$. We associate to (3.1) the two "filters" (see section 2) :

$$(3.3_+) \quad d\hat{X}_t^+ = f(\hat{X}_t^+)\,dt + \varepsilon^{-1}(dY_t - \hat{X}_t^+ dt)$$

$$(3.3_-) \quad d\hat{X}_t^- = f(\hat{X}_t^-)\,dt - \varepsilon^{-1}(dY_t + \hat{X}_t^- dt)$$

with the initial conditions $\hat{X}_0^+ = x_0^+, \hat{X}_0^- = -x_0^-$.

3.1 Detection of the zero crossing by $\{X_t\}$

We first need to detect when X_t might cross zero. For that sake, we shall make use of the :

Lemma 3.1 *For any* $0 < a < b$, $c > 0$, $0 \leq \alpha < 1/2$ *and* $0 < \beta < 1 - 2\alpha$, *there exist* $k > 0$, $\varepsilon_0 > 0$ *s.t. for any* $0 < \varepsilon \leq \varepsilon_0$,

$$P(\sup_{[a,b]} ||X_t| - \hat{X}_t^+| > c\varepsilon^\alpha) \leq exp(-k/\varepsilon^\beta)$$

Proof : It follows readily from (3.1), (3.2_+) and the Itô–Tanaka formula ($\{L_t, t \geq 0\}$ denotes the local time at 0 of $\{X_t\}$) :

$$d(|X_t| - \hat{X}_t^+) = \varepsilon^{-1}(|X_t| - \hat{X}_t^+)dt$$
$$+(\text{sign}(X_t)f(X_t) - f(\hat{X}_t^+))dt + \text{sign}(X_t)dV_t - dW_t + 2dL_t$$

$$\begin{aligned} |X_t| - \hat{X}_t^+ &= e^{-t/\varepsilon}(|X_0| - \hat{X}_0^+) \\ &+ \int_0^t e^{-(t-s)/\varepsilon}(\text{sign}(X_s)f(X_s) - f(\hat{X}_s^+))\,ds \\ &+ \int_0^{+} e^{-(t-s)/\varepsilon}\text{sign}(X_s)\,dV_s - \int_0^t e^{-(t-s)/\varepsilon}dW_s \\ &+2\int_0^t e^{-(t-s)/\varepsilon}dL_s \end{aligned}$$

It suffices to establish the result for each of the terms in the above right side. The four first terms are treated as in [5, Lemma 3.1] with the help of Lemma 3.2 below. The last term is analysed as in [15, Proposition I.3.1].◇

Lemma 3.2 *For any* $0 < a < b$, *there exists* $c > 0$ *s.t.*

$$E\left\{\sup_{a\le t\le b} \exp[cX_t^2]\right\} < \infty \ , \quad \sup_{\varepsilon>0} E\left\{\sup_{a\le t\le b} \exp[c(\hat{X}_t^+)^2]\right\} < \infty$$

Proof : The first result is well–known (see e.g. [15, Lemma I.2.7]). The second one can be established as follows :

$$d\hat{X}_t^+ = f(\hat{X}_t^+)\,dt + \varepsilon^{-1}(|X_t| - \hat{X}_t^+)\,dt + dW_t$$

For $x \in \mathbb{R}$ and $\{Z_t\}$ a bounded variation process, let $U_t(x,z)$ denote the solution of :

$$U_t(x,z) = x + \int_0^t f(U_s(x,z))\,ds + W_t + Z_t$$

Let $U_t^M = U_t(x_0, Z^M)$, where $\{Z_t^M\}$ is the smallest increasing process s.t.

$$U_t(x_0, Z^M) \ge |X_t|,\ t \ge 0,$$

and $U_t^m = U_t(x_0, Z^m)$, where $\{Z_t^m\}$ is the largest decreasing process s.t.

$$U_t(x_0, Z^m) \le |X_t|,\ t \ge 0.$$

It follows from a comparison theorems for one dimensional SDEs that

$$U_t(x_0, Z^m) \le \hat{X}_t^+ \le U_t(x_0, Z^M)$$

and

$$E(\sup_{t\in[a,b]} \exp[cU_t^2(x_0, Z^m)] + \sup_{t\in[a,b]} \exp[cU_t^2(x_0, Z^M)]) < \infty$$

follows from the same result for $\{|X_t|\}$ (possibly with a different constant c). Note that another proof of the second part of this Lemma, which carries over to higher dimension, is given in [15, Lemma I.2.8].$\diamondsuit$

With the help of Lemma 3.1, we can easily build a procedure which detects the possible zero crossings by $\{X_t\}$.

We choose $\bar{c} > 0$ and $0 \le \alpha < 1/2$. Whenever $\hat{X}_t^+ \le \bar{c}\varepsilon^\alpha$, we conclude that X_t might be zero, and we choose 0 as the estimate of X_t. Whenever $\hat{X}_t^+ > \bar{c}\varepsilon^\alpha$, we decide that $X_t \neq 0$ and we try to estimate its sign. For any $0 \leq a < b$, we define

$$C(a,b) = \{\hat{X}_t^+ > \bar{c}\varepsilon^\alpha, a \leq t \leq b\}$$

The next and last step consists in a test for deciding, conditioned upon the observation to belong to $C(a,b)$, upon the sign of $X_t, t \in [a,b]$.

There are two possible tests for this problem. The first one is an extrapolation of the test used in the case $\varepsilon = 0$, and the second one is a likelihood ratio test based on the outputs of the two filters (3.3_+) and (3.3_-).

Before presenting those two tests, let us formulate a stronger version of $(DA2)$, which will be supposed to hold throughout the rest of this section :

$$(DA2s) \quad \exists \ c, d > 0 \text{ s.t. } \inf_{|x-y| \leq c} [f(x) + f(y)] \geq d$$

3.2 Deciding the Sign of X_t : a Test based on the Increments of $\{Y_t\}$.

Define

$$F(x) = \int_0^x [f(y) + f(-y)]dy$$

$L(a,b)$, which was defined in section 2, can be rewritten as :

$$\begin{aligned} L(a,b) &= F(Z_b) - F(Z_a) + \frac{1}{2}\int_a^b [f'(-Z_t) - f'(Z_t)]dt \\ &+ \frac{1}{2}\int_a^b [f^2(-Z_t) - f^2(Z_t)]dt \end{aligned}$$

Of course, in the case $\varepsilon > 0$, $Z_t = h(X_t)$ is not observed, and $L(a,b)$ is no longer a statistics. We note that :

$$\begin{aligned} \varepsilon^{-1}(Y_{t+\varepsilon} - Y_t) &= \varepsilon^{-1}\int_t^{t+\varepsilon} Z_s ds + W_{t+\varepsilon} - W_t \\ &\simeq Z_t + W_{t+\varepsilon} - W_t \end{aligned}$$

Let $m = [\varepsilon^{-1}(b-a)]$ ($[\cdot]$ denotes the integer part of its argument), $t_k = a + k\varepsilon$, $k = 0, 1, \ldots, m$,

$$\bar{Z}_k = \varepsilon^{-1}(Y_{t_{k+1}} - Y_{t_k}),\ k = 0, 1, \ldots, m-1$$

We can then define the statistics :

$$\begin{aligned} L^\varepsilon(a,b) &= F(\bar{Z}_{m-1}) - F(\bar{Z}_0) \\ &+ \varepsilon/2 \sum_{k=0}^{m-1} [f'(-\bar{Z}_k) - f'(\bar{Z}_k) + f^2(-\bar{Z}_k) - f^2(\bar{Z}_k)] \end{aligned}$$

Assuming in addition to the above hypotheses that f' is Lipschitz, it is not hard to show that for small $\varepsilon > 0$ $L^\varepsilon(a,b)$ is close to $L(a,b)$. Hence, if $L^\varepsilon(a,b) > 0$ (resp. < 0), and provided the observation belongs to $C(a,b)$ and $\int_a^b [f(Z_t) + f(-Z_t)]^2 dt$ is large enough, there is a high probability that $X_t > 0$ (resp. < 0), $t \in [a,b]$.

Again, the details can be found in [15]. We note that the extension of this method to higher dimension requises that f be a gradient.

3.3 A Likelihood Ratio Test based on the Outputs of the two Approximate Linear Filters.

We consider again the filtering problem (3.1), to which we associate the two approximate linear filters (3.3_+) and (3.3_-). Note that, if $\mathcal{Y}_t = \sigma\{Y_s; 0 \le s \le t\}$,

$$dY_t = E(|X_t|/\mathcal{Y}_t)dt + \varepsilon d\nu_t$$

where $\{\nu_t\}$, the innovation process, is a standard Wiener process (see e.g. [9]). We expect that if $X_t > 0$ (resp. < 0), $t \in [a,b]$, then $\hat{X}_t^+$ (resp. $\hat{X}_t^-$) is very close to $E(|X_t|/\mathcal{Y}_t)$, at least after some time.

Therefore we introduce the following quantity, which we interpret as being an approximate log–likelihood ratio. Let $a < e < b$. Define

$$\hat{L}^\varepsilon = \varepsilon^{-2} \int_e^b (\hat{X}_t^+ + \hat{X}_t^-) dY_t - \varepsilon^{-2}/2 \int_e^b (|\hat{X}_t^+|^2 - |\hat{X}_t^-|^2) dt$$

We shall now show that, provided $e-a$ is not too small and $b-e$ is large enough, the conditional probability $P(\hat{L}^\varepsilon > 0/A_+(a,b))$, where

$A_+(a,b) = \{X_t > 0, a < t < b\}$, is very close to one. A similar result holds for $P(\hat{L}^\varepsilon < 0/A_-(a,b))$. let

$$M_t^+ = \exp[\varepsilon^{-1} \int_a^{a\vee t} (X_s - |X_s|)dW_s - \varepsilon^{-2}/2 \int_a^{a\vee t} (X_s - |X_s|)^2 ds]$$

and P^+ be a new probability measure given by $\frac{dP^+}{dP} = M_b^+$. From Girsanov's theorem,

$$dY_t = X_t dt + \varepsilon dW_t^+,\ 0 \le t \le b$$

where $\{W_t^+, 0 \le t \le b\}$ is a standard Wiener process under P^+. Hence, again from well–known results from nonlinear filtering, if

$$\tilde{X}_t^+ = E^+(X_t/\mathcal{Y}_t),\ 0 \le t \le b$$

then

$$dY_t = \tilde{X}_t^+ dt + \varepsilon d\nu_t^+, 0 \le t \le b$$

We have

$$\begin{aligned} \hat{L}^\varepsilon &= \varepsilon^{-1} \int_e^b (\hat{X}_t^+ + \hat{X}_t^-) d\nu_t^+ + \varepsilon^{-2}/2 \int_e^b (\hat{X}_t^+ + \hat{X}_t^-)^2 dt \\ &+ \varepsilon^{-2} \int_e^b (\hat{X}_t^+ + \hat{X}_t^-)(\tilde{X}_t^+ - \hat{X}_t^+) dt \end{aligned}$$

We shall show that on $A_+(a,b)$ the third term above is negligible, compared to the second term. Hence the sign of L^ε is given by that of

$$\begin{aligned} \hat{L}^{\varepsilon,+} &= \varepsilon^{-1} \int_e^b (\hat{X}_t^+ + \hat{X}_t^-) d\nu_t^+ + \varepsilon^{-2}/2 \int_e^b (\hat{X}_t^+ + \hat{X}_t^-)^2 dt \\ &= N_b + 1/2 < N >_b \end{aligned}$$

where $\{N_t\}$ is a P^+ martingale. Hence $P^+(\hat{L}^{\varepsilon,+} > 0/ < N >_b > r)$ is close to one when r is large. We now establish :

Lemma 3.3 *For any* $0 < \beta < 1$, $r < (b-e)d^2$, *there exists* k, $\varepsilon_0 > 0$ *s.t. for any* $\varepsilon \in [0, \varepsilon_0]$,

$$P(\varepsilon^{-2} \int_e^b (\hat{X}_t^+ + \hat{X}_t^-)^2 dt < r) \le \exp(-k/\varepsilon^\beta)$$

Proof : From lemma 3.1 (and its analogue with $\hat{X}_t^+$ replaced by $-\hat{X}_t^-$), for any $\beta < 1$, there exists k, ε_0 s.t. :

$$P(\{\sup_{[a,b]} ||X_t| - \hat{X}_t^+| > \frac{c}{2}\} \cup \{\sup_{[a,b]} ||X_t| + \hat{X}_t^-| > \frac{c}{2}\}) \leq exp(-k/\varepsilon^\beta)$$

However,

$$\frac{d}{dt}(\hat{X}_t^+ + \hat{X}_t^-) = -\varepsilon^{-1}(\hat{X}_t^+ + \hat{X}_t^-) + f(\hat{X}_t^+) + f(\hat{X}_t^-)$$

hence

$$\hat{X}_t + \hat{X}_t^- = e^{-(t-a)/\varepsilon}(\hat{X}_a^+ + \hat{X}_a^-) + \int_a^t e^{-(t-s)/\varepsilon}[f(\hat{X}_s^+) + f(\hat{X}_s^-)]ds$$

The first term in the above right side is very small for $t \geq e$. The second term is bounded below by :

$$\varepsilon(1 - e^{-(e-a)/\varepsilon}) \inf_{a \leq t \leq b}[f(\hat{X}_t^+) + f(\hat{X}_t^-)]$$

Provided $|\hat{X}^+ + \hat{X}_t^-| \leq c$, we deduce from (DA 2s) that $f(\hat{X}_t^+) + f(\hat{X}_t^-) \geq d$ The result now follows. ◊

The fact that the term

$$\varepsilon^{-2} \int_e^b (\hat{X}_t^+ + \hat{X}_t^-)(\tilde{X}_t^+ - \hat{X}_t^+)dt$$

is negligible on $A^+(a,b)$ follows from Lemma 3.3, the obvious remark that P^+ and P coïncide on $A^+(a,b)$ (since $M_b^+ = 1$ on this set) and Theorem 3 from Picard [13] which states that for any $p \geq 1$,

$$(E^+[|\hat{X}_t^+ - \tilde{X}_t^+|^p])^{1/p} = 0(\varepsilon^{3/2}).$$

We can easily conclude from the above :

Proposition 3.1 *There exists a continuous decreasing function ρ : $\mathbb{R}_+ \to \mathbb{R}_+$, with $\lim_{x \to +\infty} \rho(x) = 0$, and for any $p \geq 1$, there exist k, $\varepsilon_0 > 0$ s.t. :*

$$P(\{L^\varepsilon < 0\} \cap A_+(a,b) \cap C(a,b)) \leq k\varepsilon^p + \rho(b-a)$$

for any $\varepsilon \in (0, \varepsilon_0]$. ◊

We note that the difference with the results under (DA 1) is the appearance of the term $\rho(b-a)$: for a fixed interval [a,b], under (DA 2) the probability of making a wrong decision does not tend to zero as $\varepsilon \downarrow 0$. This is consistent with the results is the $\varepsilon = 0$ case.

4 Remarks on the problems with $\dim X > \dim Y$

Suppose that $\dim X = 2$ and $\dim Y = 1$, and that

$$dY_t = h(X_t^1)\,dt + \varepsilon dW_t$$

Assume first that h is monotone. Then one can show (see Yaesh, Bobrovsky, Schuss [16], Picard [14]) that there exist efficient linearized filters, provided that the variance of the conditional law of X_t, given $\{Y_s, 0 \le s \le t\}$ is small. But now this need not be the case in general. It is the case for the following model, which has been rigorously analysed by Milheiro de Oliveira [7].

$$\begin{cases} dX_t^1 &= f_1(X_t^1, X_t^2)dt \\ dX_t^2 &= f_2(X_t^1, X_t^2)dt + g(X_t^1, X_t^2)dV_t \\ dY_t &= h(X_t^1)dt + \varepsilon dW_t \end{cases}$$

where f is C^2, for each $x_1, x_2 \to f_1(x_1, x_2)$ is one to one and its inverse is Lipschitz, and some further regularity assumptions are satisfied. Milheiro de Oliveira [7] gives a two dimensional filter with output $\bar{X}_t$ which is such that, for $t \ge t_0 > 0$,

$$E(|X_t^1 - \bar{X}_t^1|^2) = 0(\varepsilon^{3/2}), \quad E(|X_t^2 - \bar{X}_t^2|^2) = 0(\varepsilon^{1/2})$$

It is also possible to derive approximate finite dimensional filters, even when the covariance of the conditional law is not small with ε, provided the problem has a special structure. Let us describe a problem which has been successfully treated in Roubaud [15]. Again, $\dim X = 2$ and $\dim Y = 1$. We assume that :

$$\begin{cases} dX_t &= (f(X_t^1) + bX_t^2)\,dt + GdV_t \\ dY_t &= h(X_t^1)\,dt + \varepsilon dW_t \end{cases}$$

where $f \quad : \mathbb{R} \to \mathbb{R}^2$ and $h \quad : \mathbb{R} \to \mathbb{R}$ are piecewise–linear (with say two pieces), h being non monotone, $b \in \mathbb{R}^2$, G is a 2×2 matrix, and $\{V_t\}$ is two-dimensional standard Wiener process. As in the preceding section, we associate to this problem two (linear) filters, and test procedures to decide which filter to follow. The conditional variance in the x^1 direction is small, but it is in general of order one in the x^2 direction. The fact that the system is linear in X_t^2 is crucial here. Note that one major difference with the situation of the preceding section is that the filter (or at least its second component) does not have a short memory.

Bibliography

[1] A. Bensoussan, *On some approximation techniques in nonlinear filtering,* in Stoch. Diff. Systems, Stoch. Control Th. and Applic., W.H. Fleming & P.L. Lions eds, IMA **10**, Springer 1988.

[2] B.Z. Bobrovsky, M. Zakai, *Asymptotic a priori estimates for the error in the nonlinear filtering problem,* IEEE–IT **28**, 1982, p. 371–376.

[3] W. Fleming, D. Ji, E. Pardoux, *Piecewise linear filtering with small observation noise,* in Analysis and Optimization of Systems, A. Bensoussan & J.L. Lions eds., LNCIS **111**, Springer 1988.

[4] W. Fleming, D. Ji, P. Salame, Q. Zhang, *Discrete time piecewise linear filtering with small observation noise,* Brown University Report, LCDS/CCS **88–27**, 1988.

[5] W. Fleming, E. Pardoux, *Piecewise monotone filtering with small observation noise,* SIAM J. Control **27**, 1989, p. 1156–1181.

[6] R. Katzur, B.Z. Bobrovsky, Z. Schuss, *Asymptotic analysis of the optimal filtering problem for one-dimensional diffusions measured in a low noise channel,* SIAM J. Appl. Math. **44**, 1984, Part I : 591–604, Part II : 1176–1191.

[7] P. Milheiro de Oliveira, *Etudes asymptotiques en filtrage non linéaire avec petit bruit d'observation,* Thèse, Univ. de Provence, 1990.

[8] P. Milheiro de Oliveira, M.C. Roubaud, *Filtrage linéaire par morceaux d'un système en temps discret avec petit bruit d'observation,* Rapport de recherche INRIA, 1991.

[9] E. Pardoux, *Filtrage non linéaire et équations aux dérivées partielles stochastiques associées,* in Ecole d'été de Probabilité de St-Flour 1989, LNM, Springer, to appear.

[10] J. Picard, *Nonlinear filtering of one-dimensional diffusions in the case of a high signal-to-noise ratio,* SIAM J. Appl. Math. **46**, 1986, p. 1098–1125.

[11] J. Picard, *Filtrage de diffusions vectorielles faiblement bruitées,* in Analysis and Optimization of Systems, A. Bensoussan & J.L.Lions eds, LNCIS **83**, Springer 1986.

[12] J. Picard, *Nonlinear filtering and smoothing with high signal–to–noise ratio,* in Stochastic Processes in Physics and Engineering, Reidel 1988.

[13] J. Picard, *Asymptotic study of estimation problems with small observation noise,* in Stochastic Modelling and Filtering, LNCIS **91**, Springer 1987.

[14] J. Picard, *Efficiency of the extended Kalman filter for non linear systems with small noise,* Rapport de recherche INRIA **1068**, 1989.

[15] M.C. Roubaud, *Filtrage linéaire par morceaux avec petit bruit d'observation,* Thèse, Univ. de Provence, 1990.

[16] I. Yaesh, B.Z. Bobrovsky, Z. Schuss, *Asymptotic analysis of the optimal filtering problem for two–dimensional diffusions measured in a low noise channel,* SIAM J. Appl. Math. **50**, 1990, p. 1134–1155.

A Simple Proof of Uniqueness for Kushner and Zakai Equations

B.L. Rozovskii †
Department of Mathematics and Center for Applied Mathematical Sciences
University of Southern California, Los Angeles
Ca. 90089–1113 U S A

1 Introduction

The uniqueness problem for Kushner and Zakai equations in non–linear filtering theory was addressed by many authors (see e.g. Bensoussan [1], Chaleyat-Maurel, Michel and Pardoux [2], Kallianpur, Karandikar [7], Krylov, Rozovskii [8], Kunita [9], Kurtz, Ocone [10], Rozovskii [15] etc.).

The purpose of this article is to present a simple proof of uniqueness for a generalized solutions to the Kushner and Zakai equations for diffusion processes. These solutions are considered in the space of signed measures. The result holds under rather mild assumptions on the coefficients of the corresponding processes. For example, in the case of non–degenerate signal process only boundness, continuity in t and Hölder continuity in x is assumed. In the degenerate case the coefficients are, in addition, assumed to belong to $C_b^{2+\alpha}(R^d)$. Of course, uniqueness of "classical" solutions which belong to L_1 follows from the above result.

The most general result on the uniqueness of the Kushner and Zakai equations for Markov signal process (which is the case in this article) is due to Kurtz, Ocone [10]. However, in [10] the solutions are

†The work was supported in part by NSF grant No. DMS–9002 997

ISBN 0-12-481005-5

considered in the space of probability measures. Wether it is possible to derive from [10] the uniqueness results of this article related to the case when the coefficients are only Hölder-continuous in x and continuous in t remains an open problem (even in the case of non-negative measure-valued solutions).

Another very general result about uniqueness of the Zakai equation was obtained by Chaleyat-Maurel, Michel and Pardoux [2] in the case of a non-Markov signal process. Infinite differentiability of the coefficients was the price the authors have paid to allow a feedback option.

The methods imploied in [2] and [10] differ from the one in this article. The results have substantial intersections but neither one includes another.

It is also necessary to mention that articles [?] by Bensoussan and [12] by Purtukhija, Rozovskii are conceptually related to this one. However, in both articles non-negative measure-valued solutions are considered and it is assumed in [1] that the singnal process and the noise in the observation process are independent.

In conclusion I wish to thank Dan Ocone and Etienne Pardoux for helpful discussion.

2 Setting of the Problem. Notation

Let $(\Omega, \mathcal{F}, P)$ be a probability space. Given are a number $T > 0$ and two standard Brownian motions $w_1(t)$ and $w_2(t)$ taking values in R^d and R^{d_1}, respectively. Also given are two random variables x_0 and y_0 which are independent of w_1 and w_2 and finite (P-a.s.). Denote $L_{m \times n}$ the space of $M \times n$-matrices. Let

$$b \ : [0,T] \times R^d \to R^d, \ h \ : [0,T] \times R^d \to R^{d_1},$$

$$\sigma_1 \ : [0,T] \times R^d \to L_{d \times d}, \ \sigma_2 \ : [0,T) \times R^d \to L_{d \times d_1}$$

be Borel functions. Here and below it is assumed that b, h, σ_1 and σ_2 are bounded and Hölder-continuous.

Consider the system of Ito equations

$$(1.1)\quad \begin{aligned} dx(t) &= b(t,x(t))dt + \sum_{i=1}^{2} \sigma_i(t,x(t))dw_i(t) \\ dy(t) &= h(t,x(t))dt + dw_2(t), t \in]0,T] \\ x(0) &= x_0, y(0) = y_0. \end{aligned}$$

Throughout what follows it is assumed that system (1.1) has a unique strong solution. As usual in filtering it is assumed that $y(t)$ is observable and $x(t)$ (the signal process) is not. Denote by $\mathcal{F}_t^y$ the σ-field generated by $y(s)$ for $s \leq t$ and completed with respect to the measure P. Let $f : R^d \to R^1$ be a function from C_b^2. Denote $\pi_t[f] := E[f(x(t))|\mathcal{F}_t^y]$. It is a standard fact that $\pi_t[f]$ is the best in the mean-square $\mathcal{F}_t^y$-measurable estimate for $f(x(t))$ and one can choose a modification of $\pi_t[f]$ which possesses the following stochastic differential (see Fujisaki, Kallianpur, Kunita [5] or Liptser, Shiryayev [11]) :

$$(1.2)\quad \begin{aligned} d\pi_t[f] = \pi_t[\mathcal{L}f]dt &+ (\pi_t[\mathcal{M}f] - \\ -\pi_t[f]\pi_t[h(t)])(dy(t) &- \pi_t[h(t)]dt), t \leq T, \end{aligned}$$

where $\mathcal{L}f := \sum_{i,j=1}^{d} a^{ij}(t,x)\frac{\partial^2}{\partial x_i \partial x_j} f(x) + \sum_{i=1}^{d} b^i(t,x)\frac{\partial}{\partial x_i} f(x)$,

$$\mathcal{M}f := \sum_{i=1}^{d} \sigma_2^i(t,x)\frac{\partial}{\partial x_i} f(x) + h(t,x)f(x),$$

and $a(t,x) := \frac{1}{2}[\sigma_1\sigma_1^*(t,x) + \sigma_2\sigma_2^*(t,x)]$.

In filtering the ultimate goal is to "calculate" $\pi_t[f]$. Relation (1.2) is an important step in this direction because it reduces the filtering problem to a problem in PDE's. For example if conditional distribution $P(x(t) \in dx|\mathcal{F}_t^y)$ has a smooth density $\pi(t,x)$ with respect to Lebesgue measure it is readily checked that relation (1.2) can be reduced to the stochastic PDE (see e.g. Liptser, Shiryayev [11])

$$(1.3)\quad d\pi(t,x) = \mathcal{L}^*\pi(t,x)dt + (\mathcal{M}^*\pi(t,x) - \pi(t,x)\int_{R^d} h(t,x) \times$$

$$\times \quad \pi(t,x)dx)(dy(t) - \int_{R^d} h(t,x)\pi(t,x)dxdt).$$

This equation is sometimes referred to as the Kushner equation. It is rather difficult for analysis because of its cumbersom, non-linear integro-differential structure. However, it has a much simpler linear counterpart. In his celebrated work [18] M. Zakai proved that the filtering density $\pi(t,x)$, if it exists and is smooth enough, can be represented in the form

$$\pi(t,x) = \varphi(t,x) / \int_{R^d} \varphi(t,x)dx.$$

where $\varphi(t,x)$ is a solution to the linear stochastic PDE

$$(1.4) \quad d\varphi(t,x) = \mathcal{L}^*\varphi(t,x)dt + \mathcal{M}^*\varphi(t,x)dy(t).$$

The function φ is usually referred to as the non-normalized filtering density and equation (1.4) as the Zakai equation.

Unfortunately, even for this simplified equation, analytical problems such as existence and uniqueness of a classical solution are by no means simple.

The objective of this article is to demonstrate that the uniqueness problem for the above equations can be simplified dramatically if a broader concept of a generalized (measure-valued) solution is considered. Uniqueness of classical solutions follows from this property of the measure-valued solutions almost automatically.

3 The Main Result

Denote by $\mathcal{M}(R^d)$ the set of totally finite, countably additive signed measures (see e.g. [6]) furnished with the topology of weak convergence, and let $\mathcal{M}_b(R^d)$ and $\mathcal{M}^+(R^d)$ be subspaces of $\mathcal{M}(R^d)$ consisting of all uniformly bounded signed measures on R^d and of all probability measures on $\mathbb{R}^d$, respectively. If μ belongs to $\mathcal{M}(R^d)$ we denote $|\mu|$ its total variation. We will also use the notation $< \mu, f > := \int_{R^d} f(x)\mu(dx)$ for $f \in L_1(d|\mu|)$.

Definition 3.1 (i) *An $\mathcal{F}_t^y$-adapted stochastic process P_t taking values in $\mathcal{M}_b(R^d)$ is said to be a measure–valued solution to the*

Kushner equation (1.2) corresponding to the initial condition $P_0(dx) = P(x_0 \in dx|y_0)$ *if for every* $f \in C_b^2(R^d)$*, the equality holds :*

$$(2.1) \quad < P_t, f > \; = \; < P_0, f > + \int_0^t < P_s, \mathcal{L}(s)f > ds +$$

$$+ \int_0^t < P_s, \mathcal{M}(s)f \; - \; < P_s, h(s) > f > (dy(s) -$$

$$- \; < P_s, h(s) > ds), P\text{-a.s.} \forall t \in [0,T].$$

(ii) *An* $\mathcal{F}_t^y$*-adapted stochastic process* Φ_t *taking values in* $\mathcal{M}(R^d)$ *is said to be a measure valued solution to the Zakai equation (1.4) corresponding to the initial condition* $\Phi^0(dx) = P(x_0 \in dx|y_0)$ *if* $< |\Phi \cdot|, 1 > \in L_2([0,T]) \times \Omega \,; dt \times dP)$*, for every* $t \le T$*,* $< |\Phi_t|, 1 > \in L_2(\Omega, dP)$*, and for any* $f \in C_b^2(R^d)$ *the following equality holds*

$$(2.2) \quad < \Phi_t \,, f > = < \Phi^0 \,, f > + \int_0^t < \Phi_s, \mathcal{L}(s)f > ds + \int_0^t < \Phi_s \,, \mathcal{M}(s)f > dy(s), \quad P\text{-a.s. } \forall t \in [0,T]. \quad \square$$

Remark 3.1 *It is a typical case when one can prove existence of a classical solution to the Kushner or Zakai equation, in the class of integrable functions, while the non negativity problem for the solution remains open. In this case it is very convinient to have a uniqueness theorem for a generalized solution in the space of signed measures.* □

Let $m \in C([0,T] \,; R^{d_1})$ and $\mathcal{L}_m(t,x)f(x) \; := \; \mathcal{L}(t,x)f(x) + \mathcal{M}(t,x)m(t)f(x)$. Consider the backward Cauchy problem :

$$(2.3) \quad -\frac{\partial v^m(t,x)}{\partial t} = \mathcal{L}_m v^m(t,x) \;\; t < T_0, x \in R^d$$

$$(2.4) \quad v^m(T_0, x) = \xi(x)$$

We introduce the following assumption.

(H) For all $\xi \in C_b^2(R^d)$ and $m \in C([0,T] \,; R^{d_1})$, problem (2.3), (2.4) has a solution in $C_b^{1,2}([0,T_0] \times R^d)$ for every $T_0 \le T$. □

There are two important types of sufficient conditions which guarantee that (H) holds :

(H_1) (see Eidel' man [3]). The matrix a is uniformly non–degenerate and the coefficients of the operators $\mathcal{L}$ and $\mathcal{M}$ are continuous in t and Hölder-continuous in x. Moreover, the continuity in t of the matrix a is uniform in x. □

(H_2) (see Freidlin [4]). The coefficients of the operators $\mathcal{L}$ and $\mathcal{M}$ belong to $C_b^{0,2+\alpha}([0,T]\times R^d)$, where $\alpha > 0$. □

Proposition 3.1 *Both Kushner and Zakai equations have non negative measure-valued solutions P_t and Φ_t, respectively. These solutions are connected by the relations*

$$(2.5) \quad P_t = \Phi_t / < \Phi_t, 1 >,$$

and

$$(2.6) \quad \Phi_t = P_t exp\{\int_0^t < P_s, h(s) > dy(s) \\ -\tfrac{1}{2}\int_0^t | < P_s \, , h(s) > |^2 ds\}. \quad \Box$$

Proof : It is at standard fact (see e.g. Yor [17] or Rozovskii [14] that there exists an $\mathcal{F}_t^y$–adapted process P_t in $\mathcal{M}^+(R^d)$ such that for every $f \in L_\infty(R^d), \pi_t[f] =< f, P_t >$. Thus, the existence of a solution to the Kushner equation follows immediately from (1.2).

By standard arguments, one can prove that there exists a modification of a process $P_t[f]$ such that given f, (2.1) holds for all $t \in [0,T]$ and $\omega \in \tilde{\Omega}$ where $\tilde{\Omega} \subset \Omega$ and $P(\tilde{\Omega}) = 1$. Define Φ_t by (2.6). Then by direct application of the Ito formula one can easily verify that $\Phi_t[f]$ satisfies (2.2). It is also obvious that

$\Phi_t[1] = exp\{\int_0^t < P_s \, , h(s) > dy(s) - \frac{1}{2}\int_0^t | < P_s \, , h(s) > |^2 ds\} \in L_2([0,T]\times\Omega \, ; dt \times dP)$.

Thus, existence of a measure-valued solution to the Zakai equation is proved. To complete the proof, note that

$< \Phi_t, 1 >= exp\{\int_0^t < P_s, h(s) > dy(s) - \frac{1}{2}\int_0^t | < P_s, h(s) > |^2 ds\}$ and thus is positive with probability 1. □

Theorem 3.1 *(cf.[12]) Assume (H). Then measure-valued solutions to the Zakai and Kushner equations are unique.* □

Proof : First consider the Zakai equation. By standard approximation arguments one can derive from the Definition 2.1 that a measure-valued solution Φ_t to the Zakai equation has a modification such that for each $\psi \in C_b^{1,2}([0,T] \times R^d)$ there exists an $\omega -$ set Ω' of probability 1 such that for all $t \in [0,T]$ and $\omega \in \Omega'$ the equality holds :

$$
\begin{aligned}
& < \Phi_t, \psi(t) >=< \Phi^0, \psi(0) > \\
(2.7) \quad & + \int_0^t < \Phi_s, \frac{\partial}{\partial s}\psi(s) + \mathcal{L}(s)\psi(s) > ds \\
& + \int_0^t < \Phi_s, \mathcal{M}(s)\psi(s) > dy(s).
\end{aligned}
$$

Next fix some $m \in C([0,T]\,; R^{d_1})$. Let $v(t,x)$ be a solution to problem (2.3), (2.4) for the above m. Denote $q_t := exp\{\int_0^t m(s)dy(s) - \frac{1}{2}\int_0^t |m(s)|^2 ds\}, \rho_t^{-1} := exp\{-\int_0^t h(s,x(s))dy(s) + \frac{1}{2}\int_0^t |h(s,x(s)|^2 ds\}$, and $\gamma_t := q_t\rho_t^{-1}$.
Now we differentiate by the Ito formula the product $< \Phi_t, v(t) > \gamma_t$. Making use of (2.7) we get

$$
\begin{aligned}
& < \Phi_t, v(t) > \gamma_t =< \Phi^0, v(0) > + \\
(2.8) \quad \int_0^t & < \Phi_s, \frac{\partial}{\partial s}v(s) + \mathcal{L}_m v(s) > \gamma_s ds + \int_0^t (< \Phi_s, v(s) > \times \\
(m(s) & - h(s,x(s)) + < \Phi_s, \mathcal{M}(s)v(s) >)\gamma_s w_2(s).
\end{aligned}
$$

Since $v(s)$ is a solution to problem (2.3), (2.4) the second term in the right hand part of (2.8) is zero. It is readily checked that the third term is a martingale. Therefore, taking expectations of both sides of (2.8), we get

$$E(< \Phi_{T_0}, \xi > \gamma_{T_0}) = Ev(0, x_0).$$

Let us introduce a new probability measure $\tilde{P}$ by the formula $d\tilde{P} = \rho_T^{-1}dP$. Since evidently $P(\rho_T > 0) = 1$, measures P and $\tilde{P}$ are mutually absolutely continuous and $dP/d\tilde{P} = \rho_T$. Henceforth, $\tilde{E}$ always denote the expectation with respect to $\tilde{P}$.

It is a standard fact that

$$Ev(0,x_0) = E[\xi(x_m(T_0))exp\int_0^{T_0} h(s,x_m(s))m(s)ds]$$

where $x_m(t)$ is a solution to the Ito equation

$$x_m(t) = x_0 + \int_0^t (b(s, x_m(s)) + \sigma_2(s, x_m(s)m(s))ds$$
$$+\sum_{i=1}^2 \int_0^t \sigma_i(s, x_m(s))dw_i(s).$$

By Girsanov's theorem

$$E \quad [\xi(x_m(T_0))exp\{\int_0^{T_0} h(s, x_m(s))m(s)ds\}] =$$
$$E \quad [\xi(x(T_0))q_T] = \tilde{E}[\xi(x(T_0))\rho_{T_0} q_{T_0}] =$$
$$\tilde{E} \quad [\tilde{E}[\xi(x(T_0))\rho_{T_0}|\mathcal{F}^y_{T_0}]q_{T_0}].$$

On the other hand

$$E(< \Phi_{T_0}, \xi > \gamma_{T_0}) = \tilde{E}(< \Phi_T, \xi > q_{T_0}),$$

and we arrived at the equality

$$(2.9) \quad \tilde{E}[\tilde{E}[\xi(x(T_0))\rho_{T_0}|\mathcal{F}^y_{T_0}]q_{T_0}] = E[< \Phi_{T_0}, \xi > q_{T_0}]$$

Now note that on the space $(\Omega, \mathcal{F}, \tilde{P})$ $y(t)$ is a Wiener martingale and thus the family $\{q_{T_0} := q_{T_0}(m), m \in C([0, T_0] ; R^{d_1})\}$ is total in $L_2(\Omega, \mathcal{F}^y_{T_0}, \tilde{P})$ (see e.g. Rozovskii [13]). The latter means that if $\xi \in L_2(\Omega, \mathcal{F}^y_{T_0}, P)$ and $\tilde{E}\xi q_{T_0}(m) = 0$ for all $m \in C([0, T_0] ; R^{d_1})$ then $\xi = 0$ P-a.s. Thus (2.9) implies that

$$(2.10) \quad < \Phi_{T_0}, \xi >= \tilde{E}[\xi(x(T_0))\rho_{T_0}|\mathcal{F}^y_{T_0}] \text{ P-a.s.}$$

It is readily checked that for an arbitrary measure-valued solution P_t of the Kushner equation, P_t exp $\{\int_0^t < P_s, h(s) > dy(s)-$
$\frac{1}{2}\int_0^t | < P_s, h(s) > |^2 ds\}$ is a measure-valued solution to the Zakai equation. Thus P-a.s.

$$\exp\{\int_0^t < P_s, h(s) > dy(s) - \frac{1}{2}\int_0^t | < P_s, h(s) > |^2 ds =< \Phi_t, 1 >$$

and we arrive at the formula

$$(2.11) < P_t, \xi >=< \Phi_t, \xi > / < \Phi_t, 1 > \quad \forall \xi \in L_\infty(R^d) \text{ P-a.s.}$$

The proof is complete. □

Remark 3.2 *In the process of the proof we obtained as a "free gift" (combine (2.10) and (2.11)) the famous Kallianpur–Striebel formula*

$$E[\xi(x(t))|\mathcal{F}_t^y] = \frac{\tilde{E}[\xi(x(t))\rho_t|\mathcal{F}_t^y]}{\tilde{E}[\rho_t|\mathcal{F}_t^y]} P\text{-a.s. for every } \xi \in L_\infty(R^d).$$

It was also shown in the proof that a solution Φ_t *to the Zakai's equation is always represented by the formula (2.10).* □

Bibliography

[1] A. Bensoussan, *On the Integrated Formulation of Zakai and Kushner Equations* In : "Stochastic Partial differential Equations and Appl. II". (G. DaPrato, L. Tubaro, Eds.) Lecture Notes in Math. 1390, 1989, pp. 13–23.

[2] M. Chaleyat-Maurel, D. Michel, E. Pardoux, *Un théorème d'unicité pour l'équation de Zakai.* Stochastics and Stoch. Rep. 29, 1990, pp.1–12.

[3] S.D. Eidel'man, *Parabolic Systems.* North-Holland, Amsterdam-London and Wolters–Noordhoff, Groningen, 1969.

[4] M. Freidlin, *Functional Integration and Partial Differential Equations.* Princeton University Press, Princeton (NJ), 1985.

[5] M. Fujisaki, G. Kallianpur and H. Kunita, *Stochastic Differential Equations for the Nonlinear Filtering Problem.* Osaka J. Math. 9, 1972, pp. 19–40.

[6] P.R. Halmos, *Measure Theory.* D. Van Nostrand, Princeton (NJ) etc. , 1950.

[7] G. Kallianpur, R.L. Karandikar, *The Nonlinear Filtering Problem for the Unbounded Case.* Stochastic Process Appl. 18, 1984, pp. 57–66.

[8] N.V. Krylov, B.L. Rozovskii, *On Conditional Distributions of Diffusion Processes.* Math. USSR Izvestija 12 (1978), pp. 336–356.

[9] H. Kunita, *Cauchy Problem for Stochastic Partial Differential Equations Arising in Nonlinear Filtering Theory.* Syst. and Contr. Let. 1 (1981), pp. 37–41.

[10] T.G. Kurtz, D.L. Ocone, *Unique Characterization of Conditional Distribution in Nonlinear Filtering.* The Annals of Probability 16, 1988, pp. 80–107.

[11] R. Sh. Lipster, A.N. Shiryayev, *The Statistics of Random Processes I, II.* Springer–Verlag, Berlin etc., 1977.

[12] O.G. Purtukhija, B.L. Rozovskii, *Measure-valued Solutions to Second Order Stochastic parabolic Equations.* In : "Statistics and Control of Random Processes (A.N. Shiryayev, Ed.) Nauka, Moscow, 1989, pp. 177–179 (in Russian).

[13] B.L. Rozovskii, *Lecture Notes on Linear Stochastic Partial Differential Equations.* Technical Report # 25, Departement of Mathematics, The University of North Carolina at Charlotte, 1990.

[14] B.L. Rozovskii, *Stochastic Evolution Systems. Linear Theory and Applications to Nonlinear Filtering.* D. Riedel Publ. Co., Dordrecht, 1990 (to appear).

[15] B.L. Rozovskii, *On stochastic Partial Differential Equations.* Math. USSR Sbornik 25, 1975, pp. 295–356.

[16] J. Szpirglas, *Sur l'équivalence d'équations différentielles stochastique à valeurs mesures intervenant dans le filtrage markovien non–linéaire.* Ann. Inst. Poincaré, Sect. B (N.S.) 14, 1978, pp. 33–59.

[17] M. Yor, *Sur les théories du filtrage et de la prédiction.* Séminaire de Probabilités XI, Lecture Notes in Math. 581, 1977, Springer–Verlag, Berlin etc. pp. 257–297.

[18] M. Zakai, *On Optimal Filtering of Diffusion Processes.* Z. Warsch. 11, 1969, pp. 230–243.

Itô-Wiener expansions of holomorphic functions on the complex Wiener space

Ichiro Shigekawa
Department of Mathematics
Faculty of Science
Kyoto University
Kyoto 606, JAPAN

Abstract

We show L^p-convergence of Itô-Wiener expansions for holomorphic functions on a complex abstract Wiener space and the unicity theorem.

1 Introduction

In this paper, we discuss holomorphic functions on a complex abstract Wiener space. A Hilbert space of square-integrable holomorphic functions was discussed by Bargmann [1,2] in connection with quantum physics. In particular, he considered an infinite dimensional space in [2] and this paper is closely related to his work. But we discuss L^p-holomorphic functions for $p \in (1, \infty)$. We shall show Itô-Wiener expansions for L^p-holomorphic functions. To be precise, let (B, H, μ) be a complex abstract Wiener space and J_n be a projection operator to the space of multiple Wiener integrals of order n. Setting $S_n = J_0 + J_1 + \cdots + J_n$, we show that $S_n F \to F$ strongly in $L^p(B, \mu)$ as $n \to \infty$ for a holomorphic function F. In case $p = 2$, this result is well-known as the Itô-Wiener expansion (or the Wiener chaos expansion).

The organization of this paper is as follows. We give a review

ISBN 0-12-481005-5

on a complex and an almost complex abstract Wiener space in section 2 and give a characterization of holomorphic functions by using the Cauchy-Riemann equation. In section 3, we introduce complex Hermite polynomials and define the complex Wiener chaos. We give a proof of our main theorem in section 4. We reduce our problem to convergence of a Fourier series by using a rotation operator. In addition, we show a uniqueness theorem for holomorphic functions.

2 Complex abstract Wiener space and holomorphic functions

Let us review a *complex abstract Wiener space.* A complex abstract Wiener space is a triplet (B, H, μ) where B is a complex separable Banach space, H is a complex separable Hilbert space which is densely and continuously imbedded in B and μ is a Borel probability measure on B with a characteristic function $\hat{\mu}$:

$$\begin{aligned}\hat{\mu}(\varphi) &:= \int_B \exp\{\sqrt{-1}\,\mathrm{Re}\langle z, \varphi\rangle\}\mu(dz) \\ &= \exp\left\{-\frac{1}{4}|\varphi|^2_{H^*}\right\}, \qquad \varphi \in B^* \hookrightarrow H^*,\end{aligned}$$

where B^* and H^* are dual spaces of B and H respectively and $\langle \cdot, \cdot \rangle$ is a natural paring between B and B^*.

By the above definition, it is easy to see that a family $\{\langle \cdot, \varphi\rangle\,;\varphi \in B^*\}$ is a *complex* Gaussian system, i.e., any complex linear combination is a complex Gaussian random variable (we always assume that the mean is 0). We remark that in the complex case, real and imaginary parts of a complex Gaussian random variable are independent and have identical distribution. Moreover the following identity holds. For $\varphi, \psi \in B^* \hookrightarrow H^*$,

$$\int_B \langle z, \varphi\rangle \overline{\langle z, \psi\rangle}\mu(dz) = (\varphi, \psi)_{H^*}. \tag{1}$$

Let us define an isometric mapping $J: B \to B$ by

$$Jz = \sqrt{-1}z.$$

We call a quartet (B, H, μ, J) an *almost complex abstract Wiener space*. To avoid confusion, we *forget* the complex structure of B, H and we regard B as a *real* Banach space and H a *real* Hilbert space. So the complex structure of B is introduced through the mapping J; $\sqrt{-1}z = Jz$.

From now on, we consider an almost complex abstract Wiener space (B, H, μ, J). So B is a real separable Banach space, H is a real separable Hilbert space which is densely and continuously imbedded in B, J is an isometry in B with $J^2 = -I$, and its restriction to H is also an isometry. Further, μ is a Borel probability measure on B with a characteristic function $\hat{\mu}$;

$$\begin{aligned} \hat{\mu}(\varphi) &:= \int_B \exp\{\sqrt{-1}\langle z, \varphi\rangle\}\mu(dz) \\ &= \exp\left\{-\frac{1}{4}|\varphi|_{H^*}^2\right\}, \qquad \varphi \in B^* \hookrightarrow H^*, \end{aligned}$$

where B^* and H^* are dual spaces of B and H, respectively. Of course, in this case, B^* is a set of all real valued bounded $\mathbf{R}$-linear functions on B.

Let $B^{*\mathbf{C}}$ be a complexification of B^*: $B^{*\mathbf{C}} = B^* \oplus \sqrt{-1}B^*$. We can identify $B^{*\mathbf{C}}$ with a set of bounded $\mathbf{R}$-linear functions from B to $\mathbf{C}$: $B^{*\mathbf{C}} = \mathcal{L}(B; \mathbf{C})_{\mathbf{R}}$. We set

$$\begin{aligned} B^{*(1,0)} &= \{\varphi \in B^{*\mathbf{C}}; J^*\varphi = \sqrt{-1}\varphi\} \\ B^{*(0,1)} &= \{\varphi \in B^{*\mathbf{C}}; J^*\varphi = -\sqrt{-1}\varphi\}. \end{aligned}$$

Then it is easy to see that $B^{*(1,0)}$, $B^{*(0,1)}$ are complex subspaces of $B^{*\mathbf{C}}$ and

$$B^{*\mathbf{C}} = B^{*(1,0)} \oplus B^{*(0,1)} \quad \text{(direct sum)}.$$

$B^{*(1,0)}$ is a set of all $\mathbf{C}$-linear functions on B. In fact, for $\varphi \in B^{*(1,0)}$

$$\langle\sqrt{-1}z, \varphi\rangle = \langle Jz, \varphi\rangle = \langle z, J^*\varphi\rangle = \langle z, \sqrt{-1}\varphi\rangle = \sqrt{-1}\langle z, \varphi\rangle.$$

Similarly, we can define $H^{*(1,0)}$, $H^{*(0,1)}$. Now we introduce polynomials on B.

Definition 2.1 A function $F: B \to \mathbf{C}$ is called a *polynomial* if it is of the form

$$F(z) = p(\langle z, \varphi_1 \rangle, \dots, \langle z, \varphi_n \rangle) \tag{2}$$

where $n \in \mathbf{N}$ and $p(\zeta_1, \dots, \zeta_n)$ is a complex coefficient polynomial and $\varphi_1, \dots, \varphi_n \in B^{*\mathbf{C}}$. We denote a set of all polynomials by $\mathcal{P}$.

Moreover we can define a subclass of polynomials as follows:

Definition 2.2 A function $F: B \to \mathbf{C}$ is called a *holomorphic polynomial* if it is of the form

$$F(z) = p(\langle z, \varphi_1 \rangle, \dots, \langle z, \varphi_n \rangle) \tag{3}$$

where $n \in \mathbf{N}$ and $p(\zeta_1, \dots, \zeta_n)$ is a complex coefficient polynomial and $\varphi_1, \dots, \varphi_n \in B^{*(1,0)}$. We denote a set of all holomorphic polynomials by $\mathcal{P}_h$.

It is natural that holomorphic polynomials are called holomorphic. We define a wider class of holomorphic functions. To do this, we use a characterization by the Cauchy-Riemann equation. We define operators ∂ and $\bar{\partial}$ as follows. We use notions from the Malliavin calculus (see e.g., [8,10]).

For $F \in \mathcal{P}$, H-derivative $DF(z) \in H^{*\mathbf{C}}$ is defined by

$$DF(z)[h] = \lim_{t \to 0} \frac{F(z + th) - F(z)}{t}.$$

A Sobolev space $W^{s,p}$ ($s \in \mathbf{R}$, $p \in (1, \infty)$) is a completion of $\mathcal{P}$ under a norm $\| \cdot \|_{s,p}$ defined by

$$\|F\|_{s,p} := \|(1 - L)^{s/2} F\|_p$$

where L is an Ornstein-Uhlenbeck operator and $\| \cdot \|_p$ denotes a usual L^p-norm with respect to μ.

Let $\pi^*_{(1,0)}, \pi^*_{(0,1)}: H^{*\mathbf{C}} \to H^{*\mathbf{C}}$ be projection operators to subspaces $H^{*(1,0)}$ and $H^{*(0,1)}$, respectively. Then operators ∂ and $\bar{\partial}$ are defined by

$$\partial F(z) = \pi^*_{(1,0)} DF(z) = \frac{1}{2}(DF(z) - \sqrt{-1} J^* DF(z)), \tag{4}$$

$$\bar{\partial} F(z) = \pi^*_{(0,1)} DF(z) = \frac{1}{2}(DF(z) + \sqrt{-1} J^* DF(z)). \tag{5}$$

The above operators are well-defined for $F \in W^{1,p}$, but they are not closed operators in $L^p(B, \mu)$.

Let F be a function on B of the form (2). Then it is easy to see that

$$\partial F(z) = \sum_{j=1}^{n} \frac{\partial p}{\partial \zeta^j}(\langle z, \varphi_1 \rangle, \ldots, \langle z, \varphi_n \rangle) \pi^*_{(1,0)} \varphi_j, \tag{6}$$

$$\bar{\partial} F(z) = \sum_{j=1}^{n} \frac{\partial p}{\partial \bar{\zeta}^j}(\langle z, \varphi_1 \rangle, \ldots, \langle z, \varphi_n \rangle) \pi^*_{(0,1)} \varphi_j. \tag{7}$$

Here we regard φ_j an element in $H^{*\mathbf{C}}$ by an injection $B^{*\mathbf{C}} \hookrightarrow H^{*\mathbf{C}}$.

Next we calculate adjoint operators of ∂ and $\bar{\partial}$. From the definition (4) and (5), it is easy to see that for $G \in \mathcal{P}(H^{*\mathbf{C}})$ (i.e., $H^{*\mathbf{C}}$-valued polynomial)

$$\partial^* G(z) = D^* \pi^*_{(1,0)} G(z) = \frac{1}{2} D^* \{G(z) - \sqrt{-1} J^* G(z)\}, \tag{8}$$

$$\bar{\partial}^* G(z) = D^* \pi^*_{(0,1)} G(z) = \frac{1}{2} D^* \{G(z) + \sqrt{-1} J^* G(z)\}. \tag{9}$$

Here we used the fact that $\pi^*_{(1,0)}$, $\pi^*_{(0,1)}$ are projection operators in $H^{*\mathbf{C}}$. Since ∂^* and $\bar{\partial}^*$ are densely defined, ∂ and $\bar{\partial}$ are closable in $L^p(B, \mu)$. Now we can define holomorphic functions.

Definition 2.3 A function $F \in L^p(B, \mu)$ is called a L^p-holomorphic function if $F \in \mathrm{Ker}(\bar{\partial})$, i.e.,

$$\int_B F(z) \overline{\bar{\partial}^* G(z)} \mu(dz) = 0 \quad \text{for all } G \in \mathcal{P}(H^{*\mathbf{C}}).$$

A set of all holomorphic functions is denoted by $\mathcal{H}^p(B, \mu)$.

3 Complex Hermite polynomials and complex Wiener chaos

We introduce the complex Hermite polynomials (cf. Itô [3]). For $a, b \in \mathbf{Z}_+$, the complex Hermite polynomial with index (a, b) is defined by

$$H_{a,b}(\zeta, \bar{\zeta}) = (-1)^{a+b} e^{\zeta \bar{\zeta}} \frac{\partial^{a+b}}{\partial \bar{\zeta}^a \partial \zeta^b} e^{-\zeta \bar{\zeta}}, \quad \zeta \in \mathbf{C}, \tag{10}$$

or more explicitly

$$H_{a,b}(\zeta,\bar{\zeta}) = \sum_{k=0}^{a\wedge b}(-1)^k \frac{a!b!}{k!(a-k)!(b-k)!}\zeta^{a-k}\bar{\zeta}^{b-k}. \tag{11}$$

The generating function is given by

$$\sum_{a,b=0}^{\infty} \frac{\bar{t}^a t^b}{a!b!} H_{a,b}(\zeta,\bar{\zeta}) = e^{-t\bar{t}+t\bar{\zeta}+\bar{t}\zeta}, \qquad t,\zeta \in \mathbf{C}. \tag{12}$$

The following identities are well-known:

$$H_{a,b}(\zeta,\bar{\zeta}) = \overline{H_{b,a}(\zeta,\bar{\zeta})} \tag{13}$$

$$H_{a,0}(\zeta,\bar{\zeta}) = \zeta^a \tag{14}$$

$$\frac{\partial}{\partial \zeta} H_{a,b}(\zeta,\bar{\zeta}) = aH_{a-1,b}(\zeta,\bar{\zeta}) \tag{15}$$

$$\frac{\partial}{\partial \bar{\zeta}} H_{a,b}(\zeta,\bar{\zeta}) = bH_{a,b-1}(\zeta,\bar{\zeta}) \tag{16}$$

$$\left(-\frac{\partial}{\partial \bar{\zeta}} + \zeta\right) H_{a,b}(\zeta,\bar{\zeta}) = H_{a+1,b}(\zeta,\bar{\zeta}) \tag{17}$$

$$\left(-\frac{\partial}{\partial \zeta} + \bar{\zeta}\right) H_{a,b}(\zeta,\bar{\zeta}) = H_{a,b+1}(\zeta,\bar{\zeta}). \tag{18}$$

Moreover,

$$\{\frac{1}{\sqrt{a!b!}} H_{a,b}(\zeta,\bar{\zeta})\,;\; a,b \in \mathbf{Z}_+\}$$

forms a C.O.N.S. in $L^2(\mathbf{C}, \frac{1}{\pi}e^{-|\zeta|^2}d\eta d\xi)$ $(\zeta = \eta + \sqrt{-1}\xi)$.

By using the complex Hermite polynomial, we can construct a basis in $L^2(B,H,\mu)$, where (B,H,μ,J) is an almost complex abstract Wiener space as before. Take any sequence $\{\varphi_j\}_{j=1}^{\infty} \subseteq B^{*(1,0)}$ so that $\{\varphi_j\}$ is a C.O.N.S. in $H^{*(1,0)}$, and fix it. It is evident that $\{\bar{\varphi}_j\}_{j=1}^{\infty} \subseteq B^{*(0,1)}$ is a C.O.N.S. in $H^{*(0,1)}$. Set

$$\Phi = \{\alpha = (\alpha_1, \alpha_2, \cdots) \in \mathbf{Z}_+^{\mathbf{N}}\,;\; |\alpha| = \sum \alpha_j < \infty\}.$$

Then the Fourier-Hermite function $H_{\alpha,\beta}(z)$ on B $(\alpha,\beta \in \Phi)$ is defined by

$$H_{\alpha,\beta}(z) = H_{\alpha_1,\beta_1}(\langle z,\varphi_1\rangle, \overline{\langle z,\varphi_1\rangle}) \cdot H_{\alpha_2,\beta_2}(\langle z,\varphi_2\rangle, \overline{\langle z,\varphi_2\rangle}) \cdot \cdots.$$

Then $\{\frac{1}{\sqrt{\alpha!\beta!}}H_{\alpha,\beta}(z)\,;\ \alpha,\beta\in\Phi\}$ forms a C.O.N.S. in $L^2(B,\mu)$.

Definition 3.1 The closed linear span of $\{H_{\alpha,\beta}(z)\,;\ |\alpha|=m,\ |\beta| = n\}$ in $L^2(B,\mu)$ is called a complex Wiener chaos with index (m,n) and denoted by $C_{(m,n)}$.

An element of $C_{(m,n)}$ is called a complex multiple Wiener integral with index (m,n).

We denote the (real) Wiener chaos with index l by C_l. The relation between real and complex Wiener chaos is given as follows:

Proposition 3.1 *It holds that*

$$C_l = \bigoplus_{m+n=l} C_{(m,n)} \qquad \textit{(direct sum)}. \tag{19}$$

Proof. Let $H_l(\xi)$ be a (real) Hermite polynomial defined by

$$H_l(\xi) = \frac{(-1)^l}{2^l l!}e^{\xi^2/2}\frac{d^l}{d\xi^l}e^{-\xi^2/2}.$$

(19) is easily obtained by noticing the following identities: for $\zeta = \xi+\sqrt{-1}\eta$,

$$\begin{aligned} H_{m,n}(\zeta,\bar{\zeta}) &= m!n!\sum_{l=0}^{m+n}\sum_{k=0\vee(l-n)}^{m\wedge l}(-1)^k\sqrt{-1}^l \\ &\quad\times\binom{m+n-l}{m-k}\binom{l}{k}H_{m+n-l}(\xi)H_l(\eta), \end{aligned} \tag{20}$$

$$\begin{aligned} H_m(\xi)H_n(\eta) &= \sum_{l=0}^{m+n}\sum_{k=(l-m)\vee 0}^{l\wedge m}\frac{1}{l!(m+n-l)!2^{m+n}}(-1)^k \\ &\quad\times\sqrt{-1}^n\binom{l}{k}\binom{m+n-l}{n-k}H_{l,m+n-l}(\zeta,\bar{\zeta}). \end{aligned} \tag{21}$$

□

Let J_l, $J_{(m,n)}$ be projection operators to subspaces C_l, $C_{(m,n)}$ in $L^2(B,\mu)$. On the space C_l, the L^2-norm and L^p-norm ($p\in(1,\infty)$)

are equivalent (see Sugita [8]), and J_l and $J_{(m,n)}$ are bounded operators in $L^p(B,\mu)$.

Since B has a complex structure, we can define rotation U_θ as follows. For $\theta \in \mathbf{R}$, $U_\theta : L^p(B,\mu) \to L^p(B,\mu)$ is defined by

$$U_\theta F(z) = F(e^{\sqrt{-1}\theta} z) = F(\cos\theta\, z + \sin\theta\, Jz). \tag{22}$$

Then $\{U_\theta; \theta \in \mathbf{R}\}$ is a strongly continuous 1-parameter transformation group with period 2π.

Proposition 3.2 *It holds that*

$$U_\theta J_{(m,n)} = J_{(m,n)} U_\theta = e^{\sqrt{-1}\theta(m-n)} J_{(m,n)}, \tag{23}$$

$$U_\theta J_l = J_l U_\theta. \tag{24}$$

Proof. By noting (11), we easily obtain (23). Moreover by Proposition 3.1, we have $J_l = \sum_{m+n=l} J_{(m,n)}$. Hence (24) is easily obtained from (23). □

Proposition 3.3 *For $F \in L^p(B,\mu)$, $F \in \mathcal{H}^p(B,\mu)$ if and only if for any $n \geq 1$,*

$$J_{(m,n)} F = 0.$$

Proof. First we show the necessity. Assume $F \in \mathcal{H}^p(B,\mu)$. Take any $\alpha, \beta \in \Phi$, $j \in \mathbf{N}$. Set

$$G(z) = H_{\alpha,\beta}(z)\bar{\varphi}_j \tag{25}$$

(remember that $\{\varphi_j\}$ is a fixed sequence in $B^{*(1,0)}$). By noting (8) and (18), we have

$$\bar{\partial}^* G(z) = H_{\alpha,\beta+\delta_j}(z)$$

where $\delta_j = (0,\ldots,0,\overset{j}{\check{1}},0,\ldots)$. Then for any $n \geq 1$,

$$\begin{aligned}\int_B & J_{(m,n)} F(z) \overline{H_{\alpha,\beta+\delta_j}(z)} \mu(dz) \\ &= \int_B F(z) \overline{J_{(m,n)} H_{\alpha,\beta+\delta_j}(z)} \mu(dz)\end{aligned}$$

$$\begin{aligned}
&= \delta_{m,|\alpha|}\delta_{n,|\beta|+1} \int_B F(z)\overline{H_{\alpha,\beta+\delta_j}(z)}\mu(dz) \\
&= \delta_{m,|\alpha|}\delta_{n,|\beta|+1} \int_B F(z)\overline{\bar{\partial}^* G(z)}\mu(dz) \\
&= 0.
\end{aligned}$$

Since $n \geq 1$, it is easy to see that

$$\begin{aligned}
\int_B J_{(m,n)} F(z)\overline{H_{\alpha,0}(z)}\mu(dz) &= \int_B F(z)\overline{J_{(m,n)} H_{\alpha,0}(z)}\mu(dz) \\
&= 0.
\end{aligned}$$

Thus we have $J_{(m,n)}F = 0$.

Conversely, assume that $J_{(m,n)}F = 0$ for any $n \geq 1$. Then for G given by (25), it holds that

$$\begin{aligned}
\int_B F(z)\overline{\bar{\partial}^* G(z)}\mu(dz) &= \int_B F(z)\overline{H_{\alpha,\beta+\delta_j}(z)}\mu(dz) \\
&= \int_B F(z)\overline{J_{(|\alpha|,|\beta|+1)} H_{\alpha,\beta+\delta_j}(z)}\mu(dz) \\
&= \int_B J_{(|\alpha|,|\beta|+1)} F(z)\overline{H_{\alpha,\beta+\delta_j}(z)}\mu(dz) \\
&= 0.
\end{aligned}$$

For $G(z) = H_{\alpha,\beta}(z)\varphi_j$, it holds that $\bar{\partial}^* G = 0$ by (9). Hence we also have

$$\int_B F(z)\overline{\bar{\partial}^* G(z)}\mu(dz) = 0.$$

Thus we have $F \in \mathcal{H}^p(B,\mu)$, which completes the proof. □

4 Itô-Wiener expansions of holomorphic functions

Let the notation be as before. Set $S_n = J_0 + J_1 + \cdots + J_n$, i.e., S_n is the partial sum of the Itô-Wiener expansion. In $L^2(B,\mu)$, it is well-known that $S_n \to I$ strongly as $n \to \infty$. But it is open whether $S_n \to I$ strongly as $n \to \infty$ in $L^p(B,\mu)$ for $p \in (1,\infty)$. We prove it only for $F \in \mathcal{H}^p(B,\mu)$.

Theorem 4.1 *For $F \in \mathcal{H}^p(B,\mu)$, it holds that*

$$\lim_{n\to\infty} S_n F = F \quad \text{in } L^p(B,\mu).$$

Proof. We first show that for $l \geq 1$,

$$J_l F = \frac{1}{2\pi}\int_0^{2\pi}(e^{-\sqrt{-1}l\theta} + e^{\sqrt{-1}l\theta})U_\theta F\,d\theta. \tag{26}$$

To show this, set

$$K = J_l - \frac{1}{2\pi}\int_0^{2\pi}(e^{-\sqrt{-1}l\theta} + e^{\sqrt{-1}l\theta})U_\theta\,d\theta.$$

Then by using Proposition 3.2 and Proposition 3.3, for any $m \in \mathbf{Z}_+$, $n \in \mathbf{N}$, $G \in \mathcal{P}$ we have

$$\int_B KF(z)\overline{J_{(m,n)}G(z)}\mu(dz) = \int_B KJ_{(m,n)}F(z)\overline{G(z)}\mu(dz) = 0.$$

Moreover, by using (23) we have

$$\begin{aligned}
J_{(m,0)}KF &= J_{(m,0)}J_lF - \frac{1}{2\pi}\int_0^{2\pi}(e^{-\sqrt{-1}l\theta} + e^{\sqrt{-1}l\theta})J_{(m,0)}U_\theta F\,d\theta \\
&= \delta_{l,m}J_{(m,0)}F \\
&\quad - \frac{1}{2\pi}\int_0^{2\pi}(e^{-\sqrt{-1}l\theta} + e^{\sqrt{-1}l\theta})e^{\sqrt{-1}m\theta}J_{(m,0)}F\,d\theta \\
&= \delta_{l,m}J_{(m,0)}F - \delta_{l,m}J_{(m,0)}F \\
&= 0.
\end{aligned}$$

Hence we have

$$\int_B KF(z)\overline{J_{(m,0)}G(z)}\mu(dz) = \int_B J_{(m,0)}KF(z)\overline{G(z)}\mu(dz) = 0.$$

Thus we obtain (26). Similarly, we have

$$J_0F = \frac{1}{2\pi}\int_0^{2\pi} U_\theta F\,d\theta.$$

Now S_n can be written as

$$S_nF(z) = \sum_{k=-n}^{n} \frac{1}{2\pi}\int_0^{2\pi} e^{-\sqrt{-1}k\theta}U_\theta F(z)\,d\theta.$$

Further, by noting that U_t is an isometry in $L^p(B,\mu)$, we have

$$\begin{aligned}
\|F - S_nF\|_p^p &= \frac{1}{2\pi}\int_0^{2\pi} \|U_tF - U_tS_nF\|_p^p dt \\
&= \frac{1}{2\pi}\int_0^{2\pi} dt \int_B |U_tF(z) \\
&\quad - \sum_{k=-n}^{n} \frac{1}{2\pi}\int_0^{2\pi} e^{-\sqrt{-1}k\theta}U_\theta U_tF(z)\,d\theta|^p\mu(dz) \\
&= \int_B \mu(dz)\frac{1}{2\pi}\int_0^{2\pi} |U_tF(z) \\
&\quad - \sum_{k=-n}^{n} e^{\sqrt{-1}kt}\frac{1}{2\pi}\int_0^{2\pi} e^{-\sqrt{-1}k\theta}U_\theta F(z)\,d\theta|^p dt.
\end{aligned}$$

We note that for μ-a.e. z,

$$\int_0^{2\pi} |U_tF(z)|^p dt < \infty,$$

since

$$\begin{aligned}
\int_B\int_0^{2\pi} |U_tF(z)|^p dt\,\mu(dz) &= \int_0^{2\pi} dt \int_B |U_tF(z)|^p\mu(dz) \\
&= 2\pi\|F\|_p^p.
\end{aligned}$$

Now we use the Riesz theorem, i.e., the L^p-convergence of Fourier series for L^p-functions (see e.g., [4] Chapters II and III). So for μ-a.e. z,

$$\lim_{k\to\infty}\int_0^{2\pi} |U_tF(z) - \sum_{k=-n}^{n} e^{\sqrt{-1}kt}\frac{1}{2\pi}\int_0^{2\pi} e^{-\sqrt{-1}k\theta}U_\theta F(z)\,d\theta|^p dt = 0,$$

and there exists a constant c_p such that for any $n \in \mathbf{Z}_+$,

$$\begin{aligned}
&\int_0^{2\pi} | \sum_{k=-n}^{n} e^{\sqrt{-1}kt}\frac{1}{2\pi}\int_0^{2\pi} e^{-\sqrt{-1}k\theta}U_\theta F(z)\,d\theta|^p dt \\
&\qquad\qquad \leq c_p \int_0^{2\pi} |U_tF(z)|^p\,dt.
\end{aligned}$$

Set

$$G_n(z) = \int_0^{2\pi} |U_t F(z) - \sum_{k=-n}^{n} e^{\sqrt{-1}kt} \frac{1}{2\pi} \int_0^{2\pi} e^{-\sqrt{-1}k\theta} U_\theta F(z)\, d\theta|^p dt.$$

Then we have

$$\lim_{n\to\infty} G_n(z) = 0$$

and

$$|G_n(z)| \le 2^{p-1}(1+c_p) \int_0^{2\pi} |U_t F(z)|^p\, dt.$$

By Lebesgue's dominated convergence theorem, we have

$$\lim_{n\to\infty} \|F - S_n F\|_p = 0,$$

which completes the proof. □

Proposition 4.2 *$\mathcal{H}^p(B,\mu)$ is a closure of $\mathcal{P}_h$ in $L^p(B,\mu)$.*

Proof. It is evident that $\overline{\mathcal{P}}_h \subseteq \mathcal{H}^p(B,\mu)$ by (7). We shall show the converse: $\mathcal{H}^p(B,\mu) \subseteq \overline{\mathcal{P}}_h$.

Take a sequence $\{\varphi_j\}_{j=1}^\infty$ in $B^{*(1,0)}$ so that $\{\varphi_j\}$ is a C.O.N.S. in $H^{*(1,0)}$. Define a σ-field $\mathcal{F}_k$ by

$$\mathcal{F}_k = \sigma\{\varphi_j\,;\, j = 1, 2, \ldots, k\}.$$

Then by the martingale convergence theorem, we have

$$\lim_{k\to\infty} E[F|\mathcal{F}_k] = F \quad \text{in } L^p(B,\mu)$$

since $\bigvee_{k=1}^\infty \mathcal{F}_k = \mathcal{B}(B)$. Here we denote by E the expectation with respect to μ, $E[\,\cdot\,|\mathcal{F}_k]$ stands for the conditional expectation given $\mathcal{F}_k$, and $\mathcal{B}(B)$ is a topological Borel σ-field on B.

We show $E[F|\mathcal{F}_k] \in \mathcal{H}^p(B,\mu)$. To see this, we note that D and J commute with an operation of conditional expectation $E[\,\cdot\,|\mathcal{F}_k]$. Hence for $G \in \mathcal{P}(H^{*\mathbf{C}})$

$$\begin{aligned}(E[F|\mathcal{F}_k], \bar\partial^* G) &= (F, E[\bar\partial^* G|\mathcal{F}_k]) \\ &= (F, \bar\partial^* E[G|\mathcal{F}_k]) \\ &= 0\end{aligned}$$

which implies $E[F|\mathcal{F}_k] \in \mathcal{H}^p(B,\mu)$.

On the other hand, by the definition of $\mathcal{F}_k$, we can write

$$E[F|\mathcal{F}_k] = f(\langle z,\phi_1\rangle,\ldots,\langle z,\phi_k\rangle)$$

for an $f \in \mathcal{H}^p(\mathbf{C}^n,\mu^n_{\mathbf{C}})$. Here $\mu^n_{\mathbf{C}}$ is a complex Gaussian distribution on $\mathbf{C}^n$. By Theorem 4.1 $S_n f \to f$ in $L^p(\mathbf{C}^n,\mu^n_{\mathbf{C}})$. But $S_n f$ is a holomorphic polynomial since $\mathbf{C}^n$ is of finite dimension. Now the proof is complete. □

Lastly we show the uniqueness theorem in the following form.

Theorem 4.3 *For* $F, G \in \mathcal{H}^p(B,\mu)$, *if* $\mu\{z;\ F(z) = G(z)\} > 0$, *then* $F = G$ μ-a.e.

Proof. It is enough to prove it in the case $G = 0$. Take any $\varphi \in B^*$ so that $|\varphi|_{H^*} = 1$. By the Riesz theorem, H and H^* are isomorphic, so we denote the isomorphism by $\iota\colon H^* \to H$. Define H_1 to be a linear span of $\iota(\varphi)$ and $\iota(J\varphi)$. Let H_2 be an orthogonal complement of H_1 and denote projection operators to H_1 and H_2 by π_1 and π_2, respectively.

Set $\overline{H}_2$ be a closure of H_2 in B. Here we regard H as a subspace of B: $H \hookrightarrow B$. π_1, π_2 can be extended to B so that $\pi_1\colon B \to H_1$, $\pi_2\colon B \to \overline{H}_2$. Let μ' be an image measure of μ by π_2. Then $(\overline{H}_2, H_2, \mu', J)$ is again an almost complex abstract Wiener space. Define an isomorphism $\mathbf{C} \times \overline{H}_2 \cong B$ by

$$\mathbf{C} \times \overline{H}_2 \ni (\xi + \sqrt{-1}\eta, w) \mapsto \xi\iota(\varphi) + \sqrt{-1}\eta\,\iota(J\varphi) + w \in B.$$

Then under this isomorphism, $\mu = \mu^1_{\mathbf{C}} \times \mu'$ where $\mu^1_{\mathbf{C}}$ is a complex Gaussian distribution on $\mathbf{C}$ given by

$$\mu^1_{\mathbf{C}}(d\zeta) = \frac{1}{\pi}e^{-\xi^2-\eta^2}\,d\xi\,d\eta \quad (\zeta = \xi + \sqrt{-1}\eta \in \mathbf{C}).$$

By Proposition 4.2, there exists a sequence of holomorphic polynomials $\{P_n\}_n$ so that $\|F - P_n\|_p \to 0$ as $n \to \infty$. Since $\mathbf{C} \times \overline{H}_2 \cong B$, we can write $F(z) = F(\zeta, w)$ as a function on $\mathbf{C} \times \overline{H}_2$. Then by taking a subsequence if necessary, we may assume that for μ'-a.e. w,

$$\lim_{n\to\infty} \int_{\mathbf{C}} |F(\zeta,w) - P_n(\zeta,w)|^p \mu^1_{\mathbf{C}}(d\zeta) = 0. \tag{27}$$

Note that for any fixed w, $P_n(\zeta, w)$ is a (holomorphic) polynomial in ζ.

We can easily see that for any w that satisfies (27), $\{P_n(\zeta, w)\}_n$ converges compact uniformly in ζ. Hence $F(\zeta, w)$ is holomorphic in $\zeta \in \mathbf{C}$ (or more precisely, by changing values on a $\mu^1_{\mathbf{C}}$-null set). Thus by the uniqueness theorem for holomorphic functions on $\mathbf{C}$, we have for μ'-a.e. w,

$$\mu^1_{\mathbf{C}}\{\zeta;\ F(\zeta, w) = 0\} = 0 \text{ or } 1. \tag{28}$$

Set

$$A = \{z \in B;\ F(z) = 0\}$$

and define $A + \iota(\varphi)$ to be a shift of A. By (28) we have $\mu(A \triangle (A + \iota(\varphi)) = 0$ where $\triangle$ denotes the symetric difference. Since $\varphi \in B^*$ is arbitrary, we have $\mu(A) = 0$ or 1 by the ergodicity. Now the proof is complete. □

Bibliography

[1] V. Bargmann, On a Hilbert space of analytic functions and an associated integral transform, I, *Comm. Pure Appl. Math.*, **14** (1961), 187–214

[2] V. Bargmann, Remarks on a Hilbert space of analytic functions, *Proc. Nat. Acad. Sci. USA*, **48** (1961), 199–204

[3] K. Itô, Complex multiple Wiener integral, *Japan J. Math.*, **22** (1953), 63–86.

[4] Y. Katznelson, *"An introduction to harmonic analysis,"* Dover Publications, New York, 1976.

[5] M. Krée, Propriété de trace en dimension infinie, d'espaces du type Sobolev, *C. R. Acad. Sc. Paris*, **279** (1974), 157–160.

[6] I. E. Segal, Mathematical characterisation of the physical vacuum, III, *J. Math.*, **6** (1962), 500–523.

[7] E. M. Stein, *"Singular integrals and differentiable properties of functions,"* Princeton University Press, Princeton, New Jersey, 1970.

[8] H. Sugita, Sobolev spaces of Wiener functionals and Malliavin's calculus, *J. Math. Kyoto Univ.*, **25** (1985), 31–48.

[9] H. Sugita, On a characterization of the Sobolev spaces over an abstract Wiener space, *J. Math. Kyoto Univ.*, **25** (1985), 717–725.

[10] S. Watanabe, *"Lectures on stochastic differential equations and Malliavin calculus,"* Tata Institute of Fundamental Research, Springer-Verlag, Berlin-Heidelberg-New York, 1984.

Limits of the Wong-Zakai Type with a Modified Drift Term

Héctor J. Sussmann[1]
Department of Mathematics, Rutgers University
New Brunswick, NJ 08903, U.S.A.

Abstract

We study Stratonovich stochastic differential equations driven by an m-dimensional Wiener process W, with $m \geq 2$. If W is approximated by processes W^ν with more regular sample paths, then it is known that the solutions of the equations driven by the W^ν will converge to the solution of the equation driven by W, provided that the approximations satisfy the conditions of the Wong-Zakai theorem. McShane gave an example showing that, if those conditions are not satisfied, then a different limiting equation can arise. Here we describe a large class of equations, obtained from the original one by suitably modifying the drift term, that can arise as limiting equations by some choice of the sequence $\{W^\nu\}$.

1 Introduction

Consider a stochastic differential equation

$$dx = f_0(x)dt + \sum_{i=1}^{m} f_i(x)dW_i \ , \tag{1}$$

where x is n-dimensional, $W = (W_1, \ldots, W_m)$ is a standard m-dimensional Brownian motion, the vector fields f_i satisfy appropriate

[1]Work supported in part by the National Science Foundation under NSF Grant DMS-8902994.

ISBN 0-12-481005-5

smoothness and growth conditions, and the solutions are always understood to be in the Stratonovich sense.[5]

If we approximate W by a sequence of processes W^ν with more regular (e.g. Lipschitz) sample paths, then the well known Wong-Zakai Approximation Theorem (cf. [9], [10]) says that the solutions $t \to X^\nu(t)$ of the corresponding approximating equation

$$dx = f_0(x)dt + \sum_{i=1}^{m} f_i(x)dW_i^\nu , \qquad (2)$$

with some given initial condition $X^\nu(0) = \bar{X}$, converge to the solution X of (1) with the same initial condition, *provided that the approximations* W^ν *satisfy some extra conditions*, which always hold if $m = 1$, but may fail if $m > 1$. McShane gave an example in [4] showing that, for $m = 2$, the X^ν may indeed fail to converge to X. In this note we investigate the possible limits that can be obtained by taking more general sequences of approximating processes and show that, by a suitable choice of the approximation, it is possible to make the X^ν converge to the solution of an equation

$$dx = (f_0(x) + g(x))dt + \sum_{i=1}^{m} f_i(x)dW_i , \qquad (3)$$

with a different drift term. We will show that g can be chosen to be an arbitrary element of Λ, where Λ is the linear span of all the Lie brackets of the f_i for $i = 1, \ldots, m$ that contain at least two factors. The precise statement is given below in Theorem 1. Our result is a generalization of Theorem 7.2 of [3], Chapter 6, where it is shown that, by a suitable choice of the approximating sequence, one can produce a drift term which is an arbitrary linear combination of brackets $[f_i, f_j]$, $i, j > 0$.

We remark that we only prove almost sure convergence for a fixed T and a fixed initial condition. With a more careful analysis, one can prove a.s. convergence uniformly in t for t in any bounded interval, and convergence of the stochastic flows.

2 Differential Equations with Inputs

We let $C_b^\infty(\mathbb{R}^n, \mathbb{R}^n)$ denote the class of all maps $f : \mathbb{R}^n \to \mathbb{R}^n$ of class C^∞ such that all the partial derivatives $\frac{\partial^{\alpha_1+\ldots+\alpha_n} f}{\partial x_1^{\alpha_1}\ldots\partial x_n^{\alpha_n}}$ of all orders (*including* order 0) are bounded on $\mathbb{R}^n$. (In particular, every $f \in C_b^\infty(\mathbb{R}^n, \mathbb{R}^n)$ is globally bounded and globally Lipschitz.)

We assume that $f_0, \ldots, f_m \in C_b^\infty(\mathbb{R}^n, \mathbb{R}^n)$. We let $\mathcal{U}^m$ denote the space of all locally absolutely continuous functions $U : [0, \infty) \to \mathbb{R}^m$ such that $U(0) = 0$. If $U \in \mathcal{U}^m$, then we write $u = \dot{U} = \frac{dU}{dt}$, so $u \in L^1_{\text{loc}}([0, \infty), \mathbb{R}^m)$.

Let $U \in \mathcal{U}^m$, and write $u = \dot{U}$. Write $U_0(t) \equiv t$, i.e. $u_0(t) \equiv 1$. Then the ordinary differential equation

$$\dot{x} = f_0(x) + \sum_{i=1}^{m} u_i(t) f_i(x) \ , \tag{4}$$

can also be written in the form $dx = f_0(x)\,dt + \sum_{i=1}^m f_i(x) dU_i$, or $dx = \sum_{i=0}^m f_i(x) dU_i$. It is clear that (4) satisfies the conditions of the Carathéodory existence and uniqueness theorem. Moreover, since the f_i are bounded, trajectories do not escape in finite time. So, given any $a \in [0, \infty)$, $\bar{x} \in \mathbb{R}^n$, there exists a unique solution $t \to x(t)$ of (4) such that $x(a) = \bar{x}$. For fixed $b \in [0, \infty)$, we will use $\Phi^U_{b,a}$ to denote the map that assigns to each $\bar{x}$ the value $x(b)$ of the corresponding solution. That is, $t \to \Phi^U_{t,a}(x)$ is the solution of (4) that goes through x when $t = a$. Each map $\Phi^U_{b,a}$ is a C^∞ diffeomorphism from $\mathbb{R}^n$ onto $\mathbb{R}^n$. Moreover, these diffeomorphisms satisfy $\Phi^U_{a,a} =$ identity, and $\Phi^U_{c,b}\Phi^U_{b,a} = \Phi^U_{c,a}$ for all $a, b, c \in [0, \infty)$.

We now define the *iterated integrals* $\int_a^b u_I$, where $I = (i_1, \ldots, i_r)$ is an arbitrary member of $\mathcal{I}(m)$, the space of all finite sequences of indices $i \in \{0, \ldots, m\}$. (We will write $|I|$ for the *length* of I, i.e. the number r. When $|I| = 1$, so $I = (i)$ for some i, we will just write $\int_a^b u_i$ instead of $\int_a^b u_{(i)}$. The empty sequence $\emptyset$ is a member of $\mathcal{I}(m)$.) The definition is recursive: we let $\int_a^b u_I = 1$ if $I = \emptyset$, and for a general I we write $I = (i, I')$, and define $\int_a^b u_I$ to be equal to $\int_a^b u_i(t)(\int_a^t u_{I'})dt$.

We will also write $U_I(b,a)$ for $\int_a^b u_I$. Notice that the identity $U_I(b,a) = U_i(b) - U_i(a)$ holds when $|I| = 1$, $I = (i)$, but no similar formula is true when $|I| \neq 1$, since in that case the additivity property $U_I(c,a) = U_I(c,b) + U_I(b,a)$ does not hold in general.

We are interested in the derivatives of $U_I(t,s)$ with respect to both variables t and s. Define $u_I^{+,s}(t) = \frac{\partial}{\partial t} U_I(t,s)$ and $u_I^{-,t}(s) = -\frac{\partial}{\partial s} U_I(t,s)$. Then $\int_a^b u_I = \int_a^b u_I^{+,a}(t)\,dt = \int_a^b u_I^{-,b}(s)\,ds$. The functions $u_I^{+,a}(t)$, $u_I^{-,b}(s)$ are equal, respectively, to

$$u_{i_1}(t) \int_a^t \int_a^{t_1} \cdots \int_a^{t_{k-2}} u_{i_2}(t_1) \ldots u_{i_k}(t_{k-1})\, dt_{k-1} \ldots dt_1 \tag{5}$$

and

$$u_{i_k}(s) \int_s^b \int_{t_{k-1}}^b \cdots \int_{t_2}^b u_{i_1}(t_1) \ldots u_{i_{k-1}}(t_{k-1})\, dt_{k-1} \ldots dt_1 \ . \tag{6}$$

if $I = (i_1, \ldots, i_k) \in \mathcal{I}(m)$.

Now suppose that φ is a scalar- or vector-valued function of class C^∞. Write $f_i\varphi$ to denote the result of applying f_i to φ as a first-order differential operator, i.e. $(f_i\varphi)(x) = \lim_{h\to 0} \frac{1}{h}(\varphi(x + hf_i(x)) - \varphi(x))$. More generally, if $I = (i_1, \ldots, i_r) \in \mathcal{I}(m)$, we write $f_I = f_{i_1} f_{i_2} \ldots f_{i_r}$. Then (4) implies the equation

$$\varphi(\Phi_{t,a}^U(x)) = \varphi(x) + \sum_{i=0}^m \int_a^t u_i(s)(f_i\varphi)(\Phi_{s,a}^U(x))\, ds \ , \tag{7}$$

which is the $k = 0$ case of the general formula

$$\varphi(\Phi_{t,a}^U(x)) = \sum_{|I|\leq k} U_I(t,a)(f_{I\#}\varphi)(x) + R_{k,t,a,U,\varphi,f}(x) \tag{8}$$

where, for any multiindex $I = (i_1, \ldots, i_r)$, we use $I^\#$ to denote the reversed multiindex, i.e. $I^\# = (i_r, \ldots, i_1)$, and the remainder $R_{k,t,a,U,\varphi,f}(x)$ is given by

$$R_{k,t,a,U,\varphi,f}(x) = \sum_{|I|=k+1} \int_a^t u_I^{-,t}(s)(f_{I\#}\varphi)(\Phi_{s,a}^U(x))\, ds \ . \tag{9}$$

It is easy to see that (8) is actually true for all k. (The proof is by induction, using repeated integrations by parts.) A particularly important choice of φ is $\varphi = E^n$, where $E^n : \mathbb{R}^n \to \mathbb{R}^n$ is the identity map. In that case, (8) becomes

$$\Phi_{t,a}^U(x) = \sum_{|I| \leq k} U_I(t,a) E_I^f(x) + R_{k,t,a,U,E^n,f}(x) \tag{10}$$

where $E_I^f = f_{I\#} E^n$, and

$$R_{k,t,a,U,E^n,f}(x) = \sum_{|I|=k+1} \int_a^t u_I^{-,t}(s) E_I^f(\Phi_{s,a}^U(x))\, ds \,. \tag{11}$$

We remark that all the vector-valued functions E_I^f belong to $C_b^\infty(\mathbb{R}^n, \mathbb{R}^n)$.

3 Stochastic Ordinary Inputs

Now assume that $(\Omega, \mathcal{F}, P)$ is a probability space, and $\{\mathcal{F}_t\}$ is an increasing family of sub-σ-fields of $\mathcal{F}$. An m-dimensional *ordinary input process* (OIP) on $(\Omega, \mathcal{F}, P)$ is a stochastic process $U = \{U(t) : t \geq 0\}$ such that all the sample paths $t \to U(t)(\omega)$, $\omega \in \Omega$, belong to $\mathcal{U}^m$. In that case, the derivative $\dot{U}$, the iterated integrals, the solutions of (4), and all the other U-dependent objects introduced above are well defined for each $\omega \in \Omega$.

As usual, we call a process U *adapted* if $U(t)$ is $\mathcal{F}_t$-measurable for each t. However, it is also useful to define a weaker concept, namely, that of a π-*adapted* process, where π is a partition of $[0, \infty)$. Precisely, we define a *partition* of $[0, \infty)$ to be an infinite sequence $\pi = \{t_j\}_{j=0}^\infty$ such that $0 = t_0 < t_1 < t_2 < \ldots$ and $\lim_{j\to\infty} t_j = \infty$. The *mesh* $|\pi|$ of a partition π is the number $\sup\{t_j - t_{j-1} : j = 1, 2, \ldots\}$. If π is a partition, then we call U π-*adapted* to $\{\mathcal{F}_t\}$ if, for every j, $U(t)$ is $\mathcal{F}_{t_j}$-measurable whenever $t \leq t_j$.

Now suppose that $\pi = \{t_j\}_{j=0}^\infty$ is a partition of $[0, \infty)$ and U is a π-adapted m-dimensional OIP. Let $k > 0$ be an integer, and let $C \in \mathbb{R}$, $C > 0$. We will say that U belongs to the class $OIP(m, k, C, \pi)$ if U

satisfies the following three bounds

$$|\mathbb{E}(U_I(t_j, t_{j-1}))/\mathcal{F}_{t_{j-1}}| \leq C(t_j - t_{j-1}) \,, \tag{12}$$

$$\mathbb{E}(U_I(t_j, t_{j-1})^2/\mathcal{F}_{t_{j-1}}) \leq C(t_j - t_{j-1}) \,, \tag{13}$$

and

$$|\mathbb{E}(U_I(t, t_{j-1})^2)/\mathcal{F}_{t_{j-1}}| \leq C \,, \tag{14}$$

for all choices of $I \in \mathcal{I}(m)$ such that $|I| \leq k$, all $j \in \{1, 2, 3, \ldots\}$, and all $t \in [t_{j-1}, t_j]$, as well as the bound

$$|u_I^{-,t}(s)| \leq C \tag{15}$$

for all $I \in \mathcal{I}(m)$ such that $|I| = k + 1$ and all s, t such that $t_{j-1} \leq s \leq t \leq t_j$ for some j.

If $U \in OIP(m, k, C, \pi)$ and $1 \leq |I|, |J| \leq k$, then it follows from the Schwartz inequality for conditional expectations that

$$|\mathbb{E}(U_I(t_j, t_{j-1})U_J(t_j, t_{j-1}))/\mathcal{F}_{t_{j-1}}| \leq C(t_j - t_{j-1}) \,. \tag{16}$$

We now define $D_I(f) = \sup\{||x - y||^{-1}||E_I^f(x) - E_I^f(y)|| : x, y \in \mathbb{R}^n \,, x \neq y\}$, and let $D = D(k, f) = \max\{D_I(f) : 1 \leq |I| \leq k + 1\}$.

Lemma 1 *For every k, m, f, C there exist constants K, μ, depending only on k, m, D and C, but not on the particular choice of U, X, Y, f or π, with the property that, whenever $\pi = \{t_j\}_{j=0}^{\infty}$ is a partition of $[0, \infty)$, and U is a process in $OIP(m, k, C, \pi)$, then the bound*

$$||\Phi^U_{t_j,t_{j-1}}(X) - \Phi^U_{t_j,t_{j-1}}(Y)||_{L_2} \leq (1 + Ke^{\mu|\pi|}(t_j - t_{j-1}))||X - Y||_{L^2}$$

holds whenever $X, Y : \Omega \to \mathbb{R}^n$ are $\mathcal{F}_{t_{j-1}}$-measurable and square-integrable.

PROOF. Throughout this proof, we will use the notation $\mathcal{E}(X, Y)$ to denote $\mathcal{E}(X) - \mathcal{E}(Y)$, whenever $\mathcal{E}$ is some expression that depends on X. Write $a = t_{j-1}$, $b = t_j$, and use $\mathbb{E}_a$ to denote conditional expectation with respect to $\mathcal{F}_a$. Let $a \leq t \leq b$. Using (10), we get

$$\begin{aligned} \Phi^U_{t,a}(X, Y) &= X - Y + \sum_{1 \leq |I| \leq k} U_I(t, a) E_I^f(X, Y) \\ &\quad + R_{k,t,a,U,E^n,f}(X, Y) \,. \end{aligned} \tag{17}$$

Using the bound $||E_I^f(x,y)|| \leq D||x-y||$, we get

$$\begin{aligned} \mathbb{E}(U_I(t,a)^2||E_I^f(X,Y)||^2) &= \mathbb{E}(||E_I^f(X,Y)||^2 \cdot \mathbb{E}_a(U_I(t,a)^2) \\ &\leq C^2D^2\mathbb{E}(||X-Y||^2), \end{aligned}$$

so that

$$||U_I(t,a)\cdot E_I^f(X,Y)||_{L^2} \leq CD||X-Y||_{L^2} \tag{18}$$

if $1 \leq |I| \leq k$. Similarly, if $|I| = k+1$, we have

$$||u_I^{-,t}(s)\cdot E_I^f(\Phi_{s,a}^U(X,Y))|| \leq CD||\Phi_{s,a}^U(X,Y)|| \tag{19}$$

pointwise, so

$$||u_I^{-,t}(s)\cdot E_I^f(\Phi_{s,a}^U(X,Y))||_{L^2} \leq CD||\Phi_{s,a}^U(X,Y)||_{L^2}\,. \tag{20}$$

Since $R_{k,t,a,U,E^n,f}(X,Y) = \sum_{|I|=k+1}\int_a^t u_I^{-,t}(s)\cdot E_I^f(\Phi_{s,a}^U(X,Y))\,ds$, we find

$$||R_{k,t,a,U,E^n,f}(X,Y)||_{L^2} \leq \mu\int_a^t ||\Phi_{s,a}^U(X,Y)||\,ds\,, \tag{21}$$

where $\mu = (m+1)^{k+1}CD$. Combining (17), (18) and (21), we get

$$||\Phi_{t,a}^U(X,Y)||_{L^2} \leq (1+\nu CD)||X-Y||_{L^2} + \mu\int_a^t ||\Phi_{s,a}^U(X,Y)||_{L^2}\,ds\,, \tag{22}$$

where $\nu = m+1+(m+1)^2+\ldots+(m+1)^k = \frac{(m+1)^{k+1}-m-1}{m}$.

Gronwall's inequality then yields

$$||\Phi_{t,a}^U(X,Y)||_{L^2} \leq (1+\nu CD)\mathrm{e}^{\mu|\pi|}||X-Y||_{L^2}\,. \tag{23}$$

If we now use (21) again, with $t=b$, together with (23), we get

$$||R_{k,b,a,U,E^n,f}(X,Y)||_{L^2} \leq \mu(1+\nu CD)\mathrm{e}^{\mu|\pi|}(b-a)||X-Y||_{L^2}\,. \tag{24}$$

Using (17) with $t=b$, we can write $\Phi_{b,a}^U(X,Y) = A+B$, where $A = X-Y+\sum_{1\leq|I|\leq k}U_I(b,a)(E_I^f(X,Y))$ and $B = R_{k,b,a,U,E^n,f}(X,Y)$.

We have already estimated $||B||_{L^2}$ in (24). To get a bound for $||A||_{L^2}$ write

$$\begin{aligned} ||A||^2 &= ||X-Y||^2 + 2\sum_{1\le|I|\le k} U_I(b,a)\langle X-Y, E_I^f(X,Y)\rangle \\ &+ 2\sum_{1\le|I|,|J|\le k} U_I(b,a)U_J(b,a)\langle E_I^f(X,Y), E_J^f(X,Y)\rangle . \end{aligned} \tag{25}$$

Then

$$\mathbb{E}_a(||A||^2) = ||X-Y||^2 + 2\sum_{1\le|I|\le k} \mathbb{E}_a(U_I(b,a))\langle X-Y, E_I^f(X,Y)\rangle$$

$$+2\sum_{1\le|I|,|J|\le k} \mathbb{E}_a(U_I(b,a)U_J(b,a))\langle E_I^f(X,Y), E_J^f(X,Y)\rangle , \tag{26}$$

so that

$$\mathbb{E}_a(||A||^2) \le (1 + (2\nu CD + \nu^2 CD^2)(b-a))||X-Y||^2 . \tag{27}$$

Taking expectations, we get

$$\mathbb{E}(||A||^2) \le (1 + (2\nu CD + \nu^2 CD^2)(b-a))\mathbb{E}(||X-Y||^2) , \tag{28}$$

so that

$$||A||_{L^2} \le (1 + (\nu CD + \frac{1}{2}\nu^2 CD^2)(b-a))||X-Y||_{L^2} . \tag{29}$$

Combining (29) with the bound for B, we get

$$||\Phi_{b,a}^U(X) - \Phi_{b,a}^U(Y)||_{L_2} \le (1 + Ke^{\mu|\pi|}(b-a))||X-Y||_{L^2} , \tag{30}$$

with $K = \nu CD + \frac{1}{2}\nu^2 CD^2 + \mu(1+\nu CD)$. ∎

4 The Chen-Fliess Series

It is clear from the preceding considerations that it is important to be able to analyze sums of the form $\sum_{1\le|I|\le k} U_I(b,a)(f_{I\#}\varphi)(x)$. To compute such expressions, we use the formalism of the *Chen-Fliess series* (cf. [2], [6], [7], [8]).

If $\mathcal{X}$ is a nonempty set, we use $\hat{\mathbf{A}}(\mathcal{X})$ to denote the algebra of *noncommutative formal power series* in $\mathcal{X}$, i.e. the set of all infinite linear combinations $\sum_{M\in\mathcal{M}(\mathcal{X})} s_M M$, where $\mathcal{M}(\mathcal{X})$ is the set of all *monomials* in $\mathcal{X}$, that is, the set of all finite sequences of elements of $\mathcal{X}$. The length of a monomial is its *degree*. Monomials are multiplied by just concatenating them, and then the product of two elements of $\hat{\mathbf{A}}(\mathcal{X})$ is well defined. The empty sequence is a monomial of degree 0, and is denoted by 1. Then $1.S = S.1 = S$ for all $S \in \hat{\mathbf{A}}(\mathcal{X})$. A linear combination of monomials of degree k is said to be *homogeneous of degree* k, and the set of all such combinations is denoted by $\mathbf{A}^k(\mathcal{X})$. Clearly, every $S \in \hat{\mathbf{A}}(\mathcal{X})$ has a unique decomposition $S = \sum_{k=0}^{\infty} H_k(S)$ as a sum of homogeneous components. If we regard $\hat{\mathbf{A}}(\mathcal{X})$ as a Lie algebra, with the bracket defined by $[S,T] = ST - TS$, then the Lie subalgebra of $\hat{\mathbf{A}}(\mathcal{X})$ generated by $\mathcal{X}$ is denoted by $\mathbf{L}(\mathcal{X})$ and its elements are known as *Lie polynomials* in $\mathcal{X}$. Those $S \in \hat{\mathbf{A}}(\mathcal{X})$ all whose homogeneous components $H_k(S)$ are in $\mathbf{L}(\mathcal{X})$ are known as *Lie series* in $\mathcal{X}$, and the set of all such series is denoted by $\hat{\mathbf{L}}(\mathcal{X})$. The *order* $\omega(S)$ of a series $S \in \hat{\mathbf{A}}(\mathcal{X})$ is the smallest k such that the k-th homogeneous component of S is $\neq 0$. (If $S = 0$ then $\omega(S)$ is defined to be $+\infty$.) An infinite sum $S_1 + S_2 + S_3 + \ldots$ of series in $\hat{\mathbf{A}}(\mathcal{X})$ such that $\omega(S_j) \to \infty$ as $j \to \infty$ is convergent in an obvious way since, for each k, $H_k(S_j) = 0$ for all but finitely many j's. In particular, the exponential e^S, and the logarithm $\log(1+S)$ are well defined by the usual power series if $\omega(S) \geq 1$. If S is a Lie series then $\omega(S) \geq 1$, so e^S and $\log(1+S)$ are defined. The elements of the form e^S, with $S \in \hat{\mathbf{L}}(\mathcal{X})$, are known as *exponential Lie series* in $\mathcal{X}$.

Given an input $U \in \mathcal{U}^m$, we can consider the differential equation

$$\dot{S}(t) = S(t)(X_0 + u_1(t)X_1 + \ldots + u_m(t)X_m) , \tag{31}$$

where $u_i = \dot{U}_i$ and $X_0, \ldots, X_m$ are formal noncommutative indeterminates. We can regard S as evolving in the algebra $\hat{\mathbf{A}}(X_0, \ldots, X_m)$ of noncommutative formal power series in the $m+1$ indeterminates $X_0, \ldots, X_m$. (That is, $\hat{\mathbf{A}}(X_0, \ldots, X_m)$ is the set of all formal infinite sums $S = \sum_{I\in\mathcal{I}(m)} s_I X_I$, where, if $I = (i_1, \ldots, i_r)$, $r > 0$, we define $X_I = X_{i_1} X_{i_2} \ldots X_{i_r}$, and we let $X_\emptyset = 1$.) If we solve (31) with initial condition $S(a) = 1$, then the solution is given by

$$S(t) = \sum_{I\in\mathcal{I}(m)} U_I(t,a) X_{I\#} \ . \tag{32}$$

We can also consider (31) as evolving in $\mathbf{A}_k(X_0, \ldots, X_m)$, the free nilpotent associative algebra of order k in $X_0, \ldots, X_m$, i.e. the set of all sums $S = \sum_{I\in\mathcal{I}(m), |I|\leq k} s_I X_I$, where monomials are multiplied in the usual way, and every monomial of degree $> k$ is set equal to zero. In this case, the solution is given by

$$S(t) = \sum_{I\in\mathcal{I}(m),\, |I|\leq k} U_I(t,a) X_{I\#} \ . \tag{33}$$

The value at b of this solution will be denoted by $S_{k,a,b}(U)$, or $S_{k,a,b}(u)$, and referred to as the *Chen-Fliess series of U from a to b, truncated at order k*. Formula (33) shows that $S_{k,a,b}(U)$ is just a way of coding all the iterated integrals $U_I(b,a)$, $|I| \leq k$, into one algebraic expression.

It is clear that, if a function $t \to S(t)$ is a solution of (32), and $Q \in \mathbf{A}_k(X_0, \ldots, X_m)$, then $t \to QS(t)$ is also a solution. In particular, if $a < b < c$, then $t \to S_{k,a,t}(U)$ and $t \to S_{k,a,b}(U) S_{k,b,t}(U)$ are both solutions, whose values at $t = b$ coincide. Hence the identity

$$S_{k,a,c}(U) = S_{k,a,b}(U) S_{k,b,c}(U) \tag{34}$$

holds in $\mathbf{A}_k(X_0, \ldots, X_m)$. Notice that, when $k = 1$, Formula (34) just amounts to the statement that $U_I(c,a) = U_I(c,b) + U_I(b,a)$ whenever $|I| = 1$, i.e. to the property that the integral is additive with respect to the interval. So (34) can be viewed as a generalization to high-order iterated integrals of the additivity property.

We will need the *Campbell-Hausdorff formula* (CHF), cf. [1]. To state the CHF, let A, B be indeterminates. The CHF then says that

$$e^A e^B = e^{A+B+\frac{1}{2}[A,B]+C(A,B)} \ , \tag{35}$$

where $C \in \hat{\mathbf{L}}(A, B)$ is a Lie series in A, B of order 3. Naturally, if $S, T \in \hat{\mathbf{L}}(X_0, \ldots, X_m)$, we can plug them into (35) and get $e^S e^T =$

$e^{S+T+\frac{1}{2}[S,T]+C(S,T)}$, so in particular the set of Lie series is closed under multiplication. A similar formula holds in $\mathbf{L}_k(X_0, \ldots, X_m)$ (where $\mathbf{L}_k(X_0, \ldots, X_m)$ is the truncated version of $\hat{\mathbf{L}}(X_0, \ldots, X_m)$, i.e. the Lie subalgebra of $\mathbf{A}_k(X_0, \ldots, X_m)$ generated by the X_i), and in this case the series $C(S, T)$ is actually a finite sum.

5 Construction of Approximating Processes

We now fix m and define, for each k, each formal bracket $B = [X_{i_1}, [\ldots, [X_{i_{r-1}}, X_{i_r}] \ldots]]$, $i_j \in \{1, \ldots, m\}$ for $j = 1, \ldots, r$, each interval $[a, b] \subset [0, \infty)$, and each real number $\tau > 0$, two controls $u(B, a, b, \pm, \tau) : [a, b] \to \mathbb{R}^m$, such that

$$S_{k,a,b}(u(B, a, b, \pm, \tau)) = e^{(b-a)X_0 \pm \tau^r B + Z(B,a,b,\pm,\tau)} , \tag{36}$$

where $Z(B, a, b, \pm, \tau) = P_B^{\pm}((b-a)X_0, \tau X_1, \ldots, \tau X_m)$, and $P_B^{\pm}$ are Lie polynomials of order ≥ 2 in indeterminates $Y_0, \ldots, Y_m$, that do not contain monomials in $Y_1, \ldots, Y_m$ of degree $\leq r$ (i.e. $P_B^{\pm}$ are such that $\omega(P_B^{\pm}) \geq 2$ and $\omega(P_B^{\pm}(0, Y_1, \ldots, Y_m)) > r$).

The $u(B, a, b, \pm, \tau) : [a, b] \to \mathbb{R}^m$ are constructed inductively as follows. Assume first that $r = 1$, so $B = X_i$ for some $i \in \{1, \ldots, m\}$. Then (writing $u(i, a, b, \pm, \tau)$ instead of $u(X_i, a, b, \pm, \tau)$) we define $u(i, a, b, \pm, \tau)$ to be the control whose i-th component is constant and equal to $\pm\frac{\tau}{b-a}$, while all the other components are zero. Now assume that $u(B', a, b, \pm, \tau)$ has been defined whenever B' has degree $r - 1$. Pick B of degree r, and write $B = [X_i, B']$. Divide the interval $I = [a, b]$ into four equal subintervals $I_j = [t_{j-1}, t_j]$, $j = 1, \ldots, 4$, where we let $t_j = a + j\delta$, for $j = 0, \ldots, 4$, with $\delta = \frac{1}{4}(b - a)$. Then define $u(B, a, b, +, \tau)$ to be equal to $u(i, t_0, t_1, +, \tau)$ on I_1, to $u(B', t_1, t_2, +, \tau)$ on I_2, to $u(i, t_2, t_3, -, \tau)$ on I_3, and to $u(B', t_3, t_4, -, \tau)$ on I_4. Having defined $u(B, a, b, +, \tau)$, we construct $u(B, a, b, -, \tau)$ by "changing sign and reversing time," that is, by letting $u(B, a, b, -, \tau)(t) = -u(B, a, b, +, \tau)(a + b - t)$ for $a \leq t \leq b$.

With this definition of the $u(B, a, b, \pm, \tau)$, we now show by induction on r that the Chen-Fliess series of $u(B, a, b, \pm, \tau)$ satisfies the desired properties. Consider first the case $r = 1$. In this case, it is obvious that

$$S_{k,a,b}(u(i,a,b,\pm,\tau)) = e^{(b-a)X_0 \pm \tau X_i} . \tag{37}$$

Now assume that the desired property holds for $r-1$. Let B be of degree r, and write $B = [X_i, B']$. In view of (34), we have

$$S_{k,a,b}(u(B,a,b,+,\tau)) = S_1 S_2 S_3 S_4 , \tag{38}$$

where $S_j = S_{k,t_{j-1},t_j}(u(B_j, t_{j-1}, t_j, \theta_j, \tau))$, $B_1 = B_3 = X_i$, $B_2 = B_4 = B'$, $\theta_1 = \theta_2 = +$, $\theta_3 = \theta_4 = -$. Then $S_1 = e^{\delta X_0 + \tau X_i}$ and $S_3 = e^{\delta X_0 - \tau X_i}$, where $\delta = \frac{b-a}{4}$. By the inductive hypothesis, we have

$$S_2 = e^{\delta X_0 + \tau^{r-1} B' + R^+} \qquad \text{and} \qquad S_4 = e^{\delta X_0 - \tau^{r-1} B' + R^-} ,$$

where $R^\pm = P^\pm_{B'}(\delta X_0, \tau X_1, \ldots, \tau X_m)$, and the $P^\pm_{B'}$ are Lie polynomials in $Y_0, \ldots, Y_m$ such that $\omega(P^\pm_{B'}) \geq 2$ and $\omega(P^\pm_{B'}(0, Y_1, \ldots, Y_m)) \geq r$.

We now repeatedly apply the CHF. (In our case, all the Lie series occurring in the computation are actually Lie polynomials, because we are working in a nilpotent algebra.) We get $S_1 S_2 = e^{Z^+}$, $S_3 S_4 = e^{Z^-}$, where

$$Z^\pm = 2\delta X_0 \pm \tau X_i \pm \tau^{r-1} B' + \frac{1}{2}\tau^r [X_i, B] + Q^\pm , \tag{39}$$

$$S_{k,a,b}(u(B,a,b,+,\tau)) = S_1 S_2 S_3 S_4 = e^{(b-a)X_0 + \tau^r [X_i, B] + Q} , \tag{40}$$

$$\begin{aligned} Q^\pm = {} & R^\pm \pm \frac{1}{2}\delta\tau^{r-1}[X_0, B'] \pm \frac{1}{2}\delta\tau[X_i, X_0'] + \frac{1}{2}\delta[X_0, R^\pm] \\ & \pm \frac{1}{2}\tau^r [X_i, R^\pm] + C(\delta X_0 \pm \tau X_i, \delta X_0 \pm \tau^{r-1} B' + R^\pm) , \end{aligned} \tag{41}$$

$$Q = Q^+ + Q^- + \frac{1}{2}[Z^+, Z^-] + C(Z^+, Z^-) . \tag{42}$$

It is clear that Q is a Lie polynomial in $\delta X_0, \tau X_1, \ldots, \tau X_m$, i.e. $Q = P^+_B(\delta X_0, \tau X_1, \ldots, \tau X_m)$ for some Lie polynomial in $Y_0, \ldots, Y_m$. Moreover, P^+_B clearly has order ≥ 2. We must now show that $\omega(P^+_B(0, Y_1, \ldots, Y_m)) > r$. That is, we must show that, if we plug in $\delta = 0$ in Q, then the resulting expression is divisible by τ^{r+1}. It is easy to see that, in the right-hand side of (42), the only possible terms of degree $\leq r$ in τ must come from the sum $Q^+ + Q^-$. Using (41), we conclude immediately that such terms can only arise from the sum $P^+_{B'}(0, \tau X_1, \ldots, \tau X_m) + P^-_{B'}(0, \tau X_1, \ldots, \tau X_m)$. So our conclusion will follow if we show that this sum vanishes, i.e. that $P^+_{B'}(0, Y_1, \ldots, Y_m) + P^-_{B'}(0, Y_1, \ldots, Y_m) = 0$. This in turn follows from the equality $\tilde{S}_2 \tilde{S}_4 = 1$, where the $\tilde{S}_j$ are the series obtained from the S_j by setting $X_0 = 0$. These series can be computed by setting $X_0 = 0$ in (31) and then solving on the intervals $[t_{j-1}, t_j]$ with input $u(B, a, b, \pm, \tau)$. If we let $\hat{S}_j$ denote the corresponding solutions with initial condition $\hat{S}_j(t_{j-1}) = 1$, then $\tilde{S}_j = \hat{S}_j(t_j)$. By translation invariance, we have $\tilde{S}_2 = S_+(\delta)$,, $\tilde{S}_4 = S_-(\delta)$, where $S_\pm$ is the solution of (31) on $[0, \delta]$, with input $u(B', 0, \delta, \pm, \tau)$ and initial condition $S_\pm(0) = 1$. Since, as explained above, $u(B', a, b, -, \tau)$ is obtained from $u(B', a, b, +, \tau)$ by changing sign and reversing time, it follows easily that $S_-(\delta) = S_+(\delta)^{-1}$, completing the proof of our conclusion.

We record for future use the trivial fact that

$$|u(B, a, b, \pm, \tau)_i| \leq \frac{4^{r-1}\tau}{b-a} \,. \tag{43}$$

We now let $g \in \Lambda$, so we can write $g = \sum_{\mu=1}^{p} g_\mu B_\mu(f)$, where the B_μ are Lie brackets of the form $[X_{i_1^\mu}, [\ldots, [X_{i^\mu_{r(\mu)-1}}, X_{i^\mu_{r(\mu)}}] \ldots]]$, $i_k^\mu \in \{1, \ldots, m\}$, $r(\mu) \geq 2$, and $B_\mu(f)$ is the vector field obtained by plugging in f_i for X_i for each i. We assume, without loss of generality, that all the numbers g_μ are nonzero.

We now let $(\Omega, \mathcal{F}, P)$ be a probability space and $\{\mathcal{F}_t\}$ be a filtration as above. Let $W = (W_1, \ldots, W_m)$ be an m-dimensional standard Wiener process on $(\Omega, \mathcal{F}, P)$ that has continuous sample paths and is adapted to $\{\mathcal{F}_t\}$, in the sense that W_t is $\mathcal{F}_t$-measurable and $W_t - W_s$ is independent from $\mathcal{F}_s$ whenever $s < t$. For each integer $\nu = 1, 2, \ldots$ we let π_ν be the partition $\{t_j^\nu\}_{j=0}^\infty$, where $t_j^\nu = j2^{-\nu}$. We write $\Delta W_i(j, \nu) = W_i(j2^{-\nu}) - W_i((j-1)2^{-\nu})$.

Using the $u(B, a, b, \pm, \tau)$ defined above, we will construct for each ν a π_ν-adapted input process U^ν. We define U^ν by specifying its derivative $u^\nu = \dot{U}^\nu$. Divide the interval $I_j^\nu = [(j-1)2^{-\nu}, j2^{-\nu}]$ into two equal subintervals $I_j^{\nu,-}$, $I_j^{\nu,+}$. On $I_j^{\nu,-}$, we let the component

u_i^ν be equal to $2^{\nu+1}\Delta W_i(j,\nu)$ if $|\Delta W_i(j,\nu)| \le 2^{-\frac{2\nu}{5}}$, and to zero otherwise. (It then follows, in particular, that $|u_i^\nu(t)| \le 2^{1+\frac{3\nu}{5}}$.) On $I_j^{\nu,+}$ we proceed as follows. Let $\alpha_\mu = \frac{|g_\mu|}{|g_1|+\ldots+|g_p|}$, so that $0 < \alpha_\mu$ and $\alpha_1 + \ldots + \alpha_p = 1$. Divide $I_j^{\nu,+}$ into intervals $I_j^{\nu,+,\mu}$, $\mu = 1,\ldots,p$, of length $\alpha_\mu 2^{-\nu-1}$. If $I_j^{\nu,+,\mu} = [a(j,\nu,\mu), b(j,\nu,\mu)]$, then we let u^ν be equal to $u(B_\mu, a(j,\nu,\mu), b(j,\nu,\mu), \pm, \tau_{\mu,\nu})$, where the sign is + or − depending on whether g_μ is > 0 or < 0, and the number $\tau_{\mu,\nu}$ is chosen so that $\tau_{\mu,\nu}^{r(\mu)} = |g_\mu| 2^{-\nu}$.

Then, if we apply (43) to the controls u_i^ν on an interval $I_j^{\nu,+,\mu}$, we get $|u_i^\nu(t)| \le \alpha_\mu^{-1} 2^{\nu+1} 4^{r(\mu)-1} |g_\mu|^{\frac{1}{r(\mu)}} 2^{-\frac{\nu}{r(\mu)}}$. We now pick $k \ge 3$ such that $r(\mu) \le k$ for all μ. We then have the pointwise inequality

$$|u_i^\nu(t)| \le \kappa 2^{\rho\nu}, \tag{44}$$

where $\kappa = 2\max(1, \max\{\alpha_\mu^{-1} 4^{r(\mu)-1} |g_\mu|^{\frac{1}{r(\mu)}} : \mu = 1,\ldots,p\})$ and $\rho = \frac{k-1}{k}$. (This has just been shown to be true on $I_j^{\nu,+}$, but it clearly holds on $I_j^{\nu,-}$ as well, since (a) $\kappa \ge 2$ and (b) $\rho \ge \frac{3}{5}$, because $k \ge 3$.)

We let $I_j^{\nu,-} = [a(j,\nu,0), b(j,\nu,0)]$. Then it is easy to see that

$$S_{k,a(j,\nu,\mu),b(j,\nu,\mu)}(U^\nu) = e^{2^{-\nu-1}X_0 + \sum_{i=1}^m \tilde{\Delta} W_i(j,\nu) X_i} \tag{45}$$

if $\mu = 0$, where $\tilde{\Delta} W_i(j,\nu) = \chi_{\nu,j,i} \Delta W_i(j,\nu)$, and $\chi_{\nu,j,i}$ is the indicator function of the set $\tilde{\mathcal{B}}_{\nu,j,i} = \{\omega \in \Omega : |\Delta W_i(j,\nu)| \le 2^{-\frac{2\nu}{5}}\}$. If $\mu > 0$, we have

$$S_{k,a(j,\nu,\mu),b(j,\nu,\mu)}(U^\nu) = e^{\alpha_\mu 2^{-\nu-1} X_0 + g_\mu 2^{-\nu} B_\mu + \ldots}, \tag{46}$$

where "..." denotes a Lie polynomial whose coefficients are bounded by a fixed constant times $2^{-\theta\nu}$, where $\theta = \frac{k+1}{k}$. From this, using the Campbell-Hausdorff formula, we conclude that

$$S_{k,(j-1)2^{-\nu},j2^{-\nu}}(U^\nu) = e^{2^{-\nu}X_0 + \sum_{i=1}^m \tilde{\Delta} W_i(j,\nu) X_i + 2^{-\nu} G + \ldots}, \tag{47}$$

where $G = \sum_{i=1}^p g_\mu B_\mu$.

Notice that, since $2^{\frac{\nu}{2}} \Delta W_i(j,\nu)$ is normalized Gaussian, we have

$$P(\tilde{\mathcal{B}}_{\nu,j,i}) \ge 1 - \sqrt{\frac{2}{\pi}} 2^{-\frac{\nu}{10}} e^{-2^{\frac{\nu}{5}-1}} \tag{48}$$

so that, for any fixed $T > 0$, if we let $\mathcal{B}_{T,N}$ be the event that $\chi_{\nu,j,i} = 1$ for all i, j, ν such that $i \in \{1, \ldots, m\}$, $j2^{-\nu} \leq T$, and $\nu \geq N$, then we have $P(\mathcal{B}_{T,N}) \geq 1 - T\sqrt{\frac{2}{\pi}} \sum_{\nu=N}^{\infty} 2^{\frac{9\nu}{10}} e^{-2^{\frac{\nu}{5}-1}}$, so that $P(\mathcal{B}_{T,N}) \to 1$ as $N \to \infty$.

We will need the following technical result:

Lemma 2 *The process U^ν is in $OIP(m, k, C, \pi_\nu)$, where C is a fixed constant, independent of ν.*

PROOF. The iterated integrals $U_I^\nu(t_j^\nu, t_{j-1}^\nu)$ for $1 \leq |I| \leq k$ can be obtained from the Chen-Fliess series (47) by computing the exponential. Since $|\tilde{\Delta} W_i(j, \nu)| \leq 2^{-\frac{2\nu}{5}}$, it is clear that all the coefficients of $S_{k,(j-1)2^{-\nu},j2^{-\nu}}(U^\nu) - 1$ (i.e. all the $U_I^\nu(t_j^\nu, t_{j-1}^\nu)$ with $1 \leq |I| \leq k$) are pointwise bounded by a fixed constant times $2^{-\frac{2\nu}{5}}$, so that (14) holds. Moreover, it follows from (47) that

$$S_{k,(j-1)2^{-\nu},j2^{-\nu}}(U^\nu) = 1 + \sum_{i=1}^{m} \tilde{\Delta} W_i(j, \nu) X_i + \sum_{i,i'=1}^{m} \tilde{\Delta} W_i(j, \nu) \tilde{\Delta} W_{i'}(j, \nu) X_i X_{i'} + \ldots, \tag{49}$$

where "..." denotes a finite sum of terms that are bounded by a fixed constant times $2^{-\nu}$. So the conditions of (12) and (13) will be trivially verified if we show that, if we let A be any of the variables $V_i = \tilde{\Delta} W_i(j, \nu)$ or $V_{ij} = \tilde{\Delta} W_i(j, \nu) \tilde{\Delta} W_{i'}(j, \nu)$, then $|\mathbb{E}(A)|$ and $\mathbb{E}(A^2)$ are both bounded by a constant times $2^{-\nu}$. (Since A is independent from $\mathcal{F}_{t_{j-1}^\nu}$, we can compute true expectations instead of conditional ones.) And these bounds follow trivially from the fact that $V_i = 2^{-\nu/2} H_i$, where the H_i are obtained by symmetrically truncating normalized Gaussian random variables. This completes the proof that the bounds (12), (13), (14) hold.

As for (15), recall that the components $u_i^\nu(t)$ satisfy (44). Since every integration over an interval of length $\leq 2^{-\nu}$ improves the bound by a factor of $2^{-\nu}$, we conclude that a k-th order iterated integral of u^ν is bounded by $\kappa^k 2^{(\rho-1)k\nu}$, i.e. by $\kappa^k 2^{-\nu}$. So

$$|(u^\nu)_I^{-,t}(s)| \leq \kappa^{k+1} 2^{(\rho-1)\nu}, \tag{50}$$

and (15) holds, since $\rho < 1$. ∎

It is clear from our construction that $U^\nu(j2^{-\nu}) = W(j2^{-\nu})$ on $\mathcal{B}_{T,N}$, if $j2^{-\nu} \leq T$. In view of (44), we have $||U^\nu(t) - U^\nu(j2^{-\nu})|| \leq c2^{-\frac{\nu}{k}}$ pointwise, where c is a fixed constant. Since $P(\mathcal{B}_{T,N}) \to 1$ as $N \to \infty$, and W has continuous sample paths, it follows that

$$P(\lim_{\nu\to\infty}(\sup\{||W(t) - U^\nu(t)|| : 0 \leq t \leq T\}) = 0) = 1 \qquad (51)$$

for every $T > 0$, so the U^ν are indeed approximations of W.

6 Proof of Convergence

We now fix an $\mathcal{F}_0$-measurable square-integrable initial condition $\bar{X} : \Omega \to \mathbb{R}^n$, and let $t \to X(t)$ denote the Stratonovich solution of (3) such that $X(0) = \bar{X}$. Also, let W^ν denote the ordinary input process such that $W^\nu(t_j^\nu) = W(t_j^\nu)$ for all j, and W^ν is linear on the intervals $[t_{j-1}^\nu, t_j^\nu]$ of the partition π_ν. Let $w^\nu = \dot{W}^\nu$. Define $\tilde{w}^\nu$ to be the result of truncating w^ν as before, i.e. let $\tilde{w}_i^\nu = w_i^\nu$ on $[t_{j-1}^\nu, t_j^\nu]$ if on that interval $|w_i^\nu| \leq 2^{-\frac{2\nu}{5}}$, and otherwise let $\tilde{w}_i^\nu = 0$. We then let $\tilde{W}^\nu$ be the integral of $\tilde{w}^\nu$.

It is clear that both W^ν and $\tilde{W}^\nu$ are π_ν-adapted OIP's. Moreover, the $\tilde{W}^\nu$ are in $OIP(m, k, C, \pi_\nu)$ for a fixed C, independent of ν, provided that $k \geq 2$. (The proof is analogous to, but easier than that of Lemma 2.) It is then easy to see that

$$S_{k,(j-1)2^{-\nu},j2^{-\nu}}(\tilde{W}^\nu) = e^{2^{-\nu}X_0 + \sum_{i=1}^m \tilde{\Delta}W_i(j,\nu)X_i} . \qquad (52)$$

We now want to consider the maps $\Phi_{a,b}^U$ defined for an OIP U, using the equation $dx = (f_0(x) + g(x))dt + \sum_{i=1}^m f_i(x)dU_i$ instead of (4). We will use $\hat{\Phi}_{a,b}^U$ to denote these maps, so as to avoid any confusion with the $\Phi_{a,b}^U$ that are associated to (4).

Theorem 1 *Let $(\Omega, \mathcal{F}, P)$ be a probability space endowed with a filtration $\{\mathcal{F}_t\}$. Let W be an m-dimensional standard Wiener process with respect to $\{\mathcal{F}_t\}$. Let $f_0, \ldots, f_m \in C_b^\infty(\mathbb{R}^n, \mathbb{R}^n)$, let Λ_0 be the Lie algebra of vector fields generated by $f_1, \ldots, f_m$, and let $\Lambda = [\Lambda_0, \Lambda_0]$. Let $g \in \Lambda$. Let $\bar{X} \in L^2((\Omega, \mathcal{F}_0, P); \mathbb{R}^n)$, and let $t \to X(t)$ be the*

Stratonovich solution of (3) with $X(0) = \bar{X}$. *Let* $\{U^\nu\}$ *be the ordinary input processes constructed in* §5. *Then, for every* $T > 0$,

$$\Phi^{U^\nu}_{T,0}(\bar{X}) \to X(T) \qquad \text{a.s.} \qquad \text{as} \qquad \nu \to \infty\,. \tag{53}$$

PROOF. Lemma 1 gives us estimates

$$||\Phi^{U^\nu}_{t^\nu_j,t^\nu_{j-1}}(X) - \Phi^{U^\nu}_{t^\nu_j,t^\nu_{j-1}}(Y)||_{L_2} \le (1 + K2^{-\nu})||X - Y||_{L^2}\,, \tag{54}$$

$$||\hat{\Phi}^{\tilde{W}^\nu}_{t^\nu_j,t^\nu_{j-1}}(X) - \hat{\Phi}^{\tilde{W}^\nu}_{t^\nu_j,t^\nu_{j-1}}(Y)||_{L_2} \le (1 + K2^{-\nu})||X - Y||_{L^2}\,, \tag{55}$$

valid for all ν, j, and all square-integrable $\mathcal{F}^\nu_{t_{j-1}}$-measurable X, Y. (The exponential factor that occurs in the formula of Lemma 1 is bounded independently of ν, since $|\pi_\nu| \le 1$ for all ν.)

Write $X^\nu_j = \hat{\Phi}^{\tilde{W}^\nu}_{t^\nu_j,0}(\bar{X})$, $Y^\nu_j = \Phi^{U^\nu}_{t^\nu_j,0}(\bar{X})$, and $Z^\nu_j = X^\nu_j - Y^\nu_j$. Then $X^\nu_j = \hat{\Phi}^{\tilde{W}^\nu}_{t^\nu_j,t^\nu_{j-1}}(X^\nu_{j-1})$ and $Y^\nu_j = \Phi^{U^\nu}_{t^\nu_j,t^\nu_{j-1}}(Y^\nu_{j-1})$, and therefore $Z^\nu_j = A^\nu_j - B^\nu_j$, where

$$\begin{aligned} A^\nu_j &= \hat{\Phi}^{\tilde{W}^\nu}_{t^\nu_j,t^\nu_{j-1}}(X^\nu_{j-1}) - \hat{\Phi}^{\tilde{W}^\nu}_{t^\nu_j,t^\nu_{j-1}}(Y^\nu_{j-1}) \\ B^\nu_j &= \hat{\Phi}^{\tilde{W}^\nu}_{t^\nu_j,t^\nu_{j-1}}(Y^\nu_{j-1}) - \Phi^{U^\nu}_{t^\nu_j,t^\nu_{j-1}}(Y^\nu_{j-1})\,. \end{aligned}$$

From (55) we get the bound $||A^\nu_j||_{L^2} \le (1 + K2^{-\nu})||Z^\nu_{j-1}||_{L^2}$.

We now estimate B^ν_j. Let $\hat{f}_0 = f_0 + g$, $\hat{f} = (\hat{f}_0, f_1, \ldots, f_m)$. Write $a = t^\nu_{j-1}$, $b = t^\nu_j$. We pick $k \ge \max(3, r(1), \ldots, r(p))$, and apply (10) for f with $U = U^\nu$, and for $\hat{f}$ with $U = \tilde{W}^\nu$, and let $t = b$. We get

$$\Phi^{U^\nu}_{b,a}(x) = \sum_{|I| \le k} U^\nu_I(b,a) E^f_I(x) + R_{k,b,a,U^\nu,E^n,f}(x)\,, \tag{56}$$

$$\hat{\Phi}^{\tilde{W}^\nu}_{b,a}(x) = \sum_{|I| \le k} \tilde{W}^\nu_I(b,a) E^{\hat{f}}_I(x) + R_{k,b,a,\tilde{W}^\nu,E^n,\hat{f}}(x)\,. \tag{57}$$

In view of (50), plus the analogous bound for $\tilde{W}^\nu$, and (11), the remainders $R_{k,b,a,U^\nu,E^n,f}(x)$, $R_{k,b,a,\tilde{W}^\nu,E^n,\hat{f}}(x)$ are bounded by a fixed constant times $2^{-\theta\nu}$. (Recall that $\theta = 1 + \frac{1}{k}$.) Moreover, we have

$$\sum_{|I| \le k} U^\nu_I(b,a) E^f_I(x) = (S_{k,a,b}(U^\nu)(f)E^n)(x)\,, \tag{58}$$

$$\sum_{|I| \le k} \tilde{W}^\nu_I(b,a) E^{\hat{f}}_I(x) = (S_{k,a,b}(\tilde{W}^\nu)(\hat{f})E^n)(x)\,, \tag{59}$$

where, for a noncommutative polynomial P in the X_i, $P(f)$ denotes the partial differential operator obtained by plugging in the f_i for the X_i. Using "..." to denote terms that are bounded by a fixed constant times $2^{-\theta\nu}$, we have

$$\begin{aligned} S_{k,a,b}(U^\nu) &= 1+(b-a)(X_0+G)+V+\dots, \\ S_{k,a,b}(\tilde{W}^\nu) &= 1+(b-a)X_0+V+\dots, \end{aligned}$$

where $V=\sum_{i=1}^m \tilde{\Delta}W_i(j,\nu)X_i+\sum_{i,i'=1}^m \tilde{\Delta}W_i(j,\nu)\tilde{\Delta}W_{i'}(j,\nu)X_iX_{i'}$, so that

$$\begin{aligned} \sum_{|I|\le k} U_I^\nu(b,a)E_I^f(x) &= x+(b-a)(f_0(x)+g(x))+V(f)(x)+\dots, \\ \sum_{|I|\le k} \tilde{W}_I^\nu(b,a)E_I^{\hat{f}}(x) &= x+(b-a)\hat{f}_0(x)+V(f)(x)+\dots, \end{aligned}$$

with $V(f)=\sum_{i=1}^m \tilde{\Delta}W_i(j,\nu)f_i+\sum_{i,i'=1}^m \tilde{\Delta}W_i(j,\nu)\tilde{\Delta}W_{i'}(j,\nu)f_if_{i'}E^n$.

Since $\hat{f}_0=f_0+g$, we conclude that $||B_j^\nu||\le\gamma 2^{-\theta\nu}$ pointwise, where γ is a fixed constant.

Therefore $||Z_j^\nu||_{L^2}\le(1+K2^{-\nu})||Z_{j-1}^\nu||_{L^2}+\gamma 2^{-\theta\nu}$. From this it follows easily by induction on j that $||Z_j^\nu||_{L^2}\le j e^{jK2^{-\nu}}\gamma 2^{-\theta\nu}$, i.e.

$$||\hat{\Phi}_{T,0}^{\tilde{W}^\nu}(\bar{X})-\Phi_{T,0}^{U^\nu}(\bar{X})||_{L^2}\le\gamma T e^{KT}2^{-\frac{\nu}{k}}, \tag{60}$$

if $T=t_j^\nu=j2^{-\nu}$. Actually, (60) holds for arbitrary T, with the factor γT replaced by $\gamma(T+\lambda)$ for some fixed $\lambda>0$. (To see this, let $t_{j-1}^\nu\le T<t_j$, and notice that (7) (for $U=U^\nu$, $\varphi=E^n$) together with (44) imply the pointwise bound $||\Phi_{T,t_{j-1}}^{U^\nu}(x)-x||\le\text{constant}\cdot 2^{-\frac{\nu}{k}}$. A similar bound holds for $\hat{\Phi}_{T,t_{j-1}}^{\tilde{W}^\nu}\cdot$.)

It follows from (60) that $\hat{\Phi}_{T,0}^{\tilde{W}^\nu}(\bar{X})-\Phi_{T,0}^{U^\nu}(\bar{X})\to 0$ almost surely. On the set $\mathcal{B}_{T,N}$, $\hat{\Phi}_{T,0}^{\tilde{W}^\nu}(\bar{X})=\hat{\Phi}_{T,0}^{W^\nu}(\bar{X})$ for sufficiently large ν. Since $P(\cup_N\mathcal{B}_{T,N})=1$, we conclude that $\hat{\Phi}_{T,0}^{\tilde{W}^\nu}(\bar{X})-\Phi_{T,0}^{W^\nu}(\bar{X})\to 0$ almost surely. Finally, $\hat{\Phi}_{T,0}^{W^\nu}(\bar{X})$ converges a.s. to $X(T)$ by the Wong-Zakai theorem. So (53) holds. ∎

Bibliography

[1] N. Bourbaki, *Groupes et Algèbres de Lie,* Élements de Mathématique, Fascicule XXXVII, Chap. II et III, Hermann, Paris, 1972.

[2] M. Fliess, *Réalisation locale des systèmes non linéaires, algèbres de Lie filtrées transitives et séries génératrices non commutatives,* Invent. Math. **71**, 1983, p. 521-537.

[3] N. Ikeda and S. Watanabe, *Stochastic Differential Equations and Diffusion Processes*, North-Holland, 1981.

[4] E. J. McShane, *On the use of stochastic differentials in models of random processes,* Proc. Sixth Berkeley Symp. Math. Statist. Prob. **3**, 1972, p. 263-294.

[5] R. L. Stratonovich, *A new form of representation of stochastic integrals and equations,* SIAM J. Control, 1966, p. 362-371.

[6] H. J. Sussmann, *Lie brackets and local controllability: a sufficient condition for scalar input systems,* SIAM J. Control and Optimization **21**, 1983, p. 686-713.

[7] H. J. Sussmann, *A general theorem on local controllability,* SIAM J. Control and Optimization **25**, 1987, p. 158-194.

[8] H. J. Sussmann, *Exponential Lie series and the discretization of stochastic differential equations,* in "Stochastic Differential Systems, Stochastic Control and Applications," W. H. Fleming and P. L. Lions Eds., I.M.A. vols. in Math. and its Apps. No. 10, Springer-Verlag, 1988, p. 563-582.

[9] E. Wong and M. Zakai, *On the relationship between ordinary and stochastic differential equations and applications to stochastic problems in control theory,* Proc. Third IFAC Congress, 1966, paper 3B.

[10] E. Wong and M. Zakai, *Riemann-Stieltjes approximations of stochastic integrals* Z. Wahrscheinlichkeitstheorie und Verw. Gebiete **12**, 1969, p. 87-97.

Donsker's δ-functions in the Malliavin calculus

Shinzo Watanabe
Department of Mathematics
Kyoto University, JAPAN

1 Introduction

Let $W = W_0(\mathbf{R}^r)$ be the Banach space of (r-dimensional) continuous paths $w: [0,1] \to \mathbf{R}^r$ such that $w(0) = 0$ endowed with the supremum norm and P be the (r-dim.) standard Wiener measure on W so that (W, P) be the *r-dimensional Wiener space.* The measure of Brownian bridge $P_{00}^{10}(\cdot) = P(\cdot | w(1) = 0)$ on W is singular to the Wiener measure P and yet, it is often convenient to think of its formal or fictitious density. The idea is exactly the same as considering the Dirac δ-function as a formal density of the unit point measure with respect to the Lebesgue measure. Of course, the Dirac δ-function is now a well-defined mathematical object in the Schwartz theory of distributions and therefore, it is quite natural to define this formal density (call it *Donsker's δ-function* by following Kuo [5]) to be a distribution or a generalized function on the Wiener space. Thus, it is H.-H. Kuo ([6]) who first defined this notion in the setting of Hida's theory of white noise analysis.

Being inspired by the work of Kuo, we showed in [10] that this kind of notions can be naturally and more generally defined in the setting of Malliavin's calculus. Indeed, if $F: W \to \mathbf{R}^d$ is a d-dimensional Wiener functional which is regular in the sense of Malliavin and if δ_x is the Dirac δ-function at $x \in \mathbf{R}^d$, then the *composite* $\delta_x \circ F$ can be defined as a *generalized Wiener functional* on W, more generally, $T \circ F$ can be defined for any Schwartz distribution T on $\mathbf{R}^d$, cf. [4], [11]. We call $\delta_x \circ F$ again a Donsker's δ-function. The *generalized expectation* $E[\delta_x \circ F] := p_F(x)$ coincides with the density of the

ISBN 0-12-481005-5

law $P \circ F^{-1}$ of F on $\mathbf{R}^d$ with respect to the the Lebesgue measure and, if $p_F(x) > 0$, the normalized Donsker's δ-function $\delta_x \circ F/p_F(x)$ may be regarded as a formal density of the conditional probability $P(\ \mid F = x)$. Actually, by the "quasi-sure analysis" recently developed by Airault-Malliavin [1] and Sugita [9], among others, we can define a family of Borel probabilities on W having the normalized Donsker's δ-functions as formal densities so that we can discuss a satisfactory disintegration theory on the foliation of submanifolds $\{w; F(w) = x\}$ imbedded in the Wiener space. The idea is quite similar to that the Schwartz distribution theory is used for area integrals over manifolds imbedded in the Euclidean space as is discussed, e.g., in Chapter III of Gel'fand-Shilov [2]. Such a theory of generalized expectations for generalized functionals, particularly for Donsker's δ-functions, provides us with efficient probabilistic methods in problems related to heat kernels and thereby justifies path integral methods in heat equation approaches to various problems in mathematics and mathematical physics, cf., e.g., a survey paper by Ikeda [3] or Watanabe [12].

In section 2, we give a short review on notions and results in the Malliavin calculus relevant to Donsker's δ-functions. In section 3, we discuss a problem of deciding exactly the Sobolev spaces to which these Donsker's δ-functions actually belong.

2 A survey of Sobolev spaces in the Malliavin calculus

Let (W, P) be, as above, the r-dimensional Wiener space and $H \subset W$ be the *Cameron-Martin Hilbert space* consisting of $w \in W$ which are absolutely continuous in $t \in [0, 1]$ with square integrable derivatives and endowed with, as its norm, the L^2-norm of derivatives. In the following discussions, the triple (W, H, P) may be replaced by any abstract Wiener space. As usual, a P-meaurable function on W is called a *Wiener functional*; to be more precise, it is an equivalence class of P-measurable functions coinciding with each other P-almost everywhere. Denoting by E a separable real Hilbert space, in general, let $\mathcal{L}_p(E)$ be the usual L_p-space of E-valued Wiener functionals with the norm $\| \ \|_p$, $1 \leq p < \infty$. The family of *Sobolev spaces* $\mathcal{D}_p^s(E)$, $1 < p < \infty$, $s \in \mathbf{R}$, of E-valued Wiener (and generalized Wiener) functionals has been introduced in the Malliavin calculus ([4], [11]): If L is the *Ornstein-Uhlenbeck operator* on W so that $L = -D^*D$, D being the *H-derivative* and D^* being its dual, then roughly

$$\mathcal{D}_p^s(E) = (I - L)^{-s/2}(\mathcal{L}_p(E))$$

and its norm $\| \ \|_{p,s}$ is defined by $\|F\|_{p,s} = \|(I-L)^{s/2}F\|_p$. In particular, $\mathcal{D}_p^0(E) = \mathcal{L}_p(E)$ and $\| \ \|_{p,0} = \| \ \|_p$. If $s \leq s'$ and $p \leq p'$, then $\mathcal{D}_{p'}^{s'}(E) \subset \mathcal{D}_p^s(E)$ and the dual space $\mathcal{D}_p^s(E)'$ of $\mathcal{D}_p^s(E)$ is naturally identified with $\mathcal{D}_q^{-s}(E)$, $p^{-1} + q^{-1} = 1$. Set

$$\mathcal{D}_{\infty-}^{\infty}(E) = \bigcap_{s>0} \bigcap_{1<p<\infty} \mathcal{D}_p^s(E)$$

and

$$\mathcal{D}_{1+}^{-\infty}(E) = \bigcup_{s>0} \bigcup_{1<p<\infty} \mathcal{D}_p^{-s}(E)$$

and call them the *space of test Wiener functionals* and the *space of generalized Wiener functionals*, respectively. When $E = \mathbf{R}$, we omit to write the value space E so that $\mathcal{L}_p(\mathbf{R}) = \mathcal{L}_p$, $\mathcal{D}_p^s(\mathbf{R}) = \mathcal{D}_p^s$, $\mathcal{D}_{\infty-}^{\infty}(\mathbf{R}) = \mathcal{D}_{\infty-}^{\infty}$ etc. It is convenient to *extend Sobolev norms* $\| \ \|_{p,s}$ *being defined for every* $\Phi \in \mathcal{D}_{1+}^{-\infty}(E)$ *by setting* $\|\Phi\|_{p,s} = \infty$ *if* $\Phi \notin \mathcal{D}_p^s(E)$ so that $\Phi \in \mathcal{D}_p^s(E)$ if and only if $\|\Phi\|_{p,s} < \infty$.

If $\mathbf{1}$ is the Wiener functional identically equal to 1, then $\mathbf{1} \in \mathcal{D}_{\infty-}^{\infty}$ and, for $\Phi \in \mathcal{D}_{1+}^{-\infty}$, the natural coupling $\langle \Phi, \mathbf{1} \rangle$ is called the *generalized expectation of* Φ and is denoted by $E(\Phi)$. When Φ is given by an integrable Wiener functionals, this notion obviously coincides with the usual expectation $E(\Phi) = \int_W \Phi(w)P(dw)$.

$\mathcal{D}_{\infty-}^{\infty}$ is an algebra and the product $F \cdot \Phi \in \mathcal{D}_{1+}^{-\infty}(E)$ is defined for $F \in \mathcal{D}_{\infty-}^{\infty}$ and $\Phi \in \mathcal{D}_{1+}^{-\infty}(E)$. This operation is closed in $\mathcal{D}_{\infty-}^{\infty}(E)$, i.e., $\mathcal{D}_{\infty-}^{\infty} \times \mathcal{D}_{\infty-}^{\infty}(E) \subset \mathcal{D}_{\infty-}^{\infty}(E)$.

The H-derivative D is extended naturally to a linear operator

$$D: \mathcal{D}_{1+}^{-\infty}(E) \to \mathcal{D}_{1+}^{-\infty}(H \otimes E)$$

which sends $\mathcal{D}_p^{s+1}(E)$ into $\mathcal{D}_p^s(H \otimes E)$ continuously for every $s \in \mathbf{R}$ and $1 < p < \infty$, particularly, $\mathcal{D}_{\infty-}^{\infty}(E)$ into $\mathcal{D}_{\infty-}^{\infty}(H \otimes E)$.

Let $F: W \to \mathbf{R}^d$ be a d-dimensional Wiener functional. F is called *regular in the sense of Malliavin* if

(i) $F \in \mathcal{D}^{\infty}_{\infty-}(\mathbf{R}^d)$, i.e., $F = (F^1, \ldots, F^d)$ with $F^i \in \mathcal{D}^{\infty}_{\infty-}$,

(ii) $\sigma_F = (\sigma_F^{ij})$, defined by $\sigma_F^{ij} = \langle DF^i, DF^j \rangle_{H \otimes \mathbf{R}}$, $i, j = 1, \ldots, d$, satisfies that

$$(\det \sigma_F)^{-1} \in \bigcap_{1<p<\infty} \mathcal{L}_p.$$

Note that $\sigma_F^{ij} \in \mathcal{D}^{\infty}_{\infty-}$, $\det \sigma_F \geq 0$ a.e. and we set $0^{-1} = \infty$ by convention. σ_F is called the *Malliavin covariance of* F.

Suppose that we are given a d-dim. Wiener functional $F: W \to \mathbf{R}^d$ regular in the Malliavin sense. Then for every real tempered distribution $T \in \mathcal{S}'(\mathbf{R}^d)$ on $\mathbf{R}^d$, the *composite (or pull-back)* $T \circ F \in \mathcal{D}^{-\infty}_{1+}$ can be defined in the following manner: For $\varphi \in \mathcal{S}(\mathbf{R}^d)$, $\varphi \circ F$ is the usual composite and $\varphi \circ F \in \mathcal{D}^{\infty}_{\infty-}$. By the integration by parts and chain rules for derivatives, we have for every $G \in \mathcal{D}^{\infty}_{\infty-}$ that

$$E[(1 + |x|^2 - \Delta)^k \varphi \circ F \cdot G] = E[\varphi \circ F \cdot \eta_{2k}(G)], \; k = 0, 1, 2, \ldots, \quad (1)$$

where Δ is the Laplacian on $\mathbf{R}^d$ and $\eta_{2k}(G) \in \mathcal{D}^{\infty}_{\infty-}$ is obtained as a polynomial in components of $\gamma = (\gamma^{ij}) = \sigma_F^{-1}$, F and G and their derivatives, (cf. [4], Chap. V, §9). Note that $\gamma^{ij} \in \mathcal{D}^{\infty}_{\infty-}$ by the regularity of F. By studying the expression of $\eta_{2k}(G)$ carefully, we have the following estimate of L_1-norm: for every $1 < q < \infty$

$$C_{q,k;F} := \sup\{\|\eta_{2k}(G)\|_1 \mid G \in \mathcal{D}^{\infty}_{\infty-}, \|G\|_{q,2k} \leq 1\} < \infty. \quad (2)$$

Hence, from (1), we deduce that

$$\|(1 + |x|^2 - \Delta)^k \varphi \circ F\|_{p,-2k} \leq C_{q,k;F} \cdot \sup_{x \in \mathbf{R}^d} |\varphi(x)|,$$
$$\varphi \in \mathcal{S}(\mathbf{R}^d), \; k = 0, 1, 2, \ldots \quad (3)$$

where $p^{-1} + q^{-1} = 1$. Since, for every $T \in \mathcal{S}'(\mathbf{R}^d)$, we can find $k \in \mathbf{Z}_+$ and a bounded continuous function $f(x)$ with $\lim_{|x| \to \infty} f(x) = 0$ such that $T = (1 + |x|^2 - \Delta)^k f$, we deduce from (3) in a routine way that $T \circ F$ can be uniquely defined as limit in $\mathcal{D}_p^{-2k}$ of $(1 + |x|^2 - \Delta)^k \varphi_m \circ F$

where we take any $\varphi_m \in \mathcal{S}(\mathbf{R}^d)$ such that $\varphi_m \to f$ uniformly. Since $1 < q < \infty$ is arbitrary, p is also arbitrary. Consequently,

$$T \circ F \in \bigcup_{k=0}^{\infty} \bigcap_{1<p<\infty} \mathcal{D}_p^{-2k} \quad \text{for every } T \in \mathcal{S}'(\mathbf{R}^d).$$

The formula (1) still holds if $\varphi \in \mathcal{S}(\mathbf{R}^d)$ is replaced by $T \in \mathcal{S}'(\mathbf{R}^d)$.

If δ_x is the Dirac δ-function at $x \in \mathbf{R}^d$, then $x \to \delta_x \in \mathcal{S}'(\mathbf{R}^d)$ is C^∞ and hence $x \to \delta_x \circ F$ is also C^∞. Therefore, for every $G \in \bigcap_{k=0}^{\infty} \bigcup_{1<q<\infty} \mathcal{D}_q^{2k}$, $x \to E[\delta_x \circ F \cdot G]$ is C^∞. In particular, $p_F(x) = E[\delta_x \circ F]$, which can be easily identified with the density of the law $P \circ F^{-1}$ of F, is C^∞ in x and, on the *strict support* $\mathrm{SSP}(F) = \{x; p_F(x) > 0\}$ *of the law* $p \circ F^{-1}$ *of* F, $E[\delta_x \circ F \cdot G]/p_F(x) = E[G|F = x]$ is C^∞ in x for every $G \in \bigcap_{s>0} \bigcup_{1<p<\infty} \mathcal{D}_p^s$.

3 The exact Sobolev spaces to which Donsker's δ-functions belong

Here, we would obtain the condition on $1 < p < \infty$ and $\alpha > 0$ so that $\delta_x \circ F \in \mathcal{D}_p^{-2\alpha}$ for $F: W \to \mathbf{R}^d$ regular in the Malliavin sense.

Theorem. *A necessary and sufficient condition for* $\alpha > 0$ *and* $1 < p < \infty$ *so that* $\delta_x \circ F \in \mathcal{D}_p^{-2\alpha}$ *for every d-dim. Wiener functional* $F: W \to \mathbf{R}^d$ *regular in the Malliavin sense and* $x \in \mathbf{R}^d$ *is*

$$\begin{aligned} &\text{(i)} \quad \alpha \geq \tfrac{d}{2} \text{ and } 1 < p < \infty \\ \text{or} \quad &\text{(ii)} \quad 0 < \alpha < \tfrac{d}{2} \text{ and } 1 < p < \tfrac{d}{d-2\alpha}. \end{aligned} \tag{4}$$

Proof. First we prove the necessity of (4) by considering a particular case of $x = 0$ and the following F: $F(w) = (h_1(w), \ldots, h_d(w)) := \boldsymbol{h}(w)$, $\{h_i\}$ being an orthonormal system in H where $h_i(w) = D^*[h_i](w)$ is the Wiener's stochastic integral (the first order Wiener chaos associated to h_i). Then, if $\alpha > 0$,

$$\begin{aligned} &(I - L)^{-\alpha} \delta_0 \circ F(w) \\ &\quad = \frac{1}{\Gamma(\alpha)} \int_0^\infty e^{-t} t^{\alpha-1} e^{tL} (\delta_0 \circ F)(w) dt \end{aligned}$$

$$= \frac{1}{\Gamma(\alpha)} \int_0^\infty e^{-t} t^{\alpha-1} E^\omega[\delta_0(e^{-t}\boldsymbol{h}(w) + \sqrt{1-e^{-2t}}\boldsymbol{h}(\omega))]dt$$

$$= \frac{1}{\Gamma(\alpha)(2\pi)^{d/2}} \int_0^\infty e^{-t} t^{\alpha-1}(1-e^{-2t})^{-d/2} \exp\left\{ -\frac{e^{-2t}|\boldsymbol{h}(w)|^2}{2(1-e^{-2t})} \right\} dt.$$

From this explicit expression and the fact that $\boldsymbol{h}(w)$ is a d-dim. standard Gaussian random variable, we deduce easily that $\|(I - L)^{-\alpha}\delta_0 \circ F\|_p < \infty$ if and only if (p, α) satisfies (4).

It suffices to show, therefore, that *if* $1 < p < \infty$ *and* $\alpha > 0$ *satisfy (4) then* $\delta_x \circ F \in \mathcal{D}_p^{-2\alpha}$ *for every* $x \in \mathbf{R}^d$ *and every* $F: W \to \mathbf{R}^d$ *regular in the Malliavin sense.* For this, we note by the same proof as for (2) that, for every $1 < \bar{q} < q$,

$$C_{q,\bar{q},k;F} := \sup\{\|\eta_{2k}(G)\|_{\bar{q}} \mid G \in \mathcal{D}^\infty_{\infty-}, \|G\|_{q,2k} \le 1\} < \infty. \quad (5)$$

Let p and $\bar{p}$ be the dual indices of q and $\bar{q}$; $p^{-1} + q^{-1} = 1$ and $\bar{p}^{-1} + \bar{q}^{-1} = 1$ so that $p < \bar{p}$. As before, $\mathcal{L}_p$ is the L_p-space over the Wiener space and we denote by L_p and $|||\ \ |||_p$ the usual L_p-space on $\mathbf{R}^d$ with respect to the Lebesgue measure dx and its norm. For $\alpha \ge 0$, define a linear operator $A_\alpha: \mathcal{S}'(\mathbf{R}^d) \to \mathcal{D}^{-\infty}_{1+}$ by

$$A_\alpha T = (I-L)^{-\alpha}\{(1+|x|^2 - \Delta)^\alpha T \circ F\}, \quad T \in \mathcal{S}'(\mathbf{R}^d). \quad (6)$$

Then, if $\alpha = k$ is a nonnegative *integer*, we have by (1) and (6) that

$$\begin{aligned} \|A_k T\|_p &= \sup\{E[(1+|x|^2-\Delta)^k T \circ F \cdot (I-L)^{-k} H]; \\ &\qquad\qquad H \in \mathcal{D}^\infty_{\infty-}, \|H\|_q \le 1\} \\ &= \sup\{E[T \circ F \cdot \eta_{2k}((I-L)^{-k}H)]; H \in \mathcal{D}^\infty_{\infty-}, \|H\|_q \le 1\} \\ &= \sup\{E[T \circ F \cdot \eta_{2k}(G)]; G \in \mathcal{D}^\infty_{\infty-}, \|G\|_{q,2k} \le 1\} \\ &\le \sup\{\|T \circ F\|_{\bar{p}} \cdot \|\eta_{2k}(G)\|_{\bar{q}}; G \in \mathcal{D}^\infty_{\infty-}, \|G\|_{q,2k} \le 1\} \\ &= C_{q,\bar{q},k;F} \cdot \|T \circ F\|_{\bar{p}} \end{aligned}$$

and this is dominated further by $C_{q,\bar{q},k;F}\|p_F(x)\|_\infty^{1/\bar{p}} \cdot |||T|||_{\bar{p}}$ because $P \circ F^{-1}$ has bounded smooth density $p_F(x)$. By a theorem of Stein (cf. [7] p. 163),

$$\sup_{\tau \in \mathbf{R}} \|(1+|x|^2-\Delta)^{i\tau}\|_{L_p \to L_p} < \infty \quad \text{and}$$

$$\sup_{\tau \in \mathbf{R}} \|(I-L)^{i\tau}\|_{\mathcal{L}_p \to \mathcal{L}_p} < \infty.$$

Hence, if A_ζ, $\zeta = \alpha + i\beta$, is the analytic continuation of A_α, then

$$\sup_{\tau \in \mathbf{R}} \|A_{k+i\tau}\|_{L_p \to \mathcal{L}_p} < \infty \textit{ for every } k = 0, 1, \ldots \textit{ and } 1 < p < \bar{p}. \tag{7}$$

By appealing to the Stein interpolation theorem ([8]), we can conclude, for every $N > 0$, that

$$\sup_{0 \le \alpha \le N} \|A_\alpha\|_{L_p \to \mathcal{L}_p} < \infty. \tag{8}$$

Now, if $\alpha > 0$, then $T_x = (1 + |x|^2 - \Delta)^{-\alpha}\delta_x \ge 0$ satisfies

$$\begin{array}{l} T_x(y) \textit{ is bounded and smooth on } \{y;\ |y - x| > \eta\} \\ \textit{and } \int_{|x-y|>\eta} T_x(y)dy < \infty \ \textit{ for every } \eta > 0, \end{array} \tag{9}$$

$$T_x(y) \sim \begin{cases} \text{const.}\, |x - y|^{-(d-2\alpha)} & \textit{if } d \ne 2\alpha, \\ \text{const.}\, \log \frac{1}{|x-y|} & \textit{if } d = 2\alpha, \textit{ near } x = y. \end{cases} \tag{10}$$

If $\alpha > 0$ and $1 < p < \infty$ satisfy (4), then we can choose $\bar{p} > p$ such that $(\alpha, \bar{p})$ still satisfies (4) and hence, we see immediately from (9) and (10) that

$$|||T_x|||_{\bar{p}} < \infty. \tag{11}$$

Then, by (8) and (11),

$$\|(I - L)^{-\alpha}\delta_x \circ F\|_p = \|A_\alpha T_x\|_p \le \|A_\alpha\|_{L_p \to \mathcal{L}_p} \cdot |||T_x|||_{\bar{p}} < \infty,$$

that is, $\delta_x \circ F \in \mathcal{D}_p^{-2\alpha}$. □

Acknowledgment. The problem of §3 and an idea of the proof to use an interpolation theorem have been suggested by S. Kusuoka in his lectures at Kyoto University.

Bibliography

[1] H. Airault et P. Malliavin, Intégration géometrique sur l'espace de Wiener, *Bull. Sc. math.*, 2e *série*, **112** (1988), 3–52.

[2] I. M. Gel'fand and G. E. Shilov, *Generalized functions,* Vol. I, Academic Press, 1964.

[3] N. Ikeda, Probabilistic methods in study of asymptotics, in *Ecole d'Ete de Probabilités de Saint-Flour, 1988,* to appear in **LNM.**

[4] N. Ikeda and S. Watanabe, *Stochastic differential equations and diffusion processes,* Kodansha/North-Holland, 1981, Second Edition 1989.

[5] H.-H. Kuo, *Gaussian measures in Banach spaces,* **LNM 463**, Springer, 1975.

[6] H.-H. Kuo, Donsker's delta function as a generalized Brownian functional and its application, in *Theory and Application of Random Fields, Proc. IFIP Conf. Bangalore, 1982 (ed. G. Kallianpur),* **LNCI 49** (1983), 167–178.

[7] P. A. Meyer, Retour sur la théorie de Littlewood-Paley, in *Sém. Prob. XV, 1979/80 (ed. J. Azéma and M. Yor)* **LNM 850** (1981), 151–166.

[8] E. M. Stein and G. Weiss, *Introdunction to Fourier analysis on Euclidean spaces,* Princeton Univ. Press, 1971.

[9] H. Sugita, Positive generalized Wiener functions and potential theory over abstract Wiener spaces, *Osaka J. Math.*, **25** (1988), 665–696.

[10] S. Watanabe, Malliavin's calculus in terms of generalized Wiener functionals, in *Theory and Application of Random Fields, Proc. IFIP Conf. Bangalore, 1982 (ed. G. Kallianpur),* **LNCI 49** (1983), 284–290.

[11] S. Watanabe, *Lectures on stochastic differential equations and Malliavin calculus,* Tata Institute of Fundamental Research/Springer, 1984.

[12] S. Watanabe, Short time asymptotic problems in Wiener functional integration theory, Application to heat kernels and index theorem, in *Stochastic analysis and Related Topics II, Proc. Silivri 1988 (ed. H. Korezlioglu et A. S. Ustunel),* **LNM 1444** (1990), 1–62.

Implementing Boltzmann Machines

Eugene Wong
Department of Electrical Engineering
and Computer Sciences
University of California at Berkeley
Berkeley, CA 94720

1 Introduction

Consider a discrete-time Hopfield net [3] where the state $X_i(t)$ at node i takes values ± 1 and satisfies a transition equation of the form

$$X_i\,(t+1) \;=\; sgn\,[\sum_{j=1}^{n} w_{ij}\, X_j\,(t) + \theta_i] \tag{1}$$

or

$$= \;\; X_i\,(t) \tag{2}$$

We assume that at each t only one node can change state (i.e., satisfy (1)) but that every node satisfies it infinitely often as $t \longrightarrow \infty$.

Now define an energy function

$$V(x) = -[\frac{1}{2}\sum_{i,j} w_{ij}\, x_i\, x_j \;+\; \sum_i \theta_i\, x_i] \quad x \in \{-1,1\} \tag{3}$$

and assume that the weights w_{ij} satisfy the conditions: $w_{ii} = 0$, $w_{ij} = w_{ji}$ for all i and j. Under these assumptions, it is easy to show that at every t we have

$$V(X(t+1)) \;-\; V(X(t)) = [1 - sgn\,(\Delta_i\,(t))]\,\Delta_i(t) \tag{4}$$

for some i, where

$$\Delta_i(t) = X_i(t)\,[\sum_j w_{ij}\, X_j(t) \;+\; \theta_i] \tag{5}$$

ISBN 0-12-481005-5

Equation (4) indicates that $V(X(t))$ never increases, and decreases for at most a finite number of values of t. Thus, there exists a $t_0 < \infty$ such that

$$V(X(t)) - V(X(t_0)) = 0 \quad \textit{for all } t \geq t_0$$

which implies that for every i and every $t \geq t_0$,

$$\Delta_i(t) \geq 0 \quad \textit{and} \quad X_i(t) = X_i(t_0)$$

Since every component of X is tested infinitely often, $V(X(t_0))$ cannot be decreased by a change in any single component of $X(t_0)$. Thus, $X(t_0)$ is a *local minimum* of $V(x)$.

As a tool for optimization, Hopfield nets suffer from two flaws: local minimum and restricted $V(x)$, while enjoying an implementation advantage. These nets are no mere algorithms, but models that are readily implementable as hardware. Our goal in this paper is to remove the defects while preserving the implementation advantage.

Global minimization can be achieved through simulated annealing [5] which is based on the following idea. Suppose that instead of satisfying (1,2), $X(t)$ is a stochastic process with a distribution of the form

$$Prob\,(X(t) = x) = \frac{1}{Z}\, e^{-\frac{1}{T}V(x)}\,,\ x \in \{-1,1\}^n \tag{6}$$

which is known as the *Gibbs distribution.* Further, suppose that the parameter T (interpreted as *temperature*) is reduced slowly so that the distribution of $X(t)$ remains approximately Gibbs, i.e.,

$$Prob\,(X(t) = x) \approx \frac{1}{Z(t)}\, e^{-\frac{1}{T(t)}V(x)}$$

for all t. Then, denoting the global minimum by V_m, we have

$$Prob\,(X(t) = x) \approx \frac{e^{-\frac{1}{T(t)}[V(x) - V_m]}}{\sum_x e^{-\frac{1}{T(t)}[V(x) - V_m]}}$$

$$\longrightarrow \begin{cases} 0, & V(x) \neq V_m \\ \frac{1}{K}, & V(x) = V_m \end{cases} \quad \textit{as } T(t) \to 0$$

where K is the number of values of x that attain the global minimum.

It is well known that a process with an equilibrium Gibbs distribution can be generated as a stationary Markov chain. In [1] a neural network with states forming such a process is called a *Boltzmann machine*, and in [2] it is proposed to approximate a Boltzmann machine by injecting noise in a Hopfield net. Our first objective in this paper is to show that an exact, not approximate, construction of Boltzmann machine can be achieved by injecting noise in a Hopfield net, but that this has to be done in a precise way that we shall describe in the sequel.

Our second objective is to generalize $V(x)$ to an arbitrary function. We show that even in the general case $X(t)$ can still be realized by a network (though no longer a Hopfield net). Finally, we discuss alternative architectures that are suitable for implementing the resulting network.

2 Boltzmann Machine

Boltzmann machine is a binary-state and discrete-time neural network whose states form a stationary Markov chain $X(t)$ with an equilibrium Gibbs distribution. Such a Markov chain can be constructed by controlling its transitions as follows:

1. At time t, select one component, say $X_i(t)$, for possible change.

2. Compute the energy change that would result from a state change $X_i(t) \rightarrow -X_i(t)$, viz.,

$$\Delta_i(t) = 2X_i(t) \left[\sum_j w_{ij} X_j(t) + \theta_i\right] \tag{7}$$

3. Set

$$\begin{aligned} X_i(t+1) &= -X_i(t) \quad \textit{with probability} \;\; \pi(\Delta_i(t)) \\ &= X_i(t) \quad \textit{with probability} \;\; 1 - \pi(\Delta_i(t)) \end{aligned} \tag{8}$$

The quantity π is known as the *acceptance probability* (accepting a change). In [4] it is shown that an acceptance probability of the form

$$\pi(\Delta) = e^{-\frac{\Delta}{2T}} f(|\Delta|) \tag{9}$$

suffices to ensure that $X(t)$ will have a steady state distribution given by a Gibbs distribution (4). It is also shown in [4] that

$$\pi(\Delta) = min(1, e^{-\frac{\Delta}{T}}) \tag{10}$$

is optimal in the sense of maximizing the speed of reaching equilibrium.

We note that for $T = 0$, (10) reduces to the following:

$$\begin{aligned} \pi(\Delta) &= 1 \quad \Delta < 0 \\ &= 0 \quad \Delta > 0 \end{aligned} \tag{11}$$

and (8) can then be written as

$$\begin{aligned} X_i(t+1) &= X_i(t)\, sgn[\Delta_i(t)] \\ &= X_i(t)\, sgn\{X_i(t)\, [\sum_j w_{ij}\, X_j(t) + \theta_i]\} \\ &= sgn[\sum_j w_{ij}\, X_j(t) + \theta_i] \end{aligned} \tag{12}$$

which is just (1). Thus, for $T = 0$, a Boltzmann machine is indeed a Hopfield net.

For $T > 0$, we hypothesize that the state transition equation (8) for a Boltzmann machine is re-expressible as

$$X_i(t+1) = X_i(t)\, sgn[\Delta_i(t) - Z] \tag{13}$$

where Z is a random variable with a probability density function $p_Z(z)$. From (13), we have

$$\begin{aligned} \pi(\Delta_i(t)) &= prob\,(X_i(t+1) = -X_i(t)) \qquad (14) \\ &= prob\,(\Delta_i(t) < Z) \\ &= \int_{\Delta_i(t)}^{\infty} p_Z(z)\, dz \qquad (15) \end{aligned}$$

Comparing (14) with (8) shows that if we set

$$\int_{\Delta}^{\infty} p_Z(z)\, dz = \pi(\Delta) \tag{16}$$

then (13) is indeed equivalent to (8) and our hypothesis is confirmed. We can now differentiate (16) to get

$$p_Z(z) = -\frac{d}{dz}\pi(z) \tag{17}$$

which prescribes the distribution of Z in terms of the acceptance probability π.

We have thus shown that a Boltzmann machine can be implemented by injecting noise $Z(t)$ in a Hopfield net so that the transition equation is given by

$$X_i(t+1) = sgn[\sum_j w_{ij} X_j(t) + \theta_i - X_i(t)Z(t)] \tag{18}$$

where $Z(t)$ is independent for different t's (i.e., a white noise) but not Gaussian. Instead, it distribution is governed by the choice of the acceptance probability through (17). Both (17) and (18) represent new results.

From the example given by (10), we have

$$p_Z(z) = e^{-\frac{z}{T}}\, 1(z) \tag{19}$$

where $1(z)$ is the unit-step function. A random variable with this distribution can be generated as

$$Z = Z_1^2 + Z_2^2$$

where Z_1 and Z_2 are independent $N(0, T/2)$ random variables. We note, however, no allowable choice of the acceptance probability would yield a Z that is Gaussian.

Thus, while a Boltzmann machine can be implemented by injecting noise in a Hopfield net, the noise must be multiplied by the state before added to the input as prescribed by (18), and the noise cannot be Gaussian.

3 Generalizing the Energy Function

Any function $V(x)$ on the hypercube $\{-1,1\}^n$ admits a multilinear expansion of the form:

$$V(x) = -\sum_{\underline{k}} w_{\underline{k}} x_{k_1} x_{j_2} \cdots x_{k_m} + V_0 \tag{20}$$

where $\underline{k} = (k_1, j_2, \cdots, j_m)$ denotes an ordered subset of $\{1,2,\cdots,n\}$ and the summation is taken over all $2^n - 1$ such ordered subsets. Henceforth, we shall write $x_{\underline{k}}$ to denote the product $x_{k_1} x_{j_2} \cdots x_{j_m}$. For a $V(x)$ given by (3.1), we can now generalize (7) to read

$$\Delta_i(t) = 2 \sum_{\underline{k} \in i} w_{\underline{k}} X_{\underline{j}}(t) \tag{21}$$

With Δ thus generalized, a stationary Markov chain with an equilibrium Gibbs distribution for a general $V(x)$ can again be constructed using (13) and (18). However, $X_i(t)\, \Delta_i(t)$ is no longer linear in $X(t)$. Rather, we have

$$\begin{aligned} X_i(t+1) &= sgn[X_i(t)\, \Delta_i(t) - X_i(t) Z(t)] \\ &= sgn[\sum_{\underline{k} \in i} w_{\underline{k}} X_{k_1}(t) \cdots X_i(t) \cdots X_{j_m}(t) - X_i(t)\, Z(t)] \end{aligned}$$

Even at zero temperature (hence zero noise), (3) is not realizable as a Hopfield net. To realize (3) as a network, we shall develop several alternative architectures based on the first of the equations in (3).

4 Alternative Network Architectures

We begin by rewriting the state transition equation as follows:

$$X_i(t+1) = Y_i(t)\, X_i(t) \tag{22}$$

$$\begin{aligned} Y_i(t) &= sgn[\Delta_i(t) - Z(t)] \qquad (23) \\ &= sgn[\sum_{\underline{k} \in i} w_{\underline{k}} X_{\underline{k}}(t) - Z(t)] \end{aligned}$$

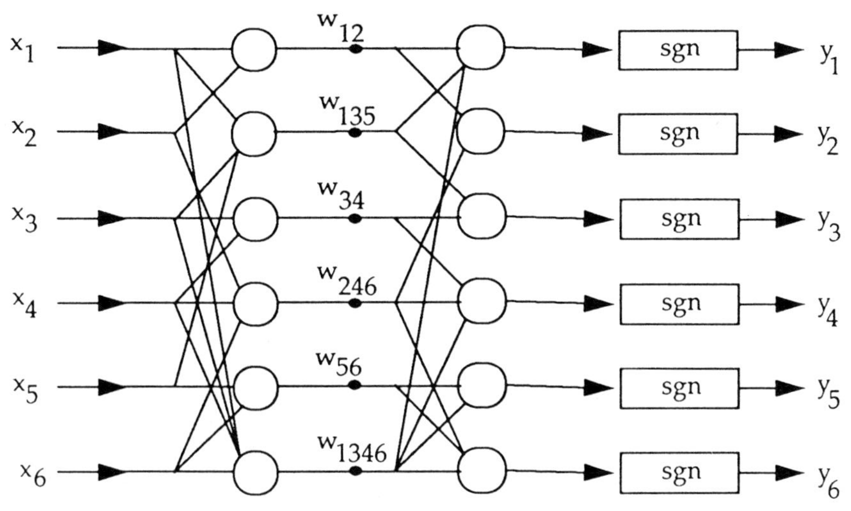

Figure 1: **Analog Network Realization**

An obvious network structure to implement (22) is to make use of multipliers to compute $w_{\underline{k}} X_{\underline{k}}$ and adders to yield Y_i. An example is given in Fig. 1.

$$\begin{aligned} V(x) &= w_{12}x_1x_2 + w_{34}x_3x_4 + w_{56}x_5x_6 + w_{135}x_1x_3x_5 \\ &+ w_{246}x_2x_4x_6 + w_{1346}x_1x_3x_4x_6 \end{aligned} \quad (24)$$

It is apparent that only the adders in the last stage represent analog operations. The rest is entirely digital. This means that a considerable reduction in interconnect-complexity can be realized by using random-access memory and indexing. In such an arrangement, the physical connections are replaced by logical connections. A hybrid architecture comprising both analog and digital circuits is shown in Figure 2.

In this arrangement, the network operates as follows:

1. A component i is chosen at random.
2. The INDEX is accessed to determine for each $\underline{k} \in i$ the address where $x_{\underline{k}} w_{\underline{k}}$ is stored.

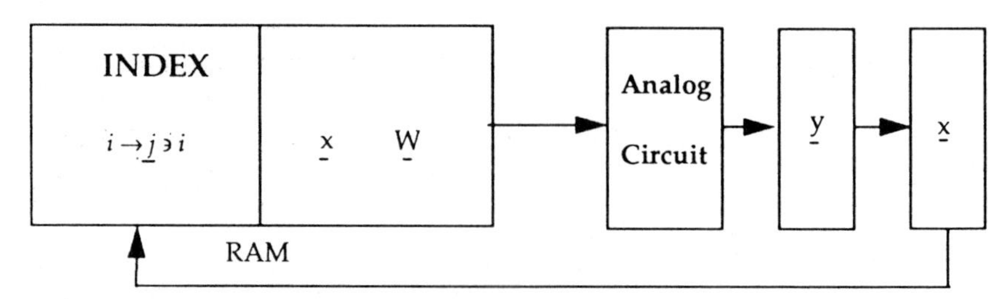

Figure 2: **A Hybrid Network**

3. Each product $x_{\underline{k}} w_{\underline{k}}$ is retrieved and converted to an analog value.

4. The sum $\sum_{\underline{k} \in i} X_{\underline{k}} w \underline{k} - Z$ is computed.

5. Y_i is computed as the "sign" of the sum computed in (d).

6. If $Y_i = -1$, change the sign of X_i and go to (g), otherwise go to (a).

7. For each $\underline{k} \in i$, change the sign of $X_{\underline{k}} w_{\underline{k}}$.

We note that the need to access the set $\{X_{\underline{k}} w_{\underline{k}}\}$ makes the operation slow. However, there is ample opportunity for *pipelining*, which means that the cycles can be overlapped. In this respect, the situation is no different from digital signal processing in general where pipelining is essential for achieving acceptable speeds.

Finally, we note that only analog-add and not analog-multiply is required. If the degree of accuracy needed is not too high, it may well be better to replace the analog adders by digital ones, thereby obviating the need for D/A conversion and affording a further opportunity to eliminate physical interconnections. We would then have an all-digital system that can be implemented in various ways, including ones that use only off-the-shelf components. In this form, the distinction between hardware and software blurs, and the use of a standard DSP (digital signal processing) system for implementation provides an attractive alternative.

5 Acknowledgement

The research reported herein was supported by U.S. Army Research Office Grant DAAL03-89-K-0128 and U.S. Air Force Office of Scientific Research Contract F49620-87-C-0041.

Bibliography

[1] D. H. Ackley, G. W. Hinton, and T. J. Sejnowski, *A learning algorithm for Boltzmann machines,* Cognitive Science 9 (1985) 147-169.

[2] J. Alspector and R. B. Allen, *A neuromorphic VLSI learning system,* in Advanced Research in VLSI, Proc. 1987 Stanford Conference, P. Losleben ed., MIT Press, Cambridge, 1987, 313-349.

[3] J. J. Hopfield, *Neural networks and physical systems with emergent collective computational abilities,* Proc. National Academy of Sciences 79 (1982) 2554-2558.

[4] G. Kesidis and E. Wong, *Optimal acceptance probability for simulated annealing,* Stochastics and Stochastic Reports, vol. 29 (1990) 221-226.

[5] S. Kirkpatrick, C. D. Gelatt, Jr., and M. P. Vecchi, *Optimization by simulated annealing,* Science 220 (1983) 671-680.

Infinite Dimensionality Results for MAP Estimation

Ofer Zeitouni
Department of Electrical Engineering
Technion - Israel Institute of Technology
Haifa 32000, Israel

Abstract

The issue of the infinite dimensionality of solutions to the maximum a posteriori estimation of trajectories of diffusions is addressed in this paper. We adapt the probabilistic methods of Ocone used in the context of nonlinear filtering to the MAP situation (i.e., nonlinear PDE set-up) and prove, at least in the one dimensional case, that the class of problems for which a finite dimensional MAP estimator exists is not richer than its filtering counterpart.

1 Introduction

The issue of the finite dimensionality of the solution to the filtering problem has been an active research field during the first half of this decade, and was settled only recently, following a program suggested first by Brockett and Clark [2] (and independently by [10]), and implemented by [4], [7], [8], to name just few of the contributions. It is by now well known that finite dimensional filters are generally hard to come by, and that in particular, in the one dimensional case, with regular coefficients, the only examples possible of finite dimensional realizations are the Kalman and Benes filters.

Recently, an alternative to the filtering problem was suggested via the Maximum a-Posteriori approach, [15]. It turns out that this approach applies to a wider class of models than the problem of the optimal filtering of diffusions, and in particular can be extended to

ISBN 0-12-481005-5

SPDE's, c.f. [18]. It is therefore of interest to check whether some of the results on finite dimensionality of the solutions to the filtering equation can be transferred to the MAP setting. The purpose of this paper is to give a partial answer to this question. It turns out that in this situation, the answer can still be formulated in terms of suitable algebraic conditions. In the one dimensional case, which is the object of study in this paper, we prove that, under some mild technical conditions, the only finite dimensional realizations for the MAP equations are those for which the filtering problem possesses a finite dimensional solution (c.f. corollary 4.1 and 4.2 below).

The techniques which we will use are borrowed from [12], who proposed a Malliavin calculus approach to the issue of finite dimensionality. Using this approach, the proof of nonexistence of finite dimensional filters is purely probabilistic, and the algebraic conditions arise only at the end of the derivation. It leads conceptually to a stronger result than the algebraic methods, for it precludes not only the existence of a finite dimensional filter of the diffusion type but rather the existence of any type of finite dimensional filter.

Ocone's results relied heavily on the linear structure of Zakai's equation and on a stochastic integration by parts formula based on the robust form of Zakai's equation. As we will see below, we deal with nonlinear PDE's and thus the method of characteristics will replace the linear structure of the equation, whereas due to the simpler form of the Ito term in the stochastic PDE, a simple Stratonovich like integration by parts formula will be enough to carry through our program. It is worthwhile to note that the same program can be carried out for a class of stochastic, nonlinear PDE's which do not arise necessarily from the MAP problem.

A main difficulty in our approach is that, unlike the filtering case, the nonlinear PDE that we deal with does not possess a globally unique solution. We will therefore work with a subset of Wiener space where we can find a common region where the equation does possess a unique solution. This will not affect the results for we will work with a local version of the Malliavin technique, due to [1].

In order to make the presentation as clear as possible, we restrict ourselves to the one dimensional case with constant diffusion coefficients. The techniques presented do carry over to the general case, at least in those cases where the MAP estimation problem is well posed and one has an existence theorem for its solution (c.f. [16] for a discussion of those cases), and whenever the growth and smoothness conditions on the coefficients are such that the method of characteristics that we use can be made to work.

The organization of the paper is as follows. In the rest of this introduction, we describe the MAP trajectory estimation problem and state our basic assumptions, together with the notations that will be used. In section 2 we recall those results from [12] and [1] that will be needed in the sequel. For a complete review of the techniques of the Malliavin calculus used, we refer the reader to [13]. In section 3, we develop the auxiliary machinery related to our class of Hamilton Jacobi equations, namely we prove the existence of a unique global solution over some strip for a subset of the Wiener space, prove the C^{∞} of the solution in the space variable and the smoothness in the Malliavin sense that we need later. Finally, in section 4, we prove theorem 4.1 and its corollaries which provide a partial answer to the issue of the finite dimensionality.

We turn now to some definitions. Let w_t, ν_t be two independent, one dimensional standard Brownian motions adapted to some filtration $\mathcal{F}_t$. Let x_t, $0 \leq t \leq T$ (the signal process) and y_t, $0 \leq t \leq T$ (the observation process) denote the solutions to the following Ito equations:

$$dx_t = f(x_t)dt + dw_t \tag{1}$$

and

$$dy_t = h(y_t)dt + d\nu_t \tag{2}$$

where x_0 possesses the density $p_0(x_0)$ and is independent of the filtration generated by the Wiener processes w_t and ν_t. Throughout, we use the following notations:

1) $C_b^k(R)$ denotes the space of functions $f : R \to R$ which are bounded together with their derivatives up to order k, with the k-th order derivative being absolutely continuous.

2) $C^k[0,T]$ denotes the space of functions which are k times continuously differentiable.

3) H denotes the usual Cameron-Martin space, i.e. the space of functions such that $||f||_H^2 \hat{=} \int_0^T f'(s)^2 ds < \infty$, with $f(0) = 0$.

4) $K \hat{=} L^2(R; e^{-x^2/2}dx)$. Note that K forms a Hilbert space when equipped with the natural scalar product associated with it and the Hermite polynomials e_i form a C^∞ orthonormal basis of K.

5) $|| \cdot ||_t$ denotes the supremum norm on $[0, t]$.

6) (Θ, B, μ) denotes the standard Wiener space, i.e. Θ is the space of continuous functions on $[0, T]$ starting at 0, B is the Borel σ-algebra and μ is Wiener's measure on $\{\Theta, B\}$. The sub σ-algebra generated by $\theta(s), 0 \le s \le t$, is denoted B_t.

7) For an arbitrary Hilbert space X, $L^q(\mu, X)$ denotes the space of X valued random functions F with $E||F||_X^q < \infty$.

8) $\mathcal{F}_t^y$ denotes the filtration generated by $y(s), 0 \le s \le t$.

Throughout, c will denote a generic constant, i.e. $c = 2c = c^3$, etc. We make the following assumptions on the coefficients of (1, 2):

A-1) $f(x) \in C_b^\infty(R), h(x) \in C^\infty(R), h'(x) \in C_b^\infty(R)$.

A-2) There exists a $c > 0$ such that $\log p_0(x) < c$ for all $x \in R$.

A-3) $p_0(x) \in C^\infty$.

Under A-1-A-3, it can be shown using the results of [15] and [17] that

$$\lim_{\epsilon \to 0} \frac{P(||\phi - x||_t < \epsilon | \mathcal{F}_t^y)}{P(||w||_t < \epsilon)} = \exp(J(\phi, t)) \tag{3}$$

where

$$\begin{aligned} J(\phi, t) &\hat{=} \log(p_0(\phi_0)) - \frac{1}{2}\int_0^t (\dot{\phi}_s - f(\phi_s))^2 ds + h(\phi_t) y_t \\ &- \frac{1}{2}\int_0^t f'(\phi_s) ds - \int_0^t y_s h'(\phi_s)\dot{\phi}_s ds - \frac{1}{2}\int_0^t h^2(\phi_s) ds \end{aligned} \tag{4}$$

and $\phi \in C^{1+\alpha}[0, t]$, any $\alpha > 0$. $J(\cdot, T)$ is interpreted as the conditional posterior density of path. Under the additional hypotheses

A-4) $|\log p_0(x)|$ is at most of quadratic growth as $|x| \to \infty$.

A-5) There exists a $c > 0$ such that $\lim_{|x|\to\infty} \frac{h^2(x)}{x^2} > c$,

it follows that a maximum a posteriori estimator, defined as

$$\hat{\phi} = \text{argmax}_{\phi \in H} J(\phi, T)$$

exists and belongs to $C^{1+\alpha}[0,T]$, all $0 < \alpha < 1/2$ (c.f. [16]).

In order to relate the MAP estimator to a stochastic PDE, we follow again [16] and make use of Bellman's optimality principle: define

$$v(t,x) \hat{=} \inf_{\phi \in H, \phi_t = x} (-J(\phi, t)) + \int_0^x f(\theta) d\theta.$$

One then suspects from optimal control theory (c.f. e.g [6]) that $v(x,t)$ satisfies whenever it is differentiable the following stochastic PDE:

$$dv(x,t) = [-\frac{1}{2}\left(\frac{\partial v(x,t)}{\partial x}\right)^2 + \frac{1}{2} l(x)]dt - h(x)dy_t \tag{5}$$

where
$l(x) \hat{=} f^2(x) + f'(x) + h^2(x)$, $v(x,0) = -\log(p_0(x)) + \int_0^x f(\theta)d\theta$.

Note that as pointed out in [16], (5) is related to a logarithmic transformation of Zakai's equation. The latter possesses however nicer analytical properties due to the existence of an additional Laplacian term in the RHS of (5). Note also that by defining $u(x,t) = v(x,t) + h(x)y_t$, one obtains a robust form of Bellman's equation, namely

$$\frac{\partial u(x,t)}{\partial t} = -\frac{1}{2}\left(\frac{\partial u(x,t)}{\partial x}\right)^2 + h'(x)y_t \frac{\partial u(x,t)}{\partial x} - \frac{1}{2} h'^2(x) y_t^2 + \frac{1}{2} l(x) \tag{6}$$

with the initial conditions $u(x,0) = v(x,0)$.

In the sequel, we concentrate on analyzing the existence of solutions, moment bounds, smoothness in the Malliavin sense and existence of densities (over function space, c.f. below) of solutions to (5). As pointed out in [12], the existence of a "density" over function space (or appropriate subspaces of it) is related to the existence of finite dimensional realizations to solutions to (5). To see that such realizations are at all possible, consider the case $h(x) = x, f(x) = -x, \log(p_0(x)) = -x^2$. Substituting $v(x,t) = a_t x^2 + b_t x + c_t$, one checks that $v(x,t)$ is indeed a solution of (5) when a_t, b_t, c_t are represented as outputs of simple recursive filters driven by the observation y_t. On the other hand, note that the non existence of finite dimensional realizations for $v(x,t)$ does not preclude the existence of such realizations for the optimal trajectory $\hat{\phi}$. This is similar to the issue of the existence of "universal" versus "non universal" finite dimensional nonlinear filters. For a discussion of this point in the context of nonlinear filtering, c.f. e.g. [4].

We often need the following assumption:

A-6) For all $x \in R$,

$$\left|\frac{\partial \log p_0(x)}{\partial x}\right| < \tilde{c}, \left|\frac{\partial^2 \log p_0(x)}{\partial x^2}\right| < \tilde{c}, |(h^2(x))''| < \tilde{c}$$

2 Stochastic Calculus of Variations in Hilbert Space

We present here those definitions and results from [12] and [1] that we need. A comprehensive survey of the techniques involved appears in [13]. Let X be a separable Hilbert space, and let $F : \theta \to X$ be measurable. If for all $\theta \in \Theta$ there exists a Hilbert Schmidt operator $DF(\theta) : H \to X$ such that $\forall \epsilon > 0$, $\forall \gamma \in H$,

$$\lim_{t \to 0} \mu\{\theta| \; ||(F(\theta + t\gamma) - F(\theta))/t - DF(\theta)(\gamma)||_X > \epsilon\} = 0,$$

and if the map $t \to F(\theta + t\gamma)$ is absolutely continuous for all θ, γ, we call the map $DF : \Theta \to HS(H; X)$ the *H differential* of F. The class of all functions F which possess an H differential is denoted W^1, and for any $\gamma \in H$, we use $D_\gamma F(\theta)$ for $DF(\theta)(\gamma)$.

Let now $F \in W^1$. The *Malliavin covariance derivative* $\nabla F^* \nabla F$: $\Theta \to X \bigotimes X$ is defined by

$$\nabla F^* \nabla F(\Theta) \hat{=} \sum_{i=1}^{\infty} D_{\gamma_i} F(\theta) \bigotimes D_{\gamma_i} F(\theta),$$

where γ_i is any orthonormal complete base in H. Various other definitions for the Malliavin covariance matrix exist. Again, we refer the reader to [13] for information concerning the Malliavin covariance.

The importance of the Malliavin covariance to our needs is captured in the following lemma, which is contained in [1], theorem 3.9:

Lemma 2.1 *1) Let $F : \Theta \to R^n$ belong to W^1. Assume that $||DF(\theta)||_{HS}^2 < \infty$, where $|| \cdot ||_{HS}$ denotes the Hilbert Schmidt norm of an operator. Further assume that $\nabla F^* \nabla F : \Theta \to X \bigotimes X > 0$ a.s. (μ). Then $d(\mu \circ F^{-1})/dx$ exists, i.e. the R^n random variable $F(\theta)$ possesses a density w.r.t. Lebesgue measure.*
2) A local version of the above result holds true, i.e. assume that the above conditions hold true only on an open set $\Lambda \subset \Theta$ with $\mu(\Lambda) > 0$. Then the conclusion of the lemma still holds for $\hat{\mu}$, the restriction of μ to Λ.

Following [12], we need to define what we mean by densities in Hilbert spaces:

Definition 2.1 Let p be a random variable with values in X. Let S denote any closed finite dimensional subspace of X, and let P_S denote the projection onto S. The cylinder sets based on S are the elements of the σ-algebra

$$B_S = (P_S^{-1} u | u \text{ is a Borel set of } S).$$

Define a Lebesgue measure m_S on (X, B_S) by $m_S(P_S^{-1} u) = m(u)$ for all $u \in S$ Borel, where m is Lebesgue's measure on S. Then $\mu \circ p^{-1}$ is said to admit a density w.r.t. cylinder sets if $\mu \circ p^{-1}|_B << m_S$ for all S as above.

As in [12], the fact that $\mu \circ p^{-1}$ admits a density w.r.t. cylinder sets implies the non existence of a finite dimensional representation for p, i.e. one cannot find a finite set of R valued random variables $x_1, \cdots, x_k$ such that $p = f(x_1, \cdots, x_k)$. That is due to the fact that if such a finite dimensional representation existed, the law of p would have been based on the finite dimensional subspace S defined by it; Therefore, for any finite subspace S' strictly including S, $\mu \circ p^{-1}|_{B_{S'}}$ could not be absolutely continuous w.r.t. to $m_{S'}$, in contradiction with the existence of densities. We note that for this argument to work, it is not necessary to prove that $\mu \circ p^{-1}$ as a random variable in X admits a density w.r.t. cylinder sets. Rather, it is enough to prove this for the restriction of this random variable to one infinite dimensional Hilbert subspace of X. Our goal in the sequel will therefore to derive conditions for $v(t,x)$ to possess a density in a subset of an infinite dimensional Hilbert space.

3 Some properties of $v(t,x)$, and a stochastic gradient representation

In this section, we develop the technical preliminaries which are necessary for the computation of the necessary condition for finite dimensionality. In particular, based on the method of characteristics, we prove local existence and smoothness results for $v(x,t)$. As a corollary, we obtain a local uniqueness result for trajectory MAP estimation (c.f. Lemma 3.2).

The property needed in the sequel is the following:

(P) There exists a set $\Gamma \in \Theta$ and a non random constant $\tau > 0$ such that:

$$a)\ \mu(\Gamma) > 0$$

$$b)\ \forall\ y \in \Gamma, (x,t) \in R \otimes [0,\tau], u(x,t) \text{ is } C^1 \text{ w.r.t. } t \text{ and } C^\infty \text{ w.r.t } x$$

We bring a short account of the method of characteristics used in the appendix. By classical results (c.f., e.g., [3], pg. 24, thm. 8.1), there exists for each $x \in R$ a neighborhood (in $R \otimes [0,T]$) of $(x,0)$ with the required properties. However, we require this neighborhood to be uniform in x for a set of μ positive measure, and for that we need the restrictions imposed on the coefficients of the diffusion process and its observation function. Indeed, we have:

Lemma 3.1 *Under A-1-A-6, property (P) above holds true.*

Proof: We show that the characteristic curves (c.f. Lemma A-1 in the appendix) define, for the set $\Gamma \hat{=} \{y_. : ||y||_T < 1\}$, a ($C^\infty$ in the space variable, C^1 in the time variable) diffeomorphism in the neighborhood $[0,\tau] \otimes R$, where τ is defined below in equation (12). Therefore, by the appendix, (P) holds true for $y \in \Gamma$.

We begin by considering the Hamiltonian for the robust equation (6), viz.

$$H(t,x,p)=\frac{1}{2}p^2-y_t h'(x)p-\frac{1}{2}l(x)+\frac{1}{2}y_t^2 h'^2(x) \tag{7}$$

The characteristic equations are therefore:

$$\begin{aligned}\dot{X} &= P-y_t h'(X), \quad X(0)=x && (8)\\ \dot{P} &= y_t h''(X)P+\frac{1}{2}l'(X)-y_t^2 h'(X)h''(X), \\ P(0) &= -\frac{\partial}{\partial x}\log p_0(x)+f(x) && (9)\end{aligned}$$

Define $\xi_t \hat{=} \frac{\partial X(t)}{\partial x}$ and $\pi_t \hat{=} \frac{\partial P(t)}{\partial x}$. Then, one has

$$\begin{aligned}\dot{\xi}_t &= \pi_t-y_t h''(X)\xi_t. \quad \xi(0)=1 && (10)\\ \dot{\pi}_t &= y_t h^{(3)}(X)P\xi_t+y_t h''(X)\pi_t+\frac{1}{2}l''(X)\xi_t-\frac{y_t^2(h'(X)^2)''}{2}\xi_t \\ \pi(0) &= -\frac{\partial^2}{\partial x^2}\log p_0(x)+\frac{\partial}{\partial x}f(x) && (11)\end{aligned}$$

Let c denote a common bound on $|l''(x)|, |f(x)|, |h'(x)|, |\frac{\partial^2 \log p_0(x)}{\partial x^2}|$, $|\frac{\partial \log(p_0(x))}{\partial x}|$, $|h''(x)|$, $|l'(x)|, |h'(x)h''(x)|$, $|(h'(x)^2)''|$. Using (9) and these bounds, it follows that $||P||_1 \leq 2(c+1)e^c$. Define now

$$m \hat{=} \sup\{\xi_s | s\in[0,1], \xi_\gamma>0 \quad \text{for all } \gamma\in[0,s]\}.$$

$$\theta \hat{=} \inf\{s|\xi_s=m\}.$$

Let $\tau' \hat{=} \inf\{s|\xi_s=0\}\wedge 1$. Clearly, $\tau'>\theta$. On the other hand, in $[0,\tau']$,

$$|\dot{\pi}_t| \leq (2(c+1)ce^c+c)m+c|\pi|$$

which implies that

$$||\pi||_{\tau'} \leq (1+2(c+1)e^c)e^c m+2ce^c \leq 4(c+1)e^{2c}m$$

where the last inequality followed from the fact that $m\geq 1$. Therefore, for t such that $(t+\theta)\wedge 1\in[0,\tau']$, one obtains that

$$\xi_{t+\theta} \geq m-4tm(c+1)e^{2c}-cmt.$$

Let

$$\tau \hat{=} \frac{1}{10(c+1)e^2 c} \wedge 1 \tag{12}$$

For $t \le \tau$, one obtains that $\xi_t > 0$, from which the C^1 diffeomorphism in space follows. The proof of the C^∞ property with respect to space is similar and therefore omitted. □

Remark If $h(x)$ is linear, assumption A-6 may be relaxed so that no bound is required on the first derivative of $p_0(\cdot)$, and still the result of Lemma 3.1 and the subsequent derivation hold true. For brevity, we do not consider this case in the sequel. Note however that this extension allows one to consider the Gaussian situation.

We recall now that, by a slight adaptation of the proof in [16], a MAP trajectory estimator with endpoint x, defined as

$$\hat{\phi}_x = \text{argmax}(J(\phi, t) | \phi \in H, \phi_t = x)$$

exists. The following lemma, which is an adaptation of [6] to our needs, provides information on the value function:

Lemma 3.2 *Under A-1-A-5, a generalized solution $v(x,t)$ which is the value function of the optimal control problem (4) (modulo the deterministic C^∞ drift shift $\int_0^x f(s)ds$) exists in K. Moreover, for all $n > 1$,*

$$E \int_R (v(x,t))^n e^{-x^2/2} dx < \infty \tag{13}$$

$$E \int_R \left(\frac{\partial v(x,t)}{\partial x}\right)^n e^{-x^2/2} dx < \infty \tag{14}$$

with the bounds being uniform in $t \in [0, \tau]$. Finally, under the additional assumption A-6, the solution $v(x,t)$ for $y \in \Gamma$ is a classical solution (and therefore unique), (13, 14) hold in $[0, \tau]$ where the expectations are restricted to the cylinder set defined by Γ, and

$$\left|\frac{\partial v(x,t)}{\partial x}\right| \le c', \quad \left|\frac{\partial^2 v(x,t)}{\partial x^2}\right| \le c'(1 + |x|^2) \tag{15}$$

for some c' which depends on $c, \tilde{c}$ but is independent of $y \in \Gamma$.

Proof: In this proof, c' denotes a generic constant which may depend on T but doesn't depend on y, and whose value varies from line to line. By A-1-A-3,

$$-K'(1+||y||_T^2+x^2) \leq J(\hat{\phi}_x,T) \leq c'(1+||y||_T^2+|x|).$$

Therefore, by the boundedness of $f(x)$, one has that

$$|v(t,x)| \leq c'(1+||y||_T^2+|x|^2)$$

which implies that, for all $n \geq 1$,

$$\int_R v(t,x)^n e^{-x^2/2}dx < \infty$$

with the same bound on the expectations of the last quantity. Therefore, the first half of the lemma is proved.

Rewriting now $J(\phi)$ as

$$\begin{aligned} J(\phi,T) &= \log(p_0(\phi_0)) + h(\phi_T)y_T - \frac{1}{2}\int_0^T f'(\phi_s)ds \\ &- \frac{1}{2}\int_0^T (\dot{\phi} - f(\phi_s) + y_s h'(\phi_s))^2 ds + \frac{1}{2}\int_0^T y_s^2 h'^2(\phi_s)ds \\ &- \int_0^T f(\phi_s)y_s h'(\phi_s)ds - \frac{1}{2}\int_0^T h^2(\phi_s)ds, \end{aligned}$$

one obtains by comparing the optimal path $\hat{\phi}_x$ with the path $\phi = x$ as in [16] that $||\hat{\phi}_x||_H \leq c'(1+||y||_T^2+|x|)$. Therefore, using the Sobolev identity, one has that

$$||\hat{\phi}_x||_T \leq c'(1+||y||_T^2+|x|).$$

Therefore, using A-1-A-5, one obtains that

$$\begin{aligned} |J(\hat{\phi}_x+x'-x,T) - J(\hat{\phi}_x,T)| &\leq c'|x-x'|(||\hat{\phi}_x||_H+1)^2 \\ &+ |\log(\frac{p_0(\hat{\phi}_x(0)+x-x')}{p_0(\hat{\phi}_x(0))})| \\ &\leq c'|x-x'|(1+||y||_T^2+|x|^2) \quad (16) \end{aligned}$$

Note now that $S(t,x) \leq -J(\hat{\phi}_x+x'-x)$, whereas $S(t,x) = -J(\hat{\phi}_x)$. Therefore,

$$S(t,x') - S(t,x) \leq c'(1+||y||_T^2+|x|^2)|x-x'|$$

Interchanging the roles of x and x', one gets the opposite inequality, and hence by the boundedness of $f(x)$ the same relation holds for $v(t,x)$. Therefore, one obtains that

$$\int_R (\frac{\partial v(x,t)}{\partial x})^n e^{-x^2/2} dx \le c'(1 + ||y||_T^{2n})$$

which proves the second part of (15).

To conclude, we use Lemma 3.1 in conjuction with the above considerations. Indeed, by (39) in the appendix, $P = \frac{\partial v}{\partial x}|_{\hat{\phi}_x}$, and since from Lemma 3.1 P is bounded uniformly in the strip $[0,\tau]$, one obtains the bound on the space derivative of v. To obtain the bound on the second spatial derivative, we proceed similarly, by defining the approximate second derivative

$$\Delta_\epsilon S(x,t) \hat{=} \frac{S(x+\epsilon,t) + S(x-\epsilon,t) - 2S(x,t)}{\epsilon^2}.$$

The rest of the proof repeats the arguments used above and is therefore omitted. □

The first step towards the stochastic gradient representation is:

Lemma 3.3 *For each* $y \in \Gamma$, $\nu \in H$, *there exists in the strip* $R \otimes [0,\tau]$ *a unique classical solution to the equation*

$$\frac{\partial \mu(x,t)}{\partial t} = -\frac{\partial v(x,t)}{\partial x}\frac{\partial \mu(x,t)}{\partial x} - h(x)\nu'(t), \quad \mu(x,0) = 0 \tag{17}$$

Moreover, for any CONS $\nu_i \in H$,

$$\sum_{i=1}^{\infty} ||\mu_{\nu_i}(x,t)||_K^2 < \infty \tag{18}$$

Proof We use again the method of characteristics. The characteristic curve for (17) satisfies

$$\frac{dX(t)}{dt} = -\frac{\partial v(X(t),t)}{\partial x}, \quad X(0) = x \tag{19}$$

Since $|\frac{\partial v(x,t)}{\partial x}| \le c(1+|x|)$, a solution to (19) exists in $R \otimes [0,\tau]$, with $|X(t)| \le c(1+|x|)$. Moreover, defining $\eta = \frac{d(\partial X(t)/\partial x)}{dt}$, one has

$$\frac{d\eta}{dt} = -\frac{\partial^2 v(X(t),t)}{\partial x^2}\eta, \quad \eta(0) = 1 \tag{20}$$

and therefore, due to our bound on $\frac{\partial^2 v(x,t)}{\partial x^2}$, $\eta(t) > 0$ for all $t \in [0, \tau]$. That implies that $X(t)$ is a diffeomorphism as a map $R \otimes [0, \tau] \rightarrow R$ and therefore $X(t)$ is a characteristic curve and a classical solution to (17) exists and is unique.

To see (18), we invoke again the characteristic representation of the solution to (17) as in (35-38) of the appendix, i.e.

$$\begin{aligned} \dot{P} &= -Pv_{xx}(X(t), t) - h'(X(t))\nu_i(t); \;\; P(0) = 0 && (21) \\ \dot{U}_{\nu_i} &= Pv_x(X(t), t) - H(X(t), t, P) + H(X(t), 0, 0) \\ &= PV_x(X(t), t) - PV_x(X(t), t) - h(X(t))\nu_i(t) \\ &= -h(X(t))\nu_i(t); \;\; U_{\nu_i}(0) = 0 && (22) \end{aligned}$$

Therefore,

$$U_{\nu_i}(x, t) = -\int_0^t h(X(x, s))\nu_i(s)ds \qquad (23)$$

where $|h(X(x, s))| \leq c|X(x, s)|$. Since, for each x, $\int_0^t h^2(X(x, s))ds < \infty$, (23) states that $U_{\nu_i}(x, t)$ is the projection of $-h(X(x, s))$ in the direction ν_i. By our bounds on $X(\cdot)$ we get therefore that

$$\sum_{i=1}^{\infty} ||U_{\nu_i}(X^{-1}(x, t), t)||_K^2 \leq \int_0^t ||h^2(X(X^{-1}(x, t), s))||_K^2 ds \leq c < \infty \qquad (24)$$

with c independent of $y \in \Gamma$. The lemma follows. □

The following Lemma is proved exactly as Lemma 3.3 was proved. We therefore omit the details.

Lemma 3.4 *Let v_ν^ϵ denote the solution to (6) with $y_t + \epsilon\nu_t$ substituted instead of y_t. Then, for all $y \in \Gamma$ and for $||\nu|| < 1$, and all ϵ small enough,*

$$\begin{aligned} |v - v_\nu^\epsilon| &\leq \epsilon c_\nu(1 + |x|) && (25) \\ \left|\frac{\partial(v - v_\nu^\epsilon)}{\partial x}\right| &\leq \epsilon c_\nu(1 + |x|) && (26) \\ \left|\left|\frac{v(x, t) - v_\nu^\epsilon(x, t)}{\epsilon} - \mu(x, t)\right|\right|_K &\rightarrow_{\epsilon \rightarrow 0} 0 && (27) \end{aligned}$$

Combining now the above Lemmas, one has:

Theorem 3.1 *$D_\nu v(x, t)$ exists and satisfies (17). Moreover, for all $y \in \Gamma$, $Dv(x, t)$ is Hilbert-Schmidt as an operator $H \rightarrow K$.*

4 Conditions for finite dimensionality

In this section, we turn to the evaluation of a criterion for the non existence of finite dimensional realizations of $v(x,t)$. By Lemma 2.1 and the remarks immediately following it, a sufficient condition for the non-existence of finite dimensional realizations is that for all ω in a set $\tilde{\Gamma}$ with $\mu(\tilde{\Gamma}) > 0$ and $\mu(\tilde{\Gamma} \cap \Gamma) > 0$,

$$\forall a \in K', \quad \sum_{i=1}^{\infty}(D_{\alpha_i}u, a)_K^2 > 0 \tag{28}$$

where $\{\alpha_i\}$ is an infinite orthonormal system in H, the scalar product in (28) is taken in K, and by the remark which follows Definition 2.1, K' is required to be an infinite dimensional subset of K.

Definition 4.1 Λ denotes the space of functions which are C^∞ w.r.t. x and possess a Stratonovich representation w.r.t. the observation $\{y_.\}$, viz. for $f(x,t) \in \Lambda$,

$$f(x,t) = \int_0^t d_R f(x,s)ds + \int_0^t d_I f(x,s) \circ dy_s$$

Definition 4.2 Define the operator L_v^t by

$$L_v^t = \frac{\partial v(x,t)}{\partial x}\frac{\partial \cdot}{\partial x}$$

Define by $\Phi_v(t)$ the set of functions $f(x,t) \in \Lambda$ such that:

1) $h(x) \in \Phi_v(t)$.

2) If $f(x,t) \in \Phi_v(t)$, so does $A_t(f) \hat{=} L_v^t f(x,t) + d_R f(x,t)$.

3) If $f(x,t) \in \Phi_v(t)$, so does $A_i(f) \hat{=} d_I f(x,t)$.

It follows by Lemma 3.1 above that the class $\Phi_v(t)$ isn't empty, i.e. all "derivatives" in the sense of (1-3) above of $h(x)$ are in Λ. Note that the definition of $\Phi_v(t)$ is obtained from the ordinary Lie algebra when the time variable is added as a state variable.

We are now ready to state our main result, namely:

Theorem 4.1 *Assume that $\Phi_v(t)\mid_{t=0}$ spans an infinite dimensional subspace of K. Then no finite dimensional representation exists for $v(x,t)$.*

Proof Assume the contrary. Then for each ω in some set $\tilde{\Gamma}$ as in the beginning of this section and for each infinite dimensional subspace K' of K there exists an $b \in K'$ such that for all $\alpha \in H$

$$(D_\alpha v(x,t), b)_K = 0. \tag{29}$$

Define the sequence of operators $\phi(t,s) : K \bigcap C^1(R) \to K \bigcap C^1(R)$ by

$$\frac{d(\phi(t,s)\circ a)}{dt} = -\frac{\partial v(x,t)}{\partial x}\frac{\partial(\phi(t,s)\circ a)}{\partial x}, \quad \phi(s,s)\circ a = a \tag{30}$$

for all $a \in K \bigcap C^1(R)$. By the method of characteristics, the operators $\phi(t,s)$ are well defined and moreover, one easily checks that

$$D_\alpha v(x,t) = \int_0^t \phi(t,s)\circ(h(x)\alpha'(s))ds.$$

Let $\phi^*(t,s)$ denote the adjoint operator of $\phi(t,s)$, which satisfies the adjoint of equation (30). By (29), one has that

$$\begin{aligned} 0 &= (D_\alpha v(x,t), b)_K \\ &= \int_0^t (\phi(t,s)\circ(h(x)\alpha'(s)), b)_K ds \\ &= \int_0^t (\phi(t,s)\circ h(x), b)_K \alpha'(s)ds \end{aligned} \tag{31}$$

for all $s \le t < \tau$. However, since $\alpha \in H$ is arbitrary, it follows that for all $s \le t < \tau$, $(\phi(t,s)\circ h(x), b)_K = (h(x), \phi^*(t,s)\circ b)_K = 0$. Expanding the RHS of this last equality, one has that

$$\begin{aligned} 0 &= (h(x), \phi^*(t,s)\circ b)_K \\ &= (h(x), \phi^*(t,0)\circ b)_K \\ &+ \int_0^t (\frac{\partial v(x,r)}{\partial x}\frac{\partial h(x)}{\partial x}, \phi^*(t,r)\circ b)_K dr \end{aligned} \tag{32}$$

which implies that for all $r \in [0,t]$,

$$A_1 \hat{=} (\frac{\partial v(x,r)}{\partial x} \frac{\partial h(x)}{\partial x}, \phi^*(t,r) \circ b)_K = 0.$$

To proceed, one would like to repeat this procedure again, as in [12]. However, when using the above equation to deduce that $dA_1 = 0$, stochastic integrals make their appearance. Moreover, one has to note that $\phi^*(t,r)$ depends also on the future observations $y_\theta, \tau \leq \theta \leq t$ and therefore the ordinary Ito calculus cannot be applied. Even if it could, for the result to follow one would need to be able to make sense of them for each $\omega \in \Gamma'$ where Γ' differs from $\tilde{\Gamma}$ only on a fixed set of measure zero (which doesn't depend on b). The tool for accomplishing that is working pathwise, using the fact that for any (not necessarily adapted) C^1 random function g_t and any adapted continuous semimartingale f_t satisfying certain boundedness conditions,

$$f_t g_t = \int_0^t f_t \dot{g}_t dt + \int_0^t g_t \circ df_t$$

where the stochastic integral is a Stratonovich integral and the equality holds a.s. with a null set which is independent of g_t (c.f. Lemma 4.1 of [5] for the details). Since $\phi^*(t,r) \circ b$ is differentiable w.r.t. r, this smooth version of Ito's lemma (where all stochastic integrals are interpreted in the Stratonovich sense) applies here. Using this formula, and repeating the argument below (31), one obtains that for each $\delta \in \Phi_v(r)$,

$$(\delta(x,r), \phi^*(t,r) \circ b)_K = 0$$

Therefore, again by the dual form,

$$(b, \phi(t,r) \circ \delta(x,r))_K = 0$$

Taking now $r = 0$, one obtains that b is orthogonal to span $\{\phi(t,0) \circ \Phi_v(0)\}$. Therefore, using the continuity of the last expression in t, one obtains by substituting $t = 0$ that span $\{\Phi_v(0)\}$ is orthogonal to b. Taking now $K' \hat{=}$span $\{\Phi_v(0)\}$, one obtains a contradiction. □

Corollary 4.1 *Finite dimensional realizations for $v(x,t)$ exist only if $h(x) = ax^2 + bx + c$.*

Proof Apply the sequence of operators $A_i^k A_t^k$ to $h(x)$ to conclude that, for any integer k, $h_x^k \frac{\partial^k h}{\partial x^k} \in \Phi_v(t)$ for all t. For this set to be finite dimensional, $h(x)$ must be a polynomial of order 2 at most. □

Since, by (A-2) above, we have assumed that $h'(x)$ is bounded, we restrict the discussion now to the case $h(x) = ax$ (any additive constant is irrelevant to the MAP problem, and in particular the case $h(x)$ =constant is not interesting). The algebraic conditions for $h(x)$ quadratic may be evaluated in a similar way, and yield additional possible singular drifts, very much as in the filtering case (c.f. [11]), although our derivation, which is based on the method of characteristics, isn't justified in that case.

Corollary 4.2 *Finite dimensional realizations for $v(x,t)$, with $h'(x) \in C_b^\infty(R)$, exist only if $l(x) = \alpha x^2 + \beta x + \gamma$.*

Proof Applying the sequence of operators $A_i A_t^2$ to $v_x h_x$, and using the fact that by corollary 4.1 $h_{xxx} = 0$, one obtains that $\mu_0(x) = l_{xx} \in \Phi_v(t)$ for all t. Let $\mu(x)$ belong to $\Phi_v(t)$ for all t. By applying $A_i A_t^2$ one obtains that $h_x^2 \mu_{xx} \in \Phi_v(t)$ for all t. Repeating this argument, one obtains that, for all $k \geq 2$,

$$(h_x)^k \frac{d^k \mu}{dx^k} \in \Phi_v(t) \tag{33}$$

which implies that $\mu(x)$ is a combination of exponentials and polynomials. On the other hand, applying on μ as above the operator $D_i^k D_t^{k+1}$ $(k \geq 2)$ yields that

$$h_x^2 \frac{d^k \mu}{dx^k} v_x(t,x)^{k-2}|_{t=0} \in \Phi_v(0) \tag{34}$$

(34) implies that unless $v_x(0,x)$=constant, $\mu(x)$ must be a polynomial, for otherwise the resulting set $\Phi_v(t)$ would be infinite dimensional. Since $v_x(0,x)$ can't be equal to a constant by the normalization condition on $p_0(x)$, we deduce that $\mu(x)$ must be a polynomial, which implies that $l(x)$ must be a polynomial of finite order.

To conclude the proof, assume that μ is a polynomial of order k', whereas $l(x)$ has order k. One has that, for $j = k' - 1$,

$$A_t^{j+2} \mu = v_x^{j+2} \frac{d^{j+2} \mu}{dx^{j+2}} + v_x^j P + \text{ lower order terms (in } v_x)$$

where P is a polynomial of order $k-1$. By our choice of j, the first term vanishes and applying A_i^j one concludes that $\Phi_v(t)$ includes a polynomial of order $k-1$. Repeating this argument, one obtains that unless μ_0 is a constant, $\Phi_v(t)$ contains polynomials of arbitrarily high order, and therefore $\Phi_v(t)|_{t=0}$ cannot be finite dimensional. However, μ_0 being a constant implies that $l(x)$ is quadratic. The Corollary follows. □

5 Appendix

In this appendix, we bring, without proof, a result from the classical method of characteristics which we use section 3. The exposition follows the deterministic case one in [9]. For a more detailed introduction and complete proofs, c.f. [14], whereas the adaptation required to deal with the Ito integrals is straightforward.

Let $H(x,t,p)$ be C^1 in x,p and assume that for fixed x,p, $dH(x,t,p)$ exists in the Ito sense. Let $\phi(x)$ denote a C^1 function. Define

$$dX(t)/dt = \frac{\partial H(X,t,P)}{\partial p}, \quad X(0)=x \tag{35}$$

$$dP(t)/dt = -\frac{\partial H(X,t,P)}{\partial x}, \quad P(0)=\phi'(x) \tag{36}$$

$$dU(t)/dt = P\frac{\partial H(X,t,P)}{\partial x} + q_t, \quad U(0)=\phi(x) \tag{37}$$

$$dq_t = -dH(X,t,P), \quad q_0 = -H(x,0,\phi'(x)) \tag{38}$$

Lemma A1 *Assume that for some strip $[0,\tau]$, (35,36) define a C^1 diffeomorphism $x \to X(t,x)$. Then*

$$u(x,t) = U(X^{-1}(x,t)) \tag{39}$$

is a solution of

$$u_t + H(x,t,u_x) = 0 \quad u(0,x)=\phi(x) \tag{40}$$

for all $t \in [0,\tau)$.

Remarks It is easy to see that actually, if all functions and boundary data are smooth and the diffeomorphism property as above holds in a C^∞ sense, the solution $u(x, t)$ is also smooth. Moreover, as long as $x \to X(t, x)$ is a diffeomorphism, the resulting solution is unique. For a general proof of these facts, c.f. [14], pg. 59.

Acknowledgements I Thank D. Ocone, O. Hijab, E. Pardoux, J. Levine and M. Cohen de Lara for helpful discussions concerning this work. This work was partially suported by the U.S. Army Research contract DAAL03-86-K-0171.

Bibliography

[1] N. Bouleau and F. Hirsch, *Proprietes d'absolue continuite dans les espaces de Dirichlet et application aux equations differentielles stochastiques,* Lecture Notes in Mathematics 1204, 1986, pp. 131-161, Springer.

[2] R.W. Brockett, *Nonlinear systems and nonlinear estimation theory,* in: Stochastic Systems M. Hazewinkel and J.C. Willems, eds., Reidel-Dordrecht, 1981.

[3] S.H. Benton, *The Hamilton Jacobi equation,* 1977, Academic Press, New-York.

[4] M. Chaleyat-Maurel and D. Michel, *Des resultats de non existence de filtres en dimension finie,* Stochastics, 13, 1984, pp. 83-102.

[5] A. Dembo and O. Zeitouni, *A change of variables formula for Stratonovich integrals and existence of solutions for stochastic two point boundary value problems,* Prob. Th. Rel. Fields, 84, 1990, pp. 411-425.

[6] W.H. Fleming, *The Cauchy problem for a nonlinear first order partial differential equation,* J. Diff. Eqns., 5, 1969, pp. 515-530.

[7] O. Hijab, *Finite dimensional causal functionals of Brownian motion in nonlinear stochastic problems,* S. Bucy and J.M.F Maura, eds., Reidel-Dordrecht, 1983.

[8] J. Levine, *Finite dimensional realizations of stochastic P.D.E.'s and application to filtering,* to appear, Stochastics.

[9] P.L. Lions, *Generalized solutions to Hamilton-Jacobi equations,* Pitman, 1982.

[10] S.K. Mitter, *On the analogy between mathematical problems of non-linear filtering and quantum physics,* Ricerche di Automatica, 10, 1981, pp. 163-216.

[11] D. Ocone, *Finite dimensional estimation algebras in nonlinear filtering,* in [2].

[12] D. Ocone, *Stochastic calculus of variations for stochastic partial differential equations,* J. Func. Anal., 79, 1988, pp. 231-288.

[13] D. Ocone, *A guide to the stochastic calculus of variations,* Lecture Notes in Mathematics 1316, 1988, pp. 1-79, Springer.

[14] H. Rund, *The Hamilton-Jacobi theory in the calculus of variations,* Van Nostrand, 1966.

[15] O. Zeitouni and A. Dembo, *A maximum a posteriori estimator for trajectories of diffusion processes,* Stochastics, 20, 1987, pp. 211-246. Errata, pp. 341.

[16] O. Zeitouni and A. Dembo, *An existence theorem and some properties of maximum a posteriori estimators of trajectories of diffusions,* Stochastics, 22, 1988, pp. 197-218.

[17] O. Zeitouni, *On the Onsager-Machlup functional of diffusion processes around non C^2 curves,* Annals of Probability, 17, 1989, pp. 1037-1054.

[18] O. Zeitouni and A. Dembo, *Map estimation of elliptic Gaussian fields observed via a nonlinear channel,* To appear, J. Multivariate analysis.